Donald S. Cohen

*Differential Equations
of
Applied Mathematics*

Differential Equations of Applied Mathematics

G. F. D. DUFF
University of Toronto

D. NAYLOR
University of Western Ontario

John Wiley & Sons, Inc. · New York · London · Sydney

Copyright © 1966 by John Wiley & Sons, Inc.

All Rights Reserved

This book or any part thereof must not
be reproduced in any form without the
written permission of the publisher.

Library of Congress Catalog Card Number: 65-26844
Printed in the United States of America

Preface

This book has developed from courses of lectures given by us over a period of years to senior undergraduates in applied mathematics and engineering science at the University of Toronto. In these lectures the methods of Fourier series and integral transforms are used to discuss the standard differential equations of mathematical physics, and it is shown that a variety of physical problems can be treated by such techniques. We have felt the need for a text book in this course which would include a unified mathematical presentation of the eigenfunction method together with the early use of generalized functions. This book is the result of our efforts to combine these features, with some indication of the wide range of physical application of the differential equations.

The first chapter contains a brief review of vector spaces, inner product spaces and matrices, a discussion of ordinary differential equations as matrix systems, and a study of equations of motion for mechanical systems with a finite number of degrees of freedom.

Chapter 2 begins with the classification of linear partial differential equations of second order in two independent variables into the elliptic, parabolic and hyperbolic types. The latter, in the form of the equation of the vibrating string is then studied in detail. Here distributions are formally introduced and Fourier series are used.

Chapter 3 is concerned with the equation of heat flow, Fourier series and integrals. Chapter 4 takes up the Laplace equation in two dimensions, and harmonic boundary value problems are studied in various cases. In each of these chapters qualitative properties of the equations, as well as finite difference methods, are considered. In Chapter 5 equations of motion are derived for various physical systems and three-dimensional problems are introduced. This chapter can be studied separately or in conjunction with any of the others.

Chapter 6 describes the general theory of eigenfunctions of a bounded smooth region. A minimizing procedure is used to define the eigenfunctions and eigenvalues and to derive their variational properties. A general discussion of the eigenfunction expansion method is also included. In Chapter 7 we study Green's functions and the representation of solutions by them. Again we refer to the distribution technique to justify the appropriate bilinear series for harmonic and parabolic Green's functions.

Following these general theories, we devote Chapters 8 and 9 to the special functions of the cylinder and the sphere. These are developed as examples of the role of special functions in mathematical physics, and emphasis is placed on the origin of the formulas in physical or geometrical relationships.

In the last chapter we return to the wave equation in two and three dimensions for which the elementary solutions are now distributions. Applications to elastic waves and diffraction by a wedge are included.

The scope of the material covered permits use of the book in a full year's program. For a one-semester course a wide choice of topics is available. It is possible, for instance, to concentrate on Chapters 2, 3, 4, 6, and 7, and parts of Chapter 8. Material from Chapters 5, 8, 9, and 10 can be used as desired when the preceding chapters have been covered.

In writing this text book we have endeavored to cover the following points. First, the wide range of application of linear differential equations to physical problems. Second, a unified mathematical background for the eigenfunction method and the employment of distributions. Third, introduction of these techniques in a gradual and accessible sequence, with illustrative exercises of varying difficulty. Fourth, the background in mechanics of the problems of motion for continuous systems.

We have attempted to maintain a close relationship between the mathematical theories and their physical applications. To present distribution theory and the eigenfunction methods, we have used a spiral technique, by which we return to each topic at increasingly higher levels.

Most of the exercises are closely related to the sections they accompany. A few more difficult or far-ranging problems are included. The exercises should be regarded as an integral part of the text, and we urge the student to make every effort to study them.

Although volumes could be written on the relation of mathematics and physics, it is the authors' belief that a willing appreciation of both is a matter of critical significance. The history of controversy, which began with Heaviside's use of operators and ultimately led to the invention of distributions, indicates the value to the mathematician of physical insight. Conversely, the physicist and applied mathematician have much to gain from the modern mathematical theories based on functional analysis. It is our hope that this book will encourage its readers to value the working alliance of mathematics and physical science for what it has already made possible and for all it may contribute in the future.

We acknowledge with thanks the assistance we have received from colleagues and students in the preparation of this book. To Professor R. R. Burnside and Mr. R. A. Adams we are indebted for many helpful suggestions. Professor B. Friedman contributed numerous comments of great value to the coherence and clarity of the work. All responsibility for error or obscurity, however, remains with us, and we welcome comment, criticism, or correction from our readers.

G. F. D. DUFF
D. NAYLOR

Toronto, Canada
January 1966

Contents

CHAPTER 1. FINITE SYSTEMS 1

   1.1. Vectors and Linear Spaces 1
   1.2. Matrices and Linear Transformations 12
   1.3. Ordinary Differential Systems 19
   1.4. Finite Mechanical Systems 26
   1.5. Lagrange's Equations and Hamilton's Principle 28
   1.6. Systems with Constant Coefficients 35
   1.7. The Response Matrix and Distributions 39
   1.8. Two-Point Boundary Conditions 44

CHAPTER 2. DISTRIBUTIONS AND WAVES 49

   2.1. Equations in Two Independent Variables 49
   2.2. Wave Motion of a String 53
   2.3. Reflection of Waves 59
   2.4. Theory of Distributions 66
   2.5. Applications to the Initial Problem 73
   2.6. The Separation of Variables 81
   2.7. Fourier Series 87

CHAPTER 3. PARABOLIC EQUATIONS AND FOURIER INTEGRALS 95

3.1. The Heat Flow Equation 95
3.2. Heat Flow on a Finite Interval 100
3.3. Fourier Integral Transforms 108
3.4. Diffusion on an Infinite Interval 113
3.5. Semi-Infinite Intervals 118
3.6. Fourier Transforms of Distributions 124
3.7. Finite Difference Calculations 129

CHAPTER 4. LAPLACE'S EQUATION AND COMPLEX VARIABLES 133

4.1. Mathematical and Physical Applications 133
4.2. Boundary Value Problems for Harmonic Functions 136
4.3. Circular Harmonics 141
4.4. Rectangular Harmonics 148
4.5. Half-Plane Problems 153
4.6. Complex Integrals 159
4.7. Fourier and Laplace Transforms 167
4.8. The Finite Difference Laplace Equation 173

CHAPTER 5. EQUATIONS OF MOTION 180

5.1. Vibrations of a Membrane 180
5.2. Lateral Vibration of Rods and Plates 187
5.3. Integral Theorems and Vector Calculus 194
5.4. Equations of Motion of an Elastic Solid 201
5.5. Motion of a Fluid 206
5.6. Equations of the Electromagnetic Field 214
5.7. Equations of Quantum Mechanics 220

CHAPTER 6. GENERAL THEORY OF EIGENVALUES AND EIGENFUNCTIONS 228

6.1. The Minimum Problem 228
6.2. Sequences of Eigenvalues and Eigenfunctions 233
6.3. Variational Properties of Eigenvalues and Eigenfunctions 239
6.4. Eigenfunction Expansions 244
6.5. The Rayleigh-Ritz Approximation Method 250
6.6. On the Separation of Variables 254
6.7. Series Expansions and Integral Transforms 260

CHAPTER 7. GREEN'S FUNCTIONS 264

7.1. Inverses of Differential Operators 264
7.2. Examples of Green's Functions 270
7.3. The Neumann and Robin Functions 277
7.4. Differential and Integral Equations 283
7.5. Source Functions for Parabolic Equations 288
7.6. Convergence of Series of Distributions 292

CHAPTER 8. CYLINDRICAL EIGENFUNCTIONS 298

8.1. Bessel Functions 298
8.2. Eigenfunctions for Finite Regions 306
8.3. The Fourier-Bessel Series 310
8.4. The Green's Function 314
8.5. Functions of Large Argument 317
8.6. Diffraction by a Cylinder 323
8.7. Modified Bessel Functions 327
8.8. The Hankel and Weber Formulas 333

CHAPTER 9. SPHERICAL EIGENFUNCTIONS 341

9.1. Legendre Functions 341
9.2. Eigenfunctions of the Spherical Surface 345
9.3. Eigenfunctions for the Solid Sphere 351
9.4. Diffraction by a Sphere: Addition Theorem 356
9.5. Interior and Exterior Expansions 362
9.6. Functions of Nonintegral Order 367

CHAPTER 10. WAVE PROPAGATION IN SPACE 372

10.1. Characteristic Surfaces 372
10.2. Source Function for the Wave Equation 378
10.3. Applications. Huyghens' Premise 385
10.4. Electromagnetic and Elastic Waves 390
10.5. Wave Fronts and Rays 397
10.6. Reflection and Diffraction 401

TABLES 411

BIBLIOGRAPHY 417

INDEX 419

CHAPTER *1*

Systems of Finite Order

1.1 VECTORS AND LINEAR SPACES

The simplest definition of a *vector* is as a quantity that has magnitude and direction. It is thereby supposed that there is a geometrical or physical underlying space. For elementary applications, this space has a Cartesian coordinate system in which the vector is represented by an ordered set of numerical components. We shall frequently use vectors in this geometrical or physical way.

Vector functions of position in space, or *vector fields*, appear in many branches of applied mathematics. The determination of electromagnetic fields, fluid flows, or elastic waves are examples of problems in which vector fields appear as dependent variables. Numerous examples and applications of this type appear throughout this book as well.

We shall also use vectors in another, and quite different, way, namely, to represent functions. For instance, a function defined at a finite set of points P_i has a sequence of values $f(P_i)$ which can be regarded as components of a vector in a suitable vector space. With these applications in view, we begin by describing vector spaces and their most significant properties.

Such a vector space, or linear space V satisfies the following axioms, in which the elements $\mathbf{x}_1, \mathbf{x}_2, \ldots$ of V are called *vectors* and the coefficients $\lambda, \mu, \ldots$ *scalars*.

Axioms of vector addition for a linear space V:

(1.1.1) $\mathbf{x}_1 + \mathbf{x}_2 = \mathbf{x}_2 + \mathbf{x}_1$ (commutative law)

(1.1.2) $\mathbf{x}_1 + (\mathbf{x}_2 + \mathbf{x}_3) = (\mathbf{x}_1 + \mathbf{x}_2) + \mathbf{x}_3$ (associative law)

(1.1.3) There exists a zero vector $\mathbf{0}$ such that $\mathbf{x} + \mathbf{0} = \mathbf{x}$ for all $\mathbf{x} \in V$.

(1.1.4) To every element $\mathbf{x}$ there corresponds an inverse element $-\mathbf{x}$, such that $\mathbf{x} + (-\mathbf{x}) = \mathbf{0}$.

Scalar multiplication is defined, for real numbers $\lambda, \mu, \ldots$ with

(1.1.5) $(\lambda \mu)\mathbf{x} = \lambda(\mu \mathbf{x})$ (associative law)

(1.1.6) $(\lambda + \mu)\mathbf{x} = \lambda \mathbf{x} + \mu \mathbf{x}$

$\lambda(\mathbf{x}_1 + \mathbf{x}_2) = \lambda \mathbf{x}_1 + \lambda \mathbf{x}_2$ (distributive laws)

(1.1.7) $1 \cdot \mathbf{x} = \mathbf{x}$

Example 1.1.1 **The Cartesian n-dimensional space R^n.** Let $\mathbf{x} = (x_1, \ldots, x_n)$ be an ordered n tuple of real numbers x_i, to which correspond a point $\mathbf{x}$ with these Cartesian coordinates and a vector $\mathbf{x}$ with these components. We define addition of vectors by component addition

(1.1.8) $\mathbf{x} + \mathbf{y} = (x_1 + y_1, \ldots, x_n + y_n)$

and scalar multiplication by component multiplication

(1.1.9) $\lambda \mathbf{x} = (\lambda x_1, \ldots, \lambda x_n)$

The Euclidean length of the vector $\mathbf{x}$ is

(1.1.10) $|\mathbf{x}| = (x_1^2 + \cdots + x_n^2)^{1/2}$

and is often called the *norm* of $\mathbf{x}$.

Example 1.1.2 **Hilbert coordinate space $H = R^\infty$.** Let $\mathbf{x} = (x_1, \ldots, x_k, \ldots)$ be an infinite sequence of real numbers such that $\sum_{k=1}^{\infty} x_k^2$ converges. The sequence $\mathbf{x}$ defines a point $\mathbf{x}$ of the Hilbert coordinate space, with kth coordinate x_k and also a vector with kth component x_k, which, as in R^n, we identify with the point. Addition and scalar multiplication are defined analogously to Example 1.1.1. The norm of the Hilbert vector $\mathbf{x}$ is the Pythagorean expression

(1.1.11) $|\mathbf{x}| = \left(\sum_{k=1}^{\infty} x_k^2 \right)^{1/2}$

By hypothesis this series converges if $\mathbf{x}$ is an element of H.

We now introduce the important concept of linear dependence of vectors (Birkhoff-MacLane, Ref. 5)*. Let $\mathbf{x}_1, \ldots, \mathbf{x}_m$ be m given vectors and $\lambda_1, \ldots, \lambda_m$ an equal number of scalars. Then we can form the linear combination or *sum*

$$\lambda_1 \mathbf{x}_1 + \cdots + \lambda_k \mathbf{x}_k + \cdots + \lambda_m \mathbf{x}_m$$

which is also an element of the vector space.

Suppose that there exist values $\lambda_1, \ldots, \lambda_m$, which are not all zero, such that the above vector sum is the zero vector. Then the vectors $\mathbf{x}_1, \ldots, \mathbf{x}_m$ are said to be *linearly dependent*. That is, there exists a nontrivial linear relation or identity among the m vectors $\mathbf{x}_1, \ldots, \mathbf{x}_m$.

Example 1.1.3 The vectors $\mathbf{x}_1 = (1, 2, 0)$, $\mathbf{x}_2 = (0, 1, 1)$ and $\mathbf{x}_3 = (1, 0, -2)$ are linearly dependent because

$$\mathbf{x}_1 - 2\mathbf{x}_2 - \mathbf{x}_3 = 0$$

Geometrically, the three vectors are coplanar.

But the vectors $\mathbf{x}_1, \ldots, \mathbf{x}_m$ may not be linearly dependent. In this case we say that they are *linearly independent*. If the vectors $\mathbf{x}_1, \ldots, \mathbf{x}_m$ are linearly independent, and if

$$\lambda_1 \mathbf{x}_1 + \cdots + \lambda_k \mathbf{x}_k + \cdots + \lambda_m \mathbf{x}_m = 0$$

then the scalars λ_k must all be zero. For, if not, the vectors would be linearly dependent. Therefore $\mathbf{x}_1, \ldots, \mathbf{x}_m$ are linearly independent if and only if every nontrivial linear combination of them is different from zero.

The vectors $\mathbf{x}_1$ and $\mathbf{x}_2$ in Example 1.13 are linearly independent.

Referring now to the Cartesian vector space R^n, we shall prove that *any $n + 1$ nonzero vectors of R^n are linearly dependent*. If the vectors are $\mathbf{x}_1, \ldots, \mathbf{x}_{n+1}$, then the condition of linear dependence is

$$\lambda_1 \mathbf{x}_1 + \cdots + \lambda_{n+1} \mathbf{x}_{n+1} = 0$$

and in component form this becomes

$$\lambda_1 x_{1i} + \cdots + \lambda_{n+1} x_{n+1\,i} = 0 \qquad i = 1, \ldots, n$$

We regard the $n + 1$ scalars $\lambda_1, \ldots, \lambda_{n+1}$ as unknowns in this system of n linear homogeneous equations with coefficients x_{ki}. To show that these equations have a nontrivial solution $\lambda_1, \ldots, \lambda_n$, we consider the determinant $\Delta_1 = |x_{ki}|$, where $i, k = 1, \ldots, n$. If Δ_1 is zero, then by the standard theory of linear equations, the system with λ_{n+1} set equal to zero has a nontrivial solution. If, on the other hand, Δ_1 is not zero, we shall transpose the term $\lambda_{n+1} x_{n+1\,i}$ to the right side of each equation and select some nonzero value for λ_{n+1}. The resulting nonhomogeneous system for $\lambda_1, \ldots, \lambda_n$ has determinant $\Delta_1 \neq 0$ and, therefore,

* References of this form are to the Bibliography on p. 417.

a unique solution exists for given λ_{n+1}. In either case the system of $n+1$ linear equations for n unknowns has a nontrivial solution $\lambda_1, \ldots, \lambda_{n+1}$. Therefore $\mathbf{x}_1, \ldots, \mathbf{x}_{n+1}$ are linearly dependent as asserted.

On the other hand, *the Hilbert space of Example 1.1.2 contains infinitely many linearly independent vectors*. For example, the vectors of the sequence

(1.1.12) $$\mathbf{e}^{(k)} = (0, \ldots, 0, 1, 0, \ldots)$$

where 1 appears in the kth place and zero elsewhere, are linearly independent.

In geometry we associate each component of a vector with a coordinate axis or a dimension of the space. It is therefore natural to describe the dimension of a linear vector space by the greatest possible number of linearly independent vectors:

DEFINITION 1.1.1 *The dimension* dim V *is the maximal number of linearly independent vectors of* V.

In this first chapter we shall use mainly finite-dimensional vector spaces, and the related algebra of finite matrices. Subsequent chapters are devoted to partial differential equations, and these lead to the study of vector spaces of infinite dimension.

A set of vectors $\mathbf{x}_{(k)}$ is said to *span* V if every vector $\mathbf{x}$ of V is equal to a linear combination of the given vectors $\mathbf{x}_{(k)}$. If, furthermore, the $\mathbf{x}_{(k)}$ are linearly independent they are said to form a *basis* for V.

THEOREM 1.1.1 *Let V be a vector space of dimension n. Then:* (1) *Every basis for V has exactly n elements.* (2) *Every linearly independent subset of V has at most n elements, and, if n is finite, is a basis for V if and only if it has exactly n elements.*

Let us establish this result for a finite-dimensional vector space V_n. Suppose $\mathbf{x}_1, \ldots, \mathbf{x}_m$ form a basis for V_n. Since the vectors $\mathbf{x}_k$ are linearly independent, their number is at most n, by the definition of dim V_n. Suppose now that $m < n$. By hypothesis there exist n linearly independent vectors $\mathbf{y}_1, \ldots, \mathbf{y}_n$. Since $\mathbf{x}_1, \ldots, \mathbf{x}_m$ form a basis, we can express every $\mathbf{y}_k$ as a linear combination

$$\mathbf{y}_k = \sum_{l=1}^{m} a_{kl} \mathbf{x}_l \qquad k = 1, \ldots, n$$

Then

$$\sum_{k=1}^{n} \lambda_k \mathbf{y}_k = \sum_{l=1}^{m} \left(\sum_{k=1}^{n} \lambda_k a_{kl} \right) \mathbf{x}_l$$

and we can make the inner sums vanish for $l = 1, \ldots, m < n$ by a suitable choice of $\lambda_1, \ldots, \lambda_n$. Indeed the requisite conditions

$$\sum_{k=1}^{n} \lambda_k a_{kl} = 0 \qquad l = 1, \ldots, m$$

form a set of $m < n$ linear equations in n unknowns. By reasoning already used, we see that a nontrivial solution $\lambda_1, \ldots, \lambda_n$ of this system can be found. But then

$$\sum_{k=1}^{n} \lambda_k \mathbf{y}_k = 0$$

which contradicts the linear independence of $\mathbf{y}_1, \ldots, \mathbf{y}_n$. Therefore m must be equal to n and part (1) of Theorem 1.1.1 is proved.

To demonstrate part (2) for n finite, consider a subset $\mathbf{x}_1, \ldots, \mathbf{x}_m$ of linearly independent vectors of V_n. Again, $m \leq n$ by the definition of dimension. We have already shown that, if the vectors $\mathbf{x}_1, \ldots, \mathbf{x}_m$ form a basis, then $m = n$. Conversely, if $m = n$, then $\mathbf{x}_1, \ldots, \mathbf{x}_m$ must form a basis, because any vector $\mathbf{x}$ of V_n is a linear combination of them. To see this, we need only remember that any $n + 1$ vectors of V_n must be linearly dependent. Thus

$$\lambda \mathbf{x} + \lambda_1 \mathbf{x}_1 + \cdots + \lambda_n \mathbf{x}_n = 0$$

for some $\lambda, \lambda_1, \ldots, \lambda_n$, where $\lambda \neq 0$ because the $\mathbf{x}_k$ are linearly independent. By dividing by λ and solving for $\mathbf{x}$ we gain our result.

This last conclusion is not valid for vector spaces of infinite dimension because such a space can contain an infinite sequence of linearly independent vectors which do not form a basis. For instance, if we omit from a basis any finite number of vectors, we obtain such a sequence.

The vector spaces to be used in this book have another property which we now introduce. They carry an *inner product* $(\mathbf{x}, \mathbf{y})$. Such a product is a real number, which satisfies the following inner product axioms:

(1.1.13) $\qquad (\mathbf{x}, \mathbf{x}) \geq 0 \qquad$ with equality only for $\mathbf{x} = 0$ (definite)

(1.1.14) $\qquad (\mathbf{x}, \mathbf{y}) = (\mathbf{y}, \mathbf{x}) \qquad\qquad$ (symmetric)

(1.1.15) $\qquad (c_1 \mathbf{x}_1 + c_2 \mathbf{x}_2, \mathbf{y}) = c_1(\mathbf{x}_1, \mathbf{y}) + c_2(\mathbf{x}_2, \mathbf{y}) \qquad$ (linear)

By means of the inner product we may define a magnitude or *norm* for $\mathbf{x}$. Thus our vector space becomes a normed linear space. The norm of $\mathbf{x}$ shall be the nonnegative square root of $(\mathbf{x}, \mathbf{x})$, and is denoted by $|\mathbf{x}|$: Thus,

(1.1.16) $\qquad |\mathbf{x}| = (\mathbf{x}, \mathbf{x})^{1/2}$

Relative to the norms of $\mathbf{x}$ and $\mathbf{y}$ the inner product satisfies the inequality of Schwarz:

(1.1.17) $\qquad |(\mathbf{x}, \mathbf{y})| \leq |\mathbf{x}| \, |\mathbf{y}|$

To establish the inequality, we consider the norm of the linear combination $\lambda \mathbf{x} + \mu \mathbf{y}$:

$$0 \leq |\lambda \mathbf{x} + \mu \mathbf{y}|^2 = (\lambda \mathbf{x} + \mu \mathbf{y}, \lambda \mathbf{x} + \mu \mathbf{y})$$
$$= \lambda^2 |\mathbf{x}|^2 + 2\lambda\mu(\mathbf{x}, \mathbf{y}) + \mu^2 |\mathbf{y}|^2$$

The quadratic form in λ and μ on the right is nonnegative. Therefore, by the theory of the quadratic equation, its discriminant cannot be positive. The discriminant is

$$(\mathbf{x}, \mathbf{y})^2 - |\mathbf{x}|^2 |\mathbf{y}|^2 \leq 0$$

Transposing the second term and taking square roots, we arrive at the Schwarz inequality.

For every pair of vectors $\mathbf{x}_{(1)}$, $\mathbf{x}_{(2)}$ in R^n (or H), we can *define* an inner product as

(1.1.18) $$(\mathbf{x}_{(1)}, \mathbf{x}_{(2)}) = \sum_{k=1}^{n} x_{(1)k} x_{(2)k}$$

Here $n = \infty$ for Hilbert space and, by (1.1.17), the infinite sum is convergent. The reader can verify that axioms (1.1.13), (1.1.14), and (1.1.15) are satisfied. Just as in ordinary vector calculus, the inner product of $\mathbf{x}$ with itself is the square of the length or norm of $\mathbf{x}$:

(1.1.19) $$(\mathbf{x}, \mathbf{x}) = \sum_{k=1}^{n} x_k^2 = |\mathbf{x}|^2$$

Two vectors are *orthogonal* if their inner product is zero. By analogy with Cartesian geometry, we can define the *angle* θ between two vectors $\mathbf{x}_{(1)}$ and $\mathbf{x}_{(2)}$ by the equation

(1.1.20) $$(\mathbf{x}_{(1)}, \mathbf{x}_{(2)}) = |\mathbf{x}_{(1)}| |\mathbf{x}_{(2)}| \cos \theta$$

By the Schwarz inequality, we shall have $|\cos \theta| \leq 1$, so that θ is a real angle which can be chosen between 0 and π.

THEOREM 1.1.2 *Any set of mutually orthogonal nonvanishing vectors is linearly independent.*

Proof. Let the vectors be $\mathbf{x}_{(k)}$, and suppose

$$\sum_k \lambda_k \mathbf{x}_{(k)} = 0$$

Taking the scalar product with $\mathbf{x}_{(l)}$ and using $(\mathbf{x}_{(l)}, \mathbf{x}_{(k)}) = 0$ for $l \neq k$, we find

$$0 = \left(\mathbf{x}_{(l)}, \sum_k \lambda_k \mathbf{x}_{(k)}\right) = \sum_k \lambda_k (\mathbf{x}_{(l)}, \mathbf{x}_{(k)}) = \lambda_l |\mathbf{x}_{(l)}|^2$$

Since $|\mathbf{x}_{(l)}|$ is not zero, we conclude that $\lambda_l = 0$ for every l. By the definition of linear independence, this proves the result.

To develop further geometric relations, we introduce the concept of an orthogonal basis of unit vectors, or *orthonormal basis*. Such a basis is composed of vectors $\mathbf{e}_{(k)}$ of unit length which are mutually orthogonal. Thus

(1.1.21) $$(\mathbf{e}_{(k)}, \mathbf{e}_{(l)}) = \delta_{kl} \qquad k, l = 1, \ldots, n$$

On the right-hand side we have introduced a symbol δ_{kl}, known as the Kronecker delta. This symbol δ_{kl} denotes unity if $k = l$ and zero if $k \neq l$. We regard these unit coordinate vectors $\mathbf{e}_{(k)}$ as a generalization of the unit vectors $\mathbf{i}, \mathbf{j}, \mathbf{k}$ of vector calculus in R^3.

Now any vector $\mathbf{x}$ of V is a linear combination of the basis vectors $\mathbf{e}_{(k)}$,

$$(1.1.22) \qquad \mathbf{x} = \sum_k x_k \mathbf{e}_{(k)}$$

The n numbers x_k so determined can be called the coefficients or coordinates of $\mathbf{x}$ with respect to the basis $\mathbf{e}_{(k)}$. To determine these coordinates, we take scalar products with $\mathbf{e}_{(l)}$ in (1.1.22). By (1.1.21), we find

$$(\mathbf{e}_{(l)}, \mathbf{x}) = \sum_k x_k (\mathbf{e}_{(l)}, \mathbf{e}_{(k)}) = x_l$$

and so

$$(1.1.23) \qquad x_l = (\mathbf{e}_{(l)}, \mathbf{x})$$

We shall encounter this formula many times and in many variations throughout this book.

THEOREM 1.1.3 *A vector space V of dimension n has an orthonormal basis of n vectors $\mathbf{e}_{(k)}$.*

Proof. Let $\mathbf{x}_{(k)}$ be an arbitrary basis, $k = 1, 2, \ldots, n$. Choose a unit vector parallel to the nonzero vector $\mathbf{x}_{(1)}$,

$$(1.1.24) \qquad \mathbf{e}_{(1)} = \frac{\mathbf{x}_{(1)}}{|\mathbf{x}_{(1)}|}$$

and construct the orthonormal sequence as follows. Let a second basis vector $\mathbf{x}_{(2)}$ independent of $\mathbf{x}_{(1)}$ be employed to form

$$(1.1.25) \qquad \mathbf{y}_{(2)} = \mathbf{x}_{(2)} - \alpha \mathbf{e}_{(1)} \qquad\qquad \mathbf{y}_{(2)} \neq 0$$

and choose α so that

$$(1.1.26) \qquad (\mathbf{e}_{(1)}, \mathbf{y}_{(2)}) = (\mathbf{e}_{(1)}, \mathbf{x}_{(2)}) - \alpha(\mathbf{e}_{(1)}, \mathbf{e}_{(1)}) = 0$$

That is, $\alpha = (\mathbf{e}_{(1)}, \mathbf{x}_{(2)})$. Now choose

$$(1.1.27) \qquad \mathbf{e}_{(2)} = \frac{\mathbf{y}_{(2)}}{|\mathbf{y}_{(2)}|}$$

Clearly $\mathbf{e}_{(2)}$ is a unit vector orthogonal to $\mathbf{e}_{(1)}$. To define $\mathbf{e}_{(3)}$ we construct

$$(1.1.28) \qquad \mathbf{y}_{(3)} = \mathbf{x}_{(3)} - \alpha_{31} \mathbf{e}_{(1)} - \alpha_{32} \mathbf{e}_{(2)}$$

and choose α_{31}, α_{32} so that $(\mathbf{y}_{(3)}, \mathbf{e}_{(1)}) = 0$, $(\mathbf{y}_{(3)}, \mathbf{e}_{(2)}) = 0$. This entails

$$0 = (\mathbf{y}_{(3)}, \mathbf{e}_{(1)}) = (\mathbf{x}_{(3)}, \mathbf{e}_{(1)}) - \alpha_{31}(\mathbf{e}_{(1)}, \mathbf{e}_{(1)}) - \alpha_{32}(\mathbf{e}_{(2)}, \mathbf{e}_{(1)})$$

so that $\alpha_{31} = (\mathbf{x}_{(3)}, \mathbf{e}_{(1)})$ and likewise $\alpha_{32} = (\mathbf{x}_{(3)}, \mathbf{e}_{(2)})$. The construction continues in a straightforward, though lengthy, manner. As an exercise, the reader can complete the proof by induction on $\mathbf{e}_{(k)}$.

The sequence $\mathbf{e}_{(1)}, \mathbf{e}_{(2)}, \ldots, \mathbf{e}_{(k)}$ found by this "Gram-Schmidt" process will terminate at the nth stage, if n is finite.

A *linear subspace* W of V is a linear space contained in V. For example, the XZ coordinate plane is a two-dimensional linear subspace of R^3. A vector $\mathbf{x}$ is *orthogonal to* W if and only if it is orthogonal to every vector $\mathbf{w}$ of W.

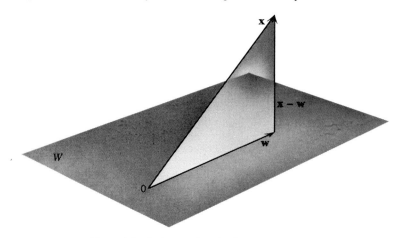

Figure 1.1 Orthogonal projection on a subspace.

The *orthogonal projection* $\mathbf{P}_W \mathbf{x}$ of $\mathbf{x}$ on W is a vector $\mathbf{w}$ of W such that $\mathbf{x} - \mathbf{w}$ is orthogonal to W. (See Figure 1.1.) It is easy to show that this projection is unique. For if two such projections $\mathbf{w}_1, \mathbf{w}_2$ exist, then $\mathbf{w}_1 - \mathbf{w}_2 \in W$, while $(\mathbf{x} - \mathbf{w}_1) - (\mathbf{x} - \mathbf{w}_2) = \mathbf{w}_2 - \mathbf{w}_1$ is orthogonal to W. Therefore $\mathbf{w}_2 - \mathbf{w}_1$ is orthogonal to itself and so is zero. Hence $\mathbf{w}_2 = \mathbf{w}_1$.

Let W_m be a linear subspace of dimension m, and let $\mathbf{e}_{(1)}, \ldots, \mathbf{e}_{(m)}$ be an orthonormal basis for W_m. We shall build an orthonormal basis $\mathbf{e}_{(1)}, \ldots, \mathbf{e}_{(n)}$ for V_n which contains this basis for W_m as a subset. To do this, we need only apply the orthogonalizing process used above to the sequence of $m + n$ vectors consisting of $\mathbf{e}_{(1)}, \ldots, \mathbf{e}_{(m)}$ together with any basis of V_n. Of these latter vectors, m will drop out by linear dependence while $n - m$ remain to augment the basis of W_m.

THEOREM 1.1.4 *Let* $\mathbf{e}_{(1)}, \ldots, \mathbf{e}_{(n)}$ *form an orthonormal basis for* V_n, *and* $\mathbf{e}_{(1)}, \ldots, \mathbf{e}_{(m)}$ *a basis for* $W_m, m < n$. *Let*

$$(1.1.29) \quad \mathbf{x} = \sum_{k=1}^{n} x_k \mathbf{e}_{(k)} \quad x_k = (\mathbf{e}_{(k)}, \mathbf{x})$$

Then the orthogonal projection of $\mathbf{x}$ *on* W_m *is the vector*

$$(1.1.30) \quad \mathbf{P}_W \mathbf{x} = \sum_{k=1}^{m} x_k \mathbf{e}_{(k)} = \sum_{k=1}^{m} (\mathbf{e}_{(k)}, \mathbf{x}) \mathbf{e}_{(k)}$$

VECTORS AND LINEAR SPACES 9

Proof. Clearly $\mathbf{P}_W \mathbf{x}$ lies in W_m, and

(1.1.31) $$\mathbf{x} - \mathbf{P}_W \mathbf{x} = \sum_{m+1}^{n} x_l \mathbf{e}_{(k)}$$

is orthogonal to W_m. Thus $\mathbf{P}_W \mathbf{x}$ is uniquely determined as required.

This result makes evident the *idempotence* property $\mathbf{P}_W \mathbf{P}_W \mathbf{x} = \mathbf{P}_W \mathbf{x}$ of the projection operator $\mathbf{P}_W$.

The power and range of our results can be better appreciated from the following example, which we shall use frequently in later chapters.

Example 1.1.4 Let D be an interval or a region, and let H be the set of real functions $\mathbf{f}$ with values $f(x)$ such that

$$\int_D f(x)^2 \, dx < \infty$$

H is a linear space, called the real Hilbert space on D. As inner product, we take the bilinear form in $f(x), g(x)$

$$(\mathbf{f}, \mathbf{g}) = \int_D f(x) g(x) \, dx$$

which satisfies the inner product axioms (1.1.13) through (1.1.15). The Schwarz inequality $|(\mathbf{f}, \mathbf{g})| \leq |\mathbf{f}| \, |\mathbf{g}|$ is valid, where now

$$|\mathbf{f}| = \left[\int_D f(x)^2 \, dx \right]^{1/2}$$

We shall return in Chapter 6 to a more detailed study of H in connection with eigenfunction expansions. Here we content ourselves with an example of the foregoing projection theorem, which is related to our subsequent use of Fourier series.

Example 1.1.5 On the interval $(0, 2\pi)$ consider the Hilbert space with inner product

$$(\mathbf{f}, \mathbf{g}) = \int_0^{2\pi} f(x) g(x) \, dx$$

Let W_m be the subspace spanned by the functions $\cos mx$, $\sin mx$, where m is a fixed integer.

Since

$$\int_0^{2\pi} \sin mx \cos mx \, dx = \frac{1}{2} \int_0^{2\pi} \sin 2mx \, dx$$

$$= \frac{1}{4m} [-\cos 2mx]_0^{2\pi} = 0$$

the two given functions are orthogonal. Since

$$\int_0^{2\pi} \sin^2 mx \, dx = \int_0^{2\pi} \cos^2 mx \, dx = \pi$$

10 SYSTEMS OF FINITE ORDER

the normalized functions are $e_1 = \pi^{-1/2} \cos mx$, $e_2 = \pi^{-1/2} \sin mx$. By Theorem 1.1.4, the projection on W_m of $f(x)$ is

$$P_m f = (e_1, f) e_1 + (e_2, f) e_2$$

$$= \int_0^{2\pi} f(s) \frac{\cos ms}{\sqrt{\pi}} ds \frac{\cos mx}{\sqrt{\pi}} + \int_0^{2\pi} f(s) \frac{\sin ms}{\sqrt{\pi}} ds \frac{\sin mx}{\sqrt{\pi}}$$

$$= \frac{1}{\pi} \int_0^{2\pi} f(s) \cos m(x-s) \, ds$$

Such integrals play a central role in Fourier series, which we encounter in Chapters 2, 3, and 6.

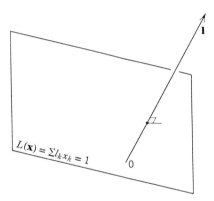

Figure 1.2 Hyperplane and normal vector.

We conclude this introductory section with a theorem which will be greatly extended in Chapter 2, and of which the finite-dimensional case shown here is a revealing example.

A *linear form*, or *functional*, on V_n is a function $L = L(x)$ from vectors x to real numbers $L(x)$, which is *additive*

$$L(x_1 + x_2) = L(x_1) + L(x_2)$$

and also *homogeneous*

$$L(\lambda x) = \lambda L(x)$$

THEOREM 1.1.5 *Let $L(x)$ be a linear form on an inner product space V_n. Then there is a unique vector $l \in V_n$ such that*

$$L(x) = (l, x)$$

Proof. The space V_n has an orthonormal basis e_k relative to the inner product. Let the number $L(e_k)$ be denoted by l_k, and let the vector l be defined as

$$l = \Sigma l_k e_k$$

Then, for any $\mathbf{x} = \Sigma x_k \mathbf{e}_k$,
$$L(\mathbf{x}) = \Sigma x_k L(\mathbf{e}_k) = \Sigma x_k l_k.$$
But also
$$(\mathbf{l}, \mathbf{x}) = \left(\sum_k l_k \mathbf{e}_k, \sum_m x_m \mathbf{e}_m \right)$$
$$= \sum_{k,m} l_k x_m (\mathbf{e}_k, \mathbf{e}_m)$$
$$= \sum_{k,m} l_k x_m \delta_{km} = \sum_k l_k x_k = L(\mathbf{x})$$
and the existence of $\mathbf{l}$ is proved.

As an exercise, show that the vector $\mathbf{l}$ is unique. *Remark.* This theorem shows that a linear form $L(\mathbf{x})$ over an inner product space V_n can be identified with a vector of the space, normal to the hyperplane $L(\mathbf{x}) = 0$. (See Figure 1.2.)

EXERCISES 1.1

1. Show that $\mathbf{x}_k = (x_{ik})$ form a basis for R^n if and only if
$$\det |x_{ik}| \neq 0$$

2. Show that $\{\mathbf{x}_k\} = (\delta_{kl})$ form a basis for R^∞. Do $\{\mathbf{y}_k\} = (\delta_{k, l+1})$, or $\{\mathbf{z}_k\} = (\delta_{k, l-1})$ form a basis for R^∞?

3. A *norm* $|\mathbf{x}|$ satisfies
$$|\mathbf{x} + \mathbf{y}| \leq |\mathbf{x}| + |\mathbf{y}|$$
$$|\lambda \mathbf{x}| = |\lambda| \, |\mathbf{x}|$$
$$|\mathbf{x}| > 0 \quad \text{unless} \quad \mathbf{x} = 0$$

Show that $|\mathbf{x}| = \max |x_l|$ defines a norm on V.

4. Does $|\mathbf{x}| = (\Sigma x_i^4)^{1/4}$ define a norm on V? Does $(\mathbf{x}, \mathbf{y}) = (\Sigma x_i^2 y_i^2)^{1/4}$ define an inner product on V?

5. Establish the Schwarz inequality for sums by examination of
$$\sum_{k,l=1}^n (a_k b_l - a_l b_k)^2 \geq 0$$

6. Find the angle between the following vectors in R^n:
$$(1, 1, 1, \ldots, 1)$$
$$(-1, 1, 1, \ldots, 1)$$

7. Find an orthonormal basis for the two-dimensional subspace of R^3 that is orthogonal to $(1, 2, 3)$.

8. Find the orthogonal projection on the subspace of Exercise 1.1.7 of the vector $(3, 2, 1)$.

9. *Weight functions.* If $\rho(x) \geq 0$, show that

$$(\mathbf{f}, \mathbf{g}) = \int_D f(x) g(x) \, \rho(x) \, dx$$

defines an inner product space.

10. Show that the Legendre polynomials

$$P_n(x) = \frac{1}{2^n n!} \frac{d^n}{dx^n} (1 - x^2)^n \qquad n = 1, 2, \ldots$$

are mutually orthogonal on the interval $(-1, 1)$ with respect to the weight function $\rho(x) = 1$. *Hint:* Use integration by parts.

11. *Complex Hilbert space.* A linear space of infinite dimension with inner product $(\mathbf{x}, \mathbf{y})$ which is a complex number satisfying:

(a) The definiteness axiom $(\mathbf{x}, \mathbf{x}) > 0$ for $\mathbf{x} \neq 0$;

(b) the linearity axiom $(c_1 \mathbf{x}_1 + c_2 \mathbf{x}_2, \mathbf{y}) = c_1 (\mathbf{x}_1, \mathbf{y}) + c_2 (\mathbf{x}_2, \mathbf{y})$;

(c) the axiom of Hermitian symmetry $(\mathbf{y}, \mathbf{x}) = \overline{(\mathbf{x}, \mathbf{y})}$, where the bar denotes the complex conjugate, is called a complex Hilbert space.

If $\rho(x)$ is a real and positive function on D, show that

$$(\mathbf{f}, \mathbf{g}) = \int_D f(x) \overline{g(x)} \rho(x) \, dx$$

defines a complex Hilbert space.

12. In a complex Hilbert space show that

$$|\text{Re } (\mathbf{f}, \mathbf{g})| \leq |\mathbf{f}| \, |\mathbf{g}| \qquad \text{where} \qquad |\mathbf{f}|^2 = (\mathbf{f}, \mathbf{f})$$

13. Show that, if $y_n(x) = (2\pi)^{-1/2} e^{inx}$, n an integer, then y_n form an orthonormal set in the complex Hilbert space on $0 \leq x < 2\pi$ with $\rho(x) \equiv 1$. Find the projection of $f(x)$ on the subspace spanned by $y_m(x)$ and $y_{-m}(x)$.

1.2 MATRICES AND LINEAR TRANSFORMATIONS

Commencing once more with a vector space V, we intend to study the linear transformations defined upon it. We consider here the case of finite dimension n, in preparation for subsequent extensions to infinite-dimensional spaces (Birkhoff-MacLane, Ref. 5; Gantmacher, Ref. 23).

A *linear transformation* T is a linear function, or mapping, which carries a vector $\mathbf{x}$ into a new vector denoted by $T\mathbf{x}$. Here T is to be additive:

(1.2.1) $$T(\mathbf{x}_1 + \mathbf{x}_2) = T\mathbf{x}_1 + T\mathbf{x}_2$$

and homogeneous with respect to scalar multiplication:

(1.2.2) $$T(\lambda \mathbf{x}) = \lambda T \mathbf{x}$$

Let $\mathbf{e}_{(k)}$ denote a basis for V, and let us express the vectors $T\mathbf{e}_{(k)}$ with respect to that basis. We have

(1.2.3) $$T\mathbf{e}_{(k)} = \sum_{l=1}^{n} T_{lk} \mathbf{e}_{(l)} \qquad k = 1, \ldots, n$$

Thus the numbers T_{lk} ($k, l = 1, \ldots, n$) are defined by the transformation T acting on the basis vectors $\mathbf{e}_{(k)}$. The array

(1.2.4)
$$\begin{pmatrix} T_{11} & \cdots & T_{1n} \\ \vdots & & \\ T_{n1} & \cdots & T_{nn} \end{pmatrix}$$

of these components is called a *matrix* and will also be denoted by the same symbol T. Rules of operation for matrices correspond to the natural properties of linear transformations. Therefore, because the order of application of linear transformations is significant, it follows that matrix multiplication is not commutative.

As T is linear, and any vector $\mathbf{x}$ depends linearly upon the $\mathbf{e}_{(k)}$, we can express the transformed vector $T\mathbf{x}$ as

(1.2.5)
$$T\mathbf{x} = T\left(\sum_k x_k \mathbf{e}_{(k)}\right) = \sum_k x_k T\mathbf{e}_{(k)} = \sum_{kl} x_k T_{lk} \mathbf{e}_{(l)}$$
$$= \sum_l \left(\sum_k T_{lk} x_k\right) \mathbf{e}_{(l)}$$

The components of $T\mathbf{x}$ with respect to $\mathbf{e}_{(k)}$ are therefore the linear combinations

(1.2.6) $$y_l = \sum_k T_{lk} x_k \qquad k, l = 1, \ldots, n$$

The action of the linear transformation T has thus been expressed by a set of linear algebraic equations, in number equal to the dimension of V. In vector-matrix notation we write

(1.2.7) $$\mathbf{y} = T\mathbf{x}$$

and drop the distinction between the matrix T and the transformation T.

As addition of transformations is defined by

(1.2.8) $$(T_1 + T_2)\mathbf{x} = T_1\mathbf{x} + T_2\mathbf{x}$$

it follows that addition of matrices is to be defined by addition of their components:

(1.2.9) $$(T_1 + T_2)_{lk} = T_{1lk} + T_{2lk}$$

The product, or repeated transformation, is

(1.2.10) $$T_2 T_1 x = T_2(T_1 x)$$

Correspondingly, the product matrix has components

(1.2.11) $$(T_2 T_1)_{lk} = \sum_h T_{2lh} T_{1hk}$$

The summation, or "contraction," runs over adjacent column and row indices.

A transformation T is said to be *nonsingular* if its *range* is the entire vector space V, that is, if the equation $\mathbf{y} = T\mathbf{x}$ has for every $\mathbf{y} \in V$ a solution $\mathbf{x} \in V$. This solution we denote by $T^{-1}\mathbf{y}$, and it is easy to verify that T^{-1} is a linear transformation, the *inverse* of T. We find

(1.2.12) $$TT^{-1} = E \qquad T^{-1}T = E$$

where E is the unit matrix

(1.2.13) $$E = (\delta_{kl}) = \begin{pmatrix} 1 & & & & \\ & 1 & & 0 & \\ & & 1 & & \\ & & & \cdot & \\ & 0 & & \cdot & \\ & & & & 1 \end{pmatrix} = \operatorname{diag}(1, 1, \ldots, 1)$$

Since $ET = TE = T$, the matrix $\lambda E = (\lambda \delta_{kl})$ acts as the operation of scalar multiplication. We remark that the determinant $|T| = \det(T_{kl})$ is not zero for a nonsingular transformation on a finite-dimensional space. However, the matrix algebra does contain divisors of zero; that is, it is possible to have $AB = 0$ with $A \neq 0$, $B \neq 0$.

For any matrix T, we define the $0th$ power T^0 as E. We have already, in effect, defined integral powers T^k, and we can construct polynomial functions and power series of a matrix T. Especially useful is the exponential series

(1.2.14) $$e^T = E + T + \frac{1}{2}T^2 + \frac{1}{3!}T^3 + \cdots + \frac{1}{m!}T^m + \cdots$$

This series, like its scalar analogue, is always convergent in the sense that its partial sums are matrices with components which tend to limits. The matrix sum of the series is formed by the limit components.

The matrix T' formed by interchanging corresponding rows and columns of T is the *transpose* of T; the corresponding transformation T' is sometimes called the *adjoint* of T. The *Hermitian conjugate*, or conjugate transposed matrix T^* is defined as the complex conjugate of T'. Thus,

(1.2.15) $$T^* = \bar{T}'$$

The transposing of a product of matrices has the effect of reversing the order of factors:

$$(T_1 T_2)' = T_2' T_1'$$

A matrix equal to its transpose $T = T'$ is called *symmetric*, and a matrix equal to its Hermitian conjugate is called *Hermitian*. These are the formal conditions for a real or complex matrix, respectively, in order to have the significant property of being *self-adjoint*. Nearly all of the matrix and differential operators we consider will be of this kind.

A real matrix T such that $TT' = E$ is called *orthogonal*; the matrix of a rotation operation has this property. Lastly, a complex matrix T, such that $TT^* = E$, is called *unitary*. The integral transforms developed in subsequent chapters will provide interesting examples of linear transformations having these properties, relative to infinite dimensional vector spaces.

Given a transformation T, we may particularly look for those nonzero vectors **x** upon which T acts as a scalar multiplying factor, that is, such that

(1.2.16) $$T\mathbf{x} = \lambda \mathbf{x}$$

Such vectors are "characteristic" for the transformation T. As well as characteristic vectors, they are known as proper vectors for T. However, we shall use the term *eigenvector* (eigen ≡ own) and the corresponding designation of the scalar λ as an *eigenvalue*.

In components we find

(1.2.17) $$(T\mathbf{x})_l = \sum_k T_{lk} x_k = \lambda x_l \qquad l = 1, \ldots, n$$

and if λ is an eigenvalue, this set of n linear algebraic equations will have a nontrivial solution. This is possible only if the *characteristic equation*

(1.2.18) $$|T - \lambda E| = \det(T - \lambda E) = 0$$

is satisfied by the parameter λ. The roots of the characteristic equation, which is a polynomial equation of degree n in λ, are the characteristic values or eigenvalues $\lambda_1, \ldots, \lambda_n$ of T. In general, therefore, a transformation T has a number of real or complex eigenvalues equal to the dimension of the underlying linear space V.

If a matrix T is symmetric or Hermitian, we can gain further information about its eigenvalues and the corresponding eigenvectors by studying a quadratic form related to T. We shall do this for symmetric matrices (Courant-Hilbert, Ref. 13, vol. 1). Consider the inner product of **x** and its transform $T\mathbf{x}$ which, for a finite-dimensional vector space, has the form

(1.2.19) $$(\mathbf{x}, T\mathbf{x}) = \Sigma x_k T_{lk} x_l = \Sigma T_{kl} x_k x_l$$

We look for extreme (i.e., maximum) values of $(\mathbf{x}, T\mathbf{x})$ for unit vectors **x**: $|\mathbf{x}|^2 = (\mathbf{x}, \mathbf{x}) = 1$. By the Lagrange multiplier theory, such a constrained maximum is attained as a free maximum for the combination

$$(\mathbf{x}, T\mathbf{x}) - \lambda(\mathbf{x}, \mathbf{x}) = (\mathbf{x}, (T - \lambda E)\mathbf{x}).$$

Differentiation of

(1.2.20) $$\sum_{k,l} (T_{kl} - \lambda \delta_{kl}) x_k x_l$$

with respect to x_k and use of the symmetric property of T, yield as equations for extremum

(1.2.21) $$\sum_l (T_{kl} - \lambda \delta_{kl}) x_l = 0 \qquad k = 1, \ldots, n$$

16   SYSTEMS OF FINITE ORDER

In matrix form this reads $T\mathbf{x} = \lambda \mathbf{x}$, so the vector $\mathbf{x}$ is a nonvanishing eigenvector. The eigenvalue λ is the corresponding value of the inner product, since

(1.2.22) $\qquad (\mathbf{x}, T\mathbf{x}) = (\mathbf{x}, \lambda \mathbf{x}) = \lambda(\mathbf{x}, \mathbf{x}) = \lambda |\mathbf{x}|^2 = \lambda$

We denote this eigenvalue and eigenvector pair by $\lambda_1, \mathbf{x}_1$.

Let us compare two eigenvectors of T, which correspond to distinct eigenvalues.

THEOREM 1.2.1 *The eigenvalues of a real symmetric matrix are real, and eigenvectors corresponding to distinct eigenvalues are orthogonal.*

Proof. We take the second statement first. Let $T\mathbf{x}_1 = \lambda_1 \mathbf{x}_1$, $T\mathbf{x}_2 = \lambda_2 \mathbf{x}_2$. Then by Exercise 1.2.6,

$$\lambda_1(\mathbf{x}_1, \mathbf{x}_2) = (\lambda_1 \mathbf{x}_1, \mathbf{x}_2) = (T\mathbf{x}_1, \mathbf{x}_2) = (\mathbf{x}_1, T\mathbf{x}_2),$$

and

$$\lambda_2(\mathbf{x}_1, \mathbf{x}_2) = (\mathbf{x}_1, \lambda_2 \mathbf{x}_2) = (\mathbf{x}_1, T\mathbf{x}_2) = \lambda_1(\mathbf{x}_1, \mathbf{x}_2)$$

by the preceding line. Therefore,

(1.2.23) $\qquad (\lambda_2 - \lambda_1)(\mathbf{x}_1, \mathbf{x}_2) = 0$

If now $\lambda_2 \neq \lambda_1$, then $(\mathbf{x}_1, \mathbf{x}_2) = 0$ as stated.

To prove the first part, let λ_1 be a nonreal eigenvalue. Then $\bar{\lambda}_1$ is a distinct eigenvalue: $T\bar{\mathbf{x}}_1 = \bar{\lambda}_1 \bar{\mathbf{x}}_1$. By (1.2.23) with $\lambda_2 = \bar{\lambda}_1$ and $\mathbf{x}_2 = \bar{\mathbf{x}}_1$, we find

(1.2.24) $\qquad (\bar{\lambda}_1 - \lambda_1)(\mathbf{x}_1, \bar{\mathbf{x}}_1) = 0$

so that $(\mathbf{x}_1, \bar{\mathbf{x}}_1) = \sum_k |x_{1k}|^2$ must be zero, and $x_{1k} \equiv 0$, or $\mathbf{x}_1 \equiv 0$. Finally, therefore, any nonreal λ_1 is not an eigenvalue.

Let us search for further eigenvectors of T among those orthogonal to $\mathbf{x}_1$. To the first maximum problem we adjoin the further condition $(\mathbf{x}_1, \mathbf{x}) = 0$, and again search for an extremum. The constraint leads to the Lagrangian expression

(1.2.25) $\qquad (\mathbf{x}, T\mathbf{x}) - \lambda(\mathbf{x}, \mathbf{x}) - 2\mu(\mathbf{x}, \mathbf{x}_1)$

where μ is a convenient multiplier. The condition for an extremum is now

(1.2.26) $\qquad (T_{kl} - \lambda \delta_{kl})x_l - \mu x_{1k} = 0$

Let us show that μ is zero. Take the inner product with $\mathbf{x}_1$ and observe that $(\mathbf{x}_1, \mathbf{x}) = 0$ by hypothesis while

$$\sum_{kl} x_{1k} T_{kl} x_l = (T\mathbf{x}_1, \mathbf{x}) = (\lambda_1 \mathbf{x}_1, \mathbf{x}) = \lambda_1(\mathbf{x}_1, \mathbf{x}) = 0$$

There remains the term $-\mu(\mathbf{x}, \mathbf{x}_1) = -\mu |\mathbf{x}_1|^2$ which must vanish. Since $\mathbf{x}_1 \neq 0$, we conclude $\mu = 0$, and so

(1.2.27) $\qquad T_{kl} x_l = \lambda x_k$

This new eigenvalue and eigenvector pair will be denoted by $\lambda_2, \mathbf{x}_2$.

As further eigenvectors are orthogonal to $\mathbf{x}_1$ and $\mathbf{x}_2$ we would proceed by adjoining the conditions $|\mathbf{x}| = 1$, $(\mathbf{x}, \mathbf{x}_1) = 0$, $(\mathbf{x}, \mathbf{x}_2) = 0$ to search for λ_3, $\mathbf{x}_3$. Proof that the procedure continues as before will be left to the reader. After n steps, we have an orthonormal basis $\mathbf{x}_1, \ldots, \mathbf{x}_n$ of V.

Denoting the components of $\mathbf{x}_k$ by x_{kl}, for $k = 1, \ldots, n$, we construct the matrix $X = (x_{kl})$ of the eigenvector components. Since

(1.2.28) $$\sum_{l=1}^{n} x_{kl} x_{ml} = \delta_{km}$$

this matrix is orthogonal:

(1.2.29) $$XX' = E$$

Furthermore, the component form of the eigenvalue relations, which are

(1.2.30) $$\sum_{l} T_{kl} x_{ml} = \lambda_m x_{mk}$$

can be written in matrix form as

(1.2.31) $$XT = DX$$

where

(1.2.32) $$D = \text{diag}(\lambda_1, \ldots, \lambda_n) = (\lambda_j \delta_{jk}) = \begin{pmatrix} \lambda_1 & & & & \\ & \lambda_2 & & 0 & \\ & & \cdot & & \\ & 0 & & \cdot & \\ & & & & \lambda_n \end{pmatrix}$$

We multiply on the right by X' and use the orthogonal property

(1.2.33) $$D = XTX' = XTX^{-1}$$

Likewise, using $X^{-1} = X'$, we have

(1.2.34) $$T = X'DX = X^{-1}DX$$

The relation now shown between T and D is known as a *similarity transformation*. Such transformations form a group which leaves invariant functional relations (see Exercise 1.2.7), i.e.,

(1.2.35) $$Xf(T)X^{-1} = f(XTX^{-1}) = f(D)$$

for any formal power series function f of the matrix T.

THEOREM 1.2.2 *The orthogonal matrix X of eigenvectors of a symmetric matrix T defines a similarity transformation by which T is transformed to the diagonal matrix of its eigenvalues.*

Now the quadratic form can be written, by Exercise 1.2.5, in the form

$$F = (\mathbf{x}, T\mathbf{x}) = (\mathbf{x}, X'DX\mathbf{x}) = (X\mathbf{x}, DX\mathbf{x})$$

18 SYSTEMS OF FINITE ORDER

Set $\mathbf{y} = X\mathbf{x}$, and note that

(1.2.36)
$$F = (\mathbf{y}, D\mathbf{y}) = \sum_{kl} y_k D_{kl} y_l$$
$$= \sum_{kl} y_k \lambda_k \delta_{kl} y_l = \sum_k \lambda_k y_k^2$$

The quadratic form has been expressed as a sum of squares, the coefficients being the eigenvalues. The new variables y_k are defined by the rotated coordinate system determined by the orthogonal matrix X. Indeed, since

(1.2.37) $\qquad (\mathbf{y}, \mathbf{y}) = (X\mathbf{x}, X\mathbf{x}) = (\mathbf{x}, X'X\mathbf{x}) = (\mathbf{x}, \mathbf{x})$

we see that for $(\mathbf{y}, \mathbf{y}) = (\mathbf{x}, \mathbf{x}) = 1$, the extreme values of F are the eigenvalues λ_k, taken at $y_l = \delta_{kl}$.

The basis $\{\mathbf{y}_k\}$ is known as the principal axis basis for F, since these vectors determine the principal axes of the ellipsoid $F = (\mathbf{x}, T\mathbf{x}) = 1$ in analytical geometry.

THEOREM 1.2.3 *The eigenvectors of T determine the principal axes of the associated quadratic form $F = (\mathbf{x}, T\mathbf{x})$. In the principal-axis coordinate system, the quadratic form F becomes a sum of squares.*

EXERCISES 1.2

1. In the V_2 spanned by $\{\cos t, \sin t\}$ $(0 \le t \le 2\pi)$, a transformation is defined by

$$Tx(t) = \int_0^{2\pi} \cos(t - \tau) x(\tau) \, d\tau$$

Find the matrix elements of T with respect to $\cos t$, $\sin t$.

2. Let V_{n+1} contain polynomials of degree n in t, with basis $1, t, t^2, \ldots, t^n$. Relative to this basis, find the matrix elements of the operations T_1, T_2, given by

$$T_1 x = t \frac{dx}{dt} \qquad T_2 x = \int_0^t \frac{x(\tau) - x(0)}{\tau} \, d\tau$$

Identify the product $T_1 T_2$.

3. Let V_n be spanned by vectors $f_1(t), \ldots, f_n(t)$ which are linearly independent over an interval $a \le t \le b$. Let

$$K(t, t_1) = \sum_{r,s=1}^{n} c_{rs} f_r(t) f_s(t_1)$$

and let

$$Tx(t) = \int_a^b K(t, t_1) x(t_1) \, dt_1$$

Express the matrix of this linear operation in terms of the matrices $C = (c_{rs})$ and $F = \left(\int_a^b f_r(t) f_s(t) \, dt \right)$.

4. Show that the eigenvalues of a Hermitian matrix are real. Show that the eigenvalues of a unitary matrix have modulus unity.

We have used the determinant rule that multiples of one row can be subtracted from another without change in the value of the determinant. The sum of terms so found is

$$\text{(1.3.13)} \qquad \frac{d}{dt} \det X(t) = \sum_{k=1}^{n} A_{kk} \det X(t)$$

and this is a first order linear homogeneous equation for $\det X(t)$. The trace of a matrix A is the sum of diagonal entries:

$$\text{(1.3.14)} \qquad \operatorname{tr} A(t) = \sum_{k=1}^{n} A_{kk}(t)$$

Solution of the first order equation gives the formula (1.3.12) stated in the Lemma.

Therefore, if $\det X(t_0) \neq 0$, it follows that $\det X(t) \neq 0$ and, hence, that the n vector solutions are linearly independent for all times t. Thus $\dim V \geq n$.

To show $\dim V \leq n$, let $\mathbf{x}(t)$ be any solution vector, and t_0 any fixed value of t. Since $\det X = \det(\mathbf{x}_k) \neq 0$, we can express $\mathbf{x}(t_0)$ in the form

$$\text{(1.3.15)} \qquad \mathbf{x}(t_0) = c_1 \mathbf{x}_1(t_0) + \cdots + c_n \mathbf{x}_n(t_0)$$

the c_k being scalars.

Because solutions of (1.3.11) are unique, it follows that for every t of our interval, we have

$$\text{(1.3.16)} \qquad \mathbf{x}(t) = c_1 \mathbf{x}_1(t) + \cdots + c_n \mathbf{x}_n(t)$$

both sides being solutions. But this is a statement that $\mathbf{x}_1(t), \ldots, \mathbf{x}_n(t)$ form a basis for V. This completes the proof of the principle of superposition.

We shall encounter numerous applications of this same principle to more general situations, such as systems of infinite order or partial differential equations. Such a principle is still valid provided that the system is linear and homogeneous.

When a basis of n linearly independent solution vectors are arranged as column vectors of a square matrix X the latter is called a *fundamental matrix*. A fundamental matrix X is nonsingular, and possesses an inverse X^{-1} defined for all times t.

THEOREM 1.3.3 *If Y is any matrix whose columns are solution vectors, then*

$$\text{(1.3.17)} \qquad Y = XC$$

where C is a matrix of constants.

Observe that Y can have any number m of columns, and then C will also have m columns.

24 SYSTEMS OF FINITE ORDER

Proof. Consider the product $X^{-1}Y$ and differentiate with respect to time t. The formula for differentiation of an inverse matrix is (Exercise 1.3.11):

(1.3.18) $$\frac{d}{dt} X^{-1} = -X^{-1} \frac{dX}{dt} X^{-1}$$

Therefore,

$$\frac{d}{dt} X^{-1}Y = -X^{-1} \frac{dX}{dt} X^{-1}Y + X^{-1} \frac{dY}{dt}$$
$$= -X^{-1}AXX^{-1}Y + X^{-1}AY$$
$$= 0$$

by the differential equations for X and Y and the property $XX^{-1} = E$. Therefore $X^{-1}Y$ is a constant matrix C, which leads at once to (1.3.17). This result is a formal generalization of (1.3.16).

Example 1.3.1 A linear equation of order n in one scalar variable y can be expressed as an n vector system. Let

$$y^{(n)} + a_1 y^{(n-1)} + \cdots + a_n y = 0$$

and set $x_1 = y$, $x_2 = y'$, ..., $x_n = y^{(n-1)}$. Then

$$\frac{d}{dt} \mathbf{x} = A\mathbf{x}$$

where A is the "companion" matrix

$$A = \begin{pmatrix} 0 & 1 & & & & \\ & 0 & 1 & & 0 & \\ & & 0 & 1 & & \\ & 0 & & \cdot & & \\ & & & & \cdot & \\ & & & & & 0 & 1 \\ -a_n, & -a_{n-1}, & \ldots, & & & -a_1 \end{pmatrix}$$

Our results all carry over to this situation, since a vector solution of the system leads back to a scalar solution of the higher-order equation. These scalar solutions also form a linear space of dimension n.

Example 1.3.2 The nonhomogeneous system

$$\frac{d}{dt} \mathbf{x} = A\mathbf{x} + \mathbf{f} \qquad \mathbf{x}(t_0) = \mathbf{x}_0$$

can be solved by the device of Theorem 1.3.3. Let X be a known fundamental matrix, and consider the vector $X^{-1}\mathbf{x}$. Thus,

$$\frac{d}{dt} X^{-1}\mathbf{x} = -X^{-1}X'X^{-1}\mathbf{x} + X^{-1}\mathbf{x}' = -X^{-1}A\mathbf{x} + X^{-1}(A\mathbf{x} + \mathbf{f})$$
$$= X^{-1}\mathbf{f}$$

By integration,
$$X^{-1}(t)\mathbf{x}(t) = X^{-1}(t_0)\mathbf{x}(t_0) + \int_{t_0}^{t} X^{-1}(s)\mathbf{f}(s)\,ds$$
and so
$$\mathbf{x}(t) = X(t)X^{-1}(t_0)\mathbf{x}(t_0) + \int_{t_0}^{t} X(t)X^{-1}(s)\mathbf{f}(s)\,ds$$

The first term on the right will be simplified if we choose X so that $X(t_0) = E$.

EXERCISES 1.3

1. Show that $y' = y^\alpha$ ($0 < \alpha < 1$) has two solutions with $y(0) = 0$. Which hypothesis of the Picard theorem is not satisfied?

2. For the linear system $\mathbf{x}' = A\mathbf{x}$, $\mathbf{x}(0) = \mathbf{x}_0$ with constant coefficient matrix A, show that the Picard approximations $\mathbf{x}_n(t)$ are the partial sums of the series for $e^{At}\mathbf{x}_0$.

3. Show how to express the single scalar equation of order n, $x^{(n)} = f(x, x', \ldots, x^{(n-1)}, t)$ as a vector system of first order, and state the existence theorem for this given equation.

4. Show that the scalar equation $x' = x^2$ has a solution singular at any specified time t_1, and express $x(0)$ in terms of t_1, for $t_1 \neq 0$.

5. If $\mathbf{x}(t)$ and $\mathbf{y}(t)$ are two solutions of (1.3.1) and $\mathbf{z}(t) = \mathbf{x}(t) - \mathbf{y}(t)$, show by subtraction of two integral equations that
$$|\mathbf{z}(t)| \leq |\mathbf{z}(t_0)| + K\int_{t_0}^{t} |\mathbf{z}(s)|\,|ds|$$
By a sequence of successive substitutions deduce that $|\mathbf{z}(t)| \leq |\mathbf{z}(t_0)|\, e^{K(t-t_0)}$. Hence show that $\mathbf{x}(t)$ depends continuously (in the sense of its norm) on $\mathbf{x}(t_0)$.

6. Assuming the matrix theory result that any nonsingular $n \times n$ matrix C can be written as e^R, for some matrix R, show that the periodic system
$$\frac{d}{dt}\mathbf{x} = A(t)\mathbf{x}, \qquad A(t + T) = A(t)$$
has a fundamental matrix of the form $P(t)e^{Rt}$, where R is a constant matrix and $P(t)$ a periodic matrix of period T. *Hint:* if $X(t)$ is a fundamental matrix, then $X(t + T)$ is also a fundamental matrix, and therefore has the form $X(t)C$.

7. If an inner product $(\mathbf{u}, \mathbf{v})$ is given, and V is the solution space of (1.3.11), show that the nonhomogeneous equation (1.3.10) has a unique solution with minimum norm, $\mathbf{u} = \mathbf{u}_1 - P_V\mathbf{u}_1$, where $\mathbf{u}_1$ is any solution of (1.3.10) and $P_V\mathbf{u}_1$ its orthogonal projection on V.

8. Show that if $(d/dt)X = AX$, $\det X \neq 0$, then the adjoint system $(d/dt)\mathbf{x} = -A(t)^T\mathbf{x}$ has as a fundamental matrix the inverse X^{-1}. A^T denotes the transpose of A.

9. If $X' = AX$, $Y' = -A^TY$, show that Y^TX is a constant matrix. Here X' denotes dX/dt and A^T the transposed matrix.

10. If $\mathbf{x}' = A(t)\mathbf{x}$, and $\mathbf{y} = P\mathbf{x}$ for a constant matrix P, show that $\mathbf{y}' = B\mathbf{y}$, where $B = PAP^{-1}$. Find the form of the system for $\mathbf{y}$ when P depends upon t.

11. By differentiating $XX^{-1} = E$ verify (1.3.18).

1.4 FINITE MECHANICAL SYSTEMS

Consider mechanical systems with a finite number of dimensions or degrees of freedom. The equations of motion are ordinary differential equations, the independent variable being the time t. The dependent variable or variables will generally be position or velocity coordinates, perhaps generalized curvilinear coordinates q_i of Lagrange.

The purpose of mechanics is to predict the position of particles and bodies and the configuration of systems as a function of the time. This is achieved by solving the differential equations of the system.

To derive our equations of motion, and also to provide a foundation for the logical structure of our subject, we state Newton's laws of motion.

1. *A body persists in its state of rest or of uniform motion in a straight line except insofar as it may be compelled by force to change that state.*

This is the law of inertia.

2. *The rate of change of momentum is proportional to the applied force and takes place in the direction of the straight line in which the force acts.*

We recall that the momentum of a given body is the product of its mass and its velocity.

3. *Action and reaction are equal and opposite.*

As this law refers to force rather than to the quantity we shall later call action, it might be stated thus: Force and counterforce are equal and opposite.

Whereas mass is a scalar, the position, velocity, and force are represented by vectors in the three-dimensional physical space of Newtonian mechanics. This is a Euclidean three-space in which the time t is a parameter having no direct geometrical significance. It is this last assumption which distinguishes Newtonian mechanics from the more modern and philosophically more sophisticated theory of relativistic mechanics.

We now express Newton's second law symbolically for the simplest system, namely, that of a single particle of mass m. Let $\mathbf{x} = (x, y, z)$ be the Cartesian position vector. Then

$$\mathbf{v} = (\dot{x}, \dot{y}, \dot{z}) \qquad \dot{x} = \frac{dx}{dt}$$

is the velocity vector. The momentum is $m\mathbf{v}$, and Newton's law states

$$\frac{d}{dt} m\mathbf{v} = \mathbf{F} \tag{1.4.1}$$

where $\mathbf{F}$ is the vector of applied force.

If the force is given in advance as a function of time, we can integrate this vector differential equation. Thus, with constant m

$$m\mathbf{v} = m\mathbf{v}_0 + \int_{t_0}^{t} \mathbf{F}\, dt' \tag{1.4.2}$$

where v_0 is the velocity vector at time $t = t_0$. Then again, after a second integration with respect to time, and division by m, we find

$$(1.4.3) \qquad \mathbf{x} = \mathbf{x}_0 + \mathbf{v}_0 t + \frac{1}{m} \int_{t_0}^{t} \int_{t_0}^{t'} \mathbf{F} \, dt'' \, dt'$$

If, for instance, $\mathbf{F}$ is the (constant) force of gravity, the integral can be evaluated at once.

In the absence of force, the first law is relevant, and we see that it is a supplement to the second law in the same sense that the choosing of a constant of integration completes the finding of an indefinite integral. We observe that two vectorial constants of integration $\mathbf{x}_0$, $\mathbf{v}_0$ are in fact needed. The genius of Newton is shown in his perception that, when the mass does not vary, force should equal the *second* time derivative of position

$$(1.4.4) \qquad m \frac{d^2 \mathbf{x}}{dt^2} = \mathbf{F}$$

The first two laws in effect state that the position vector is the solution of this differential equation, with given values of initial position and initial velocity.

The third law is needed, clearly, when we consider the effect of one body, or system, upon another.

In most problems of particle or rigid-body dynamics, however, the applied force $\mathbf{F}$ is known only as a function of the coordinates. The equations of motion can no longer be completely integrated because the unknown $\mathbf{x}$ will appear beneath the integral sign. This integral form of the equations can nonetheless be used as a basis for approximate or numerical calculation. However, a full analysis of the motion of a system will usually begin by the carrying out of all practicable exact integrations, thus reducing the order of the whole system as far as possible.

EXERCISES 1.4

1. Show that the position vector $\mathbf{x}$ can be written as

$$\mathbf{x} = \mathbf{x}_0 + \mathbf{v}_0 t + \frac{1}{m} \int_0^t (t - t') \mathbf{F}(t') \, dt$$

2. A particle of mass m falls from rest under the force of gravity g and is subject to air resistance proportional to its velocity, the coefficient of resistance being denoted by μ. Show that the equation of motion is $m\ddot{y} + \mu \dot{y} = -mg$, and find $\dot{y} = v$ and y as functions of time, given $y(0) = y_0$, $\dot{y}(0) = 0$. Show that $\lim_{t \to \infty} \dot{y}(t) = -mg/\mu$.

3. If a particle falls as in Exercise 1.4.2, but with air resistance $\dot{y}^2$, find $\dot{y}(t)$ and $y(t)$, given $y(0) = y_0$, $\dot{y}(0) = 0$. Calculate the limiting velocity.

4. A pendulum has length l and mass m, and is inclined at angle θ to the vertical. Show that its equation of motion is $l\ddot{\theta} + g \sin \theta = 0$, where g is the acceleration of gravity. Show that if the maximum angle of swing is θ_0, the period of oscillation is

$$T = \sqrt{\frac{2l}{g}} \int_{-\theta_0}^{\theta_0} \frac{d\varphi}{\sqrt{\cos \varphi - \cos \theta_0}}$$

28   SYSTEMS OF FINITE ORDER

5. A rocket of total mass m consumes c mass units of fuel per unit time while the motor runs, producing a force **F**. The rocket starts from rest under gravity and travels vertically upward. If nine-tenths of its initial mass is fuel, find its velocity at the instant the fuel is exhausted.

6. In the special theory of relativity, the mass of a particle moving at velocity v is

$$\frac{m_0}{\sqrt{1 - v^2/c^2}}$$

where m_0 is the constant rest mass and c the speed of light. A constant force F operates on the particle, which is released from rest at zero time. Show that its velocity is

$$v = \frac{Ft}{\sqrt{m_0^2 + F^2 t^2/c^2}}$$

and calculate the relativistic correction to the distance moved at the instant $v = \tfrac{1}{2}c$.

1.5 LAGRANGE'S EQUATIONS AND HAMILTON'S PRINCIPLE

The Newtonian form for the equations of motion is in theory applicable to any system containing a finite number of mass particles. However the calculation of constraints and interactions often can be greatly simplified if a more general and flexible form of these equations, due to Lagrange, is adopted. Moreover, we are then led to the general dynamical principle of least action which has a significance independent of any given coordinate system. This principle is also useful for continuous systems, which we shall later study in connection with partial differential equations.

The kinetic energy T of a moving mass point is

(1.5.1) $$T = \tfrac{1}{2}m\mathbf{v}^2 = \tfrac{1}{2}m(\dot{x}^2 + \dot{y}^2 + \dot{z}^2) = \tfrac{1}{2}m\,|\dot{x}|^2$$

and the kinetic energy for a system is found by summation over the various mass points:

(1.5.2) $$T_1 = \tfrac{1}{2}\Sigma_i m_i(\dot{x}_i^2 + \dot{y}_i^2 + \dot{z}_i^2) = \tfrac{1}{2}\Sigma_i m_i\,|\dot{x}_i|^2$$

Consider a conservative system, i.e., one possessing a potential energy function $V(x)$, from which the applied force is derived:

(1.5.3) $$\mathbf{F} = -\nabla V \equiv -\operatorname{grad} V$$

Since

(1.5.4) $$m\ddot{x}_i = \frac{d}{dt}(m\dot{x}_i) = \frac{d}{dt}\left(\frac{\partial T}{\partial \dot{x}_i}\right)$$

we see that the Newtonian equations become

(1.5.5) $$\frac{d}{dt}\left(\frac{\partial T}{\partial \dot{x}_i}\right) = -\frac{\partial V}{\partial x_i}$$

We now introduce the Lagrangian function

(1.5.6)
$$L = T - V$$

Suppose that T depends only on the $\dot{x}$ and V upon the x. Then we can write

(1.5.7)
$$\frac{d}{dt}\left(\frac{\partial L}{\partial \dot{x}_i}\right) - \frac{\partial L}{\partial x_i} = 0$$

and this is the Lagrangian form of the equations of motion in Cartesian coordinates for a particle. It can also be established for a system.

This form suggests a variational principle which will apply to any and every coordinate system (Goldstein, Ref. 26). Let the system have initial position $\mathbf{x}_1$ at time t_1, and let the position at a later time t_2 be $\mathbf{x}_2$. Consider the various paths

(1.5.8)
$$\mathbf{x} = \mathbf{x}(t) \qquad t_1 \leq t \leq t_2$$

which connect these initial and final positions.

THEOREM 1.5.1 (*Hamilton's Principle*). *The integral*

(1.5.9)
$$I = \int_{t_1}^{t_2} L \, dt$$

has a minimum (stationary) value on the path actually followed according to the equations of motion.

To find the conditions for a stationary value of this integral, we shall consider *variations* of the function $\mathbf{x}(t)$. That is, we vary $\mathbf{x}(t)$ to $\mathbf{x}(t) + \varepsilon \mathbf{x}_1(t)$, where $\mathbf{x}_1(t)$ is arbitrary except at the endpoints, where it must vanish. We insert the varied function into the integral and search for its stationary values as a function of the parameter ε. The condition for this is the vanishing of the derivative, or of the differential, with respect to ε. We shall use the differential, which we write as

$$\delta f = \frac{\partial f}{\partial \varepsilon} d\varepsilon = \frac{\partial f}{\partial \varepsilon} \delta \varepsilon$$

in order to emphasize the parametric nature of ε. The differential δf is sometimes called a *first variation* of f.

Since derivatives are commutative we have

$$\delta \frac{\partial f}{\partial x} = \frac{\partial^2 f}{\partial x \, \partial \varepsilon} d\varepsilon = \frac{\partial}{\partial x} \frac{\partial f}{\partial \varepsilon} d\varepsilon = \frac{\partial}{\partial x} \delta f$$

so that variations also commute with differentiation.

Now let $\mathbf{x}(t) + \delta\mathbf{x}(t)$ denote a comparison path, $\delta\mathbf{x}(t) = \mathbf{x}_1(t)\,\delta\varepsilon$ being regarded as a perturbation applied to a given path (Figure 1.4). The endpoints at t_1 and t_2 are held fixed so that $\mathbf{x}_1(t_1) = 0$ and $\mathbf{x}_1(t_2) = 0$. Now

(1.5.10)
$$\begin{aligned}
\delta I &= \int_{t_1}^{t_2} \frac{\partial L}{\partial \varepsilon}\,\delta\varepsilon\,dt \\
&= \int_{t_1}^{t_2} \sum_i \left(\frac{\partial L}{\partial x_i}\frac{\partial x_i}{\partial \varepsilon} + \frac{\partial L}{\partial \dot{x}_i}\frac{\partial \dot{x}_i}{\partial \varepsilon} \right) \delta\varepsilon\,dt \\
&= \int_{t_1}^{t} \sum_i \left(\frac{\partial L}{\partial x_i}\,\delta x_i + \frac{\partial L}{\partial \dot{x}_i}\,\delta \dot{x}_i \right) dt
\end{aligned}$$

The summation in the last integrand ranges over all of the coordinates of the system.

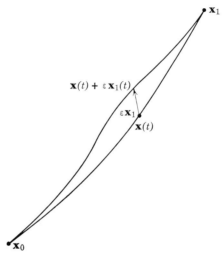

Figure 1.4 Variation of the path.

Integrating by parts, we find

(1.5.11)
$$\int_{t_1}^{t_2} \frac{\partial L}{\partial \dot{x}_i}\,\delta x_i\,dt = \left.\frac{\partial L}{\partial \dot{x}_i}\,\delta x_i\right|_{t_1}^{t_2} - \int_{t_1}^{t_2} \frac{d}{dt}\!\left(\frac{\partial L}{\partial \dot{x}_i}\right) \delta x_i\,dt$$

Therefore the "first variation" δI has the form

(1.5.12)
$$\delta I = \int_{t_1}^{t_2} \sum_i \left[\frac{\partial L}{\partial x_i} - \frac{d}{dt}\!\left(\frac{\partial L}{\partial \dot{x}_i}\right) \right] \delta x_i(t)\,dt$$

as the endpoint terms vanish. We note that, according to Lagrange's equations, this expression vanishes when $\mathbf{x}(t)$ is the actual dynamical path. That is, the path defined by the equation of motion renders the integral *stationary*. This is Hamilton's principle. Especially when the time interval $t_2 - t_1$ is short, we would expect the stationary value to be a minimum.

This variational property does not depend in any way on the coordinates used to describe the configuration of the mechanical system. Therefore, for any set of generalized coordinates q_i, the same result must hold.

THEOREM 1.5.2 *The equations of motion have the Lagrangian form*

(1.5.13) $$\frac{d}{dt}\left(\frac{\partial L}{\partial \dot{q}_i}\right) - \frac{\partial L}{\partial q_i} = 0$$

in every system of generalized coordinates q_i.

This follows because the integral I is stationary and, therefore, its first variation δI is zero for arbitrary variations δq_i. The explicit calculation would be as given above, with q_i replacing x_i. This establishes the Lagrangian form for the equations of motion in any system of generalized coordinates q_i.

In practice, the generalized coordinates most suitable for a given system are those which most directly describe the system and its constraints. For instance, a pendulum is better described by the angle coordinates of the cord rather than by the Cartesian coordinates of the weight.

A dynamical system governed by these equations of motion satisfies the law of conservation of energy, which we can deduce as a first integral of the equations themselves. We have

(1.5.14)
$$\begin{aligned}
\frac{dL}{dt} &= \sum_r \ddot{q}_r \frac{\partial L}{\partial \dot{q}_r} + \sum_r \dot{q}_r \frac{\partial L}{\partial q_r} \\
&= \sum_r \ddot{q}_r \frac{\partial L}{\partial \dot{q}_r} + \sum_r \dot{q}_r \frac{d}{dt}\left(\frac{\partial L}{\partial \dot{q}_r}\right) \\
&= \sum_r \frac{d}{dt}\left(\dot{q}_r \frac{\partial L}{\partial \dot{q}_r}\right)
\end{aligned}$$

and therefore

(1.5.15) $$\sum_r \dot{q}_r \frac{\partial L}{\partial \dot{q}_r} - L = E$$

is a constant independent of the time. To identify this constant, let us consider a system for which

(1.5.16) $$T = \sum_{r,s} \tfrac{1}{2} a_{rs} \dot{q}_r \dot{q}_s \qquad a_{rs} = a_{sr}$$

is a quadratic form in the generalized velocities $\dot{q}_r$. We also suppose $V = V(q_r)$. Then

(1.5.17) $$\frac{\partial L}{\partial \dot{q}_r} = \sum_s a_{rs} \dot{q}_s$$

so that (by Euler's theorem on homogeneous functions)

(1.5.18) $$\sum_r \dot{q}_r \frac{\partial L}{\partial \dot{q}_r} = \sum_{r,s} a_{rs} \dot{q}_r \dot{q}_s = 2T$$

Therefore (1.5.15) becomes $2T - T + V = E$ so that

(1.5.19) $$E = T + V$$

is the (constant) total energy of the system.

Example 1.5.1 **The harmonic oscillator.** One of the simplest and most widely applicable mathematical models of a vibrating system is the harmonic oscillator, a particle of mass m which, at position x, is subject to a restoring force $-kx$ proportional to its displacement. Thus

(1.5.20) $$m\ddot{x} = -kx$$

The general solution of this linear ordinary differential equation of the second order is

(1.5.21) $$x(t) = A \cos(\omega_0 t) + B \sin(\omega_0 t) \qquad A, B \text{ constants}$$

where $\omega_0^2 = k/m$. For this system

(1.5.22) $$T = \tfrac{1}{2}m\dot{x}^2, \qquad V = \tfrac{1}{2}kx^2$$

and it is easily verified that

(1.5.23) $$T + V = \tfrac{1}{2}m\dot{x}^2 + \tfrac{1}{2}kx^2 = \tfrac{1}{2}k(A^2 + B^2)$$

The total energy is constant (independent of t), and is $\tfrac{1}{2}k$ times the square of the amplitude of the vibration.

This basic mathematical model applies to many systems, some of which are:

1. Weight on a spring.
2. Longitudinal or torsional oscillations of a beam.
3. Electric circuit with inductance and capacitance but negligible resistance.
4. Quantum-mechanical oscillators.
5. Models of analogous systems in two and three dimensions.

When there is present an external force $\mathbf{F}$ not included in the potential V, such as a frictional force, we construct the differential of work done by the force,

(1.5.24) $$dW = (\mathbf{F}, d\mathbf{x})$$

and take its expression in generalized coordinates to be

(1.5.25) $$dW = \sum_r Q_r \, dq_r$$

Then the Q_r are components of force in generalized coordinates, and the equations of motion read

(1.5.26) $$\sum_r \frac{d}{dt}\left(\frac{\partial L}{\partial \dot{q}_r}\right) - \frac{\partial L}{\partial q_r} = Q_r$$

Let us consider a system wherein frictional forces proportional to the velocities are present, so that

(1.5.27) $$Q_r = -\sum_s k_{rs} \dot{q}_s$$

and let us suppose, as is commonly true in dissipative mechanical systems, that $k_{rs} = k_{sr}$. We introduce the dissipation function

(1.5.28) $$F = \frac{1}{2} \sum_{rs} k_{rs} \dot{q}_r \dot{q}_s = \tfrac{1}{2}(k\dot{q}, \dot{q})$$

and note that

(1.5.29) $$Q_r = -\frac{\partial F}{\partial \dot{q}_r}$$

Thus our equations of motion become

(1.5.30) $$\frac{d}{dt}\left(\frac{\partial L}{\partial \dot{q}_r}\right) - \frac{\partial L}{\partial q_r} + \frac{\partial F}{\partial \dot{q}_r} = Q'_r$$

where Q'_r denotes any remaining applied force.

We can no longer expect the law of conservation of energy to hold for dissipative systems. The calculation (1.5.14) now shows that

(1.5.31) $$\frac{d}{dt}\left(\sum_r \dot{q}_r \frac{\partial L}{\partial \dot{q}_r} - L\right) = -\sum_r \dot{q}_r \frac{\partial F}{\partial \dot{q}_r}$$

By noting that the left side is the rate of change of total energy, and by applying Euler's theorem on homogeneous functions to the right-hand side, we find

(1.5.32) $$\frac{d}{dt}(T + V) = -2F$$

That is, the dissipation function is half the rate of energy loss by friction.

Example 1.5.2 **The damped oscillator.** Suppose that a frictional force $-\mu\dot{x}$ acts upon an oscillating point mass m in addition to the restoring force $-kx$. The equation of motion is then

(1.5.33) $$m\ddot{x} + \mu\dot{x} + kx = f(t)$$

where $f(t)$ represents the external force applied to the system. We consider here only the case of free oscillations for which $f(t)$ is taken to vanish. The general solution of the equation is

(1.5.34) $$x(t) = A e^{\lambda_1 t} + B e^{\lambda_2 t}$$

where λ_1 and λ_2 are the roots of the characteristic equation

$$m\lambda^2 + \mu\lambda + k = 0$$

namely,

(1.5.35)
$$\lambda = -\frac{\mu}{2m} \pm \sqrt{\left(\frac{\mu}{2m}\right)^2 - \frac{k}{m}}$$

There are three cases to be considered.

1. *Weak damping*, $\mu^2 < 4km$. The roots λ_1 and λ_2 are conjugate complex and lead to the following real form of the general solution:

$$x(t) = Ce^{-\mu t/2m} \cos\left(\sqrt{\frac{k}{m} - \frac{\mu^2}{4m^2}}\, t\right) + De^{-\mu t/2m} \sin\left(\sqrt{\frac{k}{m} - \frac{\mu^2}{4m^2}}\, t\right)$$

This represents damped oscillations with an infinite number of passages through the origin. The period is $T = \dfrac{2\pi}{\sqrt{k/m - (\mu^2/4m^2)}}$ which, for $\mu = 0$, reduces to $2\pi\sqrt{m/k}$. The amplitude decreases in geometrical progression, being reduced by a factor $e^{-\mu T/2m}$, after each cycle. As μ^2 approaches $4km$, the period becomes long.

2. *Critical damping*, $\mu^2 = 4km$. The roots λ_1 and λ_2 are equal, and the general solution above degenerates. In this particular case the full solution is

$$x(t) = (A + Bt)e^{\lambda t} \qquad \lambda = -\frac{\mu}{2m}$$

This represents a motion with at most one change of sign and one maximum or minimum value. For large values of t, the displacement is exponentially small.

3. *Strong damping*, $\mu^2 > 4km$. The roots λ_1 and λ_2 are real, unequal, and negative. The general solution represents an exponentially subsiding motion with at most one change of sign of the displacement.

Another application of damped oscillations in a system with one degree of freedom is evident in the analogy between mechanical and electrical vibrations. The correspondences

Mass	$m \leftrightarrow$ inductance L,
Friction coefficient	$\mu \leftrightarrow$ resistance R,
Restoring force constant	$k \leftrightarrow$ reciprocal capacitance $1/C$,
Applied force	$f(t) \leftrightarrow$ applied emf $E(t)$,

transform the mechanical vibration equation into the electrical one, and thus permit a full description of the latter by means of the same mathematical model. This correspondence is also valid for systems with any number of degrees of freedom and, therefore, to any mechanical system there is an electrical analogue, and vice versa (Guillemin, Ref. 27). Not only does this permit a common mathematical background but also makes possible the simulation of systems of one kind by their analogues. In complicated cases this greatly facilitates numerical studies.

EXERCISES 1.5

1. By minimizing the integral $\int_{x_1}^{x_2} \sqrt{1 + (y')^2}\, dx$, show that a straight line is the shortest distance between two points.

2. *The brachistochrone.* Find the path along which a particle will slide (without friction) under gravity g, from one given higher point to another, in the least time. Show that the curve is a cycloid. *Hint:* the integral to be minimized is $\int dt = \int (1/v)\, ds$.

3. Show that the dissipation function for the damped oscillator (1.5.33) is $F = \tfrac{1}{2}\mu \dot{x}^2$ by comparing with (1.5.32).

4. A torsional double pendulum with the lower mass subject to linear friction has the equations of motion

$$M_1 \ddot{\theta}_1 + (K_1 + K_2)\theta_1 - K_2 \theta_2 = 0$$
$$M_2 \ddot{\theta}_2 + C_2 \dot{\theta}_2 + K_2(\theta_2 - \theta_1) = 0$$

Determine T, V, and F for this system.

5. For each of the three cases of the damped oscillator discussed in the text, find the paths of the motion in the phase plane of x and $y = \dot{x}$. Sketch the field of paths and find their equations, in each case.

1.6 SYSTEMS WITH CONSTANT COEFFICIENTS

As the foregoing examples in one degree of freedom suggest, linear equations with constant coefficients apply to a wide range of problems in mechanics, and especially in vibration theory. Moreover, such systems, with an infinite number of variables, will reappear in our study of partial differential equations for time-dependent problems.

Therefore let A be a matrix of constants and consider the homogeneous system

(1.6.1) $$\frac{d\mathbf{x}}{dt} = A\mathbf{x}$$

Notice that, because t does not appear explicitly in (1.6.1), the form of any solution will not be influenced by a shift of the time variable. In more technical language, the system is invariant under translation in t. Therefore, we can specify our initial time to be zero without loss of generality.

For systems with constant coefficients it is possible to determine a fundamental matrix explicitly.

THEOREM 1.6.1 *The fundamental matrix of* (1.6.1), *which is equal to the unit matrix at* $t = 0$, *is the exponential matrix*

(1.6.2) $$e^{At}$$

The verification is immediate since, by (1.2.14) with $T = tA$, we have

(1.6.3) $$\frac{d}{dt} e^{At} = A e^{At}$$

and $e^{A \cdot 0} = e^0 = E$.

36 SYSTEMS OF FINITE ORDER

We can now write down the general solution of the nonhomogeneous system

(1.6.4) $$\frac{d\mathbf{x}}{dt} = A\mathbf{x} + \mathbf{f}(t), \quad \mathbf{x}(0) = \mathbf{x}_0$$

by either using Example 1.3.2 or from first principles.
  The solution formula is

(1.6.5) $$\mathbf{x}(t) = e^{At}\mathbf{x}_0 + \int_0^t e^{A(t-s)}\mathbf{f}(s)\,ds$$

The two terms here present are each solutions of a problem with only one nonhomogeneous item or *datum*. Thus $e^{At}\mathbf{x}_0$ is the solution of the homogeneous equation (1.6.1) with the nonhomogeneous initial condition $\mathbf{x}(0) = \mathbf{x}_0$. Also, the integral in (1.6.5) is itself the solution of (1.6.4), which vanishes at $t = 0$.

The symbolic exponential matrix (1.6.2) is a power series in At. Let us now consider the effective computation of its matrix elements as functions of t. We shall need to use, and will quote without proof in two parts, a general matrix theorem which was proved for symmetric matrices in Section 1.2.

Consider a change of basis

$$\mathbf{x} = P\mathbf{y}$$

where P is a nonsingular constant matrix. With primes denoting derivatives

$$\mathbf{y}' = P^{-1}\mathbf{x}' = P^{-1}A\mathbf{x} = P^{-1}AP\mathbf{y} = B\mathbf{y}$$

say, so that the matrix of the transformed system is the *similarity transform*

$$B = P^{-1}AP$$

of A by P. We therefore attempt to diagonalize the coefficient matrix by a suitable similarity transformation. When this diagonal form has been achieved, the interlocking system of n equations will be separated or "resolved" into n separate single-component scalar equations. (Gantmacher 23)

THEOREM 1.6.2 *If A has distinct eigenvalues λ_k, or is symmetric, or is a matrix of "simple" structure, the eigenvectors of A define a diagonalizing matrix P such that*

$$D = \text{diag}(\lambda_k) = P^{-1}AP.$$

We also have $A = PDP^{-1}$. Therefore, by (1.2.35) and (1.2.32),

$$e^{At} = e^{PDP^{-1}t} = Pe^{Dt}P^{-1}$$
$$= P\,\text{diag}(e^{\lambda_k t})P^{-1}$$

from Exercise 1.2.7.

The components of e^{At} are linear combinations of the exponential functions $e^{\lambda_k t}$. In fact, by Theorem 1.3.3, the matrix

$$e^{At}P = P\,\text{diag}(e^{\lambda_k t})$$

is also a fundamental matrix. Its columns are the vectors $e^{\lambda_k t}\mathbf{y}_k$, where λ_k is the eigenvalue corresponding to the eigenvector $\mathbf{y}_k$ of A.

THEOREM 1.6.3 *If A has distinct eigenvalues, is symmetric, or is a matrix of simple structure, the general solution vector of 1.6.1 has the form*

$$\mathbf{x}(t) = \sum_{k=1}^{n} c_k \mathbf{y}_k e^{\lambda_k t}$$

where λ_k, $\mathbf{y}_k$ denote eigenvalues and eigenvectors of A and c_k denotes an arbitrary constant.

Not all matrices A, however, can be completely diagonalized by a similarity transformation. If repeated eigenvalues appear, the standard Jordan form of the matrix may contain "blocks" of several rows and columns which are "linked" by entries above the main diagonal. (Gantmacher 23)

THEOREM 1.6.4 *The standard form of a general matrix A under similarity transformations is*

$$D = \begin{pmatrix} D_1 & & & \\ & D_2 & & 0 \\ & & \cdot & \\ & 0 & & \cdot \\ & & & & D_k \end{pmatrix}$$

where each D_j is a $n_j \times n_j$ submatrix of the form

$$D_j = \begin{pmatrix} \lambda_j & 1 & & & & \\ & \lambda_j & 1 & & 0 & \\ & & \lambda_j & & & \\ & & & \cdot & & \\ & 0 & & & \cdot & 1 \\ & & & & & \lambda_j \end{pmatrix}$$

Each D_j has as its n_j diagonal entries a single eigenvalue λ_j (a given eigenvalue λ_j can appear in more than one D_j), while the superdiagonal entries are $+1$. Observe that, if all eigenvalues are distinct, then every D_j is a scalar and there are no superdiagonal positions for the additional entries. Each eigenvalue appears on the diagonal of D in a number of places equal to its multiplicity as an eigenvalue of A.

A typical block matrix D_j has the form

$$D_j = \lambda_j E_j + J_j$$

where E_j is the unit matrix of the appropriate rank n_j and

$$J_j = \begin{pmatrix} 0 & 1 & & & & \\ & 0 & 1 & 0 & & \\ & & 0 & & & \\ & & & \cdot & & \\ 0 & & & & \cdot & \\ & & & & \cdot & 1 \\ & & & & & 0 \end{pmatrix}$$

This "superdiagonal" matrix J_j has an interesting algebraic property: it is *nilpotent*, that is,

$$J_j^{n_j} = 0$$

To show this, we compute $J_j^2, J_j^3, \ldots$, and observe that each multiplication by J_j shifts the entries one row further above the diagonal. Eventually we find that $J_j^{n_j-1}$ has zeros except for a 1 in the top-right corner, and that $J_j^{n_j}$ is the zero matrix.

From this observation we can quickly compute the exponential matrix e^{At} in the general case. For we have

$$e^{At} = Pe^{Dt}P^{-1}$$

as before, where P is a matrix of suitable eigenvectors and

$$e^{Dt} = \begin{pmatrix} e^{D_1 t} & & & \\ & e^{D_2 t} & & 0 \\ & & \cdot & \\ & 0 & & \cdot \\ & & & & e^{D_h t} \end{pmatrix}$$

is also a diagonal block matrix. The blocks $e^{D_j t}$ are of rank n_j and, since E_j and J_j commute, we have

$$e^{D_j t} = e^{\lambda_j E_j t + J_j t} = e^{\lambda_j E_j t} e^{J_j t}$$
$$= e^{\lambda_j t} E_j e^{J_j t} = e^{\lambda_j t} e^{J_j t}$$
$$= e^{\lambda_j t}\left(E + J_j t + \frac{J_j^2 t^2}{2!} + \cdots + \frac{J_j^{n_j-1} t^{n_j-1}}{(n_j - 1)!} + 0 + \cdots\right)$$

The exponential series for $e^{J_j t}$ terminates after n_j terms because $J_j^{n_j}$ and therefore all higher powers are zero.

Therefore $e^{D_j t}$ has entries which are products of the scalar exponential $e^{\lambda_j t}$ and powers of t less than n_j. The coefficients of these functions are appropriate eigenvector components.

Example 1.6.1 The system

$$\frac{d}{dt} x_1 = \lambda x_1 + x_2$$

$$\frac{d}{dt} x_2 = \lambda x_2$$

has coefficient matrix $A = \begin{pmatrix} \lambda & 1 \\ 0 & \lambda \end{pmatrix}$. Thus

$$e^{At} = e^{(\lambda E + J)t} = e^{\lambda t}(E + Jt) = \begin{pmatrix} e^{\lambda t} & te^{\lambda t} \\ 0 & e^{\lambda t} \end{pmatrix}$$

The first and second columns are independent solution vectors.

EXERCISES 1.6

1. Find a fundamental matrix for the system

$$\frac{dN_1}{dt} = -k_1 N_1 \qquad \frac{dN_2}{dt} = k_1 N_1 - k_2 N_2$$

which occurs in molecular reaction theory. Consider both $k_1 \neq k_2$ and $k_1 = k_2$.

2. Express the system

$$L_1 \dot{x} + M \dot{y} + R_1 x = 0$$

$$M \dot{x} + L_2 \dot{y} + R_2 y = 0$$

for the transient currents in an electric transformer, in the form (1.6.1). Here L_1, L_2, R_1, and R_2 are positive, with $\Delta = L_1 L_2 - M^2 > 0$.
   Show that the eigenvalues are real, distinct, and negative.

3. If $P(\lambda) = (\lambda - \lambda_1)(\lambda - \lambda_2) \cdots (\lambda - \lambda_n)$, show that the nth-order linear equation

$$P(D)y = 0 \qquad\qquad D = \frac{d}{dt}$$

has solutions $e^{\lambda_k t}$. If the λ_k are distinct, show that these solutions form a basis for all solutions.

4. If $P(\lambda) = (\lambda - \lambda_1)^m$, show that

$$P(D)y = 0$$

has solutions $e^{\lambda_1 t}, te^{\lambda_1 t}, \ldots, t^{m-1}e^{\lambda_1 t}$. Hint: $F(D)(ye^{-\lambda_1 t}) = e^{-\lambda_1 t}F(D - \lambda_1)y$.

5. If $P(\lambda) = (\lambda - \lambda_1)^{m_1} \cdots (\lambda - \lambda_k)^{m_k}$, construct a basis of solutions for $P(D)y = 0$.

6. Show that the matrix solution of the system $X' = AX + XB$, $X(0) = C$; A, B, C constant, is the two-sided exponential matrix $e^{At}Ce^{Bt}$.

1.7 THE RESPONSE MATRIX AND DISTRIBUTIONS

For much of our subsequent work on time-dependent problems, it is convenient to have a modified form of the exponential fundamental matrix. Let us once more examine the general solution of Section 1.6:

(1.7.1) $$\mathbf{x}(t) = e^{At}\mathbf{x}_0 + \int_0^t e^{A(t-s)}\mathbf{f}(s)\,ds$$

Suppose that $\mathbf{x}_0 = 0$ and that the system was in equilibrium with $\mathbf{x}(t) \equiv 0$ for $t < 0$. Suppose also that at time $t = 0$ a large force $\mathbf{f}(t)$ begins to act upon the system for a short time—so short that during this time the variation of s in the exponential of (1.7.1) can be neglected. The solution is then, approximately,

(1.7.2) $$e^{At}\int_0^t \mathbf{f}(s)\,ds \qquad t > 0$$

The accuracy of this approximation improves as the duration of the force $\mathbf{f}(t)$ diminishes, and in the idealized limit of a truly impulsive force of zero duration, the solution is exact. Such a force cannot be represented by a conventional function, and we now must extend that concept of function to include impulses.

Consider a sequence of scalar functions $f_n(t)$, which, as $n \to \infty$, become concentrated near $t = 0$ in the sense that

(1.7.3) $$\int_{-\varepsilon}^{\varepsilon} f_n(s)\,ds \to 1$$

while $f_n(t)$ tend rapidly to zero for $t \neq 0$ (Figure 1.5). Such a sequence does not have a limit in a conventional sense. However, the solutions of our differential system do have a limit, and it is therefore convenient to postulate a symbolic "limit" for this sequence of force functions.

This limit we call the Dirac function and denote by $\delta(t)$. The Dirac "function" is really a generalized or ideal or symbolic function, or, as we shall say, a *distribution*. It has these properties:

(1.7.4) $$\delta(t) = 0 \qquad t \neq 0$$

(1.7.5) $$\int_{-\varepsilon}^{\varepsilon} \delta(s)\,ds = 1 \qquad \text{for every positive } \varepsilon$$

We postpone a full definition of distributions to Chapter 2. However, we give one further relation which is a formal consequence of (1.7.4) and (1.7.5). Let $\varphi(t)$ be a smooth function of t. Then

(1.7.6) $$\int_{-\infty}^{\infty} \varphi(s)\,\delta(s)\,ds = \varphi(0)$$

This is the *substitution property* of the Dirac distribution.

Formulas of this type have long been known. For example, Dirichlet's formula states (Courant-Hilbert 13, vol. 1)

(1.7.7) $$\lim_{n \to \infty} \frac{1}{\pi} \int_{-\infty}^{\infty} \varphi(s) \frac{\sin ns}{s}\,ds = \varphi(0)$$

We can express the same result in distribution language by writing

(1.7.8) $$\lim_{n \to \infty} \frac{1}{\pi} \frac{\sin nt}{t} = \delta(t)$$

Returning to (1.7.2), we set $\mathbf{f}(t) = \mathbf{f}_0 \delta(t)$, and find as solution for positive times t,

(1.7.9) $$e^{At} \mathbf{f}_0$$

The constant vector $\mathbf{f}_0$ describes the various components of the impulsive vector force. Let us employ a matrix, as in the system

(1.7.10) $$\frac{dG}{dt} = AG + E\delta(t)$$

with $G(t) \equiv 0$ for $t < 0$. The solution matrix, denoted by G, is called the response matrix or *Green's matrix* for the system.

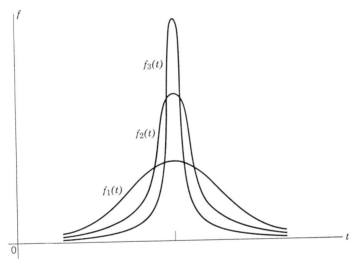

Figure 1.5 An approximating sequence.

THEOREM 1.7.1 *The Green's matrix for* (1.7.10) *is*

(1.7.11) $$G(t) = \begin{cases} 0 & t < 0 \\ e^{At} & t \geq 0 \end{cases}$$

The Green's matrix is thus a fundamental matrix truncated at the origin. By conventional standards, it does not have a derivative at $t = 0$. However, we shall write (1.7.10) with the nonhomogeneous term $E\delta(t)$ on the right. Since $G \equiv 0$ for $t < 0$, this implies

(1.7.12) $$\lim_{t \to 0+} G(t) = E$$

which we may consider as an initial condition for the Green's matrix on $0 < t < \infty$

42   SYSTEMS OF FINITE ORDER

A similar formula appears if we look for an indefinite integral of $\delta(t)$. Consider the integrals

(1.7.13) $$\int_{-\infty}^{t} f_n(s)\,ds.$$

As $n \to \infty$, they will approach zero if t is negative and, by (1.7.3), will approach unity if t is positive. Therefore define

$$H(t) = \begin{cases} 0 & t < 0 \\ 1 & t \geq 0 \end{cases}$$

This is the Heaviside unit function.

As the derivatives of (1.7.13) have the symbolic limit $\delta(t)$, we shall write

$$\frac{dH(t)}{dt} = \delta(t)$$

Interpretation of this relation between distributions will be resumed in Chapter 2.

As an exercise, the reader can show that $G(t) = H(t)e^{At}$ and verify (1.7.10).

Example 1.7.1 We can find the scalar response function $G(t)$ for an equation of order n,

$$P(D)y = (D^n + a_1 D^{n-1} + \cdots + a_n)y = \delta(t)$$

from the reduction in Example 1.3.1. The nonhomogeneous term $\delta(t)$ appears in the last component of the new system. Since $y \equiv 0$ for $t < 0$, we find $x_n(t) \to +1$ as $t \to 0+$. The remaining components are continuous at $t = 0$. Thus the response function $G(t)$ is the leading component $y \equiv x_1(t)$ and it satisfies the initial conditions

$$G(0) = x_1(0) = 0, \quad G'(0) = x_2(0) = 0, \ldots,$$
$$G^{(n-2)}(0) = x_{n-1}(0) = 0, \quad G^{(n-1)}(0+) = x_n(0+) = 1$$

That these are appropriate initial conditions for a solution of $P(D)G(t) = \delta(t)$ can be seen directly by integrating the equation over $-\varepsilon \leq t \leq \varepsilon$ and by observing that the highest order derivative is the discontinuous term on the left.

As the second order equation is most frequent in later use, we state it as a separate and important example.

Example 1.7.2 The solution of

$$P_2(D)y = y'' + k^2 y = \delta(t) \qquad y(t) \equiv 0, \quad t < 0$$

is

$$G(t) = \begin{cases} 0 & t < 0 \\ \dfrac{\sin kt}{k} & t > 0 \end{cases}$$

To demonstrate this, we need only observe that the general solution of $y'' + k^2 y = 0$ is $A \cos kx + B \sin kx$, and that the condition $G(0) = 0$ implies $A = 0$. From $G'(0+) = 1$ we have $B = 1/k$.

This formula *means* that a solution of the general nonhomogeneous second-order equation $y'' + k^2 y = f(t)$ is

$$\int_{-\infty}^{\infty} G(t - \tau) f(\tau) \, d\tau = \int_{-\infty}^{t} \frac{\sin k(t - \tau)}{k} f(\tau) \, d\tau$$

To show this, apply $P_2(D)$ on the left and use $P_2(D) G(t) = \delta(t)$ and the substitution property (1.7.6).

The following exercises illustrate the properties of various response functions $G(t)$.

EXERCISES 1.7

1. Show that $dy^{n+1}/dt^{n+1} = \delta(t)$ has the response solution

$$G(t) = t^n H(t)/n!$$

2. *The Riemann-Liouville integral.* Show that $I^n f(t) = (1/(n-1)!) \int_0^t (t - s)^{n-1} f(s) \, ds$ is an n-fold indefinite integral of $f(t)$.

3. Show that the weakly damped oscillator of Section 1.5 has the Green's function

$$G(t) = \frac{e^{-\rho t} \sin \sigma t}{\sigma} H(t)$$

where $\lambda = -\rho \pm i\sigma$.

4. For a matrix K with real nonzero eigenvalues, show that the second order system

$$Ey'' + K^2 y = E\delta(t)$$

has the Green's matrix

$$G(t) = \frac{\sin Kt}{K} H(t)$$

5. Modify the result of Exercise 1.7.4 when:
   (a) K has a block of zero eigenvalues.
   (b) K has pure imaginary eigenvalues.

6. Show that the system

$$(D^2 - k^2) y_1 - 2k \, Dy_2 = 0$$
$$2k \, Dy_1 + (D^2 - k^2) y_2 = 0$$

has the Green's matrix

$$G(t) = tH(t) \begin{pmatrix} \cos kt & \sin kt \\ -\sin kt & \cos kt \end{pmatrix}$$

7. The equations of motion of a double pendulum are

$$\ddot{\theta} + a\ddot{\varphi} + n_1^2 \theta = 0$$
$$a\ddot{\theta} + \ddot{\varphi} + n_2^2 \varphi = 0$$

Find the resonance frequencies (eigenvalues) and the Green's matrix.

1.8 TWO-POINT BOUNDARY CONDITIONS

In the following chapters we shall study partial differential equations and their solutions such as arise in specific problems of applied mathematics. In Chapter 2 we shall encounter the separation-of-variables method by which a partial differential equation can sometimes, if circumstances are just right, be "split" into two or more ordinary differential equations. However, the auxiliary conditions which specify the required solution are usually different from the "one-point" or *initial* conditions encountered so far. Therefore we now examine the simplest case of the two-point separated boundary conditions needed for these equations (Coddington-Levinson, Ref. 10).

On an interval $a \leqslant x \leqslant b$ of the independent spatial variable x we consider the second order equation of the Sturm-Liouville type

$$(1.8.1) \qquad L(y) \equiv -\frac{d}{dx}\left[p(x)\frac{dy}{dx}\right] + q(x)y = f(x)$$

Here $p(x)$ is a positive function and $f(x)$ is a given nonhomogeneous term. We wish to find a solution y which satisfies the two boundary conditions

$$(1.8.2) \qquad \begin{aligned} B_1 y(a) &\equiv a_1 y(a) + a_2 y'(a) = 0 \\ B_2 y(b) &\equiv b_1 y(b) + b_2 y'(b) = 0 \end{aligned}$$

As there is one condition at each endpoint, the boundary conditions are called *separated*.

Our equation is linear and we can superpose solutions to obtain a desired nonhomogeneous term. Let us construct a Green's function $G(x, x_1)$ for (1.8.1) corresponding to an impulsive force at x_1. By analogy with Section 1.7, we would have

$$(1.8.3) \qquad L(G) = -\frac{d}{dx}\left[p(x)\frac{dG(x, x_1)}{dx}\right] + q(x)G(x, x_1) = \delta(x - x_1)$$

while $G(x, x_1)$ also satisfies (1.8.2).

Suppose for the moment that we have found this Green's function $G(x, x_1)$. Then we have the following theorem.

THEOREM 1.8.1 *The integral*

$$(1.8.4) \qquad y(x) = \int_a^b G(x, x_1) f(x_1)\, dx_1$$

is a solution of the problem (1.8.1) *and* (1.8.2).

Proof. Apply the operator L to this function. We find, since $a < x < b$,

$$(1.8.5) \qquad Ly(x) = \int_a^b LG(x, x_1) f(x_1)\, dx_1$$

$$= \int_a^b \delta(x - x_1) f(x_1)\, dx_1 = f(x)$$

TWO-POINT BOUNDARY CONDITIONS

by the substitution property (1.7.6) of the Dirac distribution. Likewise, we apply the two boundary expressions or "operators" to the Green's function under the integral sign in (1.8.4). Since $G(x, x_1)$, as a function of x, satisfies $B_1 G(a, x_1) = 0$, $B_2 G(b, x_1) = 0$, the result is zero and the theorem is proved.

The Green's function $G(x, x_1)$ is a function of two variables. It satisfies a differential equation in x with a unit impulse on the right-hand side, concentrated at the *source* point x_1. Let us look closely at the behavior of $G(x, x_1)$ when the variable point x is near x_1 (Figure 1.6). We integrate (1.8.3) over a

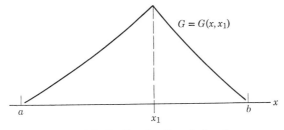

Figure 1.6 Profile of a Green's function.

small interval $(x_1 - \varepsilon, x)$ and recall that the indefinite integral of $\delta(x - x_1)$ is the Heaviside function $H(x - x_1)$. We find

$$(1.8.6) \quad -p(x)\frac{dG(x, x_1)}{dx} + \int_{x_1-\varepsilon}^{x} q(x')G(x', x_1)\, dx'$$
$$= -p(x_1 - \varepsilon)\frac{dG(x_1 - \varepsilon, x_1)}{dx} + H(x - x_1).$$

Now let x traverse the source point x_1. On the right, the Heaviside function has a unit jump discontinuity there. The other terms being continuous functions of x, we conclude that the derivative term on the left side also must have a unit jump. Therefore the derivative of $G(x, x_1)$ with respect to x has at x_1 a discontinuity of magnitude $-1/p(x_1)$.

Let $y_1(x)$ and $y_2(x)$ be linearly independent solutions of (1.8.1). We shall suppose they can be chosen so that

$$(1.8.7) \quad B_1 y_1(a) = a_1 y_1(a) + a_2 y_1'(a) = 0$$

and

$$(1.8.8) \quad B_2 y_2(b) = b_1 y_2(b) + b_2 y_2'(b) = 0$$

Then we can write

$$(1.8.9) \quad G(x, x_1) = c_1(x_1) y_1(x) + c_2(x_1) y_2(x)$$

for the interval $a \leq x \leq x_1 \leq b$ and a similar formula for the portion $a \leq x_1 \leq x \leq b$. However, since $B_1 G(a, x_1) = 0$, we must have $c_2(x_1) = 0$ in (1.8.9) and so

$$(1.8.10) \quad G(x, x_1) = c_1(x_1) y_1(x) \quad \text{for} \quad a \leq x \leq x_1 \leq b$$

In exactly the same way we find

(1.8.11) $$G(x, x_1) = c_2(x_1)y_2(x) \quad \text{for} \quad a \leq x_1 \leq x \leq b$$

Now we can determine $c_1(x_1)$ and $c_2(x_1)$. Since $G(x, x_1)$ is continuous at $x = x_1$, we have

(1.8.12) $$c_2(x_1)y_2(x_1) - c_1(x_1)y_1(x_1) = 0$$

Also we use the above jump relation

(1.8.13) $$\left.\frac{dG(x, x_1)}{dx}\right]_{x=x_1-0}^{x=x_1+0} = c_2(x_1)y_2'(x_1) - c_1(x_1)y_1'(x_1) = \frac{-1}{p(x_1)}$$

From these two formulas we find

(1.8.14) $$c_1(x_1) = \frac{-y_2(x_1)}{p(x_1)W(y_1, y_2)(x_1)} \qquad c_2(x_1) = \frac{-y_1(x_1)}{p(x_1)W(y_1, y_2)(x_1)}$$

where

(1.8.15) $$W(y_1, y_2)(x) = y_1(x)y_2'(x) - y_2(x)y_1'(x)$$

is the *Wronskian* of the two solutions.

Actually the denominator of the expressions in (1.8.14) is a *constant*, independent of x_1, as the following theorem shows.

THEOREM 1.8.2 $p(x)W(y_1, y_2)(x) = C$, *where C is a constant.*
Proof. We have
$$-(py_1'(x))' + qy_1 = 0$$
and
$$-(py_2'(x))' + qy_2 = 0$$

Multiply the first equation by y_2, the second by y_1, and subtract. The terms in $y_1 y_2$ cancel, and we are left with

(1.8.16) $$\begin{aligned} 0 &= y_1(py_2')' - y_2(py_1')' \\ &= p(y_2''y_1 - y_1''y_2) + p'(y_1y_2' - y_2y_1') \\ &= [p(y_1y_2' - y_2y_1')]' \end{aligned}$$

Upon integration, we find $p(y_1y_2' - y_2y_1')$ is a constant, as required.

We now assemble the expressions for the Green's function:

THEOREM 1.8.3 *The Green's function is given by*

(1.8.17) $$G(x, x_1) = -\frac{y_2(x_1)y_1(x)}{C} \qquad a \leq x \leq x_1 \leq b$$

and

(1.8.18) $$G(x, x_1) = -\frac{y_1(x_1)y_2(x)}{C} \qquad a \leq x_1 \leq x \leq b$$

In particular, therefore, we see that $G(x, x_1)$ is *symmetric*:

(1.8.19) $$G(x, x_1) = G(x_1, x)$$

for if we interchange x with x_1, we must change to the opposite portion of the interval (a, b) and, therefore, must use the other one of the two formulas (1.8.17) and (1.8.18) (see Figure 1.7).

Example 1.8.1 $Ly = -d^2y/dx^2$, $0 \leq x \leq 1$, $y(0) = y(1) = 0$.

Here solutions of the differential equation satisfying (1.8.7) and (1.8.8) are

$$y_1(x) = x \qquad y_2(x) = 1 - x$$

The constant C is -1. Therefore,

$$G(x, x_1) = \begin{cases} x(1 - x_1) & 0 \leq x \leq x_1 \\ x_1(1 - x) & x_1 \leq x \leq 1 \end{cases}$$

Notice that, if the solutions $y_1(x)$ and $y_2(x)$ are not linearly independent, their Wronskian $W(y_1, y_2)$ will be identically zero and so will be the constant C.

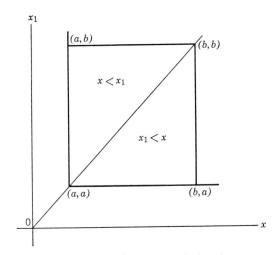

Figure 1.7 Domain of a Green's function.

In this case the Green's function does not exist. The reader will notice that, if $y_1(x)$ and $y_2(x)$ are linearly dependent, then both will satisfy each of the boundary conditions. Consequently, we say there is an *eigenfunction* of the Sturm-Liouville problem. We shall return to this question in Chapters 6 and 7.

EXERCISES 1.8

Find the Green's functions of the following two-point boundary problems:

1. $$Ly = -\frac{d^2y}{dx^2} = 0, \qquad y(0) = y'(1) = 0$$

    Answer:
    $$G(x, x_1) = \begin{cases} x & x \leq x_1 \\ x_1 & x_1 \leq x \end{cases}$$

2. $$Ly = -x\frac{d^2y}{dx^2} - \frac{dy}{dx} = 0, \qquad y(x) \text{ bounded as } x \to 0, \qquad y(1) = 0$$
Answer:
$$G(x, x_1) = \begin{cases} -\log x_1 & x \leq x_1 \\ -\log x & x_1 \leq x \end{cases}$$

3. $$Ly = -\frac{d}{dx}\left[(1 - x^2)\frac{dy}{dx}\right] = 0, \qquad y(0) = 0, \qquad y'(1) = 0$$
Answer:
$$G(x, x_1) = \begin{cases} \dfrac{1}{2} \log \dfrac{1 + x}{1 - x} & x \leq x_1 \\ \dfrac{1}{2} \log \dfrac{1 + x_1}{1 - x_1} & x_1 \leq x \end{cases}$$

4. $$Ly = -\frac{d^2y}{dx^2} - k^2 y, \qquad y(0) = y(l) = 0$$
Answer:
$$G(x, x_1) = \begin{cases} \dfrac{\sin kx \sin k(l - x_1)}{k \sin kl} & x \leq x_1 \\ \dfrac{\sin kx_1 \sin k(l - x)}{k \sin kl} & x_1 \leq x \end{cases}$$

CHAPTER *2*

Distributions and Waves

2.1 EQUATIONS IN TWO INDEPENDENT VARIABLES

Systems of a finite number of degrees of freedom are governed by ordinary differential equations. Systems containing an infinite number of discrete point masses are subject to infinite systems of ordinary differential equations. We now consider systems with a continuous mass distribution, which lead to partial differential equations, with independent variables in time and space.

This chapter and the two that follow are centred around the three types of linear partial differential equations in two independent variables. These equations are intermediate in difficulty of mathematical discussion to ordinary differential equations, on the one hand, and partial differential equations in three or more variables, on the other. These last we shall encounter later on. The restriction to linear equations is natural, at first, because nonlinear equations are more complicated to discuss. Moreover, the linear equations have a remarkably great variety of important applications (Courant-Hilbert, Ref. 13, vol. 2; Garabedian, Ref. 24).

Let $u(x, y)$ be the unknown function of coordinates x and y. The most general linear homogeneous equation of the second order is

$$(2.1.1) \quad L(u) \equiv a\frac{\partial^2 u}{\partial x^2} + 2b\frac{\partial^2 u}{\partial x \partial y} + c\frac{\partial^2 u}{\partial y^2} + 2f\frac{\partial u}{\partial x} + 2g\frac{\partial u}{\partial y} + hu = 0$$

where the coefficients $a, b, c, \ldots$ may be functions of x and y. The form of this equation resembles that of a conic section, and once again it is the coefficients of the highest order terms which determine its nature. For simplicity we suppose that these coefficients are constants.

The superposition principle holds for the linear homogeneous equation (2.1.1), i.e., its solutions form a linear space. By exhibiting an infinite sequence of independent solutions, we will show that the dimension of this solution space is infinite.

We now introduce new independent variables by an *affine* transformation

$$(2.1.2) \quad \begin{aligned} \xi &= \alpha x + \beta y \\ \eta &= \gamma x + \delta y \end{aligned} \qquad \alpha\delta - \beta\gamma \neq 0$$

Here $\alpha, \beta, \gamma, \delta$ are constants to be chosen below. It is seen that

$$(2.1.3) \quad \frac{\partial u}{\partial x} = \alpha\frac{\partial u}{\partial \xi} + \gamma\frac{\partial u}{\partial \eta} \qquad \frac{\partial u}{\partial y} = \beta\frac{\partial u}{\partial \xi} + \delta\frac{\partial u}{\partial \eta}$$

and for the second derivatives that

$$\frac{\partial^2 u}{\partial x^2} = \left(\alpha\frac{\partial}{\partial \xi} + \gamma\frac{\partial}{\partial \eta}\right)^2 u \qquad \frac{\partial^2 u}{\partial y^2} = \left(\beta\frac{\partial}{\partial \xi} + \delta\frac{\partial}{\partial \eta}\right)^2 u$$

$$\frac{\partial^2 u}{\partial x \partial y} = \left(\alpha\frac{\partial}{\partial \xi} + \gamma\frac{\partial}{\partial \eta}\right)\left(\beta\frac{\partial}{\partial \xi} + \delta\frac{\partial}{\partial \eta}\right)u$$

The second-order derivatives in the transformed equation are

$$(2.1.4) \quad (a\alpha^2 + 2b\alpha\beta + c\beta^2)\frac{\partial^2 u}{\partial \xi^2} + 2[a\alpha\gamma + b(\alpha\delta + \beta\gamma) + c\beta\delta]\frac{\partial^2 u}{\partial \xi \partial \eta} + (a\gamma^2 + 2b\gamma\delta + c\delta^2)\frac{\partial^2 u}{\partial \eta^2}$$

By a suitable choice of $\alpha, \beta, \gamma, \delta$ we can make two of these three coefficients vanish. For instance, let us suppose that $c \neq 0$, so that the roots λ_1 and λ_2 of the quadratic

$$(2.1.5) \quad a + 2b\lambda + c\lambda^2 = 0$$

are both finite. Let us set $\alpha = \gamma = 1$, $\beta = \lambda_1$, $\delta = \lambda_2$, so that

$$(2.1.6) \quad \xi = x + \lambda_1 y \qquad \eta = x + \lambda_2 y$$

Then the partial differential equation becomes

$$L(u) = [a + b(\lambda_1 + \lambda_2) + c\lambda_1\lambda_2] \frac{\partial^2 u}{\partial \xi \, \partial \eta} + \cdots = 0$$

Here some first derivative terms are not indicated explicitly.
Since

(2.1.7) $$\lambda_1 + \lambda_2 = -\frac{2b}{c} \qquad \lambda_1\lambda_2 = \frac{a}{c}$$

we find that

(2.1.8) $$\frac{2}{c}(ac - b^2)\frac{\partial^2 u}{\partial \xi \, \partial \eta} + \cdots = 0$$

Let us suppose that only second derivative terms are present in (2.1.1) and therefore also in (2.1.8). Then we find, supposing $ac - b^2 \neq 0$,

(2.1.9) $$\frac{\partial^2 u}{\partial \xi \, \partial \eta} = 0$$

This has the obvious general integral

(2.1.10) $$u = f(\xi) + g(\eta)$$

since no lower order terms are present.

As might be expected by analogy with conic sections, there are three main cases to be considered, according as the discriminant $b^2 - ac$ is positive, negative, or zero (Petrowsky, Ref. 47).

Case 1. $b^2 - ac > 0$. The roots λ_1, λ_2 are real and distinct. The standard form (2.1.9) has the general solution (2.1.10) or, by (2.1.6),

(2.1.11) $$u = f(x + \lambda_1 y) + g(x + \lambda_2 y)$$

In this case, (2.1.1) is said to be *hyperbolic*. Just as a rotation $\pi/4$ changes the rectangular hyperbola $\xi\eta = $ constant to the form $\xi^2 - \eta^2 = $ constant, so the rotation

$$s = \tfrac{1}{2}\xi + \tfrac{1}{2}\eta \qquad t = \tfrac{1}{2}\xi - \tfrac{1}{2}\eta$$

brings about the alternative standard form

(2.1.12) $$\frac{\partial^2 u}{\partial t^2} - \frac{\partial^2 u}{\partial s^2} + \cdots = 0$$

Case 2. $b^2 - ac < 0$. The roots λ_1, λ_2 are conjugate complex: $\lambda_1 = \rho + i\sigma = \bar{\lambda}_2$. Thus $\xi = x + \lambda_1 y = x + \rho y + i\sigma y$ and $\eta = x + \lambda_2 y = x + \rho y - i\sigma y = \bar{\xi}$. The standard form is

$$\frac{\partial^2 u}{\partial \xi \, \partial \bar{\xi}} = 0$$

with general integral $u = f(\xi) + g(\bar{\xi})$.

Let us now write $\xi = s + it$, where s and t are real, so that

$$s = x + \rho y, \qquad t = \sigma y$$

Also

$$s = \operatorname{Re} \xi = \tfrac{1}{2}\xi + \tfrac{1}{2}\bar{\xi} = \tfrac{1}{2}\xi + \tfrac{1}{2}\eta$$

$$t = \operatorname{Im} \xi = \frac{1}{2i}\xi - \frac{1}{2i}\bar{\xi} = \frac{1}{2i}\xi - \frac{1}{2i}\eta$$

In these variables the standard form is seen to be

(2.1.13) $$\frac{\partial^2 u}{\partial t^2} + \frac{\partial^2 u}{\partial s^2} = 0$$

In this case the equation is said to be *elliptic*, and the standard form (2.1.13) which does not involve complex quantities is preferred. The general solution now becomes

(2.1.14) $$u = f(s + it) + g(s - it)$$

That is, the solution is the sum of a formal "analytic" function of $\xi = s + it$ and a formal "antianalytic" function of $\bar{\xi} = s - it$.

Case 3. $b^2 - ac = 0$. The roots λ_1, λ_2 are real and equal: $\lambda_1 = \lambda_2$. But the transformation (2.1.6) degenerates if η becomes identical with ξ. Instead, we shall choose for η any combination of x and y not proportional to ξ. Then, because of (2.1.5), the first coefficient in (2.1.4) still vanishes. The mixed coefficient becomes

$$a\gamma + b(\delta + \gamma\lambda) + c\lambda\delta = (a + b\lambda)\gamma + (b + c\lambda)\delta$$

In this case of equal roots,

$$-\lambda = \frac{a}{b} = \frac{b}{c}$$

so this expression also vanishes. Therefore the standard form must be

(2.1.15) $$\frac{\partial^2 u}{\partial \eta^2} = 0$$

with general integral

$$u = f(\xi) + \eta g(\xi)$$

or

(2.1.16) $$u = f_1(x + \lambda y) + y g_1(x + \lambda y)$$

This is the *parabolic* case.

EXERCISES 2.1

1. Determine the type (elliptic, parabolic, hyperbolic) of each of the following equations and find the general solution of equations (a) and (c) through (f):
   (a) $u_{tt} - 3u_{tx} + 2u_{xx} = 0$.
   (b) $u_t + u_{xx} = 0$.
   (c) $u_{tt} + 3u_{xx} = 0$.
   (d) $u_{tt} + 5u_{tx} + 4u_{xx} = 0$.
   (e) $u_{tt} + 4u_{tx} + 4u_{xx} = 0$.
   (f) $u_{tt} + 3u_{tx} + 4u_{xx} = 0$.
   (g) $u_{tt} + xu_{xx} = 0$.
   (h) $u_{tt} + tu_{xx} = 0$.

2. Find the general solution of each of
   (a) $u_t - u_x = 0$.
   (b) $u_t + cu_x = 0$.
   (c) $u_{tt} - c^2 u_{xx} + a(u_t - cu_x) = 0$.
   (d) $u_{tt} - 3cu_{xt} - 10c^2 u_{xx} = 0$.

3. By a change of dependent variable $v = ue^{ax+bt}$ reduce the equation $u_{tt} = u_{xx} + \alpha u_x + \beta u_t + \gamma u$ to the form $v_{tt} = v_{xx} + hv$.

2.2 WAVE MOTION OF A STRING

Let us derive a partial differential equation as the limit of a sequence of ordinary differential equations. Suppose beads of mass $m = \rho h$ are strung

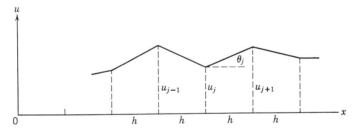

Figure 2.1 Motion of a string of beads.

at equal intervals h upon a light string which is stretched with tension T along the x-axis. Let u_j be the transverse displacement of the jth bead, and suppose u_j is small enough that we can neglect its square or higher powers. We consider a Cartesian plane of coordinates x and u, as shown in Figure 2.1. Let θ_j be the (small) angle of the string segment joining the jth and $(j + 1)$st beads to the x-axis. Thus $\sin \theta_j \approx \tan \theta_j = (u_{j+1} - u_j)/h$.

We see that two upward force components are impressed on the jth bead: from the right,
$$T \sin \theta_j \approx T \frac{u_{j+1} - u_j}{h}$$
and from the left,
$$-T \sin \theta_{j-1} \approx T \frac{u_{j-1} - u_j}{h}$$

DISTRIBUTIONS AND WAVES

Therefore the equation of motion is

(2.2.1) $$\rho h \ddot{u}_j = T \frac{u_{j+1} - 2u_j + u_{j-1}}{h}$$

We divide by $h\rho$ and let h tend to zero, so that the number of beads becomes large and the mass of each correspondingly small. The terms on the right become in the limit the second derivative with respect to length x along the string. Thus,

(2.2.2) $$\ddot{u} = c^2 u_{xx} \qquad c^2 = \frac{T}{\rho}$$

is the partial differential equation of a vibrating string of constant line density ρ. It is of the hyperbolic type.

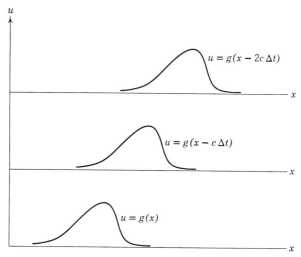

Figure 2.2 Motion of a wave form.

The general solution is

(2.2.3) $$u = f(x + ct) + g(x - ct)$$

The reader should understand the graphical interpretation of such terms as $f(x + ct)$ and $g(x - ct)$. (See Figure 2.2.) Consider the latter as a function of x for a sequence of times t. The profile of the graph of g remains the same throughout. But an increase Δt of t requires a corresponding increase $\Delta x = c\,\Delta t$ of x in order for $x - ct$ to maintain the value corresponding to any fixed place, such as a crest or singularity, on the graph. Therefore, these features and the entire graph appear to move with velocity $\Delta x/\Delta t = c$ in the positive x-direction. Likewise, the profile of the term $f(x + ct)$, as a function of x, moves with increasing time t to the left with velocity c.

These terms are interpreted physically as *waves* of arbitrary but permanent profiles, traveling with velocity c in the direction of the positive or the negative x-axis respectively.

The one-dimensional wave equation (2.2.2) arises in many other problems, some of which are:

1. Linearized supersonic airflow.
2. Sound waves in a tube or pipe.
3. Longitudinal vibrations of a bar.
4. Torsional oscillations of a rod.
5. Transmission of electricity along an insulated, low-resistance cable.
6. Long water waves in a straight canal.
7. "Whistler" waves in hydromagnetics.

Consider once more a string of density ρ, tension T, and unlimited length. Let us disregard gravity and the manner of suspension of the string. The problem of motion now is to find the displacement $u(x, t)$, given the initial position and velocity profiles

(2.2.4) $$u(x, 0) = \varphi(x) \qquad u_t(x, 0) = \psi(x)$$

We insert (2.2.3) in these equations and find, first,

(2.2.5) $$u(x, 0) = f(x) + g(x) = \varphi(x)$$

and second,

$$f'(x) - g'(x) = \frac{1}{c} \psi(x)$$

An integration with respect to x yields

(2.2.6) $$f(x) - g(x) = \frac{1}{c} \int_a^x \psi(s) \, ds$$

so that we can solve for $f(x)$ and $g(x)$. The constant of integration has been incorporated into the lower limit. We find

(2.2.7) $$f(x) = \frac{1}{2} \varphi(x) + \frac{1}{2c} \int_a^x \psi(s) \, ds$$

$$g(x) = \frac{1}{2} \varphi(x) - \frac{1}{2c} \int_a^x \psi(s) \, ds$$

When these expressions are substituted into (2.2.3), we obtain the formula of d'Alembert:

(2.2.8) $$u(x, t) = \frac{1}{2} [\varphi(x + ct) + \varphi(x - ct)] + \frac{1}{2c} \int_{x-ct}^{x+ct} \psi(s) \, ds$$

To interpret this result, we recall that the waves travel with velocity c in either direction. Now the displacement at a typical point (x, t) of space-time involves the initial displacements φ only at the two places $x + ct$, $x - ct$ from which waves of velocity c would arrive after time t. It also depends on the initial velocity profile ψ throughout this interval.

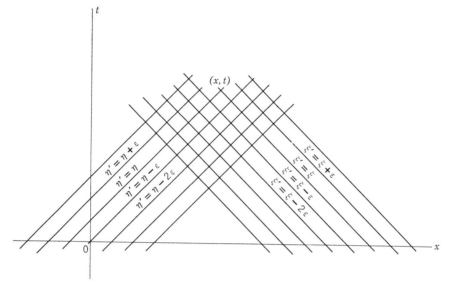

Figure 2.3 Families of characteristic lines.

In the x, t-coordinate plane we shall draw a space-time diagram of the two families of lines (see Figure 2.3)

(2.2.9) $\qquad ct + x = \text{const.} \qquad ct - x = \text{const.}$

These lines are known as *characteristic lines* or *characteristics* of the differential equation. Observe that through any fixed point (x', t') of space-time there pass two characteristic lines, namely,

$$ct' + x' = ct + x \qquad ct' - x' = ct - x$$

We shall discuss the properties of characteristic lines, and will show that, as the name suggests, they are closely connected with the nature of the hyperbolic differential equation.

The coordinates $\xi = ct + x$, $\eta = ct - x$ are called characteristic coordinates and the standard, mixed second derivative form (2.1.9) is called the characteristic form for the equation. In these new coordinates the characteristic lines are the coordinate lines themselves. From (2.1.10) we see that the general solution is composed of two terms, each of which is constant along the characteristic lines of one of the two families. Also, and this is a general property of characteristics, the differential equation is of lower (first) order in each of the characteristic variables separately.

If we compare this with our earlier description of traveling waves, we see that, in the space-time diagram, features of the wave profile are extended along characteristic lines. In fact, the locus of any feature is a characteristic line. This is especially significant if the feature is a singularity, such as a discontinuity of the function $u(x, t)$ or some of its derivatives.

THEOREM 2.2.1 *The space-time locus of a singularity consists of characteristics.*

This implies, as we see in the d'Alembert formula, that if a datum function such as $\varphi(x)$ has a singularity, then the solution function $u(x, t)$ will at any later time possess one or more related singularities lying on the same characteristics. These characteristics originate at the initial singularity.

Let us return to the role of the characteristic lines in the d'Alembert formula (2.2.8). Through the space-time point (x, t) pass two characteristic lines which, with running coordinates ξ, τ, have the equations $\xi \pm c\tau = x \pm ct$. They intersect the initial locus $\tau = 0$ in the places $\xi = x \pm ct$. These places define the domain of the datum functions φ and ψ upon which $u(x, t)$ explicitly depends.

DEFINITION 2.2.1 *The domain of dependence of the solution with respect to a given nonhomogeneous datum function is the smallest set such that the solution is independent of the values of the datum function outside this set.*

The domain of dependence of $u(x, t)$ on $\varphi(x)$ is the set of two points $x + ct$, $x - ct$, and the domain of dependence on $\psi(x)$ is the interval between the two points.

THEOREM 2.2.2 *The domain of dependence is bounded by the characteristics.*

This is a very general property of hyperbolic equations in any number of space variables, and we shall encounter it repeatedly in Chapter 10. In relativity it has the physical interpretation that waves cannot travel faster than light.

Given a particular point, we can look for the set of all points having the given point in its domain of dependence. This new set is the region of influence.

DEFINITION 2.2.2 *The region of influence of a point consists of all points at which the solution depends on data at the given point.*

The region of influence is bounded by the two characteristics through the given point and extends into the "future" $t > t_0$.

Now consider the motion of the string under external forces as represented by the equation

(2.2.10) $$\ddot{u} - c^2 u_{xx} = f(x, t)$$

Here $f(x, t)$ is the force impressed per unit of mass on the string. Let us suppose that no forces are present and that the string is in equilibrium until $t = 0$: thus $u(x, t) \equiv 0, f(x, t) \equiv 0$ for $t < 0$ while $u(x, 0) = u_t(x, 0) = 0$. We can most easily integrate (2.2.10) in the characteristic variables $\xi = ct + x$, $\eta = ct - x$, with $\partial(\xi, \eta)/\partial(x, t) = 2c$. Thus

(2.2.11) $$u_{\xi\eta} = \frac{1}{4c^2} f\left(\frac{\xi - \eta}{2}, \frac{\xi + \eta}{2c}\right)$$

We integrate this with respect to η' over the range $-\infty < \eta' \leq \eta = ct - x$, and likewise with respect to ξ' over the range $-\infty \leq \xi' \leq \xi = ct + x$. In

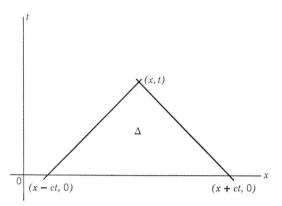

Figure 2.4 Characteristic triangle.

the x, t-plane this gives a quadrant surmounted by the triangle of Figure 2.4. Thus we find

(2.2.12)
$$u = \frac{1}{4c^2} \int_{-\infty}^{\eta} \int_{-\infty}^{\xi} f\left(\frac{\xi' - \eta'}{2}, \frac{\xi' + \eta'}{2c}\right) d\xi' \, d\eta'$$
$$= \frac{1}{2c} \iint_{\Delta} f(x', t') \, dx' \, dt'$$

where Δ denotes the *characteristic triangle* with base on the x-axis and vertex at (x, t). Again, we observe that the domain of dependence Δ of $u(x, t)$ on $f(x, t)$ is bounded by the characteristic lines through (x, t).

If the string is deflected and in motion at the initial time $t = 0$, and if external forces $f(x, t)$ also are present, we would require a superposition of expressions (2.2.8) and (2.2.12).

EXERCISES 2.2

1. Sketch the string profile at a succession of instants of time when
   (a) $u(x, 0) = 1$ if $|x| \leq a$, $u(x, 0) = 0$ if $|x| > a$; $u_t(x, 0) \equiv 0$.
   (b) $u(x, 0) \equiv 0$; $u_t(x, 0) = 1$ if $|x| \leq a$, $u_t(x, 0) = 0$ if $|x| > a$.
   (c) $u(x, 0) = a - |x|$ if $|x| \leq a$, $u(x, 0) = 0$ if $|x| > a$; $u_t(x, 0) \equiv 0$.
   (d) $u(x, 0) \equiv 0$; $u_t(x, 0) = a - |x|$ if $|x| \leq a$, $u_t(x, 0) = 0$ if $|x| > a$.
Assume no external forces.

2. Show that if the data $\varphi(x)$, $\psi(x)$ are either even or odd about $x = 0$, so is the d'Alembert solution.

3. Find the solution of the system of two first order equations for the vector $\mathbf{u}(x, t)$,

$$\frac{\partial \mathbf{u}}{\partial t} = A \frac{\partial \mathbf{u}}{\partial x} \qquad\qquad A = \begin{pmatrix} 0 & 1 \\ 1 & 0 \end{pmatrix}$$

with $\mathbf{u}(x, 0) = \mathbf{u}_0(x)$. Also find a solution of the nonhomogeneous system with the force vector $\mathbf{f}(x, t)$ present on the right-hand side.

4. A string under tension T is composed of a portion of density ρ_1 along the negative x-axis fastened at $x = 0$ to a portion of density ρ_2 along the positive x-axis. A waveform $f(x - c_1 t)$ approaches the origin from the left, f being a function which is zero except in a fixed interval, and $c_1^2 = T/\rho_1$. At the origin the solution and its first derivative with respect to x are continuous. Find the reflected and transmitted waves, and define reflection and transmission coefficients.

5. A string of density ρ under tension T lies along the x-axis. At the origin is fastened a point mass m which vibrates with the string. A wave profile $f(x - ct)$ approaches from the left, f being a function which is zero except in a fixed interval. Find the reflected and transmitted waveforms.

6. Find the general integral of $u_{\xi\eta} = au_\eta$.

7. Find the solution of $u_{tt} = u_{xx} + \alpha u_t + \alpha u_x$ which satisfies $u(x, 0) = \varphi(x)$, $u_t(x, 0) = \psi(x)$.

8. If $u(x, t)$ is a smooth solution of $\rho u_{tt} = T u_{xx}$, which vanishes for $|x|$ sufficiently large, $0 \leq t \leq t_1$, show that the energy integral

$$E(t) = \frac{1}{2} \int_{-\infty}^{\infty} [\rho u_t^2 + T(u_x)^2] \, dx$$

is constant for $0 \leq t \leq t_1$. *Hint:* Compute $E'(t)$ and integrate by parts.

9. From the result of Exercise 2.2.8 deduce that, if $u(x, 0) \equiv 0$, $u_t(x, 0) \equiv 0$, then $u(x, t) \equiv 0$ and, hence, that the initial value problem for the string equation has at most one solution. *Remark:* See Chapter 10 for further uniqueness proofs by this method.

2.3 REFLECTION OF WAVES

To establish d'Alembert's formula, we assumed the string was indefinitely long. Actually this solution is realistic only for intervals of time so short that no wave reaches an end of the string. Let us now study the effect of displacements and reflections at one end of a string.

Let a "semi-infinite" string be fastened at the origin and extend along the positive x-axis. More generally, let the end at $x = 0$ be displaced a prescribed amount

(2.3.1) $\qquad\qquad u(0, t) = h(t) \qquad\qquad t > 0$

and let the initial conditions (2.2.4) hold for $x > 0$. Then d'Alembert's formula is valid wherever no wave from the end of the string has yet arrived, that is, for $x > ct$ (see Figure 2.5). In this region we require values of $f(x)$ and $g(x)$ in (2.2.3) for positive arguments x, and these are known from the initial conditions via (2.2.7).

However, if $x < ct$, the argument of $g(x - ct)$ in (2.2.3) is negative. To find $g(-s)$ for negative arguments $-s$, we note that

(2.3.2) $$u(0, t) = h(t) = g(-ct) + f(ct)$$

so that

(2.3.3) $$g(-s) = h\left(\frac{s}{c}\right) - f(s) \qquad s > 0$$

$$= h\left(\frac{s}{c}\right) - \frac{1}{2}\varphi(s) - \frac{1}{2c}\int_a^s \psi(s_1)\, ds_1$$

With these determinations of the arbitrary functions, the general solution (2.2.3) becomes for $ct > x$,

(2.3.4) $$u(x, t) = h\left(t - \frac{x}{c}\right) + \frac{1}{2}[\varphi(ct + x) - \varphi(ct - x)] + \frac{1}{2c}\int_{ct-x}^{ct+x} \psi(s)\, ds$$

This formula contains the endpoint term $h(t - x/c)$. This represents a waveform traveling to the right from the endpoint along the family of characteristics

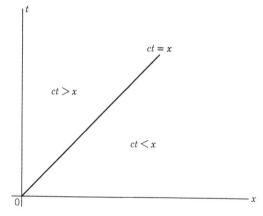

Figure 2.5 Regions of influence for the semi-infinite string.

$x - ct = $ constant. Thus the value of $h(t - x/c)$ can be found geometrically by drawing backwards in time the characteristic of this family which passes through the fieldpoint (x, t) until it meets the end $x = 0$. We then take the displacement of the end at that instant. The initial value term $\frac{1}{2}\varphi(ct - x)$ represents a wave which reached the fieldpoint along the same characteristic curve as the endpoint term, and which was the reflection, *with sign changed*, of a leftward bound wave that originated from the place $ct - x > 0$ on the x-axis. Notice also that the change of sign upon reflection has the effect of canceling the initial velocity contribution from the interval of length $ct - x$ nearest the endpoint. In this problem, the domain of dependence is bounded by the system of characteristic curves and their reflected extensions (Figure 2.6).

Now consider a *finite* length l of the string and suppose it is subject to a given motion at both ends. We have two types of problem to consider here:

1. The initial problem for the finite segment, with solution vanishing at each endpoint.
2. The boundary value problem for the finite segment, with vanishing initial values.

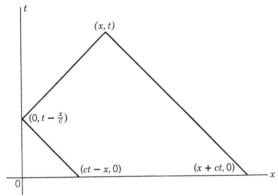

Figure 2.6 Domain of dependence for the semi-infinite string.

To obtain a solution of the full problem with given initial values as well as given motion of the ends of the string, we need to superpose suitable solutions of each type. These have been defined so that initial and boundary data do not interfere.

Turning now to the initial value problem, we shall employ a new device known as the *method of images*. Observe that an *odd* function of $x[f(-x) = -f(x)]$ will vanish at $x = 0$. Therefore, if we can arrange an odd solution on a larger interval $-l \leq x \leq l$, it will satisfy the boundary condition $u = 0$ at the origin. Likewise, if we can set up a solution which is odd with respect to the point l, the boundary condition $u = 0$ will hold at the other end $x = l$.

We can arrange this by choosing data for an "extended" initial value problem on the infinite interval, which are odd with respect to the origin *and* the point l. Such data are uniquely defined by their values on $0 \leq x \leq l$ (Figure 2.7). Thus, let

(2.3.5) $$\Phi(x) = \begin{cases} -\varphi(2nl - x) & (2n - 1)l < x < 2nl \\ \varphi(x - 2nl) & 2nl < x < (2n + 1)l \end{cases}$$

From d'Alembert's formula with datum $\Phi(x)$, we obtain a solution function for the finite string. The effect of the various successive reflections can also be computed by drawing the characteristic curves back to the ends, and then

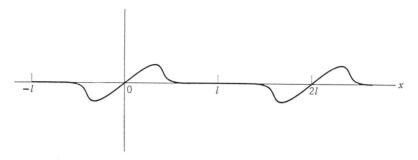

Figure 2.7 Initial data odd with respect to interval $0 \leq x \leq l$.

further back in time by repeated reflections, counting a change of sign after each reflection (Figure 2.8). The intersection with the initial segment $t = 0$, $0 \leq x \leq l$, gives the place from which the contribution began. For the construction of the solution formula, we refer to the exercises.

From the same diagram, we can also read off the solution formula for the *boundary problem*

(2.3.6) $$u(0, t) = h(t) \qquad u(l, t) = k(t) \qquad t > 0$$

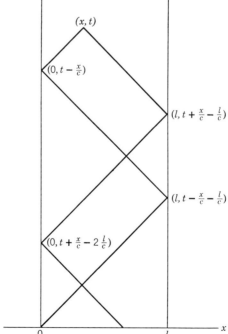

Figure 2.8 Repeated reflections.

with zero initial data. By (2.3.4), we see that the solution after one reflection at the origin is $h(t - x/c)$. The characteristic line involved here starts at the fieldpoint (x, t) and is reflected at the successive points

(2.3.7)
$$\left(0, t - \frac{x}{c}\right) \qquad \left(l, t - \frac{x}{c} - \frac{l}{c}\right)$$
$$\left(0, t - \frac{x}{c} - 2l\right) \qquad \left(l, t - \frac{x}{c} - \frac{3l}{c}\right)$$
$$\vdots \qquad \vdots$$

As there is a change of sign at each reflection, we see that the contribution to the solution at these places of reflection is

(2.3.8) $\quad h\left(t - \dfrac{x}{c}\right) - k\left(t - \dfrac{x}{c} - \dfrac{l}{c}\right) + h\left(t - \dfrac{x}{c} - \dfrac{2l}{c}\right) - k\left(t - \dfrac{x}{c} - \dfrac{3l}{c}\right) \cdots$

The number of terms in the sum is equal to the total number of reflections subsequent to $t = 0$.

The other characteristic line through the fieldpoint (x, t) is reflected at the sequence of points

(2.3.9)
$$\left(l, t + \frac{x}{c} - \frac{l}{c}\right) \qquad \left(0, t + \frac{x}{c} - \frac{2l}{c}\right)$$
$$\left(l, t + \frac{x}{c} - \frac{3l}{c}\right) \qquad \left(0, t + \frac{x}{c} - \frac{4l}{c}\right)$$
$$\vdots \qquad \vdots$$

and the contributions from this series of reflections will be

(2.3.10) $\quad k\left(t + \dfrac{x}{c} - \dfrac{l}{c}\right) - h\left(t + \dfrac{x}{c} - \dfrac{2l}{c}\right) + k\left(t + \dfrac{x}{c} - \dfrac{3l}{c}\right)$
$$- h\left(t + \frac{x}{c} - \frac{4l}{c}\right) + \cdots$$

The complete solution of the reflection problem is the sum of (2.3.8) and (2.3.10). The series terminate when the arguments of h and k become negative.

It is instructive to carry out simple numerical studies of d'Alembert's formula and of the reflection of waves at an endpoint. For this purpose we replace the partial differential equation by a difference equation. Let u_{hk} denote the value of u at the (h, k) point of a lattice in the x, t-coordinate plane, with lattice intervals Δx, Δt, respectively. Thus u_{hk} denotes $u(h\,\Delta x, k\,\Delta t)$. If we approximate second derivatives by second difference quotients, (2.2.2) is approximated by the following finite difference equation:

$$\text{(2.3.11)} \qquad \frac{u_{h,k+1} - 2u_{h,k} + u_{h,k-1}}{\Delta t^2} = c^2 \frac{u_{h+1,k} - 2u_{h,k} + u_{h-1,k}}{\Delta x^2}$$

This discrete analogue of the string equation can also be compared with (2.2.1).

Let us rearrange by solving for $u_{h,k+1}$:

$$\text{(2.3.12)} \qquad u_{h,k+1} = C(u_{h+1,k} + u_{h-1,k}) + 2(1-C)u_{h,k} - u_{h,k-1}$$

where

$$\text{(2.3.13)} \qquad C = c^2 \frac{\Delta t^2}{\Delta x^2}$$

The numerical magnitude of C depends, therefore, on the ratio $\Delta t/\Delta x$ of the mesh widths of the lattice. We achieve the simplest difference equation, and the

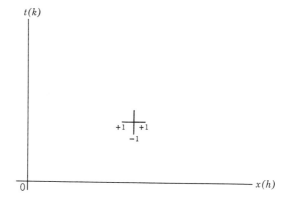

Figure 2.9 A template pattern.

one best adapted to represent wave motion along the characteristics, if we choose $C = 1$. Then (2.3.12) has the convenient form

$$\text{(2.3.14)} \qquad u_{h,k+1} = u_{h+1,k} + u_{h-1,k} - u_{h,k-1}$$

This relation can be represented on a diagram by a movable template in the form of a cross with coefficients attached to the various positions (Figure 2.9). The value at $h, k+1$ is computed by moving the template center to the h, k position and forming the suggested linear combination of values from the kth and $(k-1)$st rows.

To start the computation, we require two rows of given values, say, $k = 0$ and $k = 1$. In Figure 2.10 we show the calculation for initial values concentrated at one place, so that the two initial rows are equal. The region of influence is clearly shown by the pattern of nonzero entries.

REFLECTION OF WAVES

```
t(k)
 ·
 ·
 ·
 0  0  0 -1  0  0  0  0  0  0  0  0  1
 0  0 -1  0  0  0  0  0  0  0  0  1
 0 -1  0  0  0  0  0  0  0  0  1
 0  0  0  0  0  0  0  0  0  1
 0  1  0  0  0  0  0  0  1  0
 0  0  1  0  0  0  0  1  0  0
 0  0  0  1  0  0  1  0  0  0
 0  0  0  0  1  1  0  0  0  0
 0  0  0  0  1  1  0  0  0  0  ·  ·  ·  x(h)
```

Figure 2.10 Initial values given for $k = 0, 1$. Boundary condition $u = 0$ for $h = 0$.

If in addition to initial values we have a reflection at an end, then the end column or columns will have values assigned. For instance, if $u_x = 0$ at the end and $u = 0$, $u_t = \delta_{kl}$ initially, we have the pattern of Figure 2.11.

The simplicity and realism of these examples should not be interpreted as convincing evidence that all such finite difference patterns work well. The reader should construct for himself figures like those given here when the constant (2.3.13) is taken to be $C = 4$, and explain the results, if he can, with reference to the characteristic lines of the partial differential equation. We shall return to this question of stability in Chapter 3.

```
t(k)
 ·
 ·
 ·
 2  2  2  2  2  1  1  1  1  1  1  1  1
 2  2  2  2  1  1  1  1  1  1  1  1
 2  2  2  1  1  1  1  1  1  1  1
 2  2  1  1  1  1  1  1  1  1
 1  1  1  1  1  1  1  1  1  0
 0  0  1  1  1  1  1  1  0  0
 0  0  0  1  1  1  1  0  0  0
 0  0  0  0  1  1  0  0  0  0
 0  0  0  0  0  0  0  0  0  ·  ·  ·  x(h)
```

Figure 2.11 Initial values of u_t given for $k = 0, 1$. Boundary condition $u_x = 0$ or $u_{0,k} = u_{1,k}$.

EXERCISES 2.3

1. Show that the end condition $u_x(0, t) = h(t)$ results in reflection without change of sign and determine the solution satisfying

$$u(x, 0) = \varphi(x) \qquad u_t(x, 0) = \psi(x) \qquad 0 \le x < \infty$$
$$u_x(0, t) = h(t) \qquad \qquad 0 \le t < \infty$$

(This corresponds to a string with a looped end sliding along a perpendicular rod.)

2. Show that the even and odd extensions to $-\infty < x < \infty$ of a function $\varphi(x)$ defined for $x \geq 0$ are respectively

$$\varphi_e(x) = \varphi(|x|) \qquad \varphi_0(x) = \operatorname{sgn} x \, \varphi(|x|)$$

where $\operatorname{sgn} x = +1$ if $x > 0$, $\operatorname{sgn} x = -1$ if $x < 0$.

3. If $[x]$ denotes the largest integer less than x, show that the function $\Phi(x)$ of (2.3.5) is given by

$$\varphi_0\left(x - 2l\left[\frac{x}{2l}\right]\right).$$

4. Write down in full the solution of the initial problem $u(x, 0) = \varphi(x)$, $u_t(x, 0) = \psi(x)$, $0 \leq x \leq l$, with $u(0, t) = u(l, t) = 0$, $t > 0$.

5. With reference to characteristic lines and their reflections, construct a solution for a finite string $0 \leq x \leq l$, with zero initial conditions and end conditions either

(a) $u(0, t) = h(t) \qquad u_x(l, t) = k(t)$

or

(b) $u_x(0, t) = h(t) \qquad u_x(l, t) = k(t)$

6. Find an explicit solution for the semi-infinite string $x \geq 0$, such that $u(x, 0) = \varphi(x)$, $u_t(x, 0) = \psi(x)$, and

$$u_x(0, t) + hu(0, t) = k(t) \qquad\qquad t > 0$$

where h is a constant.

7. A semi-infinite string stretched along $x \geq 0$ is released at time zero and begins to fall under gravity, with $u_{tt} - c^2 u_{xx} = -g$. The end $x = 0$ is held fixed: $u(0, t) = 0$, $t > 0$. Find the appropriate solution of the differential equation.

8. Compute the solutions of (2.3.14) from the initial data

(a)

    0 0 0 0 0 1 0 0 0 0

    0 0 0 0 0 1 0 0 0 0

and

(b)

    0 0 0 0 0 1 0 0 0 0

    0 0 0 0 0 0 0 0 0 0

9. Prescribe a boundary condition in Exercise 2.3.8b by $u_x = 0$, i.e., $u_{0k} = u_{1k}$ for all k, and recalculate the problem of this exercise.

10. If S_k denotes the sum $\Sigma_h u_{hk}$, where u_{hk} satisfies (2.3.12), show that $S_{k+1} - 2S_k + S_{k-1} = 0$ and, hence, that $S_k = a + bk$.

2.4 THEORY OF DISTRIBUTIONS

By the conventional definition of a function, a numerical value of the function is assigned for every number of the domain. In Chapter 1, Section 7, we saw that this definition is not adequate for the description of impulsive forces, and a generalization of the concept of function is required. The need is even more urgent for partial differential equations.

Consider, for example, the two equations

(2.4.1) $$\frac{\partial}{\partial x}\left(\frac{\partial u}{\partial y}\right) = 0 \qquad \frac{\partial}{\partial y}\left(\frac{\partial u}{\partial x}\right) = 0$$

The first of these is strictly satisfied by any function $f(x)$ of x. However, if this function $f(x)$ is not differentiable, the second equation is devoid of meaning. As we should like both equations to have the same solutions, we need to look for an extension of the class of functions such that *derivatives always exist* within the new class. These generalized functions are known as distributions, and their theory was developed by L. Schwartz about 1945 (Gelfand-Shilov, Ref. 25; Schwartz, Ref. 51).

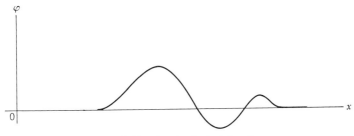

Figure 2.12 Graph of a test function.

To introduce Schwartz's generalization, we must first consider certain auxiliary functions $\varphi(x)$ of a real variable x, which we shall call *test functions*. Each test function is required to possess derivatives of every order and to vanish outside a finite interval (Figure 2.12). The test functions $\varphi(x)$ form a vector space, which we shall denote by D.

If we are given any space, we can define functions having this space as domain, by assigning to each point of the space a number. For a finite dimensional vector space V_n we thus construct scalar functions of a vector or, alternatively, scalar functions of n variables. Now we are considering a much larger vector space D of infinite dimension. Functions defined over D are functions of "infinitely many" variables—they depend on the entire set of values of the argument functions, or "test" functions φ. It is customary to call such a function of the set of values of a function a *functional*, for short. To make up for all this complexity, however, we consider the very simplest kind of functionals, namely, linear ones.

A linear functional $T = T(\varphi)$ has for each φ of D a numerical value $T(\varphi)$. Thus $T(\varphi)$ denotes a number. The numbers assigned as values of T also satisfy the linear relation

(2.4.2) $$T(c_1\varphi_1 + c_2\varphi_2) = c_1T(\varphi_1) + c_2T(\varphi_2)$$

Though we shall not emphasize the topological properties of distributions, we include the definition of continuity of T for the sake of completeness. T is *continuous* if we have $T(\varphi_n) \to 0$ as $n \to \infty$ for any sequence φ_n of D such that all $\varphi_n = 0$ outside a fixed bounded set, and φ_n together with all derivatives tend to zero uniformly as $n \to \infty$.

DEFINITION 2.4.1 *Distributions are continuous linear functionals on the vector space D.*

With this definition, the distributions $T(\varphi)$ form a linear vector space. This space is called the *dual* of the vector space D of testing functions, and it is denoted by D'. Let us show that D' is a larger space which properly contains D.

Suppose a function $f(x)$ is given. We can easily construct a linear functional representing $f(x)$, namely,

$$(2.4.3) \qquad T(\varphi) = \int_{-\infty}^{\infty} f(x)\varphi(x)\, dx$$

In fact, we shall regard the distribution T so defined as the same as the function $f(x)$. We *identify* the functional with the function. As this process of identification requires some motivation, we add a word of explanation. In Theorem 1.1.5 it was shown that in a finite-dimensional vector space every linear form L defines a vector which is also an element of the space. This is an example of such an identification, and it also shows that for a finite dimensional space of testing functions the dual space of linear functionals contains no additional elements.

For infinite dimensional vector spaces, such as are provided by testing functions, the dual space is larger, however. Therefore the identification of a particular functional of the larger space D' with a function of the smaller, conventional, class is analogous to the identification of a real number, which happens to be rational, with a member of the rational number set. We are quite accustomed to this latter identification.

Actually we have already encountered in Chapter 1 a distribution which is not a function. It is the Dirac distribution δ, defined by

$$(2.4.4) \qquad \delta(\varphi) = \varphi(0)$$

As we observed in Chapter 1, there is no classical or conventional function $\delta(x)$ with the property that for all φ of D

$$(2.4.5) \qquad \int_{-\infty}^{\infty} \delta(y)\varphi(y)\, dy = \varphi(0)$$

Therefore, to complete our analogy with functions, we introduce a *generalized function* or *symbolic function* $\delta(x)$ which is defined by this very property. We shall also *identify* the Dirac distribution δ with this Dirac "function" $\delta(x)$. This identification gives a proper and precise meaning to the symbol $\delta(x)$: it represents a specific linear functional $\delta(\varphi) = \varphi(0)$.

The translated Dirac distribution

$$(2.4.6) \qquad \delta_x(\varphi) = \varphi(x)$$

is represented by the symbolic function $\delta(x - y)$ which has the substitution property

(2.4.7) $$\int_{-\infty}^{\infty} \delta(x - y)\varphi(y)\,dy = \varphi(x)$$

mentioned in Section 1.7.

The term "distribution" arises from the notion of a distribution of mass along a line. In the case of the Dirac distribution, a unit mass is concentrated at the single point $x = 0$.

The derivatives of the smooth testing functions can also be used to define functionals. For instance, a distribution is defined by

$$T_2(\varphi) = -\varphi'(0)$$

This is called the derivative δ' of the Dirac distribution: it represents a unit "dipole" at the origin. The reason for the negative sign appears if we now consider the general problem of defining the derivative of a given distribution. Our definition must be consistent with the usual definition for the derivative of a function f, so we write

$$f'(\varphi) = \int_{-\infty}^{\infty} f'(x)\varphi(x)\,dx$$

and integrate by parts, finding

$$-\int_{-\infty}^{\infty} f(x)\varphi'(x)\,dx = -f(\varphi')$$

The integrated terms drop out because $\varphi(x)$ is zero outside a bounded interval. Now we shall *define* the derivative T' of any distribution $T = T(\varphi)$.

DEFINITION 2.4.2 *The derivative T' of the distribution $T = T(\varphi)$ is given by*

(2.4.8) $$T'(\varphi) = -T(\varphi')$$

From this definition we have immediately:

THEOREM 2.4.1 *The derivative of a distribution exists and is also a distribution.*

Example 2.4.1 The Heaviside distribution is

(2.4.9) $$H(\varphi) = \int_{-\infty}^{\infty} H(s)\varphi(s)\,ds = \int_{0}^{\infty} \varphi(s)\,ds$$

According to our definition, the derivative of the Heaviside distribution is given by

$$H'(\varphi) = -H(\varphi') = -\int_{0}^{\infty} \varphi'(s)\,ds$$
$$= -[\varphi(\infty) - \varphi(0)] = \varphi(0)$$

Therefore, for every testing function φ,

(2.4.10) $$H'(\varphi) = \varphi(0) = \delta(\varphi)$$

so that the derivative of the Heaviside distribution *is* the Dirac distribution.

Example 2.4.2 The function $|x|$ defines the distribution

$$|x|(\varphi) = \int_{-\infty}^{\infty} |s|\, \varphi(s)\, ds = \int_{0}^{\infty} s\varphi(s)\, ds - \int_{-\infty}^{0} s\varphi(s)\, ds$$

The derivative of $|x|$ is, by definition,

$$|x|'(\varphi) = -|x|(\varphi') = -\int_{0}^{\infty} s\varphi'(s)\, ds + \int_{-\infty}^{0} s\varphi'(s)\, ds$$

After integration by parts, with integrated terms which disappear, this becomes

$$|x|'(\varphi) = \int_{0}^{\infty} \varphi(s)\, ds - \int_{-\infty}^{0} \varphi(s)\, ds = \int_{-\infty}^{\infty} \operatorname{sgn} s\, \varphi(s)\, ds$$

so the derivative of the absolute value distribution is the signum distribution:

(2.4.11) $$|x|'(\varphi) = \operatorname{sgn}(\varphi)$$

Observe that

(2.4.12) $$\operatorname{sgn} x = H(x) - H(-x)$$

If we define the reversed function $\check{\varphi}(s)$ by

(2.4.13) $$\check{\varphi}(s) \equiv \varphi(-s)$$

we could now write

(2.4.14) $$\operatorname{sgn}(\varphi) = H(\varphi) - H(\check{\varphi})$$

which is the corresponding relation between the signum and Heaviside distributions.

The delta function shows that we cannot assign numerical values to a distribution at every point as we can to a function. However, we can do this at some points. In particular, we can say whether or not a distribution is zero in an interval or open set.

DEFINITION 2.4.3 *A distribution T vanishes on an interval I when $T(\varphi) \equiv 0$ for all φ that differ from zero only within I.*

Thus $T(\varphi) \equiv 0$ for all φ that vanish outside the interval I. In other words, $T(\varphi)$ does not depend on the values taken by φ within the interval I.

Example 2.4.3 Since $\delta(\varphi) = \varphi(0)$, the Dirac distribution vanishes everywhere except at the origin.

Example 2.4.4 Since $H(\varphi) = \int_0^\infty \varphi(s)\,ds$, the Heaviside distribution vanishes for $x < 0$.

DEFINITION 2.4.4 *The (closed) set of points where T does not vanish is the* support *of T*.

The support may be described as the set such that $T(\varphi)$ does depend upon the values taken by φ upon it. If T is a function $f(x)$, the support is the (closed) set whereon $f(x)$ is not zero. More technically, we can describe the support as the least closed set outside which T vanishes. The support of the Heaviside distribution is the half-axis $x \geq 0$ (Figure 2.13). Also, the support of the Dirac distribution is the origin—a single point.

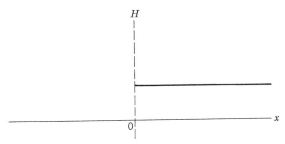

Figure 2.13 Support and singular support of Heaviside's function.

The Dirac distribution and each of its derivatives have as support one point. It is possible to prove, conversely, that only these distributions have such a support.

THEOREM 2.4.2 *A distribution T which has a support of one point (i.e., is equal to zero except at one point) is a finite linear combination of the Dirac function and its derivatives:*

(2.4.15) $$T = \sum_k a_k \delta^{(k)}$$

We omit the proof, which employs subtle properties of continuity (Schwartz, Ref. 51).

For each distribution we can define another set—the locus of singularities or the *singular support* of the distribution.

DEFINITION 2.4.5 *The singular support of a distribution T is the (closed) set of points where T is not a smooth function.*

More precisely, we can describe the singular support of T as the least closed set outside which T is equal to a function $f(x)$ possessing derivatives of all orders. The Dirac distribution has as singular support one point—the origin. The Heaviside distribution also has as singular support the origin. The distribution $|x|$ has the origin as singular support; although it is a continuous function it does not possess all derivatives there.

72   DISTRIBUTIONS AND WAVES

We conclude this discussion by introducing the very useful *power distributions* with support on the positive real axis. For any number λ with real part exceeding 0, we have:

DEFINITION 2.4.6   *The power distribution $x_+^\lambda(\varphi)$ is defined by*

$$(2.4.16) \qquad x_+^\lambda(\varphi) = \int_0^\infty s^\lambda \varphi(s)\, ds \qquad \text{Re } \lambda > 0.$$

Thus the distribution x_+^λ is to be identified with the function

$$x^\lambda(x) = \begin{cases} x^\lambda & x > 0 \\ 0 & x \leq 0 \end{cases}$$

The derivative of this function is

$$\lambda x^{\lambda-1}(x) = \begin{cases} \lambda x^{\lambda-1} & x > 0 \\ 0 & x \leq 0 \end{cases}$$

which corresponds to $\lambda x_+^{\lambda-1}$. Let us establish this relation from first principles:

$$\begin{aligned} x_+^{\lambda\prime}(\varphi) &= -x_+^\lambda(\varphi') = -\int_0^\infty s^\lambda \varphi'(s)\, ds \\ &= -[s^\lambda \varphi(s)]_0^\infty + \int_0^\infty \varphi(s)\, ds^\lambda \\ &= 0 + \lambda \int_0^\infty s^{\lambda-1} \varphi(s)\, ds = \lambda x_+^{\lambda-1}(\varphi) \end{aligned} \qquad (2.4.17)$$

Here we have used the definition of the derivative of a distribution, and integration by parts. The integrated terms vanish because Re $\lambda > 0$ and $\varphi(s)$ has bounded support. Thus we have proved:

THEOREM 2.4.3

$$(2.4.18) \qquad x_+^{\lambda\prime}(\varphi) = \lambda x_+^{\lambda-1}(\varphi)$$

The corresponding power distribution for the negative real axis is defined by

$$(2.4.19) \qquad x_-^\lambda(\varphi) = \int_{-\infty}^0 |s|^\lambda \varphi(s)\, ds$$

and its properties are referred to in the following exercise.

EXERCISES 2.4

1. Write down the symbolic or generalized functions which are identified with the linear forms $\varphi(1) + \varphi(-2)$; $\varphi(\pi) + \varphi'(2\pi) + \varphi''(b)$; $\varphi(a) - \varphi''(0)$; $\int_0^1 \varphi(s)\, ds$; $\int_a^\infty \varphi(s)\, ds$; $\int_{-\infty}^b \varphi(s)\, ds$; $\int_0^b s \varphi(s)\, ds$; $\int_0^\infty \sqrt{s} \varphi(s)\, ds$.

2. Write down the linear forms associated with the following symbolic functions:

x; x_+; x_+^2; $|x|^3$; $\sin x$; $\delta(x - a)$; $\delta'(x + b)$; $\sum_{n=-\infty}^{\infty} \delta(x + n)$; $H(x - b) - H(x - a)$.

3. Find the support and the singular support of the linear forms: $\varphi(a)$; $\varphi'(b)$;

$\int_0^1 \sin 2\pi s \varphi(s)\, ds$; $\int_0^\infty s^2 \varphi(s)\, ds$; $\int_0^\infty e^{-1/s} \varphi(s)\, ds$; $\sum_{n=-\infty}^{\infty} \varphi(n)$; $\sum_{n=0}^{\infty} \varphi^{(n)}(n)$; $\int_{-1}^1 e^{-1/(1-s^2)} \varphi(s)\, ds$.

4. Show that $x_-^\lambda(\varphi) = x_+^\lambda(\check{\varphi})$, where $\check{\varphi}(s) \equiv \varphi(-s)$. Show that $x_-^\lambda{}'(\varphi) = -\lambda x_-^{\lambda-1}(\varphi)$.

5. If a function $f(x)$ is smooth except for jumps of magnitude a_k at points x_k, show that

$$f(x) = f_0(x) + \Sigma a_k H(x - x_k)$$

where $f_0(x)$ is smooth. Deduce a formula for the derivative $f'(x)$.

6. Show that the singular support of T is contained in the support of T.

7. Show that

$$\int_{-\infty}^\infty H(x - s) H(s)\, ds = x_+(x)$$

and more generally that

$$\int_{-\infty}^\infty H(x - s) s_+^n\, ds = \frac{x_+^{n+1}(x)}{n+1} \qquad n = 1, 2, 3, \ldots$$

8. If $a(x) < b(x)$, show that

$$\frac{d}{dx} \int_{a(x)}^{b(x)} f(x, t)\, dt = \int_{-\infty}^\infty \frac{d}{dx} [H(t - a(x)) H(b(x) - t) f(x, t)]\, dt$$

Hence show that, if the limits of an integral are extended to $\pm\infty$ by inserting Heaviside factors in the integrand, then formal differentiation under the integral sign is correct.

9. Show that $e^{-1/x} H(x)$ is an infinitely differentiable function. Show that $\varphi(x) = e^{-1/(1-x^2)} H(1 - x^2)$ is a test function of D. Construct infinitely many linearly independent test functions of D with support $-1 \leq x \leq 1$.

10. Show that $\delta(ax) = (1/|a|)\, \delta(x)$.

11. The Dirac distribution in two dimensions $\delta(x, y)$ is defined by the substitution property. Show that $\delta(x, y)$ is equal, in the distribution sense, to the symbolic product $\delta(x)\delta(y)$.

12. Show that

$$\delta(ax + by)\delta(cx + dy) = \frac{1}{|ad - bc|} \delta(x)\delta(y)$$

13. If $\alpha(x)$ is a smooth function and T a distribution, show that the *product* $\alpha(x)T(x)$ defined by $\alpha T(\varphi) = T(\alpha\varphi)$ reduces to the usual product if T is a classical function. Show that $x\delta' = -\delta$.

2.5 APPLICATIONS TO THE INITIAL PROBLEM

The theory of distributions is well adapted to express solutions of initial value problems. In such a problem we are given a mechanical system, for instance a string, some initial or other data in the form of functions, and we are asked to find the value of the displacement at a given spacetime point. That value is a number. Furthermore, this number depends linearly upon the datum functions and so it is naturally a linear functional or distribution.

Therefore we may reformulate our problem: Find the solution as a distribution defined over the space of data testing functions. In this form, the problem is solved if we can display the symbolic function identified with the solution distribution. This symbolic function is often called the source function, or response function, or Green's function of the initial value problem. We repeat, this Green's function is a symbolic or generalized function identified with a distribution.

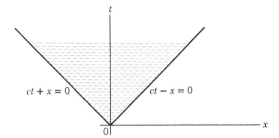

Figure 2.14 Support of the Green's function.

For the initial problem of the vibrating string it is very easy to pick out this Green's function for each of the two initial data and also for the force function. We shall also discuss the relationship of the three response functions so obtained.

Let us begin with the problem of initial velocities:

(2.5.1) $\quad u_{tt} - c^2 u_{xx} = 0 \quad u(x, 0) = 0 \quad u_t(x, 0) = \psi(x)$

By d'Alembert's formula, the solution is the definite integral

(2.5.2) $$u(x, t) = \frac{1}{2c} \int_{x-ct}^{x+ct} \psi(s)\, ds$$

We can express this as a distribution K, with symbolic function $K(x, t)$, in the form

(2.5.3) $$u(x, t) = K(\psi)(t) = \int_{-\infty}^{\infty} K(x - s, t)\psi(s)\, ds$$

provided we choose (see Figure 2.14)

(2.5.4) $$K(x - s, t) = \begin{cases} \dfrac{1}{2c} & |x - s| < ct \\ 0 & |x - s| \geq ct \end{cases}$$

Thus $K = 0$ unless ct exceeds both $x - s$ and $s - x$. This step function can be expressed by the Heaviside function:

(2.5.5) $$K(x - s, t) = \frac{1}{2c} H(ct + x - s)H(ct - x + s)$$

This is the Green's function, or source function, for the one-dimensional wave equation.

Observe that K depends on the two spatial variables x, s only through the combination $x - s$, because the string is homogeneous and the problem invariant under translation in x. The solution depends only on the distance $x - s$ between the initial point s and the field point x, not on x and s separately.

An integral of the type of (2.5.3) is known as a convolution. The factor functions K and ψ can be regarded as giving rise to a new kind of product $K * \psi$ by the operation of convolution, or "folding together."

When we regard K as a distribution, the time t enters as a parameter. Thus we can, for instance, differentiate $K(x - s, t)$ with respect to time t, and obtain

(2.5.6)
$$\frac{\partial K(x - s, t)}{\partial t} = \tfrac{1}{2}[\delta(ct + x - s)H(ct - x + s) + \delta(ct - x + s)H(ct + x - s)]$$

For $t > 0$ we can omit the Heaviside factors in each term, as a moment's consideration shows that their arguments must be positive (and their values unity) when the argument of the accompanying Dirac function is zero.

Note that when $t = 0+$ we have for K the initial "values" (i.e., distribution values)

$$K(x - s, 0) = 0 \qquad K_t(x - s, 0) = \tfrac{1}{2}\delta(x - s) + \tfrac{1}{2}\delta(s - x) = \delta(x - s)$$

Now let us examine the solution of the initial problem

(2.5.7) $$u_{tt} = c^2 u_{xx} \qquad u(x, 0) = \varphi(x) \qquad u_t(x, 0) = 0$$

This solution is

(2.5.8) $$u(x, t) = \tfrac{1}{2}[\varphi(x + ct) + \varphi(x - ct)]$$

$$= \frac{1}{2} \int_{-\infty}^{\infty} [\delta(x + ct - s) + \delta(ct - x + s)]\varphi(s)\, ds$$

which can be written as

(2.5.9) $$\int_{-\infty}^{\infty} \frac{\partial K(x - s, t)}{\partial t} \varphi(s)\, ds \qquad t > 0$$

THEOREM 2.5.1 *The source distribution for initial positions is the time derivative of the source distribution for initial velocities.*

This result, which is known as Stokes' rule, holds for all second order wave equations of the form

(2.5.10) $$u_{tt} = L(u)$$

where L is a linear operator in space variables only. The following general proof will cover all such cases. Let u satisfy (2.5.10) together with

(2.5.11) $$u(x, 0) = 0 \quad u_t(x, 0) = \psi(x)$$

Here x represents one or more spatial variables. Then the time derivative

(2.5.12) $$v(x, t) \equiv u_t(x, t)$$

itself satisfies (2.5.10), as is shown by differentiating (2.5.10) with respect to time. Also

(2.5.13) $$v(x, 0) = u_t(x, 0) = \psi(x)$$

and, finally,

(2.5.14) $$v_t(x, 0) = u_{tt}(x, 0) = L(u(x, 0)) = L(0) = 0$$

from (2.5.10) and (2.5.11).

The third problem has the force term:

(2.5.15) $$u_{tt} = c^2 u_{xx} + f(x, t)$$

A solution, as we know, is

(2.5.16)
$$u(x, t) = \frac{1}{2c} \iint_\Delta f(x', t') \, dx' \, dt'$$
$$= \frac{1}{2c} \int_{-\infty}^{t} d\tau \int_{x-c(t-\tau)}^{x+c(t-\tau)} f(s, \tau) \, ds$$
$$= \int_{-\infty}^{t} d\tau \int_{-\infty}^{\infty} K(x - s, t - \tau) f(s, \tau) \, ds$$

where K is as above.

In this integral we have set the lower limit of past time as $-\infty$. Actually we can suppose that $f(x, t)$ will be zero in the "distant past," i.e., for t sufficiently negative. If we assume that the solution $u(x, t)$ is also identically zero in the distant past, we can even drop the initial conditions at time zero. Then we have an initial problem commencing from $t = -\infty$. We shall consider this problem here as well.

The reason we have chosen $K(x, t)$ to be zero for t negative is that cause must precede effect. But this means the integrand in (2.5.16) is zero for $\tau > t$, and

APPLICATIONS TO THE INITIAL PROBLEM   77

so we can extend the interval of integration to $\tau = +\infty$. The solution (2.5.16) can now be simply and conveniently expressed as a double convolution integral

(2.5.17) $$u(x, t) = \int_{-\infty}^{\infty} \int_{-\infty}^{\infty} K(x - s, t - \tau) f(s, \tau) \, ds \, d\tau$$

THEOREM 2.5.2 (*Duhamel's Rule*). *The solution of the nonhomogeneous equation with zero initial data is the convolution (in space and time) of the Green's distribution with the forcing term.*

Proof. By direct differentiation we can show that (Exercise 2.5.10)

(2.5.18) $$\left(\frac{\partial^2}{\partial t^2} - c^2 \frac{\partial^2}{\partial x^2}\right) H(ct + x) H(ct - x) = 4c^2 \delta(ct + x) \delta(ct - x)$$

Applying the differential operator to the convolution integral, we may carry it beneath the integral sign (see Exercise 2.4.8) and find

(2.5.19) $$\left(\frac{\partial^2}{\partial t^2} - c^2 \frac{\partial^2}{\partial x^2}\right) u(x, t) = 2c \int_{-\infty}^{\infty} \int_{-\infty}^{\infty} \delta(c(t - \tau) + x - s)$$
$$\times \delta(c(t - \tau) - x + s) f(s, \tau) \, ds \, d\tau$$

To apply the substitution property of the Dirac distributions, we should change over to characteristic coordinates

(2.5.20) $$\xi' = c\tau + s \qquad \eta' = c\tau - s$$

Thus

$$s = \tfrac{1}{2}(\xi' - \eta') \qquad \tau = \frac{1}{2c}(\xi' + \eta')$$

and

(2.5.21) $$d\xi' \, d\eta' = \frac{\partial(\xi', \eta')}{\partial(s, \tau)} \, ds \, d\tau = 2c \, ds \, d\tau$$

Therefore the right side of (2.5.19) becomes

(2.5.22) $$\int_{-\infty}^{\infty} \int_{-\infty}^{\infty} \delta(ct + x - \xi') \delta(ct - x - \eta')$$
$$\times f\left(\tfrac{1}{2}(\xi' - \eta'), \frac{1}{2c}(\xi' + \eta')\right) d\xi' \, d\eta'$$
$$= f\left(\tfrac{1}{2}(\xi' - \eta'), \frac{1}{2c}(\xi' + \eta')\right)\Big|_{\substack{\xi' = ct + x \\ \eta' = ct - x}}$$
$$= f(x, t)$$

This establishes (2.5.15).

The result of these last steps of the calculation can be expressed symbolically as

(2.5.23) $$2c\delta(ct + x)\delta(ct - x) = \delta(x)\delta(t)$$

78   DISTRIBUTIONS AND WAVES

We see that under changes of coordinates the Dirac functions must transform so that their combination with the volume or area element of integration is conserved.

We can now write

(2.5.24) $$\left(\frac{\partial^2}{\partial t^2} - c^2 \frac{\partial^2}{\partial x^2}\right) K(x, t) = \delta(x)\delta(t)$$

and we regard this "point-source" equation as the fundamental property of the Green's function $K(x, t)$. Together with the causality property

(2.5.25) $$K(x, t) \equiv 0 \qquad t < 0$$

it defines $K(x, t)$ completely and uniquely.

The reader should notice that both (2.5.24) and (2.5.25) are distribution equations—they assert equality of distributions not merely of functions.

THEOREM 2.5.3 *The support of the Green's distribution $K(x, t)$ is the region of influence of the origin. The singular support of $K(x, t)$ is the pair of characteristic rays issuing from the origin.*

Proof of these statements is immediate if we recall Definition 2.2.2 of the region of influence.

Let us apply these results to find the Green's function of the telegraph equation

(2.5.26) $$\frac{\partial^2 u}{\partial t^2} = c^2 \frac{\partial^2 u}{\partial x^2} + lu \qquad l = \text{const.}$$

Any hyperbolic equation of second order in t and x with constant coefficients can be reduced to this form by a change of dependent variable (see Exercise 2.1.3).

Denote the operator of the string equation by the d'Alembertian symbol $\Box$. Then

(2.5.27) $$\Box u \equiv \frac{\partial^2 u}{\partial t^2} - c^2 \frac{\partial^2 u}{\partial x^2} = lu$$

is the equation to be studied. The Green's function $K_l(x, t)$ will satisfy

(2.5.28) $$\Box K_l(x, t) = lK_l(x, t) + \delta(x)\delta(t)$$

Let us regard both items on the right as nonhomogeneous terms like $f(x, t)$ in (2.5.15). Then, since $K(x, t) = K_0(x, t)$ is the Green's function for $\Box$, we find, by (2.5.17),

(2.5.29) $$K_l(x, t) = \int_{-\infty}^{\infty} \int_{-\infty}^{\infty} K_0(x - s, t - \tau)[lK_l(s, \tau) + \delta(s)\delta(\tau)] \, ds \, d\tau$$

$$= lK_0 * K_l(x, t) + K_0(x, t)$$

APPLICATIONS TO THE INITIAL PROBLEM   79

This is an integral equation of Volterra's type with variable limits for $K_l(x, t)$. (It can be compared with Picard's integral equation of Section 1.3.) We solve this equation by iteration: write

$$K_l = K_0 + lK_0 * K_l$$
$$= K_0 + lK_0 * (K_0 + lK_0 * K_l)$$
$$= K_0 + lK_0 * K_0 + l^2 K_0 * K_0 * K_l$$

(2.5.30)

$$= K_0 + lK_0^{(2)} + l^2 K_0^{(3)} + \cdots$$
$$= \sum_{n=0}^{\infty} l^n K_0^{(n+1)}$$

Here we have written $K_0^{(2)}$ for $K_0 * K_0$ and likewise for higher powers of convolution of K_0 with itself.

To examine these iterated Green's functions $K_0^{(n)}$, it is simplest to use the characteristic variables $\xi = ct + x$, $\eta = ct - x$ once again. Thus,

$$K_0 * K_0 = \int_{-\infty}^{\infty} \int_{-\infty}^{\infty} K_0(x - s, t - \tau) K_0(s, \tau) \, ds \, d\tau$$
$$= \frac{1}{8c^3} \int_{-\infty}^{\infty} \int_{-\infty}^{\infty} H(\xi - \xi') H(\eta - \eta') H(\xi') H(\eta') \, d\xi' \, d\eta'$$

by (2.5.5) and (2.5.21), so

(2.5.31) $$K_0^{(2)} = \frac{1}{8c^3} \xi_+ \eta_+$$

by Exercise 2.4.7 applied twice. By induction on n it is now easily proved that

(2.5.32) $$K_0^{(n)} = \frac{\xi_+^{n-1} \eta_+^{n-1}}{(2c)^{2n-1}[(n-1)!]^2}$$

Finally, therefore,

(2.5.33) $$K_l(x, t) = \sum_{n=0}^{\infty} \frac{l^n \xi_+^n \eta_+^n}{(2c)^{2n+1}(n!)^2}$$

Since

(2.5.34) $$\frac{\xi_+ \eta_+}{c^2} = \begin{cases} t^2 - \dfrac{x^2}{c^2} & \text{for } t > \dfrac{x}{c}, \ t > -\dfrac{x}{c} \\ 0 & \text{otherwise} \end{cases}$$

we can write this series as

(2.5.35) $$K_l(x, t) = I_0\left[\sqrt{l\left(t^2 - \frac{x^2}{c^2}\right)}\right] K_0(x, t)$$

Here the Bessel function $I_0(x)$ is defined by the convergent series

$$(2.5.36) \qquad I_0(x) = \sum_{n=0}^{\infty} \frac{1}{(n!)^2} \left(\frac{x}{2}\right)^{2n}$$

(see Chapter 8, Section 8.7). The support and singular support for this Green's function are the same as for the string equation itself.

These formulas for the response function of the one-dimensional wave equation have been presented in detail to illustrate the application of distributions to such problems. In the past, various devices and circumlocutions have been employed to avoid the use of derivatives that "may not exist." Oliver Heaviside (1850–1925), who was the pioneer of the "operational calculus," was criticized for lack of "rigor," whereas the real difficulty was that a suitable mathematical framework for the type of problem he considered had not yet been formulated. The delta function $\delta(x)$ was introduced by Dirac as a continuous analogue of the Kronecker delta δ_{ik}. Its interpretation as a distribution of mass was employed by L. Schwartz as a starting point for the present powerful and rigorous calculus of distributions. We shall make frequent use of this theory and therefore caution the reader that many equations should be read and understood in the distribution sense. Practice in the interpretation of such equations is an important aspect of many of our exercises.

One last word—although the distribution theory appears to overcome great difficulties, it cannot accomplish the impossible. A price must be paid for these advantages, and the price is the failure of the usual freedom to *multiply* functions freely. Thus no meaning can be attached to the square of the Dirac distribution. For linear problems, then, distributions are well suited just because we need to differentiate rather than to multiply.

EXERCISES 2.5

1. Show that

$$\delta(x) = \delta(-x)$$

$$\delta(cx) = \frac{1}{|c|} \delta(x)$$

$$\delta(f(x)) = \sum_{x_n} \frac{1}{|f'(x_n)|} \delta(x - x_n)$$

where x_n runs through the zeros of $f(x)$.

2. Show that convolution multiplication is commutative and associative: $f * g = g * f$, and $(f * g) * h = f * (g * h)$.

3. *Duhamel's rule.* If $u(x, t, \tau)$ is the solution of $\Box u = 0$, $u(x, 0, \tau) = 0$, $u_t(x, 0, \tau) = f(x, \tau)$, show that

$$\int_0^t u(x, t - \tau, \tau) \, d\tau$$

is the solution of $\Box u = f(x, t)$ with zero initial conditions.

4. Find the solution of the wave equation $\Box u = \psi(x)\,\delta(t) + \varphi(x)\,\delta'(t)$, with $u \equiv 0$ for $t < 0$.

5. Find Green's functions for the equations with constant coefficients
   (a) $\quad u_t - cu_x = \delta(x)\,\delta(t)$
   (b) $\quad u_t - cu_x - lu = \delta(x)\,\delta(t)$
   (c) $\quad u_{tt} + au_t = c^2 u_{xx} + bu_x + lu + \delta(x)\,\delta(t)$

6. Write out the full solution of the initial problem for (2.5.26) with $u(x, 0) = \varphi(x)$, $u_t(x, 0) = \psi(x)$.

7. Define initial problems for the equations of Exercise 2.5.5 and construct their solutions.

8. Carry out the induction leading to (2.5.32).

9. How does the expression $(1/2c)[H(x - s + ct) - H(x - s - ct)]$ differ from the Green's function $K(x, t)$ for the string equation?

10. Verify (2.5.18) by performing the indicated differentiations.

11. Show that $H(ct + x)H(ct - x) = H(c^2 t^2 - x^2)H(t)$.

2.6 THE SEPARATION OF VARIABLES

Our study of wave motion of a string has so far been directed toward consideration of the motion of each separate portion of the string. When the string is very long or the time intervals are quite short, so that there are few, if any, reflections from the ends, these methods have yielded explicit and convenient solutions.

Such formulas are the exception rather than the rule in applied mathematics. Even when we consider the finite string for extended periods of time, these formulas become unduly complicated, so that a less direct but more far-reaching method would be preferable. Moreover, we are aware from everyday experience, particularly of musical instruments, that the motions encountered are coherent oscillations of the entire string. The human ear will evaluate the vibration of a stretched string as a combination or superposition of various *harmonics* based upon a fundamental tone. Let us seek a mathematical description in which the motion is resolved into components of this type.

In the simplest overall motions of the string, the various parts will move in unison, the relative proportion of their displacements being unchanged with time. If the factor of proportion is denoted by $X(x)$, we have then a solution of the form

(2.6.1) $\qquad u(x, t) = X(x)T(t)$

Physically this usually represents a *standing wave*.

This is called a *separated* solution because the space and time variables appear in separate factors. What is more, we will show that the separate factors satisfy separate differential equations.

To establish this basic step of the separation of variables method, we substitute

(2.6.1) into the partial differential equation

(2.6.2) $$u_{tt} = c^2 u_{xx}$$

Let primes denote differentiation with respect to the appropriate argument variable of each factor. We find

$$X(x) \cdot T''(t) = c^2 X''(x) \cdot T(t)$$

Now divide by $c^2 X(x) T(t)$, thus "separating" the variables one to each side of the equation:

(2.6.3) $$\frac{T''(t)}{c^2 T(t)} = \frac{X''(x)}{X(x)}$$

We are now ready for the main significant step.

THEOREM 2.6.1 *The expressions on the left and right sides of* (2.6.3) *are constants, independent of both x and t.*

Proof. The left side is independent of x and so, therefore, is the right side. The right side is independent of t, and thus the left side must be as well. That is, both sides are constant with respect to both x and t, as asserted. Their common value, which we denote by $-k^2$, is called the *separation constant*. The negative sign will be required in order that k should have a real value in those solutions which satisfy the boundary conditions.

Now we obtain the two separate ordinary differential equations

(2.6.4)
$$X'' + k^2 X = 0$$
$$T'' + c^2 k^2 T = 0$$

These have as solutions the linear trigonometric combinations

(2.6.5)
$$X = A \cos kx + B \sin kx$$
$$T = C \cos kct + D \sin kct$$

provided $k \neq 0$. For $k = 0$, $X(x)$ and $T(t)$ will be linear functions.

Let the string be fixed at both ends:

(2.6.6) $$u(0, t) = 0 \qquad u(l, t) = 0 \qquad t > 0$$

Then the X factor of (2.6.1) will satisfy the same conditions, and we see that

(2.6.7) $$X(0) = A = 0 \qquad X(l) = B \sin kl = 0$$

Hence k must have one of the real values $n\pi/l$, where n is an integer. Thus we make precise the form of the standing waves:

(2.6.8) $$X(x) = B \sin \frac{n\pi x}{l}$$

The nth standing-wave profile has $n - 1$ equally spaced zeros, or nodes, in the interval $0 < x < l$ (Figure 2.15).

The solutions (2.6.8) form an infinite sequence of "harmonics" of the string, each one satisfying the end conditions (2.6.7). As (2.6.2) is linear, any superposition of these solutions is also a solution which satisfies both end conditions. We are led to inquire, therefore, whether this linear combination of the separated

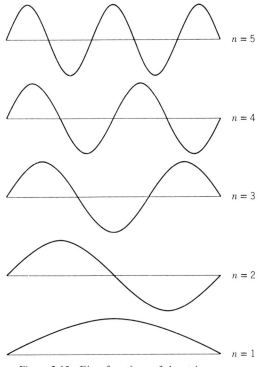

Figure 2.15 Eigenfunctions of the string.

solutions (2.6.8) will enable us to satisfy the initial conditions as well. Presumably it should, since a string can be "plucked" in any way and always gives a combination of harmonics. Let us assume a trial solution

$$(2.6.9) \qquad u(x, t) = \sum_n \sin \frac{n\pi x}{l} \left(C_n \cos \frac{n\pi ct}{l} + D_n \sin \frac{n\pi ct}{l} \right)$$

where the coefficients C_n and D_n are as yet undetermined. We can suppress the constant B of (2.6.8) in each term, because C_n and D_n are themselves quite arbitrary.

Now let us require that (2.6.9) should satisfy initial conditions

$$(2.6.10) \qquad u(x, 0) = \varphi(x) \qquad u_t(x, 0) = \psi(x) \qquad 0 \leq x \leq l$$

If we apply these conditions, we find that the constants C_n and D_n must satisfy the equations

(2.6.11)
$$\sum_n C_n \sin \frac{n\pi x}{l} = \varphi(x)$$

$$\sum_n D_n \frac{n\pi c}{l} \sin \frac{n\pi x}{l} = \psi(x)$$

From another viewpoint the problem confronting us is that of expanding the assigned functions $\varphi(x)$ and $\psi(x)$ in a series of the standing waveforms. The physical interpretation of such an expansion is that the given waveforms $\varphi(x)$ and $\psi(x)$ are expressed as a sum of harmonics of the fundamental frequency, the coefficients C_n and D_n determining the *amplitude* of the nth harmonic mode. This "expansion problem" was first successfully attacked by Fourier, and series like (2.6.11) are known as Fourier series.

The standing waveforms (2.6.8) are the simplest examples of a concept which will appear many times—that of an *eigenfunction*, or *characteristic function*. The values

(2.6.12) $$\frac{n^2\pi^2}{l^2} \qquad n = 1, 2, \ldots$$

of the separation constant k^2 are known as the *eigenvalues* (or characteristic values) of the differential equation for $X(x)$.

Let us compare these concepts with the corresponding notions of eigenvalue and eigenvector of a matrix discussed in Chapter 1, Section 1.2. A matrix A of finite order m has, in general, m eigenvalues $\lambda_1, \ldots, \lambda_m$ which are roots of a polynomial of degree m, the characteristic polynomial

(2.6.13) $$P(\lambda) = \det(A - \lambda E) = 0$$

Also the matrix A is a linear operator upon a vector space of dimension m. The eigenvectors of A are those vectors upon which A acts as a scalar multiplier of magnitude λ.

Now we have eigenvalues and eigenfunctions which pertain to a Sturm-Liouville *differential operator*, namely, the operator

(2.6.14) $$-\frac{d^2}{dx^2}$$

together with the boundary conditions $X(0) = X(l) = 0$. Given any function $X(x)$ (or distribution) on $0 \leq x \leq l$, this operator transforms it to $-X''(x)$. The eigenvalues of this operator are infinite in number, and the "eigenvectors" are functions, satisfying the boundary conditions, upon which the operator acts as a scalar multiplier of magnitude $\lambda = +k^2$:

$$-X''(x) = k^2 X(x) \qquad\qquad X(0) = X(l) = 0$$

We also have a vector space of infinite dimension, namely, the vector space of those functions on $0 \leq x \leq l$ which satisfy the boundary conditions. These are two-point boundary conditions, which determine eigenvalues and eigenfunctions. The eigenvalues are roots of a transcendental characteristic equation, in this case

(2.6.15) $$\sin(l\sqrt{\lambda}) = 0$$

which has infinitely many roots $\lambda_n = n^2\pi^2/l^2$, $n = 1, 2, \ldots$.

To explore more closely this analogy between the finite and the infinite (or continuous) system, let us return to the finite system (2.2.1) for the motion of n beads fastened upon a weightless string. We suppose the ends are held fixed. Let us separate the variables by setting

(2.6.16) $$u_i(t) = e^{ikct} y_i \qquad i = 1, \ldots, n$$

Then (2.2.1) becomes

(2.6.17) $$y_{i+1} - Cy_i + y_{i-1} = 0 \qquad i = 1, \ldots, n$$

where $y_0 = y_{n+1} = 0$ and

(2.6.18) $$C = 2 - h^2 k^2$$

where also $m = \rho l/n$ are the common masses, and $h = l/(n+1)$ is the equal interval between two adjacent beads.

The components y_i ($i = 1, \ldots, n$) define a vector $\mathbf{y}$, and the system (2.6.17) can be written

(2.6.19) $$A\mathbf{y} = \lambda\mathbf{y} \qquad\qquad \lambda = C$$

where the matrix

(2.6.20) $$A = \begin{pmatrix} 0 & 1 & 0 & 0 & \cdots \\ 1 & 0 & 1 & 0 & \\ 0 & 1 & 0 & 1 & \\ \cdot & & 1 & 0 & 1 \\ \cdot & & & & \\ \cdot & & & & \cdot & 1 \\ & & & & 1 & 0 \end{pmatrix}$$

has nonzero entries on the superdiagonal and subdiagonal only. The positions of these entries correspond to the interaction of each bead being with its nearest neighbors only.

86   DISTRIBUTIONS AND WAVES

The eigenvalues and eigenvectors of A can be found by the following device suggested by the form (2.6.8) of the standing-wave solutions. Remember that $x_i = il/(n + 1)$, and set

(2.6.21) $$y_i = B \sin i\theta$$

where θ is to be determined. Since

(2.6.22) $$y_{i+1} - Cy_i + y_{i-1} = B[\sin (i - 1)\theta - C \sin i\theta + \sin (i + 1)\theta]$$
$$= B \sin i\theta (2 \cos \theta - C)$$

by the addition formula for sines, we choose

(2.6.23) $$\cos \theta = \frac{C}{2} = 1 - \frac{h^2 k^2}{2}$$

Now (2.6.21) automatically gives $y_0 = 0$. To ensure solutions of the homogeneous system (2.6.17), we must also have $y_{n+1} = 0$, that is,

(2.6.24) $$y_{n+1} = B \sin (n + 1)\theta = 0$$

so that

(2.6.25) $$(n + 1)\theta = s\pi \qquad s = 1, 2, \ldots, n$$

From (2.6.23) and (2.6.25) we find the eigenvalue equation for k^2:

$$k^2 = k_s^2 = \frac{2}{h^2}(1 - \cos \theta) = \frac{4}{h^2} \sin^2 \frac{\theta}{2}$$

so that

(2.6.26) $$\pm k_s = \frac{2}{h} \sin \frac{s\pi}{2(n+1)} = \frac{2(n+1)}{l} \sin \frac{s\pi}{2(n+1)} \qquad s = 1, 2, \ldots, n$$

The corresponding eigenvectors $\mathbf{y}_{(s)}$ have components

(2.6.27) $$y_{(s)i} = B_s \sin \frac{is\pi}{n+1}$$

Each such eigenvector defines a "normal mode" of oscillation of the string of beads.

The most general motion is a superposition of all the normal modes, namely,

(2.6.28) $$u_i(t) = \sum_s y_{(s)i} e^{\pm i k_s ct}$$

We should allow a separate constant for each choice of the $\pm$ sign. Let us express the time exponentials by sines and cosines; then,

(2.6.29) $$u_i(t) = \sum_{s=1}^{n} \sin \frac{is\pi}{n+1} [C_s \cos (k_s ct) + D_s \sin (k_s ct)].$$

This should be compared with (2.6.9). When the limit with $h \to 0$ and $n \to \infty$ is taken, the finite normal mode sum (2.6.29) for a string of given length l goes over into the infinite sum of separated solutions (2.6.9). This is a particular case of a principle that is essentially due to Rayleigh (see p. 250).

PRINCIPLE OF RAYLEIGH *A continuous system can be approximated by a finite system, so that the eigenvalues and eigenvectors of the finite system will approximate to a finite number of the eigenvalues and eigenfunctions of the continuous system.*

EXERCISES 2.6

1. Show that the boundary conditions $u_x(0, t) = 0$ and $u_x(l, t) = 0$ representing sliding attachment at each end lead to the eigenvalues $\lambda_n = n^2\pi^2/l^2$ and eigenfunctions $\cos(n\pi x/l)$, $n = 1, 2, 3, \ldots$.

2. Show that the boundary conditions $u(0, t) = 0$, $u_x(l, t) = 0$ for a string attached at one end and sliding at the other lead to the eigenvalues $\lambda_n = (n + \tfrac{1}{2})^2\pi^2/l^2$ and eigenfunctions $\sin[(n + \tfrac{1}{2})\pi x/l]$.

3. Show that the boundary conditions $u_x(0, t) + h_1 u(0, t) = 0$ and $u_x(l, t) + h_2 u(l, t) = 0$ lead to the eigenvalue equation $(h_1 h_2 + k^2)\tan kl = (h_2 - h_1)k$, where $\lambda = k^2$.

4. Explain why the note of a violin string rises sharply by one octave when the string is clamped exactly at the centre.

5. Show that the series on the left in (2.6.11) define functions which are odd with respect to 0 and l [see $\Phi(x)$ of (2.3.5)].

6. Show that, when the (doubly odd) sine series of (2.6.11) are inserted into d'Alembert's formula (2.2.8), the resulting solution has the form (2.6.9) with

$$C_n = \frac{2}{l}\int_0^l \sin\frac{n\pi s}{l}\,\varphi(s)\,ds \qquad D_n = \frac{2}{n\pi c}\int_0^l \sin\frac{n\pi s}{l}\,\psi(s)\,ds$$

7. For fixed n, describe the spacing of the wave numbers k_s of (2.6.26). Letting $n \to \infty$, show that, for fixed s, k_s tends to $s\pi/l$.

8. Write down the equations of motion for a system of $n + 1$ equally spaced beads which approximate a string of length l fixed at the origin and free to slide at the end l [that is, $u_x(l, t) = 0$]. Find the equation for the eigenvalues and determine the eigenvectors.

9. Write down the equations for a system of $n + 2$ equally spaced beads representing a string of length l free to slide at both ends. Find the eigenvalue equation and determine the eigenvectors.

2.7 FOURIER SERIES

The solving of a partial differential equation by the eigenfunction method of the preceding section led to the problem of expanding a given datum function $\varphi(x)$ into a sum of the eigenfunctions, as expressed in (2.6.11). The eigenfunctions for the string of length l are, for integer n,

(2.7.1) $\qquad\qquad \sin\dfrac{n\pi x}{l} \quad \text{or} \quad \cos\dfrac{n\pi x}{l}$

88   DISTRIBUTIONS AND WAVES

depending on whether the string is held or is free to slide at both ends. Expansions in these eigenfunctions are known as *Fourier series*; they were the first to be discovered and are the simplest prototypes of the very general and widely applicable eigenfunction expansions we shall study later.

We shall consider the function $f(x)$ to be expanded as belonging to a vector space, and that the eigenfunctions should form a basis in that space. In fact, the vector space will be a Hilbert space, and the eigenfunctions will constitute an orthogonal basis. The eigenfunction expansion problem therefore has a geometrical interpretation: the given vector **f** shall be resolved into its components parallel to the various coordinate axes of the eigenfunction basis.

Following Fourier, we consider the interval $-l \leq x \leq l$. Let us introduce the inner product

(2.7.2) $$(\mathbf{f}, \mathbf{g}) = \int_{-l}^{l} f(x)g(x)\, dx$$

Since the integrals

(2.7.3)
$$\int_{-l}^{l} \sin\frac{n\pi x}{l} \sin\frac{m\pi x}{l}\, dx \qquad m \neq n$$

$$\int_{-l}^{l} \sin\frac{n\pi x}{l} \cos\frac{m\pi x}{l}\, dx$$

$$\int_{-l}^{l} \cos\frac{n\pi x}{l} \cos\frac{m\pi x}{l}\, dx \qquad m \neq n$$

are all in fact zero, it follows that these eigenfunctions are *orthogonal* with respect to this inner product. Here m and n are integers, which need not be distinct in the second of the three cases. Of course this orthogonality is no accident—we will show later that it follows directly from the definition of the eigenfunctions.

We shall write the expansion in the form

(2.7.4) $$f(x) = \frac{a_0}{2} + \sum_{n=1}^{\infty}\left(a_n \cos\frac{n\pi x}{l} + b_n \sin\frac{n\pi x}{l}\right) \qquad -l \leq x \leq l$$

There are now two problems:

1. Determination of the coefficients a_n, b_n.

2. Convergence of the series to $f(x)$.

The first problem we resolve by the geometrical analogy: to separate out one term in the series of orthogonal vectors on the right, we take the inner product with the vector appearing in that term. That is, to determine a_m, we multiply by $\cos(m\pi x/l)$, and integrate over $(-l, l)$. All terms on the right but one will be zero because of the orthogonality. Therefore we have

$$\int_{-l}^{l} f(x) \cos\frac{m\pi x}{l}\, dx = a_m \int_{-l}^{l} \cos^2\frac{m\pi x}{l}\, dx$$
$$= l a_m$$

That is,

(2.7.5) $$a_m = \frac{1}{l} \int_{-l}^{l} f(s) \cos \frac{m\pi s}{l} \, ds \qquad m \geq 0.$$

This formula holds for $m = 0$ as well because of the precautionary factor $\frac{1}{2}$ in the first term of (2.7.4).

In exactly the same way we deduce

(2.7.6) $$b_m = \frac{1}{l} \int_{-l}^{l} f(s) \sin \frac{m\pi s}{l} \, ds \qquad m > 0$$

and this completes the determination of coefficients for the Fourier series on the interval $-l \leq x \leq l$.

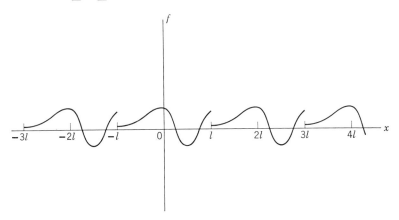

Figure 2.16 Full range Fourier series, on interval $-l \leq x \leq l$.

The Fourier series on the right in (2.7.4) is itself periodic with period $2l$, whether the given function $f(x)$ is or is not periodic. Therefore we cannot expect (2.7.4) to hold outside this interval—the series will simply repeat its range of values periodically (Figure 2.16).

To apply Fourier series to the problems of the vibrating string, we need two particular cases of the general trigonometric expansion formulated above. When the end conditions for the string are that "both ends are held fixed," the eigenfunctions of the problem are $\sin(n\pi x/l)$. We obtain a series of just these functions if we suppose that the function $f(x)$ defined on $-l \leq x \leq l$ is odd: $f(-x) = -f(x)$.

From (2.7.5) we see that a_m is the integral of an odd function, the product of the odd factor $f(x)$ and the even factor $\cos(m\pi x/l)$. This integral is therefore zero, and so all cosine terms drop out of the series.

Now the physical problem involves an arbitrary function $f(x)$ defined for $0 \leq x \leq l$. We shall extend its definition to the left half-interval by defining $f(x)$ as odd: $f(-x) = -f(x)$, $0 \leq x \leq l$. The Fourier full-range series of this odd function then contains only sine terms. This sine series we call the *half-range sine series* of the given function $f(x)$, on $0 \leq x \leq l$. Therefore the half-range sine series formulas are:

(2.7.7) $$f(x) = \sum_{n=1}^{\infty} b_n \sin \frac{n\pi x}{l} \qquad 0 \leq x \leq l.$$

where, by (2.7.6) and the oddness of $f(s)$,

(2.7.8) $$b_n = \frac{2}{l} \int_0^l f(s) \sin \frac{n\pi s}{l} ds$$

We can now determine the coefficients in the trial solution (2.6.9). The initial condition $u(x, 0) = \varphi(x)$ $(0 \leq x \leq l)$ gives

$$\varphi(x) = \sum_{n=1}^{\infty} C_n \sin \frac{n\pi x}{l}$$

that is,

(2.7.9) $$C_n = \frac{2}{l} \int_0^l \varphi(s) \sin \frac{n\pi s}{l} ds$$

Likewise the initial derivative condition $u_t(x, 0) = \psi(x)$ $(0 \leq x \leq l)$ yields

$$\psi(x) = \sum_{n=1}^{\infty} \frac{n\pi c}{l} D_n \sin \frac{n\pi x}{l}$$

that is,

(2.7.10) $$D_n = \frac{2}{n\pi c} \int_0^l \psi(s) \sin \frac{n\pi s}{l} ds$$

This completes the formal solution of the string problem with fixed ends.

The second type of half-range series, the cosine series, arises in the problem of the string with both ends free to slide along fixed parallel rails. Let the initial conditions be as above: the end conditions are

$$u_x(0, t) = 0 \qquad u_x(l, t) = 0$$

and the eigenfunctions, as mentioned in Section 2.6, are $\cos(n\pi x/l)$.

For the cosine series on $0 \leq x \leq l$ we envisage $f(x)$ defined as an *even* function $f(-x) = f(x)$ over the full range $-l \leq x \leq l$. Only the cosine terms remain in the series and we find the half-range cosine series formulas

(2.7.11) $$f(x) = \frac{1}{2} a_0 + \sum_{n=1}^{\infty} a_n \cos \frac{n\pi x}{l}$$

where

(2.7.12) $$a_n = \frac{2}{l} \int_0^l f(s) \cos \frac{n\pi s}{l} ds \qquad n = 0, 1, 2, \ldots$$

The term a_0 ($n = 0$) corresponds to the value zero of the separation constant k of (2.6.4).

A solution for the sliding-string problem must now include a contribution for $k = 0$, that is, a linear function of t. Thus,

(2.7.13) $$u(x, t) = \frac{\alpha}{2} + \frac{\beta t}{2} + \sum_{n=1}^{\infty} \cos\frac{n\pi x}{l}\left(C_n \cos\frac{n\pi ct}{l} + D_n \sin\frac{n\pi ct}{l}\right)$$

The coefficients of the Fourier series are determined as

(2.7.14) $$C_n = \frac{2}{l}\int_0^l \varphi(s) \cos\frac{n\pi s}{l}\, ds \qquad D_n = \frac{2}{n\pi c}\int_0^l \psi(s) \cos\frac{n\pi s}{l}\, ds$$

We also find from (2.7.12) that the initial position and velocity coefficients are

(2.7.15) $$\alpha = \frac{2}{l}\int_0^l \varphi(x)\, dx \qquad \beta = \frac{2}{l}\int_0^l \psi(x)\, dx$$

Let us now turn to the purely mathematical problem of the convergence of a Fourier series (Tolstov, Ref. 58). For simplicity, we shall consider the expansion of a function $f(x)$ which satisfies a fairly strong restriction. We assume that the derivative $f'(x)$ exists and is continuous throughout the interval $-\pi \leq x \leq \pi$. We can also assume, without loss of generality, that $f(x)$ is periodic of period $2\pi: f(x + 2\pi) = f(x)$.

THEOREM 2.7.1 *Let $f'(x)$ be continuous on $-\pi \leq x \leq \pi$, and let the Fourier coefficients of $f(x)$ on this interval be*

(2.7.16) $$a_n = \frac{1}{\pi}\int_{-\pi}^{\pi} f(x) \cos nx\, dx \qquad b_n = \frac{1}{\pi}\int_{-\pi}^{\pi} f(x) \sin nx\, dx$$

Then the Fourier series

(2.7.17) $$\frac{1}{2}a_0 + \sum_{n=1}^{\infty}(a_n \cos nx + b_n \sin nx)$$

converges for $-\pi < x < \pi$ to $f(x)$ as sum.

To establish convergence of the series, we must consider the partial sums

(2.7.18) $$s_n(x) = \frac{1}{2}a_0 + \sum_{k=1}^{n}(a_k \cos kx + b_k \sin kx)$$

LEMMA 2.7.1 *The partial sum $s_n(x)$ is given by Dirichlet's formula*

(2.7.19) $$s_n(x) = \frac{1}{\pi}\int_{-\pi}^{\pi} f(x + t)\frac{\sin(n + \tfrac{1}{2})t}{2\sin\tfrac{1}{2}t}\, dt$$

We prove the lemma by summing a certain finite trigonometric series. We have

(2.7.20) $$\begin{aligned}s_n(x) &= \frac{1}{2\pi}\int_{-\pi}^{\pi} f(x')\, dx' \\ &\quad + \frac{1}{\pi}\sum_{k=1}^{n}\int_{-\pi}^{\pi} f(x')(\cos kx' \cos kx + \sin kx' \sin kx)\, dx' \\ &= \frac{1}{\pi}\int_{-\pi}^{\pi} f(x')\, dx'\left(\frac{1}{2} + \sum_{k=1}^{n} \cos k(x' - x)\right)\end{aligned}$$

The sum has the form, with $t = x' - x$,

$$\frac{1}{2} + \sum_{k=1}^{n} \cos kt = \frac{1}{2}\left[1 + \sum_{k=1}^{n}(e^{ikt} + e^{-ikt})\right]$$

(2.7.21)
$$= \frac{1}{2}\sum_{k=-n}^{n} e^{ikt} = \frac{1}{2} e^{-int} \sum_{k=0}^{2n} e^{ikt}$$

$$= \frac{1}{2} e^{-int} \frac{e^{(2n+1)it} - 1}{e^{it} - 1} = \frac{e^{(n+\frac{1}{2})it} - e^{-(n+\frac{1}{2})it}}{2(e^{(\frac{1}{2})it} - e^{-(\frac{1}{2})it})}$$

$$= \frac{\sin(n + \frac{1}{2})t}{2 \sin \frac{1}{2}t}$$

Replacing x' by $x + t$ in (2.7.20), we find (2.7.19) for $s_n(x)$.

We shall require the formula obtained by integrating (2.7.21) over the interval from $-\pi$ to π. Since the cosine terms are all periodic with period 2π, they contribute zero to this integral and we have

(2.7.22)
$$\int_{-\pi}^{\pi} \frac{\sin(n + \frac{1}{2})t}{2 \sin \frac{1}{2}t} dt = \pi$$

We shall also need the following lemma which was established by Riemann.

LEMMA 2.7.2 *Let $g(t)$ be bounded and continuous in $a < t < b$, and let the derivative $g'(t)$ be continuous except at a finite number of points. Then,*

(2.7.23)
$$\lim_{n \to \infty} \int_a^b g(t) \sin nt \, dt = 0$$

For the proof, we first suppose that $g(t)$ and $g'(t)$ are continuous throughout the closed interval $[a, b]$, and therefore bounded on that interval. Integration by parts yields

(2.7.24)
$$\int_a^b g(t) \sin nt \, dt = -g(t) \frac{\cos nt}{n}\bigg]_a^b + \frac{1}{n} \int_a^b g'(t) \cos nt \, dt$$

Since $g(t)$ and $g'(t)$ are bounded and $|\cos nt| \leq 1$, the right side of (2.7.24) is smaller than $(1/n)K$, where K is a constant independent of n. The conclusion follows at once in this special case.

Suppose now that at a point t_1 the hypothesis of continuity of $g'(t)$ is not fulfilled. We then divide the interval $[a, b]$ into parts by removing an interval of length $2\delta_1$ centred at t_1. Given a positive ε, we can choose δ_1 so small that the integral (2.7.24) over $(t_1 - \delta_1, t_1 + \delta_1)$ is less than $\frac{1}{2}\varepsilon$. Having fixed δ_1, we can find bounds for $g'(t)$ in the remaining portions of $[a, b]$. According to the special case just proved, we can make the contribution from these portions less than $\frac{1}{2}\varepsilon$ by choosing n sufficiently large. Thus we can make the integral over $[a, b]$ less than any positive number ε. This completes the proof that the limit is zero. The proof for the case of a finite number of exceptional points is similar.

We are now ready for the main convergence proof. We multiply (2.7.22) by $\pi^{-1}f(x)$ and subtract from (2.7.19), obtaining

$$(2.7.25) \quad s_n(x) - f(x) = \frac{1}{\pi}\int_{-\pi}^{\pi} \frac{f(x+t) - f(x)}{2\sin\frac{1}{2}t} \sin(n + \tfrac{1}{2})t\, dt$$

Now the function

$$g(t) = \frac{f(x+t) - f(x)}{2\sin\frac{1}{2}t} \qquad t \neq 0$$

has a continuous derivative everywhere in the interval except at $t = 0$. For $t = 0$ we set

$$g(0) = \lim_{t\to 0} \frac{f(x+t) - f(x)}{2\sin\frac{1}{2}t}$$

$$= \lim_{t\to 0} \frac{f(x+t) - f(x)}{t} \lim_{t\to 0} \frac{\frac{1}{2}t}{\sin\frac{1}{2}t}$$

$$= f'(x) \cdot 1 = f'(x)$$

Thus $g(t)$ is continuous and bounded in $[-\pi, \pi]$, and satisfies the conditions of Riemann's lemma 2.7.2. Therefore, as n tends to infinity, the integral in (2.7.25) tends to zero, and the partial sums $s_n(x)$ converge to $f(x)$. This completes the proof of convergence.

EXERCISES 2.7

1. Show that any even function on the range $-l \leq x \leq l$ is orthogonal to any odd function.

2. Show that $1/\sqrt{2l}$, $(1/\sqrt{l})\cos(n\pi x/l)$, $(1/\sqrt{l})\sin(n\pi x/l)$, $n = 1, 2, \ldots$, form an orthonormal set on $-l \leq x \leq l$.

3. Calculate full range Fourier coefficients on $-l \leq x \leq l$ for the functions given by x, x^2, $|x|$, $\sin(\pi x/l)$, $\cos(3\pi x/l)$.

4. Calculate half-range sine-series coefficients for the expressions of Exercise 2.7.3 on $0 \leq x \leq l$.

5. Calculate half-range cosine-series coefficients for the expressions of Exercise 2.7.3 on $0 \leq x \leq l$.

6. Calculate full-range Fourier coefficients on $-l \leq x \leq l$ for the distributions $\delta(x)$, $\delta(x - a)$, $\delta'(x)$, $H(x)$.

7. Establish the formulas

$$\sum_{k=-\infty}^{\infty} \delta(x - a + 2kl) = \frac{1}{2l} + \frac{1}{l}\sum_{m=1}^{\infty} \cos\frac{m\pi}{l}(x - a)$$

$$\sum_{k=-\infty}^{\infty} \delta(x - a + 2kl) - \sum_{k=-\infty}^{\infty} \delta(x + a + 2kl) = \frac{2}{l}\sum_{m=1}^{\infty} \sin\frac{m\pi x}{l} \sin\frac{m\pi a}{l}$$

8. If $f(x)$ has the Fourier coefficients a_0, a_n, b_n ($n = 1, 2, \ldots$), find the Fourier coefficients of $f'(x)$ and $f''(x)$.

9. Find orthonormal eigenfunctions for the string of length l, with $u(0, t) = 0$, $u_x(l, t) = 0$, $t > 0$. *Note:* See Exercise 2.6.2.

10. Find the explicit eigenfunction expansion solution for the Cauchy problem $u(x, 0) = \varphi(x)$, $u_t(x, 0) = \psi(x)$, with the end conditions of Exercise 2.7.9.

11. Let $k(x)$ be a non-negative, even, smooth function which vanishes for $|x| > 1$, and such that
$$\int_{-\infty}^{\infty} k(x)\, dx = 1$$

Show that:

(a) $\displaystyle\lim_{\varepsilon \to 0} \frac{1}{\varepsilon} \int_{-\infty}^{\infty} k\left(\frac{x}{\varepsilon}\right) f(x)\, dx = f(0)$ if $f(x)$ is smooth;

(b) $\displaystyle\lim_{\varepsilon \to 0} \frac{1}{\varepsilon} \int_{-\infty}^{\infty} k\left(\frac{x}{\varepsilon}\right) H(x)\, dx = \frac{1}{2}$;

(c) if $f(x)$ has a jump discontinuity at the origin, then
$$\lim_{\varepsilon \to 0} \frac{1}{\varepsilon} \int_{-\infty}^{\infty} k\left(\frac{x}{\varepsilon}\right) f(x)\, dx = \frac{1}{2}[f(0+) + f(0-)].$$

12. Show that, if $f(x)$ satisfies the conditions of the convergence theorem, except that it has a jump discontinuity at a finite number of points, then at such a point x the series converges to $\frac{1}{2}[f(x + 0) + f(x - 0)]$ as sum.

13. *Parseval's theorem.* Demonstrate the formal relation
$$\int_{-l}^{l} f(x)^2\, dx = l\left[\frac{1}{2} a_0^2 + \sum_{n=1}^{\infty} (a_n^2 + b_n^2)\right]$$

Hint: Replace one factor in the integrand by the series. Of what classical theorem in two-dimensional geometry is this theorem an infinite-dimensional analogue?

CHAPTER 3

Parabolic Equations and Fourier Integrals

3.1 THE HEAT FLOW EQUATION

In the classification of linear second-order partial differential equations with two independent variables according to Section 2.1, equations for which the discriminant vanishes are termed *parabolic*. In such equations the second derivative operations form a "perfect square," and we are able to choose the coordinate system so that only one of the new coordinates, say, x, will appear in a second order derivative. The standard equation of this type is the equation of heat flow

(3.1.1) $$\frac{\partial u}{\partial t} = K \frac{\partial^2 u}{\partial x^2} \qquad K > 0$$

in which a first derivative term in the other coordinate is present.

This equation appears in a wide variety of diffusion and equalization processes, some of which are listed below:

1. Conduction of heat in bars or solids.

2. Diffusion of concentrations of liquid or gaseous substances in physical chemistry.
3. The diffusion of neutrons in atomic piles.
4. The diffusion of vorticity in viscous fluid flow.
5. Slow motion in hydrodynamics.
6. Telegraphic transmission in cables of low inductance or capacitance.
7. Equalization of charge in electromagnetic theory.
8. Long-wavelength electromagnetic waves in a highly conducting medium.
9. Thermal and dissipative effects in hydromagnetics.
10. Evolution of probability distributions in random processes.

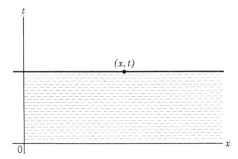

Figure 3.1 Characteristic lines of heat equation.

The parabolic equation (3.1.1) has one family of characteristic lines, namely, $t = $ constant. In this respect it is intermediate to hyperbolic equations, which have two families, and elliptic equations, which have no real characteristic lines. As we shall discover, the properties of solutions of parabolic equations are also intermediate because they share some features of each of the two other types.

Let us examine, for instance, the initial value problem for the heat flow equation. Since (3.1.1) is of first order with respect to t, we should expect to assign only one datum function at $t = 0$, say, $u(x, 0) = \varphi(x)$. Recall that for the wave equation the domain of dependence is finite and bounded by the characteristics. If in that equation the velocity c tends to infinity, then the two families of characteristic curves $t \pm x/c = $ constant coalesce to form the single family of parabolic characteristics. Thus we expect the parabolic domain of dependence to be infinite (Figure 3.1).

Let us derive the partial differential equation of heat flow in a bar extended along the x-axis, with temperature function $u(x, t)$ (Bateman, Ref. 3; Carslaw-Jaeger, Ref. 7). For simplicity we assume a constant specific heat c, so that the heat content per unit length of the bar is $\rho c u(x, t)$, where ρ is the line density, also assumed constant. We adopt Fourier's law of heat flow, which states that the amount of heat flow at any place x on the bar is proportional to the gradient (or derivative) of the temperature. By the laws of thermodynamics, this flow is in the direction of decreasing temperature.

Then the total heat in the segment $0 \leq s \leq x$ of the bar is

$$(3.1.2) \qquad H(t, x) = c\rho \int_0^x u(s, t)\, ds$$

The heat flow across the point x is, by Fourier's law,

$$(3.1.3) \qquad -\rho c K \frac{\partial u}{\partial x}$$

where $K > 0$ is the *conductivity* of the bar. Thus, if an amount of heat $f(x, t)$ is added to the bar by external heating, per unit time, then

$$(3.1.4) \qquad \frac{\partial H}{\partial t} = \rho c K \frac{\partial u}{\partial x} + \int_0^x f(s, t)\, ds$$

On the other hand, we can differentiate with respect to time under the integral sign in (3.1.2):

$$(3.1.5) \qquad \frac{\partial H}{\partial t} = \rho c \int_0^x \frac{\partial u}{\partial t}\, ds$$

Thus the right-hand sides of (3.1.4) and (3.1.5) are equal. If we then differentiate both with respect to x, we obtain the heat equation

$$(3.1.6) \qquad \frac{\partial u}{\partial t} = K \frac{\partial^2 u}{\partial x^2} + \frac{f}{\rho c}$$

with source density $f/\rho c$.

If the bar radiates heat proportionally to its temperature, then $f = -\rho h c u$, where h is a constant of proportionality, and we have the *radiation equation*

$$(3.1.7) \qquad \frac{\partial u}{\partial t} = K \frac{\partial^2 u}{\partial x^2} - hu \qquad h > 0$$

The most common end conditions for a bar are:

1. u given (assigned temperature).
2. $\partial u/\partial x$ given (assigned heat flow) (zero for *insulated* end).
3. $\partial u/\partial x + hu$ given (assigned loss by heat flow together with radiation).

By common experience it is known that the highest temperature in a bar or other conducting medium at a given instant cannot exceed both the highest initial temperature of the bar and the highest temperatures yet observed at the ends. The mathematical counterpart of this physical observation is the following theorem in which we consider only smooth solutions $u(x, t)$ of the heat flow equation.

THEOREM 3.1.1. (*Maximum Principle*). *If $u(x, t)$ satisfies (3.1.1) for $0 \leq x \leq l$, $0 \leq t \leq T$, then the maximum value of u is taken either initially (at $t = 0$) or at the endpoints $x = 0$ or $x = l$, for $0 \leq t \leq T$.*

Proof. The proof is based on the behavior of the partial derivatives u_t, u_{xx} at a hypothetical maximum place. However, in order to derive an actual contradiction, we have to consider first the differential inequality

$$\frac{\partial v}{\partial t} - K\frac{\partial^2 v}{\partial x^2} < 0 \qquad 0 \le x \le l, 0 \le t \le T \tag{3.1.8}$$

Suppose that the function $v(x, t)$ of (3.1.8) is smooth, and has a maximum at (x_0, t_0), where $0 < x_0 < l$ and $0 < t_0 \le T$. Then

$$v_{xx}(x_0, t_0) \le 0$$

since the point is a maximum. By (3.1.8) we find

$$v_t(x_0, t_0) < Kv_{xx}(x_0, t_0) \le 0$$

Therefore if δ is small enough, the time derivative $v_t(x_0, t_0 - \delta)$ is negative by continuity. Then,

$$v(x_0, t_0 - \delta) = v(x_0, t_0) - \int_0^\delta v_t(x_0, t_0 - \tau)\,d\tau > v(x_0, t_0) \tag{3.1.9}$$

But this contradicts our hypothesis that $v(x, t)$ has a maximum at (x_0, t_0). Therefore a maximum principle holds for the inequality (3.1.8).

Now consider any solution $u(x, t)$ of the heat equation (3.1.1), with $u \le M$ for $t = 0$ and $x = 0$ or $x = l$ as above. For any positive ε, let us construct

$$v = u + \varepsilon x^2 \tag{3.1.10}$$

But this v will satisfy (3.1.8) and therefore also must satisfy the maximum principle. Since

$$v \le M + \varepsilon l^2 \tag{3.1.11}$$

initially and on the boundary up to time T, we deduce that this is also true for all places in the interior of the region considered. But now

$$u = v - \varepsilon x^2 \le v \le M + \varepsilon l^2 \tag{3.1.12}$$

throughout this region. Therefore, since $\varepsilon > 0$ is arbitrary, we can conclude that

$$u \le M \tag{3.1.13}$$

throughout the region. This completes the proof.

Of course we can establish a similar result for minimum values of $u(x, t)$ by the simple remark that, if u satisfies (3.1.1), so does $-u$. Thus both maximum and minimum values are taken initially or at the endpoints.

The maximum principle has many interesting consequences. For instance, we can use it to prove uniqueness for the heat flow problem with given temperatures at the endpoints (Figure 3.2).

THEOREM 3.1.2 *A solution $u(x, t)$ of (3.1.1) subject to the initial condition*

(3.1.14) $$u(x, 0) = \varphi(x)$$

and the endpoint conditions

(3.1.15) $$u(0, t) = \mu_1(t) \qquad u(l, t) = \mu_2(t) \qquad t > 0$$

is uniquely determined.

Proof. If $u_1(x, t)$ and $u_2(x, t)$ are two such solutions, we consider their difference

(3.1.16) $$u(x, t) = u_2(x, t) - u_1(x, t)$$

Clearly $u(x, t)$ satisfies (3.1.1) with homogeneous conditions of the types (3.1.14) and (3.1.15), i.e., with $\varphi = 0$ and $\mu_1(t) = 0$, $\mu_2(t) = 0$. By the maximum

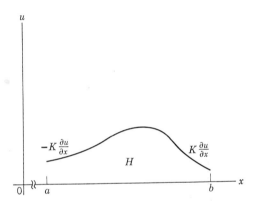

Figure 3.2 Heat distribution in a bar.

principle, u cannot exceed zero within the domain and, likewise, $-u$ cannot exceed zero. Therefore u is everywhere equal to zero, and the two given solutions are identical.

Another deduction from the maximum principle is that two solutions which differ by less than ε on the initial and boundary lines will differ by less than ε within the domain. This enables us to assert the following theorem:

THEOREM 3.1.3 *The solution $u(x, t)$ of the initial and boundary problem depends continuously on the data.*

Thus the solution of this flow problem has a certain stability—it cannot fluctuate radically because of small random disturbances. For a problem in partial differential equations to have realistic physical interpretation, it should have these properties:

1. There exists a solution.
2. The solution is unique.
3. The solution depends continuously on the data of the problem.

100 PARABOLIC EQUATIONS AND FOURIER INTEGRALS

Such a problem is said to be *correctly set* or *correctly posed*. Even before we compute or make use of a solution formula, we should like to know that the problem is correctly posed so that the specific results are stable and reliable.

Theorems 3.1.2 and 3.1.3, together with construction of a solution, would mean that the initial and boundary problem is *correctly set* for the heat flow equation. We shall see how to construct solution formulas for heat flow problems in the following section.

EXERCISES 3.1

1. By a transformation $v = ue^{\alpha x + \beta t}$ reduce the equation with constant coefficients

$$v_t = av_{xx} + bv_x + cv$$

to the form (3.1.1).

2. Establish and interpret the following equation of heat balance for a radiating bar $0 \leq x \leq l$:

$$\frac{\partial H}{\partial t} = -hH + \rho c K u_x(l, t) - \rho c K u_x(0, t) + Q$$

where

$$Q = \int_0^l f(s, t)\, ds \quad \text{and} \quad H = \rho c \int_0^l u(s, t)\, ds$$

3. If $u(x, t)$ satisfies (3.1.1), for $0 \leq x \leq l$ and $u(0, t) = u(l, t) = 0$, $t > 0$, show that the integral

$$\int_0^l u^2(x, t)\, dx$$

is a nonincreasing function of t.

4. Show that if three heat solutions u_1, u_2, u_3 satisfy $u_1 \leq u_2 \leq u_3$ on the initial line and boundary, then they satisfy the same inequalities throughout the domain.

5. Show that a bounded solution of (3.1.1) for $-\infty < x < \infty$, $t > 0$, is unique. *Hint:* Compare in finite regions with the solution $C(\frac{1}{2}x^2 + Kt)$.

6. Derive the partial differential equation

$$\frac{\partial u}{\partial t} = K \frac{\partial^2 u}{\partial x^2} - v \frac{\partial u}{\partial x}$$

for diffusion in a rod which moves with velocity v along the x-axis.

7. Formulate a problem for the heat equation that is not correctly set. *Hint:* Reverse the time.

3.2 HEAT FLOW ON A FINITE INTERVAL

Let us apply the technique of the separation of variables to the equation of heat flow

(3.2.1) $$u_t = Ku_{xx}$$

for the purpose of solving initial value problems (Figure 3.3). Adopting an appropriate time unit, we can arrange for unit conductivity: $K = 1$.

Suppose the bar $0 \leq x \leq l$ has each end maintained at zero temperature:

(3.2.2) $\qquad u(0, t) = u(l, t) = 0 \qquad t > 0$

Consider now the product solutions

(3.2.3) $\qquad u(x, t) = X(x)T(t)$

and substitute into (3.2.1). After division by (3.2.3), we find the separated form

(3.2.4) $\qquad \dfrac{T'}{T} = \dfrac{X''}{X}$

By Theorem 2.6.1, both sides of this relation are constants, independent of x and t. Comparing with Section 2.6 once more, we see that it is convenient to

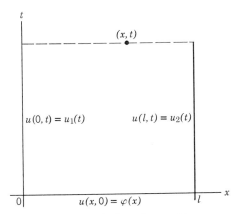

Figure 3.3 The heat flow problem.

denote the common value of the two sides in (3.2.4) by $-k^2$. The two ordinary differential equations now obtained are

(3.2.5) $\qquad T' + k^2 T = 0 \qquad t > 0$

and

(3.2.6) $\qquad X'' + k^2 X = 0 \qquad 0 \leq x \leq l$

With (3.2.6) are prescribed the boundary conditions that follow from (3.2.2), namely

(3.2.7) $\qquad X(0) = X(l) = 0$

As in Section 2.6, we find for k^2 the eigenvalues $k^2 = n^2\pi^2/l^2$, and the eigenfunctions

(3.2.8) $\qquad X_n(x) = \sin \dfrac{n\pi x}{l}$

102 PARABOLIC EQUATIONS AND FOURIER INTEGRALS

The corresponding solutions of (3.2.5) are the exponentials

(3.2.9) $$T(t) = Ce^{-k^2 t} = Ce^{-n^2\pi^2 t/l^2}$$

Because (3.2.1) is both linear and homogeneous, superposition to form a more general solution is indicated, and we write

(3.2.10) $$u(x, t) = \sum_{n=1}^{\infty} B_n e^{-n^2\pi^2 t/l^2} \sin \frac{n\pi x}{l}$$

This trial solution should satisfy the initial condition of Theorem 3.1.2:

(3.2.11) $$u(x, 0) = \varphi(x) \qquad 0 \leq x \leq l$$

Setting $t = 0$ in (3.2.10), we find that this leads to

(3.2.12) $$u(x, 0) = \varphi(x) = \sum_{n=1}^{\infty} B_n \sin \frac{n\pi x}{l} \qquad 0 \leq x \leq l$$

Exactly as in Section 2.7, we can determine the Fourier sine series coefficients as

(3.2.13) $$B_n = \frac{2}{l} \int_0^l \varphi(s) \sin \frac{n\pi s}{l} \, ds \qquad n = 1, 2, 3, \ldots$$

Substitution of these coefficients into (3.2.10) completes the finding of the formal solution. For positive values of t, the solution series (3.2.10) converges strongly because of the rapid decrease of the exponential time factors as n becomes large. The decrease of these factors as $t \to \infty$ shows that the solution $u(x, t)$ becomes small after a long period of time, i.e., it tends to zero as $t \to \infty$. Such terms are known as *transient*.

Other end conditions will lead to other eigenvalues and eigenfunctions for the spatial equation (3.2.6). For instance, if both ends of the bar are insulated so that heat cannot flow past them, the temperature gradients there must vanish. But the eigenvalue problem

(3.2.14) $$X'' + k^2 X = 0 \qquad X'(0) = X'(l) = 0$$

has the eigenvalues $k^2 = (n\pi/l)^2$ and eigenfunctions

(3.2.15) $$X_n(x) = \cos \frac{n\pi x}{l} \qquad n = 0, 1, 2, \ldots$$

as in Section 2.7. For the insulated bar we set up the trial solution

(3.2.16) $$u(x, t) = \frac{A_0}{2} + \sum_{n=1}^{\infty} A_n e^{-n^2\pi^2 t/l^2} \cos \frac{n\pi x}{l}$$

The initial condition (3.2.11) now yields

(3.2.17) $$\varphi(x) = \frac{A_0}{2} + \sum_{n=1}^{\infty} A_n \cos \frac{n\pi x}{l}$$

Therefore, as in (2.7.12), we find

(3.2.18) $$A_n = \frac{2}{l}\int_0^l \varphi(s) \cos \frac{n\pi s}{l} ds \qquad n = 0, 1, 2, \ldots$$

Observe that the terms for $n \geq 1$ in the series (3.2.16) will each tend to zero as $t \to \infty$; and their total contribution to the solution is also a transient. The limit as $t \to \infty$ of the temperature is therefore the constant term

(3.2.19) $$\frac{A_0}{2} = \frac{1}{l}\int_0^l \varphi(s) ds$$

This value is the *average* of the initial values $\varphi(x)$ on the interval: no heat can escape from the bar so the temperature is ultimately equalized about its mean value.

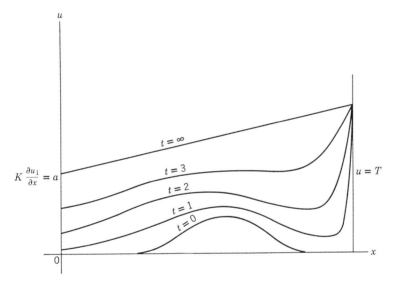

Figure 3.4 Approach of solution to stationary limit.

A more intricate problem can arise if nonhomogeneous boundary conditions are specified. Postponing until later the more difficult problem with variable endpoint temperatures, we consider the following example.

To the bar $0 \leq x \leq l$ is applied a given flow q of heat at the origin while the end $x = l$ is maintained at temperature T. The initial heat distribution is $u(x, 0) = \varphi(x)$, $0 \leq x \leq l$. To solve this problem, we should first look for the *stationary* or *equilibrium* part of the solution, that is, the part remaining after the transients have become negligible (Figure 3.4). Such a stationary term is independent of t and, therefore,

(3.2.20) $\qquad\qquad u_{xx} = 0 \qquad$ that is $\qquad u = \alpha + \beta x$

From Fourier's law (3.1.3) and the given end conditions, we find that the particular solution required is, since $K = 1$,

(3.2.21) $$u_1 = \frac{-q}{\rho c}(x - l) + T$$

We need to write the full solution u as the sum of the stationary term u_1 and a transient part u_2:

(3.2.22) $$u = u_1 + u_2$$

Let us emphasize again that this is possible only because the heat flow equation is *linear*. To determine $u_2 = u - u_1$, we have the homogeneous boundary conditions

(3.2.23) $$\frac{d}{dx}u_2(0, t) = 0 \qquad u_2(l, t) = 0 \qquad t > 0$$

to which a Fourier series expansion will now apply. The initial values are

(3.2.24) $$u_2(x, 0) = \varphi(x) - u_1 = \varphi(x) + \frac{q(x - l)}{\rho c} - T$$

The appropriate eigenvalue problem is

(3.2.25) $$X'' + k^2 X = 0 \qquad X'(0) = X(l) = 0$$

the eigenvalues are $[(n + \tfrac{1}{2})\pi/l]^2$ and the eigenfunctions are

(3.2.26) $$X_n(x) = \cos(n + \tfrac{1}{2})\frac{\pi x}{l} \qquad n = 0, 1, 2, \ldots$$

These eigenfunctions are orthogonal, since

(3.2.27) $$\begin{aligned}
&\int_0^l \cos(n + \tfrac{1}{2})\frac{\pi x}{l} \cos(m + \tfrac{1}{2})\frac{\pi x}{l}\, dx \\
&= \frac{1}{2}\int_0^l \left[\cos(n - m)\frac{\pi x}{l} + \cos(n + m + 1)\frac{\pi x}{l}\right] dx \\
&= \frac{l}{2\pi}\left[\frac{\sin(n - m)\pi x/l}{n - m} + \frac{\sin(n + m + 1)\pi x/l}{n + m + 1}\right]_0^l = 0 \qquad m \neq n
\end{aligned}$$

The eigenfunction expansion of (3.2.24) is

(3.2.28) $$u_2(x, 0) = \varphi(x) - u_1 = \sum_{n=0}^{\infty} A_n \cos\left(n + \frac{1}{2}\right)\frac{\pi x}{l}$$

where

(3.2.29) $$\begin{aligned}
A_n &= \frac{2}{l}\int_0^l [\varphi(s) - u_1(s)] \cos\left[\left(n + \frac{1}{2}\right)\frac{\pi s}{l}\right] ds \\
&= \frac{2}{l}\int_0^l \varphi(s) \cos\left[\left(n + \frac{1}{2}\right)\frac{\pi s}{l}\right] ds - \frac{(-1)^n}{n + \tfrac{1}{2}}\frac{2T}{\pi} - \frac{2ql}{\rho c(n + \tfrac{1}{2})^2 \pi^2}
\end{aligned}$$

With this formula we can complete the calculation of the solution, which is

(3.2.30)
$$u(x, t) = \frac{-q}{\rho c}(x - l) + T + \sum_{n=0}^{\infty} A_n \exp\left[-\left(n + \frac{1}{2}\right)^2 \frac{\pi^2}{l^2} t\right] \cos\left(n + \frac{1}{2}\right)\frac{\pi x}{l}$$

Now let us consider a problem with sources distributed along the bar. The nonhomogeneous equation is

(3.2.31)
$$u_t - u_{xx} = f(x, t)$$

whereas the initial condition shall be

(3.2.32) $\qquad u(x, 0) \equiv 0 \qquad 0 \leq x \leq l$

and the boundary conditions shall be

(3.2.33) $\qquad u(0, t) = u(l, t) = 0 \qquad t > 0$

Let us expand the solution function $u(x, t)$ of the nonhomogeneous equation in a series of the eigenfunctions $\sin(n\pi x/l)$ which are associated with the same boundary conditions (3.2.33). The Fourier coefficients will now be functions of the additional variable t:

(3.2.34)
$$u(x, t) = \sum_{n=1}^{\infty} c_n(t) \sin \frac{n\pi x}{l}$$

Here, according to the half-range sine series formulas,

(3.2.35)
$$c_n(t) = \frac{2}{l}\int_0^l u(x, t) \sin \frac{n\pi x}{l} dx$$

We now convert the partial differential equation (3.2.31) into a system of equations for the Fourier coefficients (3.2.35). To do this, we expand the partial differential equation itself into a Fourier sine series and equate coefficients of each eigenfunction on either side of the equation. The method of doing this is, as usual, to multiply (3.2.31) by $\sin(n\pi x/l)$ and integrate over the interval $0 \leq x \leq l$.

To evaluate the second derivative term in this calculation, we shall use the following lemma:

LEMMA 3.2.1 *If b_n is the Fourier sine coefficient of $f(x)$ on $0 \leq x \leq l$, then the Fourier sine coefficient of the second derivative $f''(x)$ is*

(3.2.36)
$$\frac{2}{l}\int_0^l f''(x) \sin \frac{n\pi x}{l} dx = -\left(\frac{n\pi}{l}\right)^2 b_n + \frac{2n\pi}{l^2}[f(0) + (-1)^{n+1} f(l)]$$

106   PARABOLIC EQUATIONS AND FOURIER INTEGRALS

Proof. Integration by parts yields

$$\frac{2}{l}\int_0^l f''(x) \sin \frac{n\pi x}{l}\, dx = \frac{2}{l}\left[f'(x) \sin \frac{n\pi x}{l}\right]_0^l - \frac{2n\pi}{l^2}\int_0^l f'(x) \cos \frac{n\pi x}{l}\, dx$$

$$= 0 - \frac{2n\pi}{l^2}\left[f(x) \cos \frac{n\pi x}{l}\right]_0^l - \frac{2}{l}\left(\frac{n\pi}{l}\right)^2 \int_0^l f(x) \sin \frac{n\pi x}{l}\, dx$$

$$= \frac{2n\pi}{l^2}[f(0) - (-1)^n f(l)] - \left(\frac{n\pi}{l}\right)^2 b_n$$

as required to complete the proof.

We apply the lemma to the second derivative term u_{xx} in the nonhomogeneous equation (3.2.31). Thus, multiplying (3.2.31) by $(2/l) \sin(n\pi x/l)$ and integrating over $(0, l)$, we have

$$\frac{2}{l}\int_0^l [u_t - u_{xx} - f(x, t)] \sin \frac{n\pi x}{l}\, dx$$

(3.2.37)
$$= \frac{2}{l}\int_0^l u_t \sin \frac{n\pi x}{l}\, dx + \left(\frac{n\pi}{l}\right)^2 c_n - \frac{2n\pi}{l^2}[u(0, t) - (-1)^n u(l, t)]$$

$$- \frac{2}{l}\int_0^l f(x, t) \sin \frac{n\pi x}{l}\, dx$$

By (3.2.33) the boundary terms vanish.

The integral over u_t is equal to $dc_n(t)/dt$, so we obtain for $c_n(t)$ the equation

(3.2.38) $$\frac{dc_n(t)}{dt} + \left(\frac{n\pi}{l}\right)^2 c_n(t) = f_n(t)$$

where $f_n(t)$ denotes the Fourier sine coefficient of $f(x, t)$:

(3.2.39) $$f_n(t) = \frac{2}{l}\int_0^l f(x, t) \sin \frac{n\pi x}{l}\, dx$$

The infinite system of ordinary differential equations (3.2.38) now replaces the partial differential equation. Because we have made appropriate choice of the eigenfunctions $\sin(n\pi x/l)$, the equations of this infinite system are separate and independent. This corresponds to the diagonal resolution of an n vector system as discussed in Section 1.6.

From (3.2.32) we have $c_n(0) = 0$ and the solution of (3.2.38) is

(3.2.40)
$$c_n(t) = \int_0^t f_n(\tau) e^{-\pi^2 n^2 (t-\tau)/l^2}\, d\tau$$

$$= \int_0^t f_n(t - \tau) e^{-\pi^2 n^2 \tau/l^2}\, d\tau$$

The solution (3.2.34) is now determined by (3.2.40) since $f_n(t)$ is given by (3.2.39). The reader should reconstruct and examine the explicit solution, which we shall consider again in Section 3.4.

EXERCISES 3.2

1. A rod of length l is initially at temperature T_0; its ends are held at zero temperature. Show that its temperature distribution after time t is

$$\frac{4T_0}{\pi} \sum_{n=1}^{\infty} \frac{e^{-(2n-1)^2\pi^2 t/l^2}}{2n-1} \sin \frac{(2n-1)\pi}{l} x$$

where x is distance measured from one end.

2. Find formulas for solutions of $u_t = K u_{xx}$ on $0 \le x \le l$ with $u(x, 0) = \varphi(x)$ and either

    (a) $u(0, t) = 0$ $u_x(l, t) + h_2 u(l, t) = 0$ $t > 0$

    or

    (b) $u_x(0, t) = 0$ $u_x(l, t) + h_2 u(l, t) = 0$ $t > 0$

    or

    (c) $u_x(0, t) + h_1 u(0, t) = 0$ $u_x(l, t) + h_2 u(l, t) = 0$ $t > 0$

3. Show that the initial value problems have solutions of the form $\int_0^l G(x, \xi, t) \varphi(\xi)\, d\xi$, where

$$G(x, t) = \frac{2}{l} \sum e^{-(n\pi/l)^2 t} \sin \frac{n\pi x}{l} \sin \frac{n\pi \xi}{l}$$

for the boundary conditions (3.2.2).

4. Find solutions of $u_t = u_{xx} + f(x, t)$ and $u(x, 0) = 0$, $0 \le x \le l$, for each of the boundary conditions (a), (b), and (c) of Exercise 3.2.2.

5. Modify the solutions of Exercise 3.2.2 as needed to satisfy the radiation equation

$$u_t = u_{xx} - hu$$

6. The initial temperature of a radiating rod of length l is zero. For $t > 0$, the two ends are maintained at a temperature of 100°. Find the temperature distribution for $t > 0$ and, in particular, find $\lim_{t \to \infty} u(x, t)$, $0 < x < l$. The partial differential equation is

$$u_t = u_{xx} - hu$$

7. Modify the solution of Exercise 3.2.6 if the end of the rod at the origin is insulated.

8. Modify the solution of Exercise 3.2.6 if the end of the rod at the origin is maintained at a temperature of 20° instead.

9. Find the solution of $u_t = u_{xx}$, $t > 0$, $0 < x < l$, such that $u(0, t) = 0$, $u(l, t) = \cos \omega t$, $t > 0$, and $u(x, 0) = 0$, $0 \le x \le l$. Describe the behavior of the solution as $t \to \infty$.

10. Find the solution of $u_t = u_{xx}$, $0 \le x \le l$, $t > 0$ with $u(x, 0) = 0$, $u(0, t) = g(t)$, $u(l, t) = 0$, $t > 0$. *Hint:* Use (3.2.34) and Lemma 3.2.1.

11. If a_n is the Fourier half-range cosine coefficient of $g(x)$, $0 \leq x \leq l$, show that the cosine coefficient of the second derivative $g''(x)$ is

$$\frac{2}{l}\int_0^l g''(x) \cos\frac{n\pi x}{l}\,dx = -\left(\frac{n\pi}{l}\right)^2 a_n - \frac{2}{l}[g'(0) + (-1)^{n+1}g'(l)]$$

12. Find the temperature in a bar with an insulated end at $x = l$, to which given amounts of heat are supplied at the end $x = 0$: $u_t = u_{xx}$, $u(x,0) = 0$, $u_x(0,t) = f(t)$, $u_x(l,t) = 0$, $t > 0$. *Hint:* Use Exercise 3.2.11.

13. If

$$\alpha_n = \frac{2}{l}\int_0^l h(x) \sin\left(n + \frac{1}{2}\right)\frac{\pi x}{l}\,dx$$

show that

$$\frac{2}{l}\int_0^l h''(x)\sin\left(n+\frac{1}{2}\right)\frac{\pi x}{l}\,dx = -\left[\left(n+\frac{1}{2}\right)\frac{\pi}{l}\right]^2\alpha_n + \frac{2}{l}\left[\left(n+\frac{1}{2}\right)\frac{\pi}{l}h(0) + (-1)^n h'(l)\right]$$

14. Find the temperature in a bar with the end $x = l$ insulated, and the end $x = 0$ maintained at a given temperature $f(t)$: $u_t = u_{xx}$, $u(x,0) = 0$, $u(0,t) = f(t)$, $u_x(l,t) = 0$, $t > 0$.

3.3 FOURIER INTEGRAL TRANSFORMS

The wave motion and heat flow problems have been resolved for a finite interval by use of the Fourier series. We now undertake the extension of this method to infinite intervals $(0, \infty)$ or $(-\infty, \infty)$. First we shall reformulate the full-range Fourier series in complex exponential form. On the interval $(-l, l)$ the complex exponentials

(3.3.1) $\qquad e^{-in\pi x/l} = \cos\dfrac{n\pi x}{l} - i\sin\dfrac{n\pi x}{l} \qquad n = 0, \pm 1, \pm 2, \ldots$

are orthogonal functions, since

(3.3.2) $\qquad \displaystyle\int_{-l}^{l} e^{-in\pi x/l} e^{im\pi x/l}\,dx = \frac{l}{\pi}\left[\frac{e^{i(m-n)\pi x/l}}{m-n}\right]_{-l}^{l} = 0$

for $m \neq n$. As these functions are complex, we must use the Hermitian scalar product

(3.3.3) $\qquad (f, g) = \displaystyle\int f(x)\overline{g(x)}\,dx$

with the corresponding norm $|f|$, where

(3.3.4) $\qquad |f|^2 = \displaystyle\int f(x)\bar{f}(x)\,dx = \int |f(x)|^2\,dx$

Now if $m = n$ in (3.3.2), the integrand is unity and the integral equal to $2l$. Therefore the exponentials

$$\frac{1}{\sqrt{2l}} e^{-in\pi x/l} \qquad n = 0, \pm 1, \pm 2, \ldots$$

form an orthonormal set for the complex Hilbert space on $-l \leq x \leq l$. (See Exercise 1.1.11.)

The complex Fourier series of a function $f(x)$ is

(3.3.5) $$f(x) = \sum_{n=-\infty}^{\infty} a_n e^{in\pi x/l} \qquad -l \leq x \leq l$$

where the coefficients a_n are, by (3.3.2), determined as

(3.3.6) $$a_n = \frac{1}{2l} \int_{-l}^{l} f(x) e^{-in\pi x/l} \, dx$$

Now let l tend to infinity. Then the eigenvalues $n\pi/l$ become crowded together and in the limit $l \to \infty$ form a dense set of values. Let us denote $n\pi/l$ by s, with $\Delta s = \pi/l$, and regard s in the limit as a continuous variable.

Assuming that the integral in (3.3.6) has a limit as $l \to \infty$, we observe that la_n will also have a limit, and so we define

(3.3.7) $$F(s) = \sqrt{\frac{2}{\pi}} \lim_{l \to \infty} la_n$$

Then (3.3.5) becomes

(3.3.8) $$f(x) = \frac{1}{\sqrt{2\pi}} \sum_{-\infty}^{\infty} \sqrt{\frac{2}{\pi}} la_n e^{in\pi x/l} \frac{\pi}{l}$$
$$\to \frac{1}{\sqrt{2\pi}} \int_{-\infty}^{\infty} F(s) e^{isx} \, ds$$

while (3.3.6) becomes

(3.3.9) $$F(s) \sim \sqrt{\frac{2}{\pi}} la_n \to \frac{1}{\sqrt{2\pi}} \int_{-\infty}^{\infty} f(x) e^{-isx} \, dx$$

The numerical factor in (3.3.7) has been chosen so that the coefficients in (3.3.8) and (3.3.9) will be equal.

These formulas, which we rewrite in a symmetrical form, imply the following theorem:

THEOREM 3.3.1 (*Fourier Integral Transforms*). *If*

(3.3.10) $$F(s) = \frac{1}{\sqrt{2\pi}} \int_{-\infty}^{\infty} f(x) e^{-isx} \, dx$$

then

(3.3.11) $$f(x) = \frac{1}{\sqrt{2\pi}} \int_{-\infty}^{\infty} F(s) e^{isx} \, ds$$

These reciprocal formulas were discovered by Fourier and Cauchy, and they are probably the most important of all integral transforms. Thus $F(s)$ is called the *Fourier transform* of $f(x)$, and $f(x)$ the *inverse Fourier transform* of $F(s)$. A table of Fourier transforms may be found in the appendix, p. 411.

Example 3.3.1 The Fourier transform of

$$f(x) = \frac{1}{1+x^2}$$

is

$$F(s) = \sqrt{\frac{\pi}{2}} e^{-|s|}$$

Example 3.3.2 The Fourier transform of

$$f(x) = \frac{x}{1+x^2}$$

is

$$F(s) = \sqrt{\frac{\pi}{2}}\, \text{sgn}\, s\, e^{-|s|}$$

(These examples are discussed in Exercise 4.6.13.)

The combined formula obtained by substituting (3.3.10) into (3.3.11), namely

(3.3.12) $$f(x) = \frac{1}{2\pi}\int_{-\infty}^{\infty} e^{isx}\,ds \int_{-\infty}^{\infty} f(y)e^{-isy}\,dy$$

is *Fourier's double integral formula*. Conditions for the rigorous validity of (3.3.12) have been the subject of many studies over the past century and a half. For a numerical (classical) function $f(x)$ such that the integral

$$\int_{-\infty}^{\infty} |f(x)|\,dx$$

is convergent, and which is continuous and of bounded variation with a finite number of *finite* discontinuities, Fourier's double integral converges to the function $f(x)$, where $f(x)$ is continuous, and to $\tfrac{1}{2}f(x+0) + \tfrac{1}{2}f(x-0)$ at a finite discontinuity. In Section 3.6 we shall discuss the Fourier integral theorem for distributions as well as for functions.

The complex Fourier transforms are analogues for an infinite interval of a complex Fourier series. There are also analogues of the half-range sine and cosine series, and these are found by considering even and odd functions. Suppose $f(x)$ even: $f(-x) = f(x)$. Then from (3.3.10)

(3.3.13)
$$F(s) = \frac{1}{\sqrt{2\pi}}\int_0^{\infty} f(x)e^{-isx}\,dx + \frac{1}{\sqrt{2\pi}}\int_{-\infty}^0 f(x)e^{-isx}\,dx$$
$$= \frac{1}{\sqrt{2\pi}}\int_0^{\infty} f(x)e^{-isx}\,dx + \frac{1}{\sqrt{2\pi}}\int_0^{\infty} f(-x')e^{isx'}\,dx'$$

where $x' = -x$, and, since $f(x)$ is even,

(3.3.14)
$$F(s) = \frac{1}{\sqrt{2\pi}}\int_0^{\infty} f(x)(e^{-isx} + e^{isx})\,dx$$
$$= \sqrt{\frac{2}{\pi}}\int_0^{\infty} f(x)\cos sx\,dx$$

But cos sx is an even function of s and of x. Hence $F(s)$ is an even function of s. Repeating the calculation, we find that (3.3.11) now yields

$$(3.3.15) \qquad f(x) = \sqrt{\frac{2}{\pi}} \int_0^\infty F(s) \cos sx \, ds$$

THEOREM 3.3.2 (*Fourier Cosine Transforms*). *If $f(x)$ is defined for $0 \leq x < \infty$, and*

$$(3.3.16) \qquad F(s) = \sqrt{\frac{2}{\pi}} \int_0^\infty f(x) \cos sx \, dx \qquad\qquad s > 0$$

then

$$(3.3.17) \qquad f(x) = \sqrt{\frac{2}{\pi}} \int_0^\infty F(s) \cos sx \, ds \qquad\qquad x > 0$$

Observe the complete symmetry of the direct and inverse cosine transforms.

Example 3.3.3 The cosine transform of $1/(1 + x^2)$ is $\sqrt{(\pi/2)}e^{-s}$. As we consider only $s > 0$, the absolute bars in Example 3.3.1 may be omitted from the exponential.

The Fourier sine transforms appear in connection with an odd function $f(x)$. Let $f(-x) = -f(x)$, and observe from (3.3.13) that

$$(3.3.18) \qquad \begin{aligned} F(s) &= \frac{1}{\sqrt{2\pi}} \int_0^\infty f(x)(e^{-isx} - e^{isx}) \, dx \\ &= -\sqrt{\frac{2}{\pi}} i \int_0^\infty f(x) \sin sx \, dx \end{aligned}$$

THEOREM 3.3.3 (*Fourier Sine Transform*). *If $f(x)$ is defined for $0 \leq x < \infty$ and*

$$(3.3.19) \qquad F_s(s) = \sqrt{\frac{2}{\pi}} \int_0^\infty f(x) \sin sx \, dx \qquad\qquad s > 0$$

is suitably convergent, then

$$(3.3.20) \qquad f(x) = \sqrt{\frac{2}{\pi}} \int_0^\infty F_s(s) \sin sx \, ds \qquad\qquad x > 0$$

The proof is easily completed if we observe from (3.3.18) that $F_s(s) = iF(s)$, where $F(s)$ is the Fourier transform of the odd extension of $f(x)$ to $-\infty < x < \infty$. Since $F_s(s)$ is an odd function of s, its inverse Fourier transform $if(x)$ will be $-i$ times its sine transform. Therefore $f(x)$ will be $-i^2 = 1$ times the sine transform of $F_s(s)$.

Example 3.3.4 The sine transform of $x/(1 + x^2)$ is $\sqrt{(\pi/2)}e^{-s}$.

EXERCISES 3.3

1. Find the Fourier transforms of

$$xe^{-x}H(x) \qquad H(x-a) - H(x-b) \qquad a < b$$

2. Find the Fourier cosine transforms of

$$x^2 e^{-x}, \qquad H(b-x)$$

3. Find the Fourier sine transforms of

$$e^{-x}\sin x, \qquad xH(b-x)$$

4. Establish Theorems 3.3.2 and 3.3.3 directly from the half-range Fourier series by making $l \to \infty$.

5. Letting $l \to \infty$ in the real form (2.7.4) of the full-range Fourier series for $-l < x < l$, verify Fourier's double real integral

$$f(x) = \frac{1}{\pi} \int_0^\infty ds \int_{-\infty}^\infty f(t) \cos s(x-t) \, dt$$

6. Show that the transform of $f(ax)$ is $(1/a)F(s/a)$.

7. Show that, *formally*, the Fourier transform of $f'(x)$ is $isF(s)$. In what sense might this formal result fail in actual validity?

8. Let $g(t)$ and $g'(t)$ be bounded, except at a finite number of points, and suppose $\int_{-\infty}^\infty |g(t)| \, dt$ is finite. Show that

$$\lim_{n \to \infty} \int_{-\infty}^\infty \sin nt \, g(t) \, dt = 0$$

Hint: Compare with Riemann's Lemma 2.7.2.

9. Show that

$$\lim_{n \to \infty} \int_a^b \frac{\sin nt}{t} \, dt$$

is equal to π if the origin lies within the interval (a, b), and is equal to zero if the origin lies outside the interval. Deduce that

$$\int_{-\infty}^\infty \frac{\sin nt}{t} \, dt = \pi \qquad n > 0$$

10. Show that, if $f(x)$ and $f'(x)$ are bounded and continuous except at a finite number of points, and if $\int_{-\infty}^\infty |f(x)| \, dx$ is finite, then

$$f(x) = \lim_{R \to \infty} \frac{1}{2\pi} \int_{-R}^R ds \int_{-\infty}^\infty f(x') \cos s(x' - x) \, dx'$$

3.4 DIFFUSION ON AN INFINITE INTERVAL

Our first application of the Fourier transform will be to the initial problem of heat flow on an infinite bar $-\infty < x < \infty$. Let

(3.4.1) $$u(x, 0) = \varphi(x) \qquad -\infty < x < \infty$$

be the initial data for a solution of

(3.4.2) $$u_t = u_{xx} \qquad t > 0$$

The Fourier transform of the solution is

(3.4.3) $$U(s, t) = \frac{1}{\sqrt{2\pi}} \int_{-\infty}^{\infty} e^{-isx} u(x, t)\, dx$$

As in Section 3.2, we transform the partial differential equation. Thus the transform of u_{xx} is

(3.4.4) $$\frac{1}{\sqrt{2\pi}} \int_{-\infty}^{\infty} e^{-isx} u_{xx}(x, t)\, dx = \frac{1}{\sqrt{2\pi}} \int_{-\infty}^{\infty} e^{-isx}\, du_x(x, t)$$
$$= \frac{1}{\sqrt{2\pi}} e^{-isx} u_x(x, t) \Big]_{-\infty}^{\infty} - \frac{1}{\sqrt{2\pi}} \int_{-\infty}^{\infty} u_x(x, t)\, de^{-isx}$$

We shall *assume* that our solution is small for large $|x|$ and, therefore, that the integrated terms vanish. Thus we have, by another partial integration

(3.4.5) $$\frac{is}{\sqrt{2\pi}} \int_{-\infty}^{\infty} u_x e^{-isx}\, dx = \frac{is}{\sqrt{2\pi}} \int_{-\infty}^{\infty} e^{-isx}\, du(x, t)$$
$$= \frac{is}{\sqrt{2\pi}} e^{-isx} u(x, t) \Big]_{-\infty}^{\infty} - \frac{is}{\sqrt{2\pi}} \int_{-\infty}^{\infty} u(x, t)\, de^{-isx}$$
$$= \frac{i^2 s^2}{\sqrt{2\pi}} \int_{-\infty}^{\infty} e^{-isx} u(x, t)\, dx = -s^2 U(x, t)$$

The equation for the transform $U(s, t)$ is therefore

(3.4.6) $$\frac{\partial U(s, t)}{\partial t} = -s^2 U(s, t)$$

since the Fourier transform of $u_t(x, t)$ is $U_t(s, t)$.

This is really a continuous system with one equation for each value of the parameter s. It can be compared with the nonhomogeneous system (3.2.38).

The solution of (3.4.6) is surely

(3.4.7) $$U(s, t) = U(s, 0) e^{-s^2 t}$$

But by transforming (3.4.1), we obtain

(3.4.8) $$U(s, 0) = \frac{1}{\sqrt{2\pi}} \int_{-\infty}^{\infty} e^{-isx} \varphi(x)\, dx$$

Let us now reconstruct the full solution, by the inverse Fourier transform, as

$$u(x,t) = \frac{1}{\sqrt{2\pi}} \int_{-\infty}^{\infty} e^{isx} U(s,t)\, ds$$

(3.4.9)
$$= \frac{1}{\sqrt{2\pi}} \int_{-\infty}^{\infty} e^{isx} U(s,0) e^{-s^2 t}\, ds$$

$$= \frac{1}{2\pi} \int_{-\infty}^{\infty} e^{isx-s^2 t}\, ds \int_{-\infty}^{\infty} e^{-isx'} \varphi(x')\, dx'$$

We shall make a formal interchange of the orders of integration of s and x'; this could be easily justified if, for instance, $\varphi(x)$ is a test function. Then

(3.4.10)
$$u(x,t) = \frac{1}{2\pi} \int_{-\infty}^{\infty} \varphi(x') \int_{-\infty}^{\infty} e^{is(x-x')-s^2 t}\, ds\, dx'$$

$$= \int_{-\infty}^{\infty} \varphi(x') K(x-x', t)\, dx' = \varphi * K(x,t)$$

say, where

(3.4.11) $$K(x-x', t) = \frac{1}{2\pi} \int_{-\infty}^{\infty} e^{is(x-x')-s^2 t}\, ds \qquad t > 0$$

is the Green's function or elementary solution of the heat equation for the infinite interval. This integral can be explicitly evaluated (see Exercise 3.4.2), and we find

(3.4.12) $$K(x,t) = \frac{1}{2\sqrt{\pi t}} e^{-x^2/4t} \qquad t > 0$$

For $t > 0$ this heat kernel $K(x,t)$ is an analytic function of x and t. Note that K is positive for all x when $t > 0$, so that the region of influence and also the domain of dependence for the heat equation includes the entire x-axis. However as $|x| \to \infty$, the solution K becomes very small, so that the amount of heat transported to great distances is exponentially small.

It is not difficult to show, either from (3.4.12) directly (see Exercise 3.4.3) or from the partial differential equation, that

(3.4.13) $$\int_{-\infty}^{\infty} K(x,t)\, dx = 1 \qquad t > 0$$

The total amount of heat remains fixed. Also, as t becomes small, the Gaussian profile of $K(x,t)$ becomes concentrated near $x = t$ (Figure 3.5). Note that in the limit $t \to 0+$, formula (3.4.10) becomes, formally

(3.4.14) $$u(x,0) = \varphi(x) = \int_{-\infty}^{\infty} \varphi(x') \lim_{t \to 0+} K(x-x', t)\, dx'$$

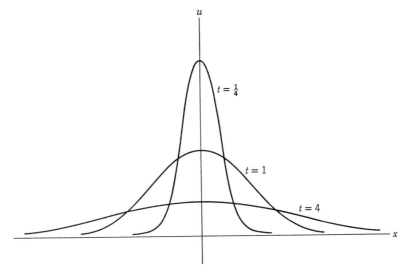

Figure 3.5 The Gaussian heat profile.

Therefore the limit is the Dirac distribution:

(3.4.15) $$\lim_{t \to 0+} K(x, t) = \delta(x)$$

Here we have another representation for the Dirac "function:"

(3.4.16) $$\delta(x) = \lim_{t \to 0+} \frac{1}{2\sqrt{\pi t}} e^{-x^2/4t}$$

This formula has a meaning in the classical sense for all $x \neq 0$, when both sides are zero. Of course it has a meaning in the distribution sense as well, namely, that the solution (3.4.10) has for initial value

(3.4.17) $$\lim_{t \to 0+} \frac{1}{2\sqrt{\pi t}} \int_{-\infty}^{\infty} \varphi(x') e^{-(x-x')^2/4t} \, dx' = \varphi(x)$$

for every testing function $\varphi(x)$. Since these functions $\varphi(x)$ are smooth, (3.4.17) is easily established for them.

Let us also examine the related problem with heat source function $f(x, t)$: thus,

(3.4.18) $$u_t - u_{xx} = f(x, t) \qquad t > 0$$

and

(3.4.19) $$u(x, 0) = 0 \qquad t > 0$$

As in the case of the wave equation in Section 2.5, we shall even consider a problem for (3.4.18) with initial time t_0 at "minus infinity."

The transformed equation is now

(3.4.20) $$\frac{\partial U(s, t)}{\partial t} + s^2 U(s, t) = F(s, t)$$

where

(3.4.21) $$F(s, t) = \frac{1}{\sqrt{2\pi}} \int_{-\infty}^{\infty} e^{-isx} f(x, t)\, dx$$

The "system" (3.4.20) is closely analogous to (3.2.38) for the finite interval. The solution of (3.4.20) with initial condition $U(s, 0) = 0$ is

(3.4.22) $$U(s, t) = \int_0^t e^{-s^2(t-t')} F(s, t')\, dt'$$

If we wish to consider $t_0 = -\infty$, then we can replace the lower limit in (3.4.22) by $-\infty$. In addition, if we recall the Green's function of Section 1.7, we see that insertion of a Heaviside factor will allow us to extend the time integration interval to $+\infty$. Therefore, we shall write

(3.4.23) $$U(s, t) = \int_{-\infty}^{\infty} e^{-s^2(t-t')} H(t - t') F(s, t')\, dt'$$

The solution of (3.4.18) is now, formally,

$$u(x, t) = \frac{1}{\sqrt{2\pi}} \int_{-\infty}^{\infty} e^{isx} U(s, t)\, ds$$

(3.4.24)
$$= \frac{1}{\sqrt{2\pi}} \int_{-\infty}^{\infty} e^{isx}\, ds \int_{-\infty}^{\infty} e^{-s^2(t-t')} H(t-t') F(s, t')\, dt'$$

$$= \frac{1}{2\pi} \int_{-\infty}^{\infty} \int_{-\infty}^{\infty} e^{isx - s^2(t-t')} H(t-t') \int_{-\infty}^{\infty} e^{-isx'} f(x', t')\, dx'\, dt'\, ds$$

Integration over s again yields the heat kernel (3.4.12) and we find

$$u(x, t) = \frac{1}{2\sqrt{\pi}} \int_{-\infty}^{\infty} \int_{-\infty}^{\infty} \frac{e^{-(x-x')^2/4(t-t')}}{\sqrt{(t-t')}} H(t-t') f(x', t')\, dx'\, dt'$$

(3.4.25)
$$= \int_{-\infty}^{\infty} \int_{-\infty}^{\infty} K(x - x', t - t') f(x', t')\, dx'\, dt'$$

$$= K * f(x, t)$$

Here we have written

(3.4.26) $$K(x, t) = \frac{1}{2\sqrt{\pi t}} e^{-x^2/4t} H(t)$$

The cutoff factor $H(t)$ in (3.4.25) ensures that no heat can arrive at any other place of the bar until after it has been supplied to the bar according to the source-density function $f(x', t')$.

DIFFUSION ON AN INFINITE INTERVAL   117

THEOREM 3.4.1 *The solution of the point-source heat flow equation*

(3.4.27) $$LK \equiv \frac{\partial K}{\partial t} - \frac{\partial^2 K}{\partial x^2} = \delta(x)\delta(t)$$

is (3.4.26).

Proof. We have

$$LK = L\left[H(t)\frac{e^{-x^2/4t}}{2\sqrt{\pi t}}\right]$$

$$= \frac{\partial}{\partial t}[H(t)]\frac{e^{-x^2/4t}}{2\sqrt{\pi t}} + H(t)L\left(\frac{e^{-x^2/4t}}{2\sqrt{\pi t}}\right)$$

$$= \delta(t)\frac{e^{-x^2/4t}}{2\sqrt{\pi t}} + H(t)0$$

since (3.4.12) is a solution of $Lu = 0$ for $t > 0$. Finally, by (3.4.16) and the general relation $\delta(t)f(t) = \delta(t)f(0)$, we find

$$LK = \delta(t)\left[\frac{e^{-x^2/4t}}{2\sqrt{\pi t}}\right]_{t=0} = \delta(t)\delta(x)$$

THEOREM 3.4.2 *The solution of the heat flow initial problem is the convolution*

(3.4.28) $$u(x, t) = \int_{-\infty}^{\infty} K(x - x', t)\varphi(x')\, dx'$$

over the space variable. The solution of the nonhomogeneous heat flow problem with $t_0 = -\infty$ is the double convolution

$$u(x, t) = \int_{-\infty}^{\infty}\int_{-\infty}^{\infty} K(x - x', t - t')f(x', t')\, dx'\, dt'$$

over space and time.

Proof. Apply the operator L and use (3.4.15) and (3.4.27) respectively.

EXERCISES 3.4

1. Show that $\int_0^{\infty} e^{-x^2}\, dx = \frac{1}{2}\sqrt{\pi}$, by considering the double integral $\int_0^{\infty}\int_0^{\infty} e^{-x^2-y^2}\, dx\, dy$ in polar coordinates.

2. Show that

$$\int_{-\infty}^{\infty} e^{isx-s^2t}\, ds = \frac{\sqrt{\pi}}{\sqrt{t}}e^{-x^2/4t} \qquad t > 0$$

assuming that certain changes of variable are justified.

3. Show that the heat kernel (3.4.26) satisfies

$$\int_{-\infty}^{\infty} K(x, t)\, dx = H(t)$$

4. *Duhamel's rule:* If $u(x, t, \tau)$ denotes the solution of $u_t = Ku_{xx}$, $u(x, 0) = f(x, \tau)$, show that $\int_0^t u(x, t - \tau, \tau) \, d\tau$ is the solution of $u_t - Ku_{xx} = f(x, t)$, $u(x, 0) = 0$.

5. Find Green's functions for these equations:
   (a) The equation of radiation $u_t - u_{xx} + hu = \delta(x)\delta(t)$.
   (b) The equation for a steadily moving medium

   $$u_t - u_{xx} + vu_x = \delta(x)\delta(t) \qquad v = \text{const.}$$

   (c) The Schrödinger equation for a free particle in one dimension

   $$i\hbar u_t + \frac{\hbar^2}{2m} u_{xx} = \delta(x)\delta(t)$$

   where $\hbar = h/2\pi$ and h is Planck's constant. (See Section 5.7.)

3.5 SEMI-INFINITE INTERVALS

For a semi-infinite bar extended along the positive x-axis, we assign one boundary condition at the end $x = 0$. The condition at the "infinite end" would be a condition of boundedness or smallness as $x \to \infty$.

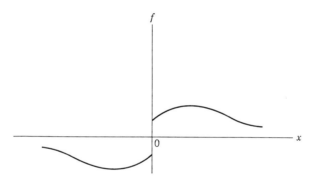

Figure 3.6 Odd extension of $f(x)$.

Let us examine the initial problem with zero temperature at the origin $x = 0$. We could attempt this in several ways, two of which are mentioned in Exercise 3.5.1, but the neatest procedure available in this instance is the method of images. Therefore, we extend the given data function $\varphi(x)$, $x > 0$, to be an odd function for $-\infty < x < \infty$. (See Figure 3.6.) This sets up a full-line initial problem that was solved in (3.4.10) and (3.4.12). In the solution, which is

$$(3.5.1) \qquad u(x, t) = \frac{1}{2\sqrt{\pi t}} \int_{-\infty}^{\infty} \varphi(x') e^{-(x-x')^2/4t} \, dx'$$

we use the fact that $\varphi(-x') = -\varphi(x')$:

$$(3.5.2) \qquad \begin{aligned} u(x, t) &= \frac{1}{2\sqrt{\pi t}} \int_0^{\infty} \varphi(x') e^{-(x-x')^2/4t} \, dx' + \frac{1}{2\sqrt{\pi t}} \int_0^{\infty} \varphi(-x') e^{-(x+x')^2/4t} \, dx' \\ &= \frac{1}{2\sqrt{\pi t}} \int_0^{\infty} \varphi(x')(e^{-(x-x')^2/4t} - e^{-(x+x')^2/4t}) \, dx' \end{aligned}$$

The Green's function for this initial problem is

(3.5.3) $$K_0(x, x', t) = \frac{1}{2\sqrt{\pi t}} (e^{-(x-x')^2/4t} - e^{-(x+x')^2/4t})H(t)$$

It represents a unit source of heat at $x = x'$, $t = 0$, together with a unit negative source, or unit "sink" of heat, at the image point $x = -x'$, $t = 0$.

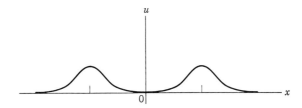

Figure 3.7 Source and image.

In much the same way we can verify that a bar with an insulated end at the origin, $u_x(0, t) = 0$, $t > 0$, will lead to the *even* Green's function

(3.5.4) $$K_e(x, x', t) = \frac{1}{2\sqrt{\pi t}} (e^{-(x-x')^2/4t} + e^{-(x+x')^2/4t})H(t)$$

This represents unit sources at $x = x'$ and at $x = -x'$, and being even (and smooth) satisfies the boundary condition $\partial K_e/\partial x = 0$ at $x = 0$, $t > 0$ (Figure 3.7).

We could also use these Green's functions for problems with a continuous source distribution $f(x, t)$ in the quadrant $x > 0$, $t > 0$, but we leave details of this to the reader.

The more difficult problem here is connected with the nonhomogeneous boundary condition, say

(3.5.5) $$u_x(0, t) = h(t) \qquad t > 0$$

for the semi-infinite bar. This represents a bar to the end of which a given supply of heat is furnished.

It is not obvious which transform should be used in connection with (3.5.5). Presumably either the Fourier sine or cosine transform is relevant, but at this stage, it is hard to predict which. Actually it is the cosine transform, as we shall verify by means of Lemma 3.5.1 below. We shall discuss the reason for this in Chapters 5 and 6, but for the present, the reader should observe that the cosine transform is associated with *even* functions of x, and that even functions naturally satisfy the corresponding *homogeneous* boundary condition $u_x = 0$.

Consider, then, the heat flow problem

(3.5.6) $$u_t - u_{xx} = 0 \qquad x > 0, t > 0$$

with

(3.5.7) $$u(x, 0) = 0, \qquad x > 0$$

120   PARABOLIC EQUATIONS AND FOURIER INTEGRALS

and

(3.5.8) $$u_x(0, t) = h(t) \qquad t > 0$$

We also suppose that $u(x, t)$ and $u_x(x, t)$ tend to zero as $x \to \infty$.
The cosine transform of $u(x, t)$ is

(3.5.9) $$U_c(s, t) = \sqrt{\frac{2}{\pi}} \int_0^\infty u(x, t) \cos sx \, dx$$

To "transform" the partial differential equation, we must calculate the cosine transform of u_{xx}. As in Lemma 3.2.1, we shall state the result of this calculation explicitly.

LEMMA 3.5.1 *Let $F_c(s)$ be the Fourier cosine transform of a function $f(x)$ which tends to zero, together with its first derivative, as $x \to \infty$. Then the cosine transform of the second derivative $f''(x)$ is*

(3.5.10) $$\sqrt{\frac{2}{\pi}} \int_0^\infty f''(x) \cos sx \, dx = -s^2 F_c(s) - \sqrt{\frac{2}{\pi}} f'(0)$$

Proof. Integration by parts gives

$$\sqrt{\frac{2}{\pi}} \int_0^\infty f''(x) \cos sx \, dx = \sqrt{\frac{2}{\pi}} [f'(x) \cos sx]_0^\infty + \sqrt{\frac{2}{\pi}} s \int_0^\infty f'(x) \sin sx \, dx$$

$$= -\sqrt{\frac{2}{\pi}} f'(0) + \sqrt{\frac{2}{\pi}} s[f(x) \sin sx]_0^\infty$$

$$- \sqrt{\frac{2}{\pi}} s^2 \int_0^\infty f(x) \cos sx \, dx$$

$$= -\sqrt{\frac{2}{\pi}} f'(0) - s^2 F_c(s)$$

as required.

Applying this formula to the differential equation, we write

(3.5.11) $$\sqrt{\frac{2}{\pi}} \int_0^\infty u_t(x, t) \cos sx \, dx = \sqrt{\frac{2}{\pi}} \int_0^\infty u_{xx}(x, t) \cos sx \, dx$$

The left side is $\partial U_c(s, t)/\partial t$. By (3.5.8) and Lemma 3.5.1, the right side is $-\sqrt{(2/\pi)} h(t) - s^2 U_c(s, t)$. For the cosine transform, we therefore find the equation or rather, the continuous system

(3.5.12) $$\frac{\partial U_c(s, t)}{\partial t} + s^2 U_c(s, t) = -\sqrt{\frac{2}{\pi}} h(t) \qquad -\infty < s < \infty$$

The solution of this equation, which by (3.5.7) should vanish for $t = 0$, is

(3.5.13) $$U_c(s, t) = -\sqrt{\frac{2}{\pi}} \int_0^t h(\tau) e^{-s^2(t-\tau)} \, d\tau$$

This determines the cosine transform of $u(x, t)$.

The solution itself will be formed by inverting that transform. From Theorem 3.3.2 we calculate

$$u(x, t) = \sqrt{\frac{2}{\pi}} \int_0^\infty U_c(s, t) \cos sx \, ds$$

(3.5.14)
$$= -\frac{2}{\pi} \int_0^\infty \cos sx \, ds \int_0^t h(\tau) e^{-s^2(t-\tau)} \, d\tau$$

$$= -\frac{2}{\pi} \int_0^t h(\tau) \, d\tau \int_0^\infty e^{-s^2(t-\tau)} \cos sx \, ds$$

The inner integral over s is related to the Green's function integral of Section 3.4, and has the value

(3.5.15) $$\int_0^\infty e^{-s^2(t-\tau)} \cos sx \, ds = \frac{1}{2}\sqrt{\frac{\pi}{t-\tau}} \, e^{-x^2/[4(t-\tau)]}$$

Consequently our solution is

(3.5.16) $$u(x, t) = -\frac{1}{\sqrt{\pi}} \int_0^t \frac{h(\tau)}{\sqrt{t-\tau}} e^{-x^2/[4(t-\tau)]} \, d\tau$$

Observe that the domain of dependence of $u(x, t)$ upon $h(t)$ includes all earlier times τ (Figure 3.8). For a solution of the problem beginning at $t_0 = -\infty$, we need only adjust the lower limit in the integral of (3.5.16). Inspection of the integral (3.5.16) shows that it will tend rapidly to zero as $x \to \infty$, and that its derivatives with respect to x do so likewise. This justifies the hypothesis used in the calculations of (3.5.10) and (3.5.11).

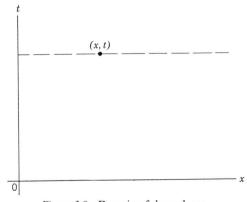

Figure 3.8 Domain of dependence.

In any application of the Fourier transform we are, in effect, expressing our solution as a superposition of exponentials, and we shall have more to say about this in the next section. However there are some problems in which the solution is a single exponential term that can be calculated in a quite elementary way. Such a problem, early studied by Fourier, is the determination of periodic temperature waves. Let x be the depth below the surface of a flat earth, and let the surface temperature be

(3.5.17) $$u(0, t) = A \cos \omega t = \operatorname{Re}(Ae^{i\omega t}) \qquad \omega > 0$$

This is a periodic or oscillatory problem, the oscillations existing throughout all time.

To determine the temperature variation for $x > 0$, we shall try the real part of

(3.5.18) $$u(x, t) = Ae^{\alpha x + i\omega t} \qquad x > 0$$

This will satisfy our heat flow equation (3.5.6) provided

(3.5.19) $$i\omega = \alpha^2$$

Therefore

(3.5.20) $$\alpha = \pm\sqrt{i\omega} = \pm \frac{1+i}{\sqrt{2}} \sqrt{\omega} = \pm\left(\sqrt{\frac{\omega}{2}} + i\sqrt{\frac{\omega}{2}}\right)$$

Our solution is now the real part of

$$Ae^{\alpha x + i\omega t} = A \exp\left[\pm\sqrt{\tfrac{1}{2}\omega}\,x + i(\pm\sqrt{\tfrac{1}{2}\omega}\,x + \omega t)\right]$$

and the real part is

(3.5.21) $$Ae^{\pm\sqrt{\frac{1}{2}\omega}\,x} \cos(\pm\sqrt{\tfrac{1}{2}\omega}\,x + \omega t)$$

The choice of the plus or minus sign is now governed by the requirement that the solution be small for large x. We choose minus, and our solution can be written

(3.5.22) $$u(x, t) = Ae^{-\sqrt{\frac{1}{2}\omega}\,x} \cos(\omega t - \sqrt{\tfrac{1}{2}\omega}\,x)$$

This formula shows that the amplitude of the temperature waves diminishes exponentially with depth x, that there is a time lag $x/\sqrt{2\omega}$ of the waves, and that the depth of penetration varies as the square root of the period $T = 2\pi/\omega$.

EXERCISES 3.5

1. Show that the initial problem $u_t = u_{xx}$, $x > 0$, $t > 0$; $u(x, 0) = \varphi(x)$, $x > 0$; $u(0, t) = 0$, $t > 0$, has solutions

$$u(x, t) = \int_0^\infty G(x, \xi, t) \varphi(\xi)\, d\xi$$

where

$$G(x, \xi, t) = \frac{2}{\pi} \int_0^\infty e^{-s^2 t} \sin sx \sin s\xi\, ds$$

in two ways:
(a) Letting $l \to \infty$ in Exercise 3.2.3.
(b) Using Fourier sine transforms as in Section 3.4.
Evaluate the integral for G and check the result.

2. Derive the results and formulas of Exercise 3.5.1 for the problem with the boundary condition $u_x = 0$.

3. Solve the problem $u_t = u_{xx}$, $x > 0$, $t > 0$; $u(x, 0) = \varphi(x)$, $x > 0$; $u_x(0, t) - hu(0, t) = 0$, $t > 0$, $h > 0$, by first finding $v(x, t) \equiv u_x(x, t) - hu(x, t)$ as a solution of a similar problem with $v(0, t) = 0$, $t > 0$, and then showing

$$u(x, t) = -\int_x^\infty e^{-h(x'-x)} v(x', t) \, dx'$$

4. Show that

$$\int_0^\infty s e^{-s^2(t-\tau)} \sin sx \, ds = \frac{x \sqrt{\pi}}{4(t-\tau)^{3/2}} e^{-x^2/[4(t-\tau)]}$$

either by:
(a) Comparing with (3.4.11) and (3.4.12).
(b) Comparing with (3.5.15).

5. Show that the solution of the boundary problem $u_t = u_{xx}$, $x > 0$, $t > 0$; $u(x, 0) = 0$, $x > 0$; $u(0, t) = g(t)$, $t > 0$ is

$$u(x, t) = \frac{x}{2\sqrt{\pi}} \int_0^t \frac{g(\tau)}{(t-\tau)^{3/2}} e^{-x^2/[4(t-\tau)]} \, d\tau$$

in two ways:
(a) By using sine transforms and Exercise 3.5.4.
(b) By comparing with (3.5.16).

6. By combining Exercises 3.5.3, 3.5.4, and 3.5.5 find the solution of the boundary problem $u_t = u_{xx}$, $x > 0$, $t > 0$; $u(x, 0) = 0$, $x > 0$; $u_x(0, t) - hu(0, t) = g(t)$, $t > 0$, $h > 0$.

7. Compare the depths of penetration of annual temperature waves of amplitude $40°C$ and daily temperature waves of amplitude $10°C$.

8. Given that K for the earth is 4.10^{-3} cm^2/sec, find the time lag of annual temperature waves at a depth of 2 meters.

9. Construct the solution of the steady-state problem $u_t = u_{xx}$, $0 \le x \le l$; $u(0, t) = A \cos \omega t$; $u(l, t) = 0$, $-\infty < t < \infty$.

10. Show that the finite bar problem $u_t = u_{xx}$, $t > 0$, $0 \le x \le l$; $u(x, 0) = 0$, $0 \le x \le l$, $u_x(0, t) = 0$, $t > 0$; $u(l, t) = g(t)$, $t > 0$, has the solution

$$u(x, t) = \frac{\pi}{l^2} \sum_{n=0}^\infty (-1)^n (2n + 1) \cos\left[\left(n + \frac{1}{2}\right)\frac{\pi x}{l}\right]$$

$$\times \int_0^t g(t - \tau) \exp\left[-\left(n + \frac{1}{2}\right)^2 \frac{\pi^2 \tau}{l^2}\right] d\tau$$

Hint: Follow the method of Section 3.5, formulas (3.5.13) through (3.5.16) but with a cosine series or finite transform adapted to the homogeneous boundary conditions.

11. Why, in (3.5.10), can we *not* say "the cosine transform of $f''(x)$ is $-s^2$ times the cosine transform of f?" For what class of functions would this statement be true? Is it true if $f(x)$ is an even function on $-\infty < x < \infty$ with continuous first derivative?

12. If $F_s(s)$ is the sine transform of $f(x)$, show that the sine transform of $f''(x)$ is

$$\sqrt{\frac{2}{\pi}} \int_0^\infty f''(x) \sin sx\, dx = \sqrt{\frac{2}{\pi}} sf(0) - s^2 F_s(s)$$

3.6 FOURIER TRANSFORMS OF DISTRIBUTIONS

In Section 3.4 we used a Fourier transform to solve a partial differential equation by reducing it to an ordinary equation. This method works because the transform in some sense replaced the operation of differentiation with respect to x by a factor is [see (3.4.3)–(3.4.6)]. A further look into this process leads us to the Fourier transforms of distributions.

Let us formally apply the Fourier transform formulas (3.3.10) and (3.3.11) to the Dirac distribution $\delta = \delta(x)$. Its transform is

(3.6.1) $$\frac{1}{\sqrt{2\pi}} \int_{-\infty}^\infty \delta(x) e^{-isx}\, dx = \frac{1}{\sqrt{2\pi}} e^{-is0} = \frac{1}{\sqrt{2\pi}}$$

which is a constant. The inverse Fourier transform formula (3.3.11) now is

(3.6.2) $$\delta(x) = \frac{1}{2\pi} \int_{-\infty}^\infty e^{isx}\, ds$$

Similarly, the translated Dirac distribution $\delta(x - a)$ has the Fourier integral transform

(3.6.3) $$\frac{1}{\sqrt{2\pi}} \int_{-\infty}^\infty \delta(x - a) e^{-isx}\, dx = \frac{1}{\sqrt{2\pi}} e^{-isa}$$

According to (2.4.8), the derivative distribution $\delta'(x)$ has the Fourier transform

(3.6.4) $$\frac{1}{\sqrt{2\pi}} \int_{-\infty}^\infty \delta'(x) e^{-isx}\, dx = -\frac{1}{\sqrt{2\pi}} \frac{d}{dx} e^{-isx} \bigg]_{x=0} = \frac{is}{\sqrt{2\pi}}$$

Likewise, the second derivative $\delta''(x)$ has the transform

(3.6.5) $$\frac{1}{\sqrt{2\pi}} \int_{-\infty}^\infty \delta''(x) e^{-isx}\, dx = +\frac{1}{\sqrt{2\pi}} \frac{d^2}{dx^2} e^{-isx} \bigg]_{x=0} = \frac{-s^2}{\sqrt{2\pi}}$$

which we may again compare with (3.4.5).

Now consider any distribution T which has compact (i.e., bounded) support S. Recall that $T(\varphi)$ does not depend on the values of φ taken outside the set S. We have, then,

(3.6.6) $$T(\varphi) = \int_{-\infty}^\infty f(x)\varphi(x)\, dx = \int_S f(x)\varphi(x)\, dx$$

where $f(x)$ is the symbolic function identified with T. To obtain the Fourier transform of $f(x)$, we should replace $\varphi(x)$ by $\sqrt{(1/2\pi)}e^{-isx}$. However, this exponential is not a testing function because its support is not bounded. Therefore let us replace $\varphi(x)$ by a truncated exponential

$$\frac{1}{\sqrt{2\pi}} e^{-isx} \varphi_1(x)$$

where $\varphi_1(x)$ is a testing function equal to 1 in the support of T. We obtain now a *function* of s:

$$F(s) = \frac{1}{\sqrt{2\pi}} \int_{-\infty}^{\infty} f(x)e^{-isx}\, dx$$

$$= \frac{1}{\sqrt{2\pi}} \int_{S} f(x)e^{-isx}\, dx$$

(3.6.7)
$$= \frac{1}{\sqrt{2\pi}} \int_{S} f(x)e^{-isx}\varphi_1(x)\, dx$$

$$= \frac{1}{\sqrt{2\pi}} \int_{-\infty}^{\infty} f(x)e^{-isx}\varphi_1(x)\, dx$$

$$= T\left(\frac{e^{-isx}}{\sqrt{2\pi}} \varphi_1(x)\right) = T\left(\frac{e^{-isx}}{\sqrt{2\pi}}\right)$$

THEOREM 3.6.1 *The Fourier transform of a distribution T with compact support is*

(3.6.8)
$$F(s) = T\left(\frac{e^{-ixs}}{\sqrt{2\pi}}\right)$$

The examples (3.6.1) through (3.6.4) can be verified by this definition. A further example is the Fourier transform of $\delta^{(n)}(x-a)$, which is

(3.6.9)
$$F(s) = \int_{-\infty}^{\infty} \delta^{(n)}(x-a) \frac{e^{-ixs}}{\sqrt{2\pi}}\, dx = -1^n \frac{d^n}{dx^n} \frac{e^{-ixs}}{\sqrt{2\pi}}\bigg]_{x=a}$$

$$= \frac{(is)^n}{\sqrt{2\pi}} e^{-ias}$$

The convolution $f * g$ of two functions was introduced in Chapter 2:

(3.6.10)
$$f * g(x) = \int_{-\infty}^{\infty} f(x-y)g(y)\, dy$$

The importance of convolutions is further emphasized by the following theorem:

THEOREM 3.6.2 *The Fourier transform of the convolution $f * g(x)$ is the product $\sqrt{2\pi}F(s)G(s)$ of the separate transforms.*

Proof. We calculate the transform as

$$\frac{1}{\sqrt{2\pi}} \int_{-\infty}^{\infty} f * g(x) e^{-isx} \, dx$$

$$= \frac{1}{\sqrt{2\pi}} \int_{-\infty}^{\infty} e^{-isx} \, dx \int_{-\infty}^{\infty} f(x-y) g(y) \, dy$$

(3.6.11)
$$= \frac{1}{\sqrt{2\pi}} \int_{-\infty}^{\infty} g(y) e^{-isy} \, dy \int_{-\infty}^{\infty} f(x-y) e^{-is(x-y)} \, d(x-y)$$

$$= \sqrt{2\pi} G(s) F(s)$$

since the inner integral for $F(s)$ is independent of y after the variable of integration $x - y$ is adopted.

A similar result will hold for inverse Fourier transforms. This convolution theorem will also be valid for distributions of compact support. For instance, the substitution property

(3.6.12) $$\delta * f(x) = \int_{-\infty}^{\infty} \delta(x-y) f(y) \, dy = f(x)$$

shows that the Dirac distribution acts as a *unit* in convolution "multiplication." Correspondingly, the transform of $\delta * f(x)$ is $\sqrt{2\pi}(1/\sqrt{2\pi}) F(s) = F(s)$. Again, the convolution of $\delta'(x-a)$ and $\delta''(x-b)$ is

$$\delta'_a * \delta''_b(x) = \int_{-\infty}^{\infty} \delta'(x-a-y) \delta''(y-b) \, dy = \delta'''(x-a-b)$$

while its Fourier transform is

$$\sqrt{2\pi} \, \frac{is}{\sqrt{2\pi}} e^{-isa} \, \frac{(is)^2}{\sqrt{2\pi}} e^{-isb} = \frac{(is)^3}{\sqrt{2\pi}} e^{-is(a+b)}$$

Another significant formula is found if we consider the norm of a function $f(x)$. Since the Fourier integrals are complex, we should use the Hermitian norm

(3.6.13) $$|f|^2 = (f, f) = \int_{-\infty}^{\infty} |f(x)|^2 \, dx = \int_{-\infty}^{\infty} f(x) \bar{f}(x) \, dx$$

Recall that a linear matrix transformation A, $\mathbf{y} = A\mathbf{x}$, is called *unitary* if

(3.6.14) $$|\mathbf{y}|^2 = (\mathbf{y}, \mathbf{y}) = (\mathbf{x}, \mathbf{x}) = |\mathbf{x}|^2$$

The matrix A is unitary if $(A\mathbf{x}, A\mathbf{x}) = (\mathbf{x}, A^*A\mathbf{x}) = (\mathbf{x}, \mathbf{x})$ for all $\mathbf{x}$, i.e., if $A^*A = E$. In component form this unitary condition reads

(3.6.15) $$\sum_k \bar{a}_{kl} a_{km} = \delta_{lm} \qquad A = (a_{kl})$$

THEOREM 3.6.3 *The Fourier transform is unitary:*

$$(3.6.16) \quad |F|^2 = (F, F) = \int_{-\infty}^{\infty} |F(s)|^2 \, ds = \int_{-\infty}^{\infty} |f(x)|^2 \, dx = (f, f) = |f|^2$$

Proof. We have

$$|f|^2 = \int_{-\infty}^{\infty} f(x)\bar{f}(x) \, dx$$

$$= \frac{1}{\sqrt{2\pi}} \int_{-\infty}^{\infty} \left[\int_{-\infty}^{\infty} e^{+isx} F(s) \, ds \right] \bar{f}(x) \, dx$$

(3.6.17)

$$= \int_{-\infty}^{\infty} F(s) \, ds \, \frac{1}{\sqrt{2\pi}} \overline{\int_{-\infty}^{\infty} e^{-isx} f(x) \, dx}$$

$$= \int_{-\infty}^{\infty} F(s) \overline{F(s)} \, ds = |F|^2$$

We have stated here the formal part of Plancherel's theorem, one of the leading results of Fourier analysis. The complete theorem, which we do not attempt to prove, shows that the *existence* of either one of the integrals implies the existence of the transformed integral. That is, if f is an element of the Hilbert space on $-\infty < x < \infty$ (i.e., of finite "length"), then its transform is also an element of the Hilbert space on $-\infty < s < \infty$, and has equal magnitude.

When the Fourier transform is regarded as an infinite and continuous matrix, its matrix elements are $e^{-isx}/\sqrt{2\pi}$, where s is a row label and x a column label. The infinite and continuous unit matrix is $E(x, y) = (\delta(x - y))$, since matrix multiplication here means integration from $-\infty$ to $+\infty$, and δ is the *unit* under this operation. The unitary condition (3.6.15) now has the continuous analogue

$$(3.6.18) \quad \int_{-\infty}^{\infty} \frac{e^{-isx}}{\sqrt{2\pi}} \frac{e^{isy}}{\sqrt{2\pi}} \, ds = \frac{1}{2\pi} \int_{-\infty}^{\infty} e^{-is(x-y)} \, dy = \delta(x - y)$$

which is valid according to the complex conjugate of (3.6.2).

There is a further relation satisfied by a unitary matrix A, which has a significant interpretation in the case of the Fourier transform. From the unitary condition $A^*A = E$ follows $A^* = A^{-1}$ and then $AA^* = E$. As well as (3.6.15), therefore, the components of A satisfy the second condition

$$\sum_l a_{kl} \bar{a}_{hl} = \delta_{kh}$$

Here summation runs over the column label l.

For the Fourier transform this relation implies integration over the space variable x:

$$\int_{-\infty}^{\infty} \frac{e^{-isx}}{\sqrt{2\pi}} \frac{e^{itx}}{\sqrt{2\pi}} \, dx = \delta(s - t)$$

This formula is equivalent to (3.6.18). We interpret it as a statement of orthogonality and normalization for the matrix elements $e^{-isx}/\sqrt{2\pi}$ of the Fourier transform. These elements are the orthonormal eigenfunctions of the differential operator $i(d/dx)$ on the interval $(-\infty, \infty)$ with the condition of boundedness:

$$i\frac{d}{dx}\frac{e^{-isx}}{\sqrt{2\pi}} = s\frac{e^{-isx}}{\sqrt{2\pi}}$$

Comparing with Theorem 1.6.2 we see that the Fourier transform *diagonalizes the derivative operator*.

If a linear transformation preserves lengths, it will also preserve scalar products. Thus, if $f(x)$ and $g(x)$ have transforms $F(s)$ and $G(s)$, we find

(3.6.19) $\quad (f, g) = \int_{-\infty}^{\infty} f(x)\bar{g}(x)\,dx = \int_{-\infty}^{\infty} F(s)\bar{G}(s)\,ds = (F, G)$

Direct proof of this formula follows (3.6.17) and is left as an exercise.

This relation can be used to define the Fourier transform of any distribution. Let φ be a test function, and Φ its Fourier transform. The Fourier transform of a distribution f is now a distribution F defined over the linear space of functions Φ by the formula

(3.6.20) $\quad F(\Phi) = f(\varphi)$

When f and F are functions, this becomes

(3.6.21) $\quad F(\Phi) = \int_{-\infty}^{\infty} F(s)\overline{\Phi(s)}\,ds = (F, \Phi) = (f, \varphi) = f(\varphi)$

by (3.6.19), and so the definition is consistent with the usual definition of the Fourier transform of a function.

EXERCISES 3.6

1. Show that the convolution of $\delta^{(m)}(x - a)$ and $\delta^{(n)}(x - b)$ is $\delta^{(m+n)}(x - a - b)$, and verify Theorem (3.6.2) for the corresponding Fourier transforms.

2. Show that (3.6.2) is *formally* equivalent to Fourier's double integral formula (3.3.12).

3. Show that, for two functions $f(x)$, $g(x)$ of finite norm, with transforms $F(s)$, $G(s)$,

$$\int_{-\infty}^{\infty} F(s)G(s)e^{ixs}\,ds = \int_{-\infty}^{\infty} f(x - y)g(y)\,dy$$

4. Show that the Fourier sine and cosine transforms are unitary, with the norm

$$|f|^2 = \int_0^{\infty} |f(x)|^2\,dx$$

Write down the matrix elements and the relations corresponding to (3.6.18) for the sine and cosine transforms.

5. If $f(x)$ has sine and cosine transforms $F_s f(s)$ and $F_c f(s)$ respectively, show that

$$F_s f'(s) = -sF_c f(s) \qquad F_c f'(s) = sF_s f(s) - \sqrt{\frac{2}{\pi}} f(0)$$

6. Find the orthonormal eigenfunctions of $i(d/dx)$ on $(-l, l)$ with the periodic boundary condition $y(-l) = y(l)$. Construct the transform which diagonalizes this operator.

7. Find the real orthonormal eigenfunctions of $-d^2/dx^2$ on $(-l, l)$ with the periodic boundary conditions $y(-l) = y(l)$, $y'(-l) = y'(l)$, and discuss the transition $l \to \infty$.

3.7 FINITE DIFFERENCE CALCULATIONS

In Chapter 2 we gave some simple examples in numerical calculation of wave motion solutions by means of a finite difference pattern which is a discrete approximation to the partial differential equation. Let us pursue this method for the heat flow equation, using a grid or lattice in the x, t-plane with intervals Δx, Δt in the x and t directions. For brevity, we write

(3.7.1) $$u_{hk} = u(h \Delta x, k \Delta t).$$

To set up a discrete analogue of the partial differential equation, we replace the derivatives appearing in it by suitably chosen differences of the function (3.7.1) defined at points of the grid. As no finite difference combination can represent derivatives with perfect accuracy, every template pattern on a finite grid involves approximation to the terms of the differential equation and also to its solution. In choosing a finite difference scheme, we must balance accuracy of approximation to the solution against simplicity and ease of calculation.

It is natural to replace the time derivative u_t by the first order difference

$$\frac{u_{h,k+1} - u_{h,k}}{\Delta t}$$

which is the first approximation to it. Likewise, the second derivative u_{xx} may be approximated by the second order difference as in (2.3.11). The equation of diffusion (3.2.1) is then replaced by

(3.7.2) $$\frac{u_{h,k+1} - u_{h,k}}{\Delta t} = \frac{u_{h+1,k} - 2u_{h,k} + u_{h-1,k}}{(\Delta x)^2}$$

We solve for the "leading term" $u_{h,k+1}$ and find

(3.7.3) $$u_{h,k+1} = C(u_{h+1,k} + u_{h-1,k}) + (1 - 2C)u_{h,k}$$

where

(3.7.4) $$C = \frac{\Delta t}{(\Delta x)^2}$$

We start our calculation with a single row of given numbers—the initial data—for $k = 0$, say. The finite difference pattern, or template, is shown in Figure 3.9 together with the coefficients at each lattice point. Moving the template from point to point, we construct the array of values for $k = 1$, $k = 2, \ldots$ in succession.

130 PARABOLIC EQUATIONS AND FOURIER INTEGRALS

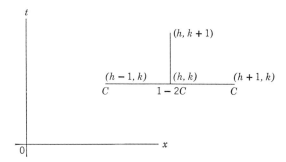

Figure 3.9 Finite difference pattern for heat equation.

For a numerical example, we must specify C numerically. Figure 3.10 shows the simplest possible calculation with $C = 1$. It is evident that the solution shown is completely unsatisfactory. It oscillates with ever-increasing amplitude and violates the maximum principle almost everywhere.

This template pattern is *unstable*. The presence of increasingly large values appears to be due to the negative coefficient in the finite difference template. Let us reduce the coefficient C so that this negative sign is removed. For good measure, choose $C = \frac{1}{4}$ so that the template is as shown in Figure 3.11. This means, incidentally, that the new time interval Δt must be a quarter the previous size if the interval Δx remains unchanged. It can be seen that the new calculation is quite realistic, and represents the evolving profile of a Gaussian fundamental solution reasonably well.

No finite difference solution with this template can represent accurately the domain of dependence of a point-source solution, since the nonzero entries would have to spread at once over the entire line $t = $ constant. However, the error at large distances is exponentially small, and the values of the solution shown decrease exponentially along the diagonal lines through the source point.

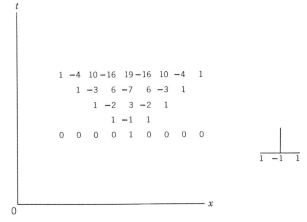

Figure 3.10 An unstable solution, $C = 1$.

For $C \leq \frac{1}{2}$, this finite difference template can be shown to be *stable*. This implies two separate results:

1. Numerical oscillation or instability is absent.
2. By decreasing the intervals of the grid, any desired accuracy of approximation to the true solution of the partial differential equation can be secured.

Proof of this second property, which is well beyond our scope, can be found in books on the numerical solution of partial differential equations (Richtmyer,

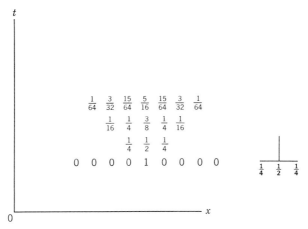

Figure 3.11 A stable solution, $C = \frac{1}{4}$.

Ref. 49). We shall add here one further remark, which leads to an apparent paradox. Suppose we have a solution pattern which we wish to improve by subdividing the grid into smaller intervals. If we divide both space and time intervals Δx, Δt in half, we obtain a new grid, and a new template pattern with it. But from (3.7.4), the new value of C will be *twice* the old one. If this carries us up into unstable values of C, then any accuracy hitherto achieved will be completely lost. Therefore, to maintain a stable pattern, we must divide the time interval Δt in four when Δx is divided in two. To cover a given region of space and time with our new calculations, we now have eight times as many points to compute. This larger volume of computation must be considered when improvement in accuracy is desired.

Finally, let us indicate why the template should be unstable when C and Δt are too large. The heat flow equation represents an equalizing, or averaging, process. At a maximum temperature, for instance, $u_t = Ku_{xx}$ is negative, so the solution decreases. As soon as the convexity represented by u_{xx} has been equalized, however, the decrease terminates. If the finite difference intervals are too long, the corrections will overshoot by a larger amount than their original size and will set up ever-increasing oscillations. For stability, therefore, the intervals Δt must be sufficiently short.

EXERCISES 3.7

1. Show that $\Sigma_h u_{hk} = $ constant is an "integral" of the difference equation (3.7.3).

2. With $C = \frac{1}{4}$ in (3.7.3), compute the solution of the boundary problem $u_{h0} = 0$, $h > 0$; $u_{0k} = 0$, $k > 0$, $u_{00} = 1$, up to $k = 4$.

3. With $C = \frac{1}{4}$ in (3.7.3), compute the solution of the boundary problem $u_{h0} = 0$, $h \geq 0$; $u_{0k} = u_{1k} = 0$, $k \leq 2$; $u_{03} = 0$, $u_{13} = 1$; $u_{0k} = u_{1k}$, $k \geq 4$, up to $k = 6$.

4. With $C = \frac{1}{8}$ in (3.7.3), compute the solution of (3.7.3) with $u_{h0} = 0$, $h \neq 0$; $u_{00} = 1$, up to $k = 4$.

5. With $C = \frac{1}{2}$ in (3.7.3), find a formula for u_{hk} when $u_{h0} = 0$, $h \neq 0$; $u_{00} = 1$.

6. Derive the template formula

$$\tfrac{3}{2} u_{hk+1} - \tfrac{1}{4}(u_{h+1,k+1} + u_{h-1,k+1}) = u_{hk}$$

and explain how it could be used for a bar of finite length. What is the domain of dependence of the lattice point $(h, 0)$?

CHAPTER *4*

Laplace's Equation and Complex Variables

4.1 MATHEMATICAL AND PHYSICAL APPLICATIONS

Section 2.1 exhibited the standard form

(4.1.1) $$\Delta u = u_{xx} + u_{yy} = 0$$

of an equation with constant coefficients of the elliptic type. This is the equation of Laplace which is the focal point of this chapter. Some of our observations will apply also to Laplace's equation in three dimensions

(4.1.2) $$\Delta u = u_{xx} + u_{yy} + u_{zz} = 0$$

but we shall postpone to Chapter 5 and beyond any systematic consideration of three-dimensional problems.

Laplace's equation is the most famous and most universal of all partial differential equations. No other single equation has so many deep and diverse mathematical relationships and physical applications. A few leading cases are given in the following:

1. Theory of functions $f(z)$ of a complex variable $z = x + iy$, and the associated conformal mapping theory (Ahlfors, Ref. 1).
2. Theory of gravitational or Newtonian potentials (Kellogg, Ref. 37).
3. Electrostatic potentials.
4. Potentials of steady current flows (magneto statics).
5. Stationary heat flow problems.
6. Potentials of incompressible inviscid fluid flow.
7. Probability density in random-walk problems.
8. Harmonic and biharmonic potentials in two-dimensional elasticity.
9. Water-wave potentials for unsteady motion.

In addition, the Laplacian *operator* Δ appears in many other partial differential equations. We shall see many examples in Chapter 5, but note here that both the wave equation of Chapter 2 and the diffusion equation of Chapter 3 contain the one-dimensional Laplace operator, which is the second derivative $\partial^2/\partial x^2$.

The Laplace equation also has many significant generalizations which lead into theories of several complex variables, harmonic vectors, elliptic differential operators, and abstract potential theory. Rather than pursue such matters, we shall consider here those special properties of the Laplacian Δ which single it out from all other differential operators.

In physical problems we imagine a Euclidean background space within which rigid-body motions are permitted. At any point of a medium, the physical properties are often not dependent on direction or orientation of the medium itself. Such a medium is *isotropic*, or invariant under rotations. Likewise, a medium may be *homogeneous*, or invariant under translations of coordinates. We would therefore expect the same rotation and translation invariance in the differential equation describing properties of the medium. The Laplacian operator has constant coefficients and thus is clearly invariant under translations. Also, it is invariant under rotations and is essentially the only operator with this property.

A function $u = u(x, y)$ which satisfies the Laplace equation (4.1.1) is called *harmonic*. Harmonic functions have the significant property of being *analytic*. Harmonic functions in one dimension of space satisfy $d^2u/dx^2 = 0$: they are linear functions. Observe also that a linear function is uniquely determined by its values at the two endpoints of an interval. In one dimension, the value of the second derivative d^2u/dx^2 is a measure of the convexity of the graph of the function $u(x)$. Likewise, in two dimensions, the value of the Laplacian Δu is a measure of the two-dimensional convexity of a function $u(x, y)$. We may therefore say that harmonic functions have zero convexity: they are as flat as possible.

Because the Laplace equation is linear and homogeneous, a principle of superposition is valid for harmonic functions. If functions $u_k(x, y)$ are harmonic in a domain R, then all constant multiples and linear combinations of these functions are also harmonic. That is, the harmonic functions in a domain R

form a linear vector space. For the Laplace equation there also holds the following theorem:

THEOREM 4.1.1 (*Maximum Principle*). *The maximum and minimum values of a function $u(x, y)$ harmonic in a domain R are attained on the boundary of R.*

To prove this result, we shall assume that $u(x, y)$ is differentiable within R; this assumption will later be independently justified. Suppose we first consider the inequality (see Section 3.1)

(4.1.3) $$v_{xx} + v_{yy} > 0$$

If such a *subharmonic* function $v = v(x, y)$ had a maximum within R, we would have, at that maximum point, $v_{xx} \leq 0$ and $v_{yy} \leq 0$ simultaneously. This contradicts (4.1.3), so the maximum value of v is attained on the boundary of R.

Suppose now that u is harmonic within R, and that the maximum of u on the boundary is M. Let

(4.1.4) $$v = u + \varepsilon r^2 = u + \varepsilon(x^2 + y^2) \qquad \varepsilon > 0$$

Then $\Delta v = 4\varepsilon > 0$, so that v attains its maximum on the boundary. This maximum, by the definition of v, cannot exceed $M + \varepsilon \max_R r^2$. Assuming that R is bounded, we have, in R

$$u = v - \varepsilon r^2$$

(4.1.5) $$< v$$
$$< M + \varepsilon \max_R r^2$$

since the maximum of v is taken on the boundary. But $\varepsilon > 0$ is arbitrary, so in R

(4.1.6) $$u \leq M$$

Thus the maximum of u is the value M which is attained on the boundary. Note that $u =$ constant is harmonic, and attains the same value inside the domain as on the boundary.

The corresponding result for minimum values follows if we consider the harmonic function which is the negative of u. This completes the proof of the maximum principle for harmonic functions. We will be able to strengthen this result slightly with the mean value theorem of Section 4.3.

From the maximum principle we are led to consider the first boundary value problem for Laplace's equation. A region R is given, and we ask for a harmonic function $u = u(x, y)$, which is equal to a given function $f(s)$ of arc length s on the boundary curve C of R (Figure 4.1). This *Dirichlet problem* would arise, for example, in finding the steady temperature distribution throughout a two-dimensional or cylindrical medium with a given temperature distribution at its boundary.

The two theorems which follow next are similar to results in Section 3.1.

THEOREM 4.1.2 *A solution of the Dirichlet problem for harmonic functions is unique.*

THEOREM 4.1.3 *A solution of the Dirichlet problem for harmonic functions depends continuously on the assigned boundary values.*

Proofs are suggested as an exercise. To show that the Dirichlet problem is correctly set, we have now to construct solutions.

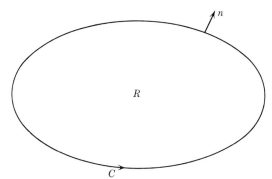

Figure 4.1 Plane region and boundary curve.

EXERCISES 4.1

1. If $\mathbf{x}' = A\mathbf{x}$, where A is the rotation matrix

$$\begin{pmatrix} \cos\theta & \sin\theta \\ -\sin\theta & \cos\theta \end{pmatrix} \quad \text{and} \quad \mathbf{x} = \begin{pmatrix} x \\ y \end{pmatrix} \quad \mathbf{x}' = \begin{pmatrix} x' \\ y' \end{pmatrix}$$

verify that $\Delta' u'(x', y') = \Delta u(x, y)$.

2. Prove Theorems 4.1.2 and 4.1.3. *Hint:* In each case, compare two suitable harmonic functions by estimating the maximum value of their difference.

3. Show that a harmonic function positive on the boundary curve C is also positive within R.

4. Show that three harmonic functions u_1, u_2, u_3 which satisfy $u_1 < u_2 < u_3$ on C also satisfy these inequalities throughout R.

5. If a harmonic function u and a subharmonic function v are equal on C, show that $v \leq u$ throughout R. (*Note:* $\Delta v \geq 0$).

6. Show that a general homogeneous polynomial $P(x, y)$ of degree n in x and y has $n + 1$ independent coefficients. Show that there are exactly two harmonic polynomials of degree n, $n = 1, 2, \ldots$. Enumerate these for $n = 1, 2, 3, 4$, and plot their nodal lines (zeros) in the x, y-plane.

4.2 BOUNDARY VALUE PROBLEMS FOR HARMONIC FUNCTIONS

All the classical boundary value problems we intend to study occur in steady-state heat-conduction problems. We therefore begin by again deriving the heat flow equation, this time for a space region of two (or more) dimensions. The Fourier heat transfer law is

(4.2.1) $$\mathbf{J} = -K_1 \nabla u$$

where $\mathbf{J}$ is the heat current vector, u is the temperature, and ∇ is the gradient operator. Recall that ∇u is the vector whose Cartesian components are the partial derivatives of u with respect to Cartesian coordinates. The diffusion coefficient K_1 is a constant of the medium. The total heat content of a region R is now

(4.2.2) $$H = c\rho \int_R u \, dA \qquad dA = dx\, dy$$

where c is the specific heat and ρ the density, both assumed constant. We obtain our equation by equating the rate of heat increase with the rate of flow across the boundary curve C:

(4.2.3) $$\frac{dH}{dt} = c\rho \int_R u_t \, dA = K_1 \int_C \nabla u \cdot \mathbf{n} \, ds$$

The normal component $\mathbf{n} \cdot \nabla u$ of ∇u integrated over C gives the inward heat flow, $\mathbf{n}$ being the outward unit normal vector. We now use Gauss' integral theorem for two dimensions (see Section 5.3):

(4.2.4) $$\int_R \nabla \cdot \mathbf{V} \, dA = \int_C \mathbf{V} \cdot \mathbf{n} \, ds$$

which expresses the boundary integral of the normal component of a vector field $\mathbf{V}$ as the area integral of the divergence of $\mathbf{V}$. This yields, in the present circumstances,

(4.2.5) $$\int_C \nabla u \cdot \mathbf{n} \, ds = \int_R \nabla \cdot \nabla u \, dA = \int_R \Delta u \, dA$$

since

(4.2.6) $$\Delta u = \nabla \cdot \nabla u = \text{div}(\text{grad } u) = (u_x)_x + (u_y)_y$$

Assembling (4.2.3) and (4.2.5), we find

(4.2.7) $$\int_R (c\rho u_t - K_1 \Delta u) \, dA = 0$$

This integral vanishes for every choice of the region R. If the integrand, which we shall assume is continuous, is not zero throughout R, we can choose a portion R_1 of R so as to contradict (4.2.7). Therefore,

(4.2.8) $$u_t = K \Delta u \qquad K = \frac{K_1}{c\rho}$$

138   LAPLACE'S EQUATION AND COMPLEX VARIABLES

is the general equation for heat conduction in two or more dimensions. For a steady or stationary heat flow, the time derivative vanishes and then

(4.2.9) $$\Delta u = 0$$

This shows that an equilibrium temperature function u is harmonic.

For such a temperature distribution, we would expect the amount of heat flow to be as small as possible consistent with the surface temperatures. It turns out that this will be true if we measure the vector $\mathbf{J}$ of (4.2.1) by its integral norm

(4.2.10) $$\int_R |\mathbf{J}|^2 \, dA = K_1^2 \int_R (\nabla u)^2 \, dA$$
$$= K_1^2 \int_R (u_x^2 + u_y^2) \, dA$$

THEOREM 4.2.1 (*Dirichlet's Principle*). *Of all functions u defined on R with given boundary values on C, the function with the least value of the Dirichlet integral*

(4.2.11) $$D(u) \equiv \int_R (\nabla u)^2 \, dA \equiv \int_R (u_x^2 + u_y^2) \, dx \, dy$$

is harmonic.

To discuss this famous theorem, we will assume that u is a harmonic function with the given boundary values. Let v be any other function with the same values on C. Then the difference

(4.2.12) $$w = v - u$$

will be zero on the boundary C. We observe that $v = u + w$ and construct the Dirichlet integral for v:

(4.2.13) $$D(v) = D(u + w) = \int_R (\nabla u + \nabla w) \cdot (\nabla u + \nabla w) \, dA$$
$$= \int_R (\nabla u)^2 \, dA + 2 \int_R \nabla u \cdot \nabla w \, dA + \int_R (\nabla w)^2 \, dA$$
$$= D(u) + 2 \int_R \nabla u \cdot \nabla w \, dA + D(w)$$

We examine the second of these three terms and "integrate by parts," that is, we apply Green's first integral theorem

(4.2.14) $$\int_R \nabla u \cdot \nabla w \, dA + \int_R w \, \Delta u \, dA = \int_R \nabla \cdot (w \, \nabla u) \, dA$$
$$= \int_C w \, \nabla u \cdot \mathbf{n} \, ds = \int_C w \, \frac{\partial u}{\partial n} \, ds$$

In these steps we have shown how Green's theorem is deduced from the Gauss divergence theorem by setting $\mathbf{V} = w\,\nabla u$. For the case at hand, w vanishes on C so the integral on the right in (4.2.14) will be zero. Also u is harmonic, so $\Delta u = 0$ and the second term on the left side of (4.2.14) disappears. Therefore the integral of the cross terms in (4.2.13) itself vanishes. This leaves

$$D(v) = D(u) + D(w) > D(u)$$

since $D(w) > 0$ unless w is a constant. As w vanishes on C, this could only be the constant zero, which we excluded at the outset by requiring $v \not\equiv u$. This concludes the formal proof of Dirichlet's principle as stated above.

The reader may notice that we have not proved that there is any harmonic function with the given boundary values. We have only shown that, if there is such a function, it has the minimum property. It is beyond our scope in this book to enter into the existence theorems for harmonic functions, so we refer the reader to Ahlfors, Ref. 1; Courant-Hilbert, Ref. 13, vol. 2. For the problems we are confronted with in applied mathematics, we can almost always safely assume that a solution of Dirichlet's problem exists.

Let us now consider the *Neumann problem*, or second boundary value problem of potential theory. This arises when given amounts of heat are supplied to various parts of the boundary C. The heat flow across C is the normal component $J_n = \mathbf{J} \cdot \mathbf{n}$ of the current vector $\mathbf{J}$, so by (4.2.1) the boundary condition for u is

(4.2.15) $$K_1 \frac{\partial u}{\partial n} = K_1 \nabla u \cdot \mathbf{n} = -\mathbf{J} \cdot \mathbf{n} = -J_n.$$

We wish to find the harmonic function u representing the stationary temperature distribution throughout R. The heat supplied to the boundary will be conducted throughout the medium in such a way as to maintain the stationary (and in general nonconstant) temperature profile. However if the situation is stationary, the total heat content of R must be stationary, i.e., constant in time. Therefore the net supply of heat at the boundary must be *zero*. We find the mathematical counterpart of this remark when we set $w = 1$ in (4.2.14) and obtain

(4.2.16) $$\int_R \Delta u \, dA = \int_C \frac{\partial u}{\partial n} \, ds$$

For a harmonic function, $\Delta u = 0$, so the boundary integral of the normal derivative also vanishes. If u is not harmonic we can interpret values of Δu as source density of heat within R, and (4.2.16) then shows that the total heat source in R is equal to the total heat flow across C.

The mathematical formulation of the Neumann problem is then: Find a function $u = u(x, y)$ which satisfies

(4.2.17) $$\Delta u = 0 \quad \text{in} \quad R$$

i.e., u is harmonic, and

(4.2.18) $$\frac{\partial u}{\partial n} = f(s) \quad \text{on} \quad C$$

where

(4.2.19) $$\int_C f(s)\, ds = 0$$

A third kind of boundary condition is needed if we allow for radiation of heat from the boundary into the surrounding medium. Let us assume Newton's linear law of cooling: the heat lost is hu, where u is the temperature *difference* from the surrounding medium, and $h > 0$ is a constant of proportionality depending on the medium. The heat g supplied at a point of the boundary is partly conducted into the medium and partly lost by radiation to the outside. Equating these amounts, we find the third boundary condition

(4.2.20) $$\frac{\partial u}{\partial n} + hu = g$$

If heat is lost by radiation from any part of the boundary, the net influx condition (4.2.19) becomes unnecessary. The temperature will so adjust itself that the heat lost by radiation is equal to the total heat input and a stationary situation becomes possible.

The third boundary problem therefore is: Find the harmonic function u which satisfies

(4.2.21) $$\frac{\partial u}{\partial n} + h(s)u = g(s) \qquad h(s) \geq 0,\ h(s) \not\equiv 0$$

on S. Here $h = h(s)$ can be a function of arc length s on the contour.

A solution of this problem is *unique*. For if two solutions u_1 and u_2 are present, their difference $u = u_2 - u_1$ satisfies

(4.2.22) $$\frac{\partial u}{\partial n} + h(s)u = 0$$

Therefore, by Green's theorem

(4.2.23) $$\int_R (\nabla u)^2\, dA = -\int_R u\, \Delta u\, dA + \int_C u\, \frac{\partial u}{\partial n}\, ds$$
$$= 0 - \int_C u(h(s)u)\, ds = -\int_C h(s)u^2\, ds$$

The integral on the right is inherently positive, unless $u = 0$ wherever $h(s) \neq 0$ on C. But the Dirichlet integral on the left is inherently positive, unless $\nabla u \equiv 0$. Both integrals must therefore be zero, and u accordingly is a constant. For the third boundary problem, the constant value of u must be zero. Therefore $u_2 \equiv u_1$ in R and on C, which completes the uniqueness proof.

For the Neumann problem (4.2.18), we cannot take the last step of this calculation, since $h \equiv 0$. Therefore u_1 and u_2 may differ by a constant in this problem.

THEOREM 4.2.2 *Solutions of the first (Dirichlet) and third boundary value problems for harmonic functions are unique. A solution of the second (Neumann) boundary value problem is determined up to an additive constant, provided the condition (4.2.19) is satisfied.*

EXERCISES 4.2

1. Derive the equation $u_t = K \Delta u + f$, for heat flow with source density $c\rho f$ within R.
2. If $\Delta u = -f$ in R and $u_n = g$ on S, show that

$$\int_R f\, dA + \int_C g\, ds = 0$$

and interpret as a heat flow condition.

3. If $\Delta u = -f$ in R and $u_n + hu = g$ on C, show that

$$\int_C hu\, ds = \int_R f\, dA + \int_C g\, ds$$

and interpret as a heat flow condition.

4. Show that $D(w) > 0$, unless w is a constant. Hence show that the solution of a Dirichlet problem is unique, using the Dirichlet integral.

5. Discuss the mixed boundary condition, where

$$u = f \text{ on } C_1 \qquad \frac{\partial u}{\partial n} + hu = g \text{ on } C_2 \qquad C = C_1 + C_2$$

6. Show how to express any of the three types of boundary condition, or any mixed condition of Exercise 4.2.5, as a condition $au + bu_n = f$.

7. Show that, with the Dirichlet scalar product,

$$(u, v) = \int_R \nabla u \cdot \nabla v\, dA$$

harmonic functions are orthogonal to functions which vanish on the boundary

8. Formulate and establish the theorems of this section in three dimensions.

4.3 CIRCULAR HARMONICS

The simplest examples of solution formulas for the harmonic boundary value problems described above arise if R is a circle (Figure 4.2). We shall first determine separated solutions for the harmonic equation in polar coordinates r, θ, where

(4.3.1) $\qquad x = r \cos \theta \qquad y = r \sin \theta$

142   LAPLACE'S EQUATION AND COMPLEX VARIABLES

Relative to the coordinates r, θ, Laplace's equation (4.1.1) becomes

(4.3.2) $$u_{rr} + \frac{1}{r} u_r + \frac{1}{r^2} u_{\theta\theta} = 0$$

If we assume
$$u = R(r)\Theta(\theta)$$
substitute into (4.3.2), and divide by $R\Theta$, we find

(4.3.3) $$\frac{1}{R}(r^2 R'' + rR') = -\frac{\Theta''}{\Theta}$$

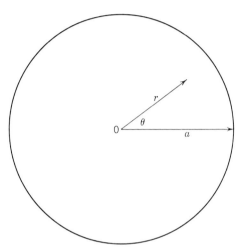

Figure 4.2   A circular region.

In this relation the first group of terms, depending on r alone, must equal a function of θ. Therefore both are constant, say, with common value K. Then we have

(4.3.4) $$r^2 R'' + rR' - KR = 0$$

(4.3.5) $$\Theta'' + K\Theta = 0$$

If the separation constant K is zero, we find at once the solutions

(4.3.6) $$R = A + B \log r$$
$$\Theta = C\theta + D$$

and, therefore,

(4.3.7) $$(A \log r + B)(C\theta + D)$$

is a harmonic function.

If the separation constant K is taken to be a real and positive number, say $K = \lambda^2$, then the equations determining $R(r)$ and $\Theta(\theta)$ become

(4.3.8)
$$r^2 R'' + rR' - \lambda^2 R = 0$$
$$\Theta'' + \lambda^2 \Theta = 0$$

From the second equation we have the familiar solution

(4.3.9) $$\Theta(\theta) = A \cos \lambda\theta + B \sin \lambda\theta$$

The equation for R is of a type called *homogeneous* by Euler. Its solutions are powers of r:

(4.3.10) $$R(r) = Ar^\lambda + Br^{-\lambda}$$

Combining these results, we see that

(4.3.11) $$(Cr^\lambda + Dr^{-\lambda})(A \cos \lambda\theta + B \sin \lambda\theta)$$

is a harmonic function.

Example 4.3.1 A harmonic function vanishing on the sides of a wedge $0 < \theta < \alpha$ is

$$r^{n\pi/\alpha} \sin \frac{n\pi\theta}{\alpha}$$

However, we may encounter a restriction on λ which is in the nature of an eigenvalue problem. Suppose the origin lies strictly inside the region or is encircled by it. In polar coordinates an increase of 2π in θ denotes a full revolution, and so the pairs $(r, \theta + 2n\pi)$, $n = 0, \pm 1, \pm 2$, all refer to the same point. Therefore, if we wish to have a single-valued harmonic function of points of the x, y-plane, the factor $\Theta(\theta)$ must be periodic of period 2π. From (4.3.11) we see that this condition is satisfied if and only if λ is an integer n. In (4.3.7) we must clearly have $C = 0$ to satisfy this condition.

THEOREM 4.3.1 *The most general single-valued harmonic function in a region enclosing the origin has the expansion*

(4.3.12) $$u = A_0 + B_0 \log r + \sum_{n=1}^{\infty} (C_n r^n + D_n r^{-n})(A_n \cos n\theta + B_n \sin n\theta)$$

Proof. Laplace's equation is linear homogeneous, superposition of solutions is permitted, and this expression gives the general element of the linear space spanned by the foregoing single-valued harmonic functions.

To illustrate the application of the general solution (4.3.12), we now consider Dirichlet and Neumann problems for the interior of the circle $r \leq a$. The solutions must be well behaved at $r = 0$, and this at once excludes the logarithmic term and the negative powers of r from (4.3.12). We can then set $C_n = 1$ with no loss of generality.

For the Dirichlet problem, the boundary condition will be $u(a, \theta) = f(\theta)$, $0 \leq \theta \leq 2\pi$. The trial solution is

$$(4.3.13) \qquad u(r, \theta) = \frac{1}{2} A_0 + \sum_{n=1}^{\infty} r^n (A_n \cos n\theta + B_n \sin n\theta)$$

On the boundary $r = a$ we apply the given condition

$$(4.3.14) \qquad u(a, \theta) = \frac{1}{2} A_0 + \sum_{n=1}^{\infty} a^n (A_n \cos n\theta + B_n \sin n\theta) = f(\theta)$$

In effect, we must expand $f(\theta)$ in a full-range Fourier series over the range $-\pi \leq \theta \leq \pi$. Evaluating the coefficients by the usual orthogonal function method, we find

$$(4.3.15) \qquad A_n = \frac{1}{\pi a^n} \int_{-\pi}^{\pi} f(\varphi) \cos n\varphi \, d\varphi \qquad n = 0, 1, 2, \ldots$$

$$(4.3.16) \qquad B_n = \frac{1}{\pi a^n} \int_{-\pi}^{\pi} f(\varphi) \sin n\varphi \, d\varphi \qquad n = 1, 2, \ldots$$

We insert these values into (4.3.13), and use the addition formula

$$\cos n(\theta - \varphi) = \cos n\theta \cos n\varphi + \sin n\theta \sin n\varphi$$

under the integral sign. Then we find

$$(4.3.17) \qquad u(r, \theta) = \frac{1}{\pi} \int_{-\pi}^{\pi} f(\varphi) \left[\frac{1}{2} + \sum_{n=1}^{\infty} \left(\frac{r}{a} \right)^n \cos n(\theta - \varphi) \right] d\varphi$$

By writing the cosine in terms of exponentials, we can express the sum as a pair of geometric series. The summation of these (Exercise 4.3.4) yields

$$(4.3.18) \qquad \frac{1}{2} + \sum_{n=1}^{\infty} \left(\frac{r}{a} \right)^n \cos n(\theta - \varphi) = \frac{1}{2} \frac{a^2 - r^2}{a^2 - 2ar \cos(\theta - \varphi) + r^2}$$

POISSON'S THEOREM 4.3.2 *The harmonic function $u(r, \theta)$ with $u(a, \theta) = f(\theta)$ is*

$$(4.3.19) \qquad u(r, \theta) = \frac{a^2 - r^2}{2\pi} \int_{-\pi}^{\pi} \frac{f(\varphi) \, d\varphi}{a^2 - 2ar \cos(\theta - \varphi) + r^2} \qquad r < a$$

This important formula of Poisson has many applications, and deserves careful consideration by the reader. Some of the exercises below are intended to bring out features of the integrand.

The *mean value property* of harmonic functions is deduced at once from (4.3.19) by setting $r = 0$.

COROLLARY 4.3.1 *The value of a harmonic function at the center of a circle is equal to the average of its values on the circumference.*

This is a statement in words of the formula

$$u(0, \theta) = u(0) = \frac{1}{2\pi} \int_{-\pi}^{\pi} f(\varphi)\, d\varphi \tag{4.3.20}$$

This corollary makes possible the following theorem which is a converse for harmonic functions of the maximum principle of Section 4.1.

THEOREM 4.3.3 *If a harmonic function in a region R assumes its maximum value at a point within R, then that function is a constant.*

Proof. Let the maximum value be attained at Q, and let P be any other point of R. Join PQ by a smooth curve lying in R, and construct a set of circles lying in R with centers on the curve, such that each center lies inside the preceding circle. The first center shall be Q and the last circle should contain P.

In the first circle the maximum is attained at the center: it is the average of values on the circumference which cannot be greater and thus cannot be less. The function is then constant on this circumference and therefore also throughout the first circle, its value being the maximum.

For the second, third, and following circles, we can repeat this reasoning. Consequently, the function is the same constant on all the circles and in particular at P. Since P is any point of R, the result is proved.

The *exterior* Dirichlet problem for a circle leads to a very similar formula, which it is interesting to compare with Poisson's formula. For the exterior problem we require u to be bounded as $r \to \infty$, and so discard from the general solution (4.3.12) the logarithmic term and the positive powers of r. The general solution now reads

$$u(r, \theta) = \frac{1}{2} A_0 + \sum_{n=1}^{\infty} r^{-n}(A_n \cos n\theta + B_n \sin n\theta) \qquad r > a \tag{4.3.21}$$

The boundary condition to determine the constants is

$$f(\theta) = \frac{1}{2} A_0 + \sum_{n=1}^{\infty} a^{-n}(A_n \cos n\theta + B_n \sin n\theta)$$

Therefore, we find the combined result

$$\begin{matrix}A_n \\ B_n\end{matrix} = \frac{a^n}{\pi} \int_{-\pi}^{\pi} f(\varphi) \begin{matrix}\cos \\ \sin\end{matrix} n\varphi\, d\varphi \qquad n = 0, 1, 2, \ldots \tag{4.3.22}$$

Returning to the trial solution, we obtain

$$u(r, \theta) = \frac{1}{\pi} \int_{-\pi}^{\pi} f(\varphi) \left[\frac{1}{2} + \sum_{n=1}^{\infty} \left(\frac{a}{r}\right)^n \cos n(\theta - \varphi) \right] d\varphi \tag{4.3.23}$$

Comparing with (4.3.17), we find that r and a have been interchanged. The final result is therefore

$$u(r, \theta) = \frac{r^2 - a^2}{2\pi} \int_{-\pi}^{\pi} \frac{f(\varphi)\, d\varphi}{a^2 - 2ar \cos(\theta - \varphi) + r^2} \qquad r > a \tag{4.3.24}$$

146   LAPLACE'S EQUATION AND COMPLEX VARIABLES

Now let us consider the interior Neumann problem for the circle $r \leq a$. Here the value of the normal derivative is given for $r = a$, that is,

(4.3.25) $$\frac{\partial u}{\partial n} = \frac{\partial u}{\partial r} = f(\theta)$$

From Section 4.2 we recall that

(4.3.26) $$\int_C \frac{\partial u}{\partial n}\, ds = a\int_{-\pi}^{\pi} f(\theta)\, d\theta = 0$$

This means that the absolute coefficient in the Fourier series for $f(\theta)$ must be zero.

As a possible form of solution we can again select (4.3.13). To apply the new boundary condition, we differentiate with respect to r and set $r = a$:

(4.3.27) $$\left.\frac{\partial u}{\partial r}\right|_{r=a} = \sum_{n=1}^{\infty} na^{n-1}(A_n \cos n\theta + B_n \sin n\theta) = f(\theta)$$

Notice that the constant term has disappeared, and the expansion of $f(\theta)$ in a Fourier series of the form (4.3.27) is possible only because of (4.3.26). The coefficients for $n \geq 1$ are

(4.3.28) $$\begin{matrix}A_n\\B_n\end{matrix} = \frac{1}{\pi n a^{n-1}} \int_{-\pi}^{\pi} f(\varphi)\begin{matrix}\cos\\\sin\end{matrix} n\varphi\, d\varphi \qquad n = 1, 2, \ldots$$

Substitution of these formulas in (4.3.13) yields

(4.3.29) $$u(r, \theta) = \frac{1}{2} A_0 + \frac{a}{\pi} \int_{-\pi}^{\pi} f(\varphi) \left[\sum_{n=1}^{\infty} \frac{1}{n}\left(\frac{r}{a}\right)^n \cos n(\theta - \varphi)\right] d\varphi$$

The series to be summed here is

(4.3.30) $$\sum_{n=1}^{\infty} \frac{1}{n} \rho^n \cos n\alpha = -\tfrac{1}{2} \log(1 + \rho^2 - 2\rho \cos \alpha)$$

and the solution formula becomes

(4.3.31) $$u(r, \theta) = \frac{1}{2} A_0 - \frac{a}{2\pi} \int_{-\pi}^{\pi} f(\varphi) \log [a^2 - 2ar \cos(\theta - \varphi) + r^2]\, d\varphi$$

where we have used (4.3.26) to discard a constant factor a^2 in the argument of the logarithm.

EXERCISES 4.3

1. Find the eigenvalues and eigenfunctions of the Sturm-Liouville problem with periodic boundary conditions

$$\Theta'' + k^2 \Theta = 0 \qquad \Theta(-\pi) = \Theta(\pi) \qquad \Theta'(-\pi) = \Theta'(\pi)$$

2. Show that $r^n \cos n\theta$, $r^n \sin n\theta$ are harmonic by considering the function $f(z) = z^n$ of the complex variable $z = re^{i\theta}$.

3. Find the steady temperature distribution within the wall $a < r < b$ of an infinitely long, hollow, circular cylinder. The inner wall temperature is $f(\theta)$, the outer wall temperature $g(\theta)$. (You need not attempt to sum the series.) *Hint:* Can we omit any terms from the trial solution (4.3.12)?

4. Show that
$$\sum_{n=1}^{\infty} x^n \cos n\alpha = \frac{x(\cos \alpha - x)}{1 - 2x \cos \alpha + x^2}$$

and deduce (4.3.18). *Hint:* Compare with (2.7.21).

5. With the help of the formula
$$\int_{-\pi}^{\pi} \frac{d\varphi}{A + B \cos \varphi} = 2\pi \frac{\operatorname{sgn} A}{\sqrt{A^2 - B^2}} \qquad |B| < |A|$$
show that
$$\lim_{r \to a} \frac{1}{2\pi} \frac{r^2 - a^2}{r^2 - 2ar \cos \varphi + a^2} = \delta(\varphi)$$

6. Show that a solution of the interior Neumann problem for $r \leq a$ can be expressed as
$$u(r, \theta) = \frac{1}{2} A_0 - \frac{a^2 r}{\pi} \int_{-\pi}^{\pi} \frac{F(\varphi) \sin (\theta - \varphi) \, d\varphi}{a^2 - 2ar \cos (\theta - \varphi) + r^2}$$
where
$$F(\varphi) = \int_0^{\varphi} f(\gamma) \, d\gamma$$

7. Show that a solution of the exterior Neumann problem for the circle is
$$u(r, \theta) = \frac{1}{2} A_0 + \frac{a}{2\pi} \int_{-\pi}^{\pi} f(\varphi) \log [a^2 - 2ar \cos (\theta - \varphi) + r^2] \, d\varphi$$

8. Find a series solution of the third boundary problem for the circle $r \leq a$, with $\partial u / \partial r + hu = f(\theta)$ for $r = a$. Determine the coefficients in (4.3.12) but do not attempt to sum the series.

9. Given the mean value formula (4.3.20), establish the maximum principle of Section 4.1.

10. If u is subharmonic ($\Delta u \geq 0$), show that its value at the center of a circle does not exceed its average on the circumference.

11. *Inversion in a circle.* Show that the general plane expansion (4.3.12) has the following property: if $u(r, \theta)$ has an expansion of this form and a is a constant, then $u(a^2/r, \theta)$ also has an expansion of this form.

12. Show that the substitution $\sigma = \log (a/r)$ converts the polar Laplace equation (4.3.2) into $u_{\sigma\sigma} + u_{\theta\theta} = 0$. Hence construct the harmonic function in the sector

$0 \leq r \leq a$, $0 \leq \theta \leq \alpha < 2\pi$ subject to the conditions $u(r, 0) = 0$, $u(a, \theta) = 0$, $u(r, \alpha) = f(r)$. Show that

$$u = \frac{2}{\pi} \int_0^a \frac{1}{r'} f(r') \, dr' \int_0^\infty \sin\left(s \log \frac{a}{r}\right) \sin\left(s \log \frac{a}{r'}\right) \frac{\sinh s\theta}{\sinh s\alpha} \, ds$$

13. By applying the transformation stated in the preceding exercise, obtain the harmonic function in the region $0 < a \leq r \leq b$, $0 \leq \theta \leq \pi$ with the boundary values $u(r, 0) = u(r, \pi) = u(a, \theta) = 0$, $u(b, \theta) = g(\theta)$.

14. Construct the harmonic function in the annulus $0 < a \leq r \leq b$, $0 \leq \theta \leq 2\pi$, with the conditions $u(a, \theta) = 0$, $u(b, \theta) = g(\theta)$.

15. Construct the harmonic function in the region $0 < a \leq r \leq b$, $0 \leq \theta \leq \pi$, with the conditions $u(a, \theta) = u(b, \theta) = 0$, $u_\theta(r, 0) = 0$, $u_\theta(r, \pi) = f(r)$. (Use a cosine transform and Exercise 3.2.11.)

16. Solve the harmonic Dirichlet problem for the semicircle $0 \leq r \leq a$, $0 \leq \theta \leq \pi$, with the boundary values $u(a, \theta) = g(\theta)$, $u(r, 0) = 0$, $u(r, \pi) = f(r)$. (Use a suitable sine transform and Lemma 3.2.1.)

4.4 RECTANGULAR HARMONICS

Next in simplicity to the harmonic functions in a circle are the harmonic functions in a rectangle. We choose the rectangle $0 \leq x \leq a$, $0 \leq y \leq b$ (Figure 4.3).

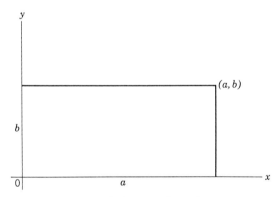

Figure 4.3 Rectangular region.

The separated solutions natural for this configuration are

(4.4.1) $$u(x, y) = X(x) Y(y)$$

When we substitute these into (4.1.1) and divide through by $X(x) Y(y)$, we have

(4.4.2) $$\frac{X''(x)}{X(x)} + \frac{Y''(y)}{Y(y)} = 0$$

The separation of variables principle shows that each of these terms is a constant.

If we denote a suitable separation constant by K, we find the ordinary differential equations

(4.4.3) $$X'' - KX = 0$$

(4.4.4) $$Y'' + KY = 0$$

When $K = 0$, the general solution is clearly

(4.4.5) $$(Ax + B)(Cy + D)$$

When $K = k^2$, a positive constant, the associated solution is

(4.4.6) $$(A \cosh kx + B \sinh kx)(C \cos ky + D \sin ky)$$

When $K = -k^2$, a negative value, the corresponding solution is

(4.4.7) $$(A \cosh ky + B \sinh ky)(C \cos kx + D \sin kx)$$

To illustrate the superposition of these solutions, consider the Dirichlet problem for the rectangle $0 \leq x \leq a$, $0 \leq y \leq b$, with the boundary values of u given in the following "elementary" form:

(i) $u = 0$ when $x = 0$ $0 \leq y \leq b$
(ii) $u = 0$ when $x = a$ $0 \leq y \leq b$
(iii) $u = 0$ when $y = 0$ $0 \leq x \leq a$
(iv) $u = g(x)$ when $y = b$ $0 \leq x \leq a$

The conditions (i) and (ii) can be met only by separated solutions of the type (4.4.7), which contain trigonometric functions of x. From condition (i) we see that $C = 0$ in any solution (4.4.7), and from (ii) we find $k = n\pi/a$, where n is an integer. The term in $\cosh ky$ fails to satisfy condition (iii), so we shall set $A = 0$. Our condition (iv) is therefore to be met by superposed solutions of the type

(4.4.8) $$u(x, y) = \sum_{n=1}^{\infty} B_n \sinh \frac{n\pi y}{a} \sin \frac{n\pi x}{a}$$

The condition is actually

(4.4.9) $$u(x, b) = g(x) = \sum_{n=1}^{\infty} B_n \sinh \frac{n\pi b}{a} \sin \frac{n\pi x}{a}$$

so that the Fourier sine series formulas give

(4.4.10) $$B_n \sinh \frac{n\pi b}{a} = \frac{2}{a} \int_0^a g(x) \sin \frac{n\pi x}{a} dx$$

The actual solution is, therefore,

(4.4.11) $$u(x, y) = \frac{2}{a} \sum_{n=1}^{\infty} \frac{\sinh (n\pi y/a)}{\sinh (n\pi b/a)} \left(\int_0^a g(x') \sin \frac{n\pi x'}{a} dx' \right) \sin \frac{n\pi x}{a}$$

The complete solution of a Dirichlet problem for the rectangle, with non-homogeneous data assigned on all four sides, will be a *sum of four expressions* like (4.4.11), each referring to a nonhomogeneous datum on one side of the rectangle. When the datum is assigned on one of the sides $x = $ constant, we should use the solutions (4.4.6) in which the roles of x and y are interchanged.

Various modifications of the boundary conditions can be made for one or more sides of the rectangles, with consequent adjustments in the solution. As an example of a "mixed" boundary problem, consider:

(i) $u = 0$ for $x = 0$ $0 \leq y \leq b$
(ii) $u_x = 0$ for $x = a$ $0 \leq y \leq b$
(iii) $u_y + hu = 0$ for $y = 0$ $0 \leq x \leq a$, $h = $ const.
(iv) $u = g(x)$ for $y = b$ $0 \leq x \leq a$

The physical interpretation of this problem is that a long solid beam of rectangular cross section is (i) kept frozen on the left side, (ii) insulated on the right side, (iii) allowed to radiate on the bottom face, and (iv) maintained at given temperature along the top.

To build up the solution, we use (4.4.7), subject to (i) and (ii), which we have encountered previously in connection with Fourier series and which lead to the eigenfunctions

$$\sin\left(n + \frac{1}{2}\right)\frac{\pi x}{a}$$

with eigenvalues $k = (n + \frac{1}{2})\pi/a$. Condition (iii), when applied to the hyperbolic factor in (4.4.7) at $y = 0$, yields

$$Bk + Ah = 0$$

and our solutions to be superposed are

$$B_n(k \cosh ky - h \sinh ky) \sin kx$$

with the above eigenvalues for k.

The trial solution can now be expressed as

(4.4.12)
$$u(x, y) = \sum_{n=1}^{\infty} B_n \left[\left(n + \frac{1}{2}\right)\frac{\pi}{a} \cosh\left(n + \frac{1}{2}\right)\frac{\pi y}{a} \right.$$
$$\left. - h \sinh\left(n + \frac{1}{2}\right)\frac{\pi y}{a} \right] \sin\left(n + \frac{1}{2}\right)\frac{\pi x}{a}$$

For this to equal $g(x)$ when $y = b$, we must have

(4.4.13)
$$g(x) = \sum_{n=1}^{\infty} B_n \left[\left(n + \frac{1}{2}\right)\frac{\pi}{a} \cosh\left(n + \frac{1}{2}\right)\frac{\pi b}{a} \right.$$
$$\left. - h \sinh\left(n + \frac{1}{2}\right)\frac{\pi b}{a} \right] \sin\left(n + \frac{1}{2}\right)\frac{\pi x}{a}$$

This yields the following Fourier coefficient relation for B_n:

(4.4.14)
$$B_n\left[\left(n+\frac{1}{2}\right)\frac{\pi}{a}\cosh\left(n+\frac{1}{2}\right)\frac{\pi b}{a} - h\sinh\left(n+\frac{1}{2}\right)\frac{\pi b}{a}\right]$$
$$= \frac{2}{a}\int_0^a g(s)\sin\left(n+\frac{1}{2}\right)\frac{\pi s}{a}\,ds$$

The formal solution of this mixed problem, with nonhomogeneous data along the top, has now been determined. Though we shall not carry this example any further here, we suggest as a test of comprehension that the reader should construct the formal solution for the same problem when the only nonhomogeneous datum is given as $u_x = f(y)$ on the side $x = a$, $0 \le y \le b$.

This method of dividing the problem into parts succeeds for two reasons. First, the separated solutions can be superposed to satisfy given data on each separate boundary segment in turn. Second, the data on every other boundary segment are zero, so that subsequent addition of the part solutions does not destroy the boundary conditions. Observe that in each of these stages of construction we have used the linear property of superposition of solutions.

If heat is supplied within the rectangle, so that the linear differential equation itself is nonhomogeneous,

(4.4.15)
$$\Delta u = -f(x, y)$$

then our solution would also contain a fifth part. This part solution will satisfy homogeneous boundary conditions on all sides, together with the nonhomogeneous differential equation. The problem of internal sources is best approached by means of Green's functions, which we shall study in Chapter 7.

Suppose now that the rectangle of length a becomes infinitely long: $a \to \infty$. We can find harmonic solutions on such a semi-infinite strip either by examining the finite solutions in the limit $a \to \infty$ or by starting anew. We consider the Dirichlet problem for a semi-infinite strip of width b. As there are two somewhat different parts to the solution, we shall find one of them in each way.

For the part problem with data on $y = b$, we require

(i)	$u = 0$	when $x = 0$	$0 \le y \le b$
(ii)	u bounded	when $x \to \infty$	$0 \le y \le b$
(iii)	$u = 0$	when $y = 0$	$0 < x < \infty$
(iv)	$u = g(x)$	when $y = b$	$0 < x < \infty$

Let us take the limit $a \to \infty$ in (4.4.11). We set $n\pi/a = s$, with $\pi/a = \Delta s$, and let $a \to \infty$, $\Delta s \to 0$. We find

(4.4.16)
$$u(x, y) = \frac{2}{\pi}\int_0^\infty \frac{\sinh sy}{\sinh sb}\sin sx\,ds \int_0^\infty g(x')\sin sx'\,dx'$$

If we invert the order of integration, we encounter the integral

$$(4.4.17) \qquad \frac{2}{\pi}\int_0^\infty \frac{\sinh sy}{\sinh sb}\sin sx \sin sx'\, ds \qquad 0 < y < b$$

which we shall later evaluate by a complex variable technique.

The part of the semi-infinite strip solution arising from data on the finite end will be constructed directly from separated solutions. The conditions to be satisfied are

(i) $u = f(y)$ when $x = 0$ $0 \le y \le b$
(ii) u bounded   as $x \to \infty$ $0 \le y \le b$
(iii) $u = 0$ when $y = 0$ $0 < x < \infty$
(iv) $u = 0$ when $y = b$ $0 < x < \infty$

Conditions (iii) and (iv) require us to use (4.4.6), that is, trigonometric functions of y. Indeed, by a now familiar process we find the y factor is $\sin(n\pi y/b)$, with $k = n\pi/b$. Now condition (ii) will be satisfied in (4.4.6) only if $A + B = 0$, that is, we must choose a decreasing exponential function of x. The trial solution now appears as

$$(4.4.18) \qquad u(x,y) = \sum_{n=1}^\infty B_n e^{-n\pi x/b} \sin\frac{n\pi y}{b}$$

Setting $x = 0$, we find the condition

$$(4.4.19) \qquad f(y) = u(0,y) = \sum_{n=1}^\infty B_n \sin\frac{n\pi y}{b}$$

Therefore, by the half-range sine-series theorem,

$$(4.4.20) \qquad B_n = \frac{2}{b}\int_0^b f(\eta)\sin\frac{n\pi\eta}{b}\,d\eta$$

and, finally

$$(4.4.21) \qquad u(x,y) = \frac{2}{b}\sum_{n=1}^\infty e^{-n\pi x/b}\int_0^b f(\eta)\sin\frac{n\pi\eta}{b}\,d\eta \sin\frac{n\pi y}{b}$$

The exponential factors will improve the convergence rate of this series for x large.

EXERCISES 4.4

1. Show that $\int_C f(\alpha)e^{\alpha x + i\alpha y}\,d\alpha$ is harmonic, for every contour C in the complex plane of α, and for every distribution $f(\alpha)$ defined on C.

2. Show how to find a harmonic function $u(x,y)$ on the rectangle $0 \le x \le a, 0 \le y \le b$, if the normal derivative of u on each side is given, subject to the integral condition of the Neumann problem. *Hint:* Use (4.4.5).

3. Construct, in four parts, the harmonic function $u(x, y)$ on the rectangle $0 \leq x \leq a$, $0 \leq y \leq b$, with:

(i) $u_x + hu = f_1(y)$ for $x = 0$ $0 \leq y \leq b$
(ii) $u_x = f_2(y)$ for $x = a$ $0 \leq y \leq b$
(iii) $u = f_3(x)$ for $y = 0$ $0 \leq x \leq a$
(iv) $u_y = f_4(x)$ for $y = b$ $0 \leq x \leq a$

4. An isosceles right-angled triangle is given. Show how to find the harmonic function which vanishes on the hypotenuse and has given values on the other two sides.

5. An isosceles right-angled triangle is given. Show how to find the harmonic function which has vanishing normal derivative on the hypotenuse and on one other side. The function itself takes given values on the third side.

6. By applying the finite transform stated in Exercise 3.2.13, find the harmonic function in the semi-infinite strip $0 \leq x < \infty$, $0 \leq y \leq b$, with the mixed boundary conditions $u = 0$ on $y = 0$, $u_y = 0$ on $y = b$, and $u = f(y)$ on $x = 0$. Show that

$$u = \frac{2}{b} \sum_{n=0}^{\infty} e^{-(n+\frac{1}{2})\pi x/b} \sin\left[\left(n + \frac{1}{2}\right) \frac{\pi y}{b}\right] \int_0^a f(y') \sin\left[\left(n + \frac{1}{2}\right) \frac{\pi y'}{b}\right] dy'$$

7. Solve the problem formulated in the preceding exercise by means of a Fourier sine transform on $0 \leq x < \infty$ and the result given in Exercise 3.5.12. Reconcile your result with that stated above and compare the two methods.

8. By means of Lemma 3.2.1 obtain the solution of the equation $u_{xx} + u_{yy} + 2 = 0$ in the rectangle $0 \leq x \leq a$, $0 \leq y \leq b$, with the condition of vanishing around the boundary. Show that

(i) $$u = \frac{8a^2}{\pi^3} \sum_{n=0}^{\infty} \left[1 - \frac{\sinh (2n+1)\frac{\pi}{a}(b-y) + \sinh (2n+1)\frac{\pi}{a} y}{\sinh (2n+1)\frac{\pi}{a} b} \right] \frac{\sin (2n+1)\frac{\pi}{a} x}{(2n+1)^3}$$

and also

(ii) $$u = \frac{32}{\pi^2} \sum_{n,m=0}^{\infty} \frac{\sin \frac{(2n+1)\pi x}{a} \sin \frac{(2m+1)\pi y}{b}}{(2n+1)(2m+1)\left[\frac{(2n+1)^2 \pi^2}{a^2} + \frac{(2m+1)^2 \pi^2}{b^2}\right]}$$

4.5 HALF-PLANE PROBLEMS

For infinite regions it is often necessary to apply a suitable integral transform. Let us consider the various possible methods of solution of the harmonic Dirichlet problem for the half-plane $y > 0$, when u is given along the x-axis (Figure 4.4). Here we are at liberty to try either a complex Fourier transform with respect to the variable x or a Fourier sine transform in the variable y. We shall describe both of these methods.

To the harmonic equation $u_{xx} + u_{yy} = 0$ we apply the exponential form of the Fourier integral, namely,

(4.5.1) $$U(s, y) = \frac{1}{\sqrt{2\pi}} \int_{-\infty}^{\infty} e^{-isx} u(x, y) \, dx$$

154   LAPLACE'S EQUATION AND COMPLEX VARIABLES

We shall *assume* that u and u_x vanish for $x \to \pm\infty$. Then the equation for $U(s, y)$ is

(4.5.2) $$\frac{\partial^2 U}{\partial y^2} = s^2 U$$

which has the solution

(4.5.3) $$U(s, y) = A(s)e^{sy} + B(s)e^{-sy}$$

We are also going to assume that u is bounded as $y \to +\infty$, and we see that the term in (4.5.3) that is large for large y depends on the sign of s. For s positive, we set $A(s) = 0$ and choose

(4.5.4) $$U(s, y) = B(s)e^{-sy} \qquad s > 0$$

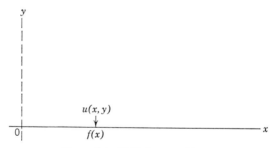

Figure 4.4   Half-plane problem.

When we put $y = 0$, we find $B(s) = U(s, 0)$, and so we can rewrite (4.5.4) as

(4.5.5) $$U(s, y) = U(s, 0)e^{-sy} \qquad s > 0$$

Likewise, for $s < 0$, we should set $B(s) = 0$ in (4.5.3) and write

(4.5.6) $$U(s, y) = A(s)e^{sy} \qquad s < 0$$

Again, we put $y = 0$ and identify $A(s)$ as $U(s, 0)$. Hence,

(4.5.7) $$U(s, y) = U(s, 0)e^{sy} \qquad s < 0$$

The two formulas (4.5.5) and (4.5.7) can be combined in one:

(4.5.8) $$U(s, y) = U(s, 0)e^{-|s|y}$$

Since $U(s, 0)$ is the Fourier transform of $u(x, 0)$ and the given values for the Dirichlet problem are $u(x, 0) = f(x)$, say, we have

(4.5.9) $$U(s, y) = \frac{1}{\sqrt{2\pi}} \int_{-\infty}^{\infty} f(x')e^{-isx'}\, dx' e^{-|s|y}$$

The desired harmonic function is the inverse transform of (4.5.9):

(4.5.10)
$$u(x, y) = \frac{1}{\sqrt{2\pi}} \int_{-\infty}^{\infty} U(s, y)e^{isx}\, ds$$
$$= \frac{1}{2\pi} \int_{-\infty}^{\infty} e^{isx}\, ds \int_{-\infty}^{\infty} f(x')e^{-isx'}\, dx'\, e^{-|s|y}$$
$$= \frac{1}{2\pi} \int_{-\infty}^{\infty} f(x')\, dx' \int_{-\infty}^{\infty} e^{is(x-x')-|s|y}\, ds$$

The inner integral can be written as

(4.5.11)
$$\int_0^{\infty} e^{is(x-x')-sy}\, ds + \int_{-\infty}^0 e^{is(x-x')+sy}\, ds$$
$$= \left[\frac{e^{is(x-x')-sy}}{i(x-x')-y}\right]_0^{\infty} + \left[\frac{e^{is(x-x')+sy}}{i(x-x')+y}\right]_{-\infty}^0$$
$$= \frac{-1}{i(x-x')-y} + \frac{1}{i(x-x')+y} = \frac{2y}{(x-x')^2 + y^2}$$

Finally, therefore, we have

(4.5.12) $$u(x, y) = \frac{y}{\pi} \int_{-\infty}^{\infty} \frac{f(x')\, dx'}{(x-x')^2 + y^2} \qquad y > 0$$

Let us show how the same result is found by use of the Fourier sine transform in y. Set

(4.5.13) $$U_s(x, s) = \sqrt{\frac{2}{\pi}} \int_0^{\infty} u(x, y) \sin sy\, dy$$

Laplace's equation, together with the boundary condition, converts into

(4.5.14) $$\frac{\partial^2 U_s}{\partial x^2} - s^2 U_s = -s\sqrt{\frac{2}{\pi}} f(x)$$

We need the solution of this equation which satisfies the "boundary" conditions $U_s \to 0$ as $x \to \pm\infty$. These are two-point conditions, and the appropriate Green's function (see Section 1.8) is

(4.5.15) $$G(x, x') = \frac{1}{2s} e^{-s|x-x'|}$$

Therefore,

(4.5.16)
$$U_s(x, s) = \frac{\sqrt{2}}{\sqrt{\pi}} s \int_{-\infty}^{\infty} f(x') G(x, x')\, dx'$$
$$= \frac{1}{\sqrt{2\pi}} \int_{-\infty}^{\infty} f(x') e^{-s|x-x'|}\, dx'$$

156   LAPLACE'S EQUATION AND COMPLEX VARIABLES

We substitute this expression into the inverse sine transform for the solution

(4.5.17)
$$u(x, y) = \sqrt{\frac{2}{\pi}} \int_0^\infty U_s(x, s) \sin sy \, ds$$
$$= \frac{1}{\pi} \int_{-\infty}^\infty f(x') \, dx' \int_0^\infty e^{-s|x-x'|} \sin sy \, ds$$

But for $\rho > 0$,

$$\int_0^\infty e^{-\rho s} \sin \alpha s \, ds = \frac{\alpha}{\alpha^2 + \rho^2}$$

This leads back to (4.5.12).

When we let y tend to zero in this solution formula, we should recover the boundary values $f(x)$ as limit on the left. This suggests the formula

(4.5.18)
$$\delta(x - x') = \frac{1}{\pi} \lim_{y \to 0} \frac{y}{(x - x')^2 + y^2}$$

We can verify this by observing the following:

1. The limit on the right is zero for $x \neq x'$.
2. The area under the curve is unity:

$$\frac{1}{\pi} \int_{-\infty}^\infty \frac{y \, dx}{(x - x')^2 + y^2} = \frac{y}{\pi} \left[\frac{1}{y} \tan^{-1} \frac{x - x'}{y} \right]_{-\infty}^\infty = \frac{1}{\pi} (\tan^{-1} \infty - \tan^{-1} - \infty)$$
$$= \frac{1}{\pi} \left[\frac{\pi}{2} - \left(-\frac{\pi}{2} \right) \right] = 1$$

With these facts in hand, it is easy to prove that

(4.5.19)
$$\frac{1}{\pi} \lim_{y \to 0} y \int_{-\infty}^\infty \frac{f(x') \, dx'}{(x - x')^2 + y^2} = f(x)$$

if $f(x)$ is any testing function or even if $f(x)$ is a bounded and continuous function. We leave the proof as an exercise for the mathematically inclined reader: we wish mainly to emphasize here that the symbolic formula (4.5.18) *means* (4.5.19)!

The Neumann problem for a half-plane can be solved by a device which is sometimes useful for plane harmonic functions, and we shall deduce the solution of this problem from the solution of the Dirichlet problem.

Let $v(x, y)$ be the required harmonic function. Thus $v_{xx} + v_{yy} = 0$ for $y > 0$, v is bounded in this upper half-plane, and

(4.5.20)
$$v_y(x, 0) = f(x) \qquad\qquad -\infty < x < \infty$$

Now define

$$u(x, y) = v_y(x, y)$$

and observe that

(i) $$\Delta u(x, y) = \Delta v_y(x, y) = \frac{\partial}{\partial y} \Delta v(x, y) = 0$$

(ii) $$u(x, 0) = v_y(x, 0) = f(x)$$

By the preceding solution we can take $u(x, y)$ as in (4.5.12). Since

(4.5.21) $$v(x, y) = \int^y u(x, \eta) \, d\eta$$

we have

(4.5.22) $$\begin{aligned} v(x, y) &= \int^y \frac{\eta}{\pi} \int_{-\infty}^{\infty} \frac{f(x') \, dx'}{(x - x')^2 + \eta^2} \, d\eta \\ &= \frac{1}{\pi} \int_{-\infty}^{\infty} f(x') \, dx' \int^y \frac{\eta \, d\eta}{(x - x')^2 + \eta^2} \\ &= \frac{1}{2\pi} \int_{-\infty}^{\infty} f(x') \log\left[(x - x')^2 + y^2\right] dx' \end{aligned}$$

An arbitrary constant can be added to this solution.

We conclude this section with a somewhat more elaborate example, illustrating the solution of boundary value problems in which the function is given on one part of the boundary and the normal derivative on the remaining part. Such problems are called "mixed." In the quadrant $x > 0$, $y > 0$, let $u(x, y)$ be harmonic, with

(4.5.23) $\quad u(x, 0) = f(x) \quad$ for $\quad y = 0 \quad\quad 0 \leq x < \infty$

(4.5.24) $\quad \dfrac{\partial u}{\partial x}(0, y) = g(y) \quad$ for $\quad x = 0 \quad\quad 0 \leq y < \infty$

It is appropriate to use a Fourier cosine transform, and we define

(4.5.25) $$U_c(s, y) = \sqrt{\frac{2}{\pi}} \int_0^{\infty} u(x, y) \cos sx \, dx$$

The ordinary differential equation for this transform is

(4.5.26) $$\frac{\partial^2 U_c}{\partial y^2} - s^2 U_c = \sqrt{\frac{2}{\pi}} g(y)$$

as is shown on integration by parts of the u_{xx} term in Laplace's equation according to Lemma 3.5.1. We have to solve this equation with the boundary conditions

(4.5.27) $$U_c(s, 0) = \sqrt{\frac{2}{\pi}} \int_0^{\infty} f(x') \cos sx' \, dx'$$

(4.5.28) $\quad\quad\quad U_c$ bounded $\quad$ for $\quad y \to +\infty$

158   LAPLACE'S EQUATION AND COMPLEX VARIABLES

The solution of the homogeneous form of (4.5.26) vanishing at $y = 0$ is $\sinh sy$, and the solution which satisfies (4.5.28) is e^{-sy}. From Section 1.8 we see that the Green's function for these boundary conditions must be

(4.5.29) $$G(y, y') = \begin{cases} \dfrac{1}{s} e^{-sy'} \sinh sy & y' \geq y \\ \dfrac{1}{s} e^{-sy} \sinh sy' & y \geq y' \end{cases}$$

For $s > 0$, the solution of (4.5.26) we require is

(4.5.30) $$U_c(s, y) = U_c(s, 0) e^{-sy} - \sqrt{\dfrac{2}{\pi}} \int_0^\infty g(y') G(y, y') \, dy'$$

By the inverse transform

(4.5.31) $$u(x, y) = \sqrt{\dfrac{2}{\pi}} \int_0^\infty U_c(s, y) \cos sx \, ds$$

We insert (4.5.30) together with (4.5.27) into (4.5.31). The combined result is, after rearrangement,

(4.5.32) $$\begin{aligned} u(x, y) = {} & \dfrac{2}{\pi} \int_0^\infty f(x') \, dx' \int_0^\infty e^{-sy} \cos sx \cos sx' \, ds \\ & - \dfrac{2}{\pi} \int_0^y g(y') \, dy' \int_0^\infty \dfrac{1}{s} e^{-sy} \cos sx \sinh sy' \, ds \\ & - \dfrac{2}{\pi} \int_y^\infty g(y') \, dy' \int_0^\infty \dfrac{1}{s} e^{-sy'} \cos sx \sinh sy \, ds \end{aligned}$$

We quote the integrals required, as their calculation is straightforward and also lengthy. We have

(4.5.33)
$$\int_0^\infty \dfrac{1}{s} e^{-sy} \cos sx \sinh sy' \, ds = \dfrac{1}{4} \log \left[\dfrac{x^2 + (y + y')^2}{x^2 + (y - y')^2} \right] \quad \text{if } y > y' \geq 0$$

$$\int_0^\infty e^{-sy} \cos sx \cos sx' \, ds = \dfrac{y(x^2 + x'^2 + y^2)}{[y^2 + (x - x')^2][y^2 + (x + x')^2]} \quad \text{if } y > 0$$

Therefore the solution of the quadrant problem is

(4.5.34)
$$\begin{aligned} u(x, y) = {} & \dfrac{2y}{\pi} \int_0^\infty \dfrac{(x^2 + x'^2 + y^2) f(x') \, dx'}{[y^2 + (x - x')^2][y^2 + (x + x')^2]} \\ & - \dfrac{1}{2\pi} \int_0^\infty g(y') \log \left[\dfrac{x^2 + (y + y')^2}{x^2 + (y - y')^2} \right] dy' \end{aligned}$$

EXERCISES 4.5

1. Show that
$$\delta(x) = -\frac{1}{\pi} \lim_{\varepsilon \to 0} \operatorname{Im} \frac{1}{x + i\varepsilon}$$

2. Find a formula for the harmonic function $u(x, y)$ in the infinite strip $-\infty < x < \infty$, $0 < y < b$, with $u(x, 0) = 0$, $u(x, b) = f(x)$. (You need not attempt to evaluate integrals.)

3. Find the solution of the Dirichlet problem for the half-plane $y < 0$, and compare with (4.5.12).

4. Solve the Dirichlet problem $u(x, 0) = f(x), 0 < x < \infty, u(0, y) = g(y), 0 < y < \infty$, for the quadrant $x > 0, y > 0$.

5. Solve the Neumann problem $u_y(x, 0) = f(x), 0 < x < \infty, u_x(0, y) = g(y), 0 < y < \infty$, for the same quadrant $x > 0, y > 0$.

6. Construct the solution (4.5.34) in two parts, each being the solution of a Dirichlet or Neumann problem in a suitable half-plane with appropriate even or odd data.

7. Show by means of an exponential Fourier transform that the harmonic function in the infinite strip $-\infty < x < \infty, 0 \leq y \leq b$, with $u(x, 0) = f(x), u_y(x, b) = 0$, is given by
$$u(x, y) = \frac{1}{b} \sin \frac{\pi y}{2b} \int_{-\infty}^{+\infty} \frac{f(x') \cosh \frac{\pi(x - x')}{2b}}{\cosh \frac{\pi(x - x')}{b} - \cos \frac{\pi y}{b}} dx'$$

By letting $b \to \infty$ deduce the half-plane Dirichlet integral (4.5.12)

8. Solve the problem of the preceding exercise by applying a finite cosine transform for $0 \leq y \leq b$ and using the result of Exercise 3.2.13. Reconcile the results.

9. Show that, if $0 < \rho < 1$,
$$\sum_{n=0}^{\infty} \rho^{n+\frac{1}{2}} \sin\left(n + \frac{1}{2}\right)\theta = \frac{\cosh(\frac{1}{2} \log \rho) \sin \frac{1}{2}\theta}{\cosh(\log \rho) - \cos \theta}$$

10. In polar coordinates the function $u(r, \theta)$ satisfies (4.3.2) in the half-plane $0 \leq r < \infty, 0 \leq \theta \leq \pi$, together with the *mixed* boundary conditions $u(r, 0) = f(r)$ and $u_\theta(r, \pi) = 0$. By applying a suitable finite sine transform to remove the θ-dependence and using Exercise 3.2.13, show that
$$u(r, \theta) = \frac{1}{\pi} \sin \frac{1}{2} \theta \int_0^\infty \frac{\cosh\left[\frac{1}{2} \log(r/r')\right] f(r') \, dr'}{r' \{\cosh[\log(r/r')] - \cos \theta\}}$$

4.6 COMPLEX INTEGRALS

Hitherto we have taken the parameter s of the Fourier integral to be real. The inversion integral is then extended over the real axis $-\infty < s < \infty$. We may, however, regard s as a complex variable, $s = \xi + i\eta$, and interpret the inversion integral as extended along the real axis of the complex plane of s.

160   LAPLACE'S EQUATION AND COMPLEX VARIABLES

Subsequently we shall use other integral transforms, the inversion of which also utilizes complex or contour integrals. We insert here a brief discussion of contour integrals and their evaluation (Copson, Ref. 11).

Let x and y be real variables, and define the complex variable $z = x + iy$, where $i^2 = -1$. We shall assume that the reader is familiar with the elementary properties of the complex number field. Consider functions $w(x, y)$ with complex values, which can be expressed as functions of the single combination $z = x + iy$. The condition for w to be of this special form is the existence of a relation between w and z which is independent of x and y. That is, the Jacobian

$$(4.6.1) \qquad \frac{\partial(w, z)}{\partial(x, y)} = w_x z_y - w_y z_x = 0$$

Let us distinguish the real and imaginary parts u, v of w by writing

$$(4.6.2) \qquad w = u + iv$$

and let us assume hereafter that u and v are twice differentiable functions of x and y. Then, since $z_x = 1$, $z_y = i$ in (4.6.1), we find that the two real-valued functions u, v satisfy a pair of equations.

THEOREM 4.6.1   *The real and imaginary parts u and v of a function $w = u + iv$ of a complex variable $z = x + iy$ satisfy the Cauchy-Riemann equations*

$$(4.6.3) \qquad u_y + v_x = 0 \qquad u_x - v_y = 0$$

From the Cauchy-Riemann equations we can deduce the following:

COROLLARY 4.6.1   *The real and imaginary parts u and v are harmonic functions of x and y.*

Proof. Differentiating the second of (4.6.3) with respect to x, the first with respect to y, we find

$$u_{xx} = v_{yx} = v_{xy} = -u_{yy}$$

Similarly, the result can be proved for v.

Example 4.6.1 The function z^n, where in polar form $z = re^{i\theta}$, is

$$z^n = r^n e^{in\theta} = r^n(\cos n\theta + i \sin n\theta)$$

The real and imaginary parts are the harmonic functions

$$r^n \cos n\theta \qquad r^n \sin n\theta$$

Example 4.6.2 The logarithm of $z = re^{i\theta}$ is $\log z = \log r + i\theta$.

The harmonic function $\log r$ is singular at the origin, while θ is a many-valued function, associating the values $\theta \pm 2n\pi$ to a point of the plane.

We write

$$(4.6.4) \qquad w = f(z) = u + iv$$

and refer to $f(z)$ as a function of a complex variable.

We now define the complex derivative of $f(z)$ with respect to the complex variable z.

THEOREM 4.6.2 *The derivative $f'(z)$ defined by the limit*

$$f'(z_0) = \lim_{z \to z_0} \frac{f(z) - f(z_0)}{z - z_0}$$

exists and is independent of the direction of approach of z to z_0.

Proof. If $z \to z_0$ along a line parallel to the real axis, then $z - z_0 = x - x_0$ is real. Then

$$\frac{f(z) - f(z_0)}{z - z_0} = \frac{u - u_0}{x - x_0} + i\frac{v - v_0}{x - x_0} \to u_x + iv_x$$

where the partial derivatives are evaluated at $z_0 = x_0 + iy_0$. Similarly, if $z \to z_0$ along a path parallel to the y-axis, then $z - z_0 = i(y - y_0)$, and

$$\frac{f(z) - f(z_0)}{z - z_0} = \frac{u - u_0}{i(y - y_0)} + \frac{i(v - v_0)}{i(y - y_0)} \to v_y - iu_y$$

By the Cauchy-Riemann equations, the two limits are equal. Likewise, if $z \to z_0$ along any path, we find a combination of the two equal limits, and it follows that the limit is independent of the direction of approach of z to its limit z_0.

If $f(z)$ has a derivative $f'(z)$ at every point of a domain D, then $f(z)$ is said to be regular analytic in D. Points at which a function $f(z)$ of a complex variable z ceases to be regular are termed *singularities*.

Next we introduce complex integrals in the complex plane, defining them by means of ordinary line integrals in the x, y-plane. Let

$$z_0 = x_0 + iy_0 \quad \text{and} \quad z_1 = x_1 + iy_1$$

be two points and L a curve joining them. Let

(4.6.5)
$$\int_L f(z)\, dz = \int_L (u + iv)(dx + i\, dy)$$
$$= \int_L [u\, dx - v\, dy + i(v\, dx + u\, dy)]$$

We now show that this complex integral is independent of the path L. If L_1 and L_2 are two paths joining z_0 to z_1 then $C = L_2 - L_1$ is a closed contour (Figure 4.5). By the negative sign, we indicate that L_1 is traversed from z_1 to z_0. We recall Green's theorem for a domain D with boundary a closed contour C:

(4.6.6)
$$\int_C (P\, dx + Q\, dy) = \iint_D (Q_x - P_y)\, dx\, dy$$

Here we suppose that P, Q, Q_x, P_y, are continuous in D and on C.

162   LAPLACE'S EQUATION AND COMPLEX VARIABLES

CAUCHY'S THEOREM 4.6.3   *Let C be a closed contour and $f(z)$ a function of a complex variable which is regular analytic within and on C. Then*

(4.6.7) $$\int_C f(z)\, dz = 0$$

Proof. From the definition of the contour integrals and Green's theorem, we have

(4.6.8)
$$\int_C f(z)\, dz = \int_C [u\, dx - v\, dy + i(v\, dx + u\, dy)]$$
$$= \iint_D [-(v_x + u_y) + i(u_x - v_y)]\, dx\, dy$$

where D is the domain interior to C. But the Cauchy-Riemann equations (4.6.3) are satisfied throughout D, since $f(z)$ is regular analytic there, and accordingly, the integrals in (4.6.8) are all zero.

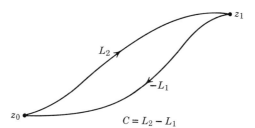

Figure 4.5   Equivalent paths and closed contour.

From Cauchy's theorem we deduce that

(4.6.9) $$\int_{L_2} f(z)\, dz - \int_{L_1} f(z)\, dz = \int_{L_2 - L_1} f(z)\, dz = 0$$

This demonstrates that the contour integral (4.6.5) is independent of the path L and depends only on the points z_0 and z_1. To indicate this, we may use an integral with upper and lower limits z_1, z_0. We see that the contour L can be deformed, with no change in the value of the integral, provided that the endpoints remain fixed. However, the path L must remain within a domain D, wherein $f(z)$ is regular analytic.

Next we define the complex indefinite integral of $f(z)$. From the Cauchy-Riemann equation $u_x = v_y$ it follows that there exists a function $\psi(x, y)$ such that

(4.6.10) $$\psi_x = v \qquad \psi_y = u$$

Likewise from the equation $u_y = -v_x$ follows the existence of a function $\varphi(x, y)$ with

(4.6.11) $$\varphi_x = u \qquad \varphi_y = -v$$

The complex function
$$g = \varphi + i\psi$$
now clearly satisfies the Cauchy-Riemann equations
$$\varphi_x = \psi_y \qquad \varphi_y = -\psi_x$$
and so is a function of the complex variable z. We calculate its derivative, as in Theorem 4.6.2.

(4.6.12) $\qquad g'(z) = \varphi_x + i\psi_x = \psi_y - i\varphi_y = u + iv = f(z)$

By analogy with real variable calculus, we say that $g(z)$ is an indefinite integral of $f(z)$.

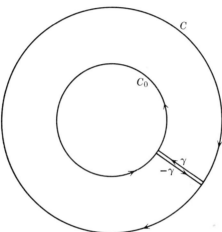

Figure 4.6 Annulus with cut.

THEOREM 4.6.4 *The contour integral* (4.6.5) *satisfies the fundamental formula of the integral calculus*

(4.6.13) $\qquad \displaystyle\int_L f(z)\,dz = \int_{z_0}^{z_1} f(z)\,dz = g(z_1) - g(z_0)$

The reader can verify this formula by separation into real and imaginary parts.

The Cauchy theorem is often used to justify the deformation of a closed contour C. Suppose, for instance, that $f(z)$ has irregularities within C but not in the annulus between C and C_0 (Figure 4.6). We then insert a cut γ connecting C and C_0, and apply the Cauchy theorem to the domain D with boundary $C + \gamma - C_0 - \gamma$, in which the cut γ is traversed twice in opposite senses. Since $f(z)$ is regular analytic in D,

$$0 = \int_{C+\gamma-C_0-\gamma} f(z)\,dz = \int_C f(z)\,dz + \int_\gamma f(z)\,dz - \int_{C_0} f(z)\,dz - \int_\gamma f(z)\,dz$$
$$= \int_C f(z)\,dz - \int_{C_0} f(z)\,dz$$

The integrals over C and C_0 are equal, so we speak of *deforming* C into C_0. We adopt the convention that $\int_C$ denotes integration over C once in a counterclockwise or "positive" sense, with the interior domain D lying on the left.

In textbooks on functions of a complex variable, it is shown that every function $f(z)$ which is regular analytic near the origin has a power series expansion, or Taylor series

(4.6.14) $$f(z) = \sum_{n=0}^{\infty} a_n z^n$$

convergent in some circle $|z| \leq R$ with the center as the origin. Likewise, as we saw in Section 4.3, every harmonic function regular at the origin has a series expansion

(4.6.15) $$u(r, \theta) = \sum_{n=0}^{\infty} r^n (A_n \cos n\theta + B_n \sin n\theta)$$

If $f(z)$ is regular analytic and single-valued near the origin, but has a singularity at $z = 0$, then $f(z)$ has a Laurent series

(4.6.16) $$f(z) = \sum_{n=-\infty}^{\infty} a_n z^n$$

If the coefficients in this sum vanish for $n < -p$, where p is a positive integer, and $a_{-p} \neq 0$, then $f(z)$ is said to have a pole of order p at the origin.

Similarly, a function $u(r, \theta)$ harmonic near the origin, but with a singularity there, has a series expansion

(4.6.17) $$u(r, \theta) = \sum_{n=-\infty}^{\infty} r^n (A_n \cos n\theta + B_n \sin n\theta)$$

Comparing this expression with the series (4.3.12) for a general plane harmonic function, we see that they are equivalent except for the logarithmic term. The complex logarithm is not a single-valued function, and so cannot be present in (4.6.16).

Suppose $f(z)$ in (4.6.16) is regular analytic in a domain D except at the origin where $f(z)$ has a pole of order p. Suppose also that D contains a closed contour C which encircles the origin once in the positive sense. Then we have the following theorem:

RESIDUE THEOREM 4.6.5 $\quad \int_C f(z)\, dz = 2\pi i a_{-1}$

To prove this "residue formula," we deform the contour C to a circle $|z| = R$, where the series (4.6.16) converges. Then on C, $z = Re^{i\theta}$ and

(4.6.18) $$\begin{aligned} \int_C f(z)\, dz &= \int_0^{2\pi} \sum_{n=-p}^{\infty} a_n R^n e^{in\theta} i R e^{i\theta}\, d\theta \\ &= i \sum_{n=-p}^{\infty} a_n R^{n+1} \int_0^{2\pi} e^{i(n+1)\theta}\, d\theta \\ &= 2\pi i a_{-1} \end{aligned}$$

since all integrals are zero except for $n = -1$.

COMPLEX INTEGRALS   165

The coefficient a_{-1} is known as the *residue* of $f(z)$ at the singularity. It is the coefficient of z^{-1} in the series expansion of the function about the origin.

If $f(z)$ has a number of singularities at points z_k and C is a closed contour which encircles them in the positive sense, then

(4.6.19) $$\int_C f(z)\, dz = 2\pi i \sum_k a_{k,-1}$$

Here $a_{k,-1}$ is the residue of $f(z)$ at $z = z_k$, defined as the coefficient of $(z - z_k)^{-1}$ in the power series expansion centred at $z = z_k$.

EXERCISES 4.6

1. If $f = u + iv$, $z = x + iy$, $\bar{z} = x - iy$, show that the Cauchy-Riemann equations are equivalent to $\partial f / \partial \bar{z} = 0$. Show that sums, products, quotients, and functions of analytic functions are analytic.

2. *Conjugate harmonic function.* If u is harmonic, show that the line integral

$$\int (u_x\, dy - u_y\, dx)$$

is path-independent and defines a function $v(x, y)$ of its upper limit. Show that $u + iv$ is a function $f(z)$.

3. Find the harmonic functions conjugate in the sense of Exercise 4.6.2 to $x^2 - y^2$, $x^3 - 3xy^2$, $r^n \cos n\theta$, $e^{\alpha x} \sin \alpha y$.

4. Show directly that φ, ψ of (4.6.10) and (4.6.11) are harmonic.

5. *Cauchy's integral formula.* If $f(z)$ is regular analytic within C, and z_0 is within C, show that

$$f(z_0) = \frac{1}{2\pi i} \int_C \frac{f(z)\, dz}{z - z_0}$$

Deduce the mean value theorem for harmonic functions.

6. Show that in (4.6.14)

$$a_n = \frac{1}{2\pi i} \int_C \frac{f(z)\, dz}{z^{n+1}}$$

where C is a closed contour encircling the origin.

7. If

$$u = \sum_{n=0}^{\infty} r^n (A_n \cos n\theta + B_n \sin n\theta)$$

show that any harmonic conjugate of u has the form

$$v = \sum_{n=0}^{\infty} r^n (-B_n \cos n\theta + A_n \sin n\theta)$$

8. In Exercises 4.6.6 and 4.6.7, show that $a_n = A_n - iB_n$, $n > 0$ if $f(z) = u + iv$.

166   LAPLACE'S EQUATION AND COMPLEX VARIABLES

9. Evaluate the following integrals taken over the circumference $|z| = 1$ of the unit circle:

$$\int \frac{dz}{z} \qquad \int \frac{dz}{1 + 4z^2} \qquad \int \frac{dz}{2 + 5z^2 + 2z^4}$$

10. Evaluate by residues

$$\int_0^{2\pi} \frac{d\theta}{5 + 4\cos\theta} \qquad \int_{-\infty}^{\infty} \frac{dz}{1 + z^2}$$

11. If $f(z)$ has a pole of order 1 (simple pole) at $z = z_1$, show that its residue there is $\lim_{z \to z_1} (z - z_1)f(z)$.

12. If $f(z_1) \neq 0$ and $g(z_1) = 0$ but $g'(z_1) \neq 0$, show that the residue of the quotient $f(z)/g(z)$ at $z = z_1$ is $f(z_1)/g'(z_1)$. Show that the residue of $\cot z$ at $z = n\pi$ is 1.

13. (a) By integrating the function $e^{isz-az}\cosh bz$ around the boundary of the quarter-circle $z = re^{i\theta}$, $0 \leq r \leq R$, $0 \leq \theta \leq \pi/2$, and letting $R \to \infty$, show that

$$\int_0^\infty e^{isx-ax} \cosh bx \, dx = i \int_0^\infty e^{-sy-iay} \cos by \, dy$$

where $s > 0$, and $a > b > 0$.

(b) By a similar method applied to the function $(z^2 + a^2)^{-1} e^{ibz}$, show that, if a, $b > 0$,

$$\int_0^\infty \frac{\cos bx \, dx}{x^2 + a^2} = \frac{\pi}{2a} e^{-ab}$$

$$\int_0^\infty \frac{\sin bx \, dx}{x^2 + a^2} = \int_0^\infty \frac{e^{-by} \, dy}{a^2 - y^2}$$

14. Show that

$$\int_0^\infty \frac{\sinh ax \sinh bx \, dx}{x \sinh cx} = \frac{1}{4} \log \left[\frac{1 + \cos \frac{\pi(a-b)}{c}}{1 + \cos \frac{\pi(a+b)}{c}} \right]$$

Assume that a, b, c are positive and that $c > a + b$.

15. Suppose that $0 < y < a$ and $x > 0$. Show that

$$\int_0^\infty \frac{\cosh[\alpha(a - y)] \cos xs \, ds}{\alpha \sinh \alpha a} = \frac{\pi e^{-xk}}{2ak} + \frac{\pi}{a} \sum_{n=1}^\infty \frac{e^{-x\lambda_n}}{\lambda_n} \cos \frac{n\pi y}{a}$$

where

$$\alpha = \sqrt{k^2 + s^2} \qquad \lambda_n = \sqrt{k^2 + \frac{n^2\pi^2}{a^2}} \qquad k \text{ real and positive}$$

16. Show that, if $0 < a < b$,

$$\int_0^\infty \frac{\cosh ax \cos cx \, dx}{\cosh bx} = \frac{\pi}{b} \frac{\cosh \frac{\pi c}{2b} \cos \frac{\pi a}{2b}}{\cosh \frac{\pi c}{b} + \cos \frac{\pi a}{b}}$$

17. By integrating the function $z^{n-1}(z+1)^{-1}$ around a circular contour indented along the positive real axis to $z = 0$, show that

$$\int_0^\infty \frac{x^{n-1}\,dx}{1+x} = \frac{\pi}{\sin n\pi}$$

where $0 < n < 1$.

18. Integrate e^{-z^2} around the rectangle $-R \le x \le R$, $0 \le y \le b$, where $z = x + iy$, and deduce that

$$\int_0^\infty e^{-x^2} \cos 2bx\,dx = \frac{\sqrt{\pi}}{2} e^{-b^2}$$

4.7 FOURIER AND LAPLACE TRANSFORMS

The Laplace transform which we now introduce is very closely related to the complex Fourier transform. In the combined Fourier integral (3.3.12), replace $f(x)$ by the product $e^{-cx}f(x)$, so that

$$(4.7.1) \qquad e^{-cx}f(x) = \frac{1}{2\pi} \int_{-\infty}^\infty e^{ixs}\,ds \int_{-\infty}^\infty e^{-cy}f(y)e^{-isy}\,dy$$

We transpose the exponential factor and rearrange:

$$(4.7.2) \qquad f(x) = \frac{1}{2\pi} \int_{-\infty}^\infty e^{(c+is)x}\,ds \int_{-\infty}^\infty f(y) e^{-(c+is)y}\,dy$$

Let us choose the complex variable $p = c + is$ as variable of integration replacing s. Then $dp = i\,ds$ and p ranges on the vertical line with abscissa c in the complex plane. Then (4.7.2) becomes

$$(4.7.3) \qquad f(x) = \frac{1}{2\pi i} \int_{c-i\infty}^{c+i\infty} e^{px}\,dp \int_{-\infty}^\infty f(y)e^{-py}\,dy$$

Alternatively, we may state this result as the inversion of a *Laplace transform*.

THEOREM 4.7.1 *The Laplace integral*

$$(4.7.4) \qquad L(p) = \int_{-\infty}^\infty f(y) e^{-py}\,dy$$

has the formal inversion integral

$$(4.7.5) \qquad f(x) = \frac{1}{2\pi i} \int_{c-i\infty}^{c+i\infty} e^{px} L(p)\,dp$$

The values of p for which (4.7.4) converges and defines a regular analytic function of p form a vertical strip in the complex plane. If $f(y) \sim e^{\alpha y}$ as $y \to \infty$ and $f(y) \sim e^{\beta y}$ as $y \to -\infty$, where α, β are real, then the integral (4.7.4) is convergent for $\alpha < \operatorname{Re} p < \beta$. In (4.7.5) we can choose c to be any real number with $\alpha < c < \beta$. The integral will converge and be independent of c provided

$L(p)$ tends to zero, as $\operatorname{Im} p \to \pm\infty$. To show this, we can deform the contour of integration. By Cauchy's theorem, we have only to show that the integrals over the short ends of a rectangle (Figure 4.7) tend to zero as the ends recede to infinity. This is clearly true when $L(p) \to 0$ as $\operatorname{Im} p \to \pm\infty$ in the strip.

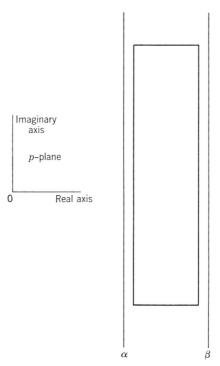

Figure 4.7 Rectangular contour in strip of convergence.

By means of the theory of residues, we can verify that the inversion integral (4.7.5) implies the validity of (4.7.4). Assuming that (4.7.5) defines $f(x)$ when $L(p)$ is given, we construct the Laplace integral, with $\alpha < \operatorname{Re} p < \beta$:

(4.7.6)
$$\begin{aligned}
\int_{-\infty}^{\infty} f(y) e^{-py}\, dy &= \int_{0}^{\infty} f(y) e^{-py}\, dy + \int_{-\infty}^{0} f(y) e^{-py}\, dy \\
&= \int_{0}^{\infty} e^{-py}\, dy \, \frac{1}{2\pi i} \int_{c-i\infty}^{c+i\infty} e^{sy} L(s)\, ds \\
&\quad + \int_{-\infty}^{0} e^{-py}\, dy \, \frac{1}{2\pi i} \int_{c-i\infty}^{c+i\infty} e^{sy} L(s)\, ds \\
&= \frac{1}{2\pi i} \int_{c_1-i\infty}^{c_1+i\infty} L(s)\, ds \int_{0}^{\infty} e^{-(p-s)y}\, dy \\
&\quad + \frac{1}{2\pi i} \int_{c_2-i\infty}^{c_2+i\infty} L(s)\, ds \int_{-\infty}^{0} e^{(s-p)y}\, dy
\end{aligned}$$

The inner integral of the first term converges if $\operatorname{Re} p > \operatorname{Re} s$, so we choose $\alpha < c_1 < \operatorname{Re} p$. Likewise, the inner integral of the second term converges if $\operatorname{Re} p < \operatorname{Re} s$, and we shall place c_2 between $\operatorname{Re} p$ and β: $\operatorname{Re} p < c_2 < \beta$.

Evaluation of the inner integrals yields

(4.7.7) $$\frac{1}{2\pi i} \int_{c_1-i\infty}^{c_1+i\infty} \frac{L(s)\,ds}{p-s} + \frac{1}{2\pi i} \int_{c_2-i\infty}^{c_2+i\infty} \frac{L(s)}{s-p}\,ds$$

Since $L(s)$ is bounded in the strip, we can write these expressions as an integral over a closed contour C by the addition of vanishingly small integrals along the short ends (Figure 4.7). We find

(4.7.8) $$\frac{1}{2\pi i}\int_{c_2-i\infty}^{c_2+i\infty} \frac{L(s)}{s-p}\,ds - \frac{1}{2\pi i}\int_{c_1-i\infty}^{c_1+i\infty} \frac{L(s)}{s-p}\,ds = \lim_{R\to\infty} \frac{1}{2\pi i}\int_C \frac{L(s)}{s-p}\,ds$$

where C is the rectangle $(c_1 \pm iR, c_2 \pm iR)$, containing p. By Cauchy's integral formula (Exercise 4.6.5), the value is $L(p)$. This proves the converse relation for the bilateral Laplace transform.

The Laplace integral (4.7.4) is often referred to as the two-sided or *bilateral* Laplace transform. The corresponding half-range or one-sided transform is sometimes used in connection with initial value problems. If in Theorem 4.7.1 we replace $f(x)$ by $f(x)H(x)$, we obtain the following:

COROLLARY 4.7.1 *The one-sided Laplace integral*

(4.7.9) $$L_1(p) = \int_0^\infty e^{-px} f(x)\,dx$$

has the inversion integral

(4.7.10) $$\frac{1}{2\pi i}\int_{c-i\infty}^{c+i\infty} e^{px} L_1(p)\,dp = \begin{cases} f(x) & x > 0 \\ 0 & x \le 0 \end{cases}$$

Observe that we can take $\beta = +\infty$ for $L_1(p)$; the integral (4.7.9) converges in the half-plane $\operatorname{Re} p > \alpha$.

Example 4.7.1 $f(x) = H(x)\,e^{ax}$, $L(p) = 1/(p-a) = L_1(p)$.

The matrix analogue of this example is: If $f(t) = e^{At}H(t)$, then $L(p) = (pE - A)^{-1}$. Referring to Sections 1.6 and 1.7, we have the following result:

THEOREM 4.7.2 *The Laplace transform of the Green's matrix $G(t) = e^{At}H(t)$ of the system $x' = Ax$ is the 'resolvent' matrix $(pE - A)^{-1}$.*

The evaluation of contour integrals of the Fourier or Laplace inversion type is sometimes carried out by the calculus of residues. To be specific, consider the integral

$$I = \int_{-\infty}^{\infty} f(x)\,dx$$

170   LAPLACE'S EQUATION AND COMPLEX VARIABLES

taken along the real axis. Let the function $f(z)$ be regular analytic for $y \geq 0$, except for singularities at points $z_1, z_2, \ldots, z_k, \ldots$, where $f(z)$ has residues $a_1, a_2, \ldots, a_k, \ldots$. We draw a large semicircular contour consisting of the real interval $-R < x < R$, and the upper semicircle $|z| = R$.

From the residue theorem (Theorem 4.6.5), we find

$$\int_{-R}^{R} f(x)\,dx + \int_0^\pi f(Re^{i\theta})iRe^{i\theta}\,d\theta = 2\pi i \sum_{k=1}^{n} a_k$$

where the summation of residues is taken over the singularities within the semicircle. Now let $R \to \infty$ and suppose that $f(z)$ tends to zero as $R \to \infty$ in such a way that the second integral on the left tends to zero. Then we find

(4.7.11) $$\int_{-\infty}^{\infty} f(x)\,dx = 2\pi i \sum_{k=1}^{\infty} a_k$$

As an example of this technique, which we shall use shortly, we find the Fourier transform of the quotient

$$\frac{\sinh ax}{\sinh bx} \qquad 0 < a < b$$

We must evaluate the real integral

(4.7.12) $$\int_{-\infty}^{\infty} e^{isx} \frac{\sinh ax}{\sinh bx}\,dx$$

and therefore, we consider the contour integral

(4.7.13) $$\int_C e^{isz} \frac{\sinh az}{\sinh bz}\,dz$$

where C is a semicircular contour.

The integrand is a regular analytic function of z except for singularities (first order poles) at the zeros $z = \pm i(n\pi/b)$ of the denominator.

Suppose $s > 0$, and consider the integrand on the semicircular contour, say with $R = (n + \tfrac{1}{2})\pi/b$, passing between the zeros. It can be shown that on this contour

(4.7.14) $$\left|\frac{\sinh az}{\sinh bz}\right| \leq 4e^{(a-b)R|\cos\theta|}$$

Therefore the semicircle integral is no greater than

$$4R\int_0^\pi e^{-sR\sin\theta - (b-a)R|\cos\theta|}\,d\theta$$

The exponent now contains the factor $s\sin\theta + (b-a)|\cos\theta|$. It is easy to show that, for $0 < \theta < \pi$, the minimum m of this expression is equal to the smaller of the two positive numbers s and $b - a$. Therefore $m > 0$, and the semicircle integral is less than

$$4R\pi e^{-mR} \to 0 \quad \text{as} \quad R \to \infty$$

Therefore we can apply the residue series formula.
The residue of the integrand at $z = i(n\pi/b)$ is

(4.7.15) $$\frac{i(-1)^n}{b} e^{-n\pi s/b} \sin \frac{n\pi a}{b}$$

(See Exercise 4.6.11.) Therefore, by (4.7.11),

(4.7.16) $$\int_{-\infty}^{\infty} e^{isx} \frac{\sinh ax}{\sinh bx} dx = \frac{-2\pi}{b} \sum_{n=1}^{\infty} (-1)^n e^{-n\pi s/b} \sin \frac{n\pi a}{b}$$

This series is convergent and can be summed as a pair of geometric series. The final result, in closed form, is

(4.7.17) $$\int_{-\infty}^{\infty} e^{isx} \frac{\sinh ax}{\sinh bx} dx = \frac{\pi}{b} \frac{\sin(\pi a/b)}{\cosh(\pi s/b) + \cos(\pi a/b)}$$

EXERCISES 4.7

1. Find the bilateral Laplace integrals of $\delta(x - a)$, x_+^λ.
2. Find the inverse Laplace transforms of e^{p^2}, pe^{-pa}.
3. Find the one-sided Laplace transforms of $H(x - a)$, $a > 0$; $\cos x$.
4. Find the inverse Laplace transforms of $(1 + p^2)^{-1}$, $p^{-\alpha}$.
5. Show that the Green's function $G(t)$ for the ordinary equation with constant coefficients, $P(D)G(t) = \delta(t)$, $G(t) \equiv 0$, $t < 0$, has the one-sided Laplace transform $L_1(p) = 1/P(p)$.
6. Show that $G(t)$ of Exercise 4.7.5 is given for $t > 0$ by

$$G(t) = \frac{1}{2\pi i} \int_C \frac{e^{\lambda t}}{P(\lambda)} d\lambda$$

where C encircles the zeros of $P(\lambda)$. Evaluate the integral by residues.

7. Show the Laplace transform $L_1(f') = pL_1(f) - f(0)$. By induction on n, show

$$L_1(f^{(n)}) = p^n L_1(f) - \sum_{k=1}^{n} p^{n-k} f^{(k-1)}(0)$$

8. (a) Show that the harmonic function $u(x, y)$ in the half-plane $-\infty < x < \infty$, $-\infty < y \le 0$, which vanishes at infinity and satisfies the boundary condition $gu_y + p^2 u = pg(x)$ on $y = 0$, is given by

$$u(x, y) = \frac{p}{\pi} \int_{-\infty}^{\infty} g(x') dx' \int_0^{\infty} \frac{e^{ys} \cos s(x' - x) ds}{p^2 + gs}$$

172   LAPLACE'S EQUATION AND COMPLEX VARIABLES

(b) The function $\varphi(x, y, t)$ satisfies $\varphi_{xx} + \varphi_{yy} = 0$ in the half-plane $-\infty < x < \infty$, $-\infty < y \leq 0$, together with the boundary condition $g\varphi_y(x, 0, t) + \varphi_{tt}(x, 0, t) = 0$ and the initial conditions $\varphi_t(x, 0, 0) = 0$, $\varphi(x, 0, 0) = g(x)$. By applying a one-sided Laplace transform, reduce the problem to that formulated in part (a) of this question and deduce that, for $t > 0$,

$$\varphi(x, y, t) = \frac{1}{\pi} \int_{-\infty}^{\infty} g(x') \, dx' \int_0^{\infty} e^{ys} \cos(t\sqrt{gs}) \cos s(x' - x) \, ds$$

The function $\varphi(x, y, t)$ is a linearized water wave potential.

9. The function $\varphi(x, y, t)$ satisfies the Laplace equation $\varphi_{xx} + \varphi_{yy} = 0$ in the infinite strip $-\infty < x < \infty$, $-h \leq y \leq 0$, with the boundary conditions $\varphi_{tt} + g\varphi_y = 0$ for $y = 0$ and $\varphi_y = 0$ for $y = -h$. If the initial conditions are $\varphi(x, 0, 0) = g(x)$ and $\varphi_t(x, 0, 0) = 0$, show by means of a Laplace transform that

$$\varphi = \frac{1}{\pi} \int_{-\infty}^{\infty} g(x') \, dx' \int_0^{\infty} \frac{\cos \mu t \cdot \cos s(x' - x) \cdot \cosh s(y + h) \, ds}{\cosh sh}$$

where $\mu = \sqrt{gs \tanh sh}$.

10. Solve the initial value problem for $\varphi_{xx} + \varphi_{yy} = 0$ in the region $0 \leq x \leq a$, $-\infty < y \leq 0$, when $\varphi_t(x, y, 0) = 0$, $\varphi(x, 0, 0) = g(x)$, $\varphi_{tt}(x, 0, t) + g\varphi_y(x, 0, t) = 0$, and $\varphi_x(a, y, t) = \varphi_x(0, y, t) = 0$. Show that

$$\varphi = \frac{1}{a} \int_0^a g(x') \, dx' \left[1 + 2 \sum_{n=0}^{\infty} e^{n\pi y/a} \cos \frac{n\pi x}{a} \cos \frac{n\pi x'}{a} \cos \left(t \sqrt{\frac{n\pi g}{a}} \right) \right]$$

11. For $0 \leq x \leq a$, establish the Fourier series

$$\frac{x}{a} = -\frac{2}{\pi} \sum_{n=1}^{\infty} \frac{(-1)^n}{n} \sin \frac{n\pi x}{a}$$

Show that, if $x > 0$,

$$\frac{1}{2\pi i} \int_L \frac{e^{px} \, dp}{p \sinh ap} = \frac{x}{a} + \frac{2}{\pi} \sum_{n=1}^{\infty} \frac{(-1)^n}{n} \sin \frac{n\pi x}{a}$$

Verify that the right-hand side represents the step function $f(x)$ defined by

$$f(x) = \begin{cases} 0 & 0 < x < a \\ 2n & (2n - 1)a < x < (2n + 1)a \end{cases}$$

where $n = 1, 2, 3, \ldots$. Show also that

$$\int_0^{\infty} e^{-px} f(x) \, dx = \frac{1}{p} \left[\frac{1}{ap} + 2 \sum_{n=1}^{\infty} \frac{(-1)^n ap}{n^2 \pi^2 + a^2 p^2} \right]$$

Deduce that

$$\frac{1}{\sinh ap} = \frac{1}{ap} + 2 \sum_{n=1}^{\infty} \frac{(-1)^n ap}{n^2 \pi^2 + a^2 p^2}$$

4.8 THE FINITE DIFFERENCE LAPLACE EQUATION

A heat flow problem or electrostatic potential distribution in a region of irregular shape will not be tractable by the methods so far treated in this chapter. Even for regions of the particular shapes we have discussed, the series and integral formulas for solutions are cumbersome, and it may be best in specific cases to use numerical calculations. We present here the method of difference equations designed to give numerical solutions of Laplace's equation. As in previous examples, we shall construct a mesh, or grid, in which the intervals are of a fixed size Δx. For this isotropic Laplace operator, we can choose the same grid interval in the y-direction: $\Delta y = \Delta x$. Let $u_{h,k}$ denote $u(h\,\Delta x, k\,\Delta y) = u(h\,\Delta x, k\,\Delta x)$ the value at the point (h, k) of the grid.

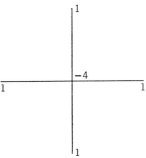

Figure 4.8 Template for Laplace equation.

We replace the Laplace equation by the simplest of the many possible finite difference approximations, namely,

$$(4.8.1) \qquad \frac{u_{h+1,k} - 2u_{h,k} + u_{h-1,k}}{(\Delta x)^2} + \frac{u_{h,k+1} - 2u_{h,k} + u_{h,k-1}}{(\Delta y)^2} = 0$$

Since $\Delta y = \Delta x$, we can write this Laplace difference equation as

$$(4.8.2) \qquad u_{h,k} = \tfrac{1}{4}(u_{h+1,k} + u_{h,k+1} + u_{h-1,k} + u_{h,k-1})$$

The value at any grid point is the average of the values at the four vertically and horizontally adjacent points. Observe that this relation is an analogue not only of the differential equation $u_{xx} + u_{yy} = 0$ but also of the mean value formula of Section 4.3.

Corresponding to (4.8.2) we construct a template pattern (Figure 4.8), which we visualize as moving from point to point of the grid.

Boundary values will be assigned at those grid points which lie closest to the boundary. For the Dirichlet problem we should assign boundary values to just enough points near the boundary to separate the inner and outer regions of the grid (Figure 4.9). We choose these points as is best to describe the shape of the region, which we assume is bounded.

The Laplace difference equation satisfies a maximum principle:

THEOREM 4.8.1 *A solution of the Laplace difference equation takes its maximum and minimum values at a boundary point.*

Proof. Suppose that the maximum value is taken at an interior point (h, k). The entries on the right of (4.8.2) cannot be greater, so (4.8.2) will not hold if any one of them is less. Therefore the maximum value is taken at the neighboring points. Continuing this argument, we find that the maximum is taken at every

174   LAPLACE'S EQUATION AND COMPLEX VARIABLES

point, including the boundary points. A similar result holds for minimum values, and this completes the proof.

This proof also shows that a solution which takes a maximum or minimum value at an interior point must be a constant.

The mean value relations (4.8.2) apply at every interior point of the region. The relations among the values $u_{h,k}$ so found form a system of linear algebraic equations of order equal to the number of interior points. For points (h, k) next to one or more boundary points, we insert the given boundary values in the appropriate terms on the right of (4.8.2). Therefore the linear algebraic system is nonhomogeneous. The mathematical problem now presented is to

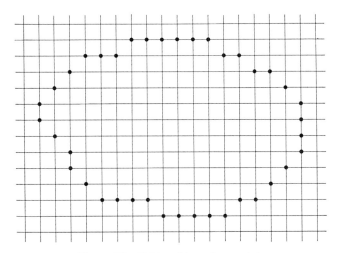

Figure 4.9 Grid and boundary points.

solve this system, which will often be of very large order. Our first question must be: Does the system have a solution, and if so, is the solution unique? Indeed this would be a very necessary property of any algebraic system which claims to simulate the Dirichlet problem.

THEOREM 4.8.2 *The Dirichlet problem for the Laplace difference equation has one and only one solution.*

Proof. To show that a system $A \cdot \mathbf{x} = \mathbf{f}$ of linear algebraic equations has a unique solution, it is sufficient to establish that its determinant is not zero. But if the determinant is zero, the corresponding homogeneous system $A \cdot \mathbf{x} = 0$ will have a nontrivial solution. We therefore prove our result by showing that the homogeneous Laplace difference system has only the zero solution. The nonhomogeneous terms are the boundary values, so the homogeneous system represents the Dirichlet problem with boundary values zero. But by Theorem 4.8.1, all the values of such a solution must be zero. This completes the existence and uniqueness proof for the Laplace difference equation.

THE FINITE DIFFERENCE LAPLACE EQUATION   175

The order of the algebraic system is too large for direct computation of determinants, and it is therefore necessary to use successive approximations for the numerical work. Let $u_{h,k}^{(n)}$ be the value of the nth approximation at (h, k) defined at interior points by the formula

(4.8.3) $$u_{h,k}^{(n)} = \tfrac{1}{4}[u_{h+1,k}^{(n-1)} + u_{h,k+1}^{(n-1)} + u_{h-1,k}^{(n-1)} + u_{h,k-1}^{(n-1)}]$$

We start with an arbitrarily chosen first approximation $u_{h,k}^{(1)}$. At every boundary point we set $u_{h,k}^{(n)}$ equal to the boundary value.

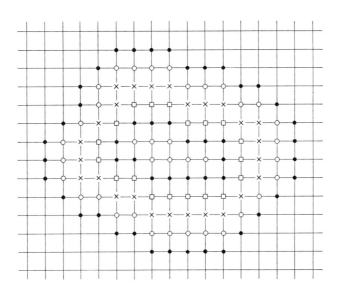

Figure 4.10 Boundary and inner layers.

THEOREM 4.8.3 *The sequence of approximations $u_{h,k}^{(n)}$ converges geometrically to the solution $u_{h,k}$ as $n \to \infty$.*

Proof. Form the difference

(4.8.4) $$v_{h,k}^{(n)} = u_{h,k} - u_{h,k}^{(n)}$$

Subtracting (4.8.3) from (4.8.2) we find that $v_{h,k}^{(n)}$ satisfies the difference formula

(4.8.5) $$v_{h,k}^{(n)} = \tfrac{1}{4}[v_{h+1,k}^{(n-1)} + v_{h,k+1}^{(n-1)} + v_{h-1,k}^{(n-1)} + v_{h,k-1}^{(n-1)}]$$

which has the same form as (4.8.3). Note also that the values of $v_{h,k}^{(n)}$ at all boundary points are zero.

We can take advantage of the vanishing of $v_{h,k}^{(n-1)}$ at boundary points to estimate $v_{h,k}^{(n)}$ on the layer of grid points next to the boundary. Working inward step by step we can estimate the differences $v_{h,k}^{(n)}$ at all grid points. For this purpose let us define layers $L_1, L_2, \ldots, L_N$ as follows (Figure 4.10). L_1 consists of all grid points having a boundary point as an immediate neighbor. L_2 consists of all further grid points neighboring a point of L_1, and so on, to the innermost layer L_N.

Since the greatest and least values of the solution are taken at boundary points, we can choose the first approximation so that $v_{h,k}^{(1)}$ is at most half the difference of these least and greatest values. Now let

$$(4.8.6) \qquad M_n = \max_{h,k} |v_{h,k}^{(n)}| \qquad n = 1, 2, \ldots$$

be the maximum error of the nth approximation. From (4.8.5) we see that $|v_{h,k}^{(n)}| \leq \tfrac{1}{4} \cdot 4 \max |v_{h,k}^{(n-1)}| = M_{n-1}$ and hence that $M_n \leq M_{n-1}$.

We now show that the sequence M_n is geometrically convergent to zero. At points of L_1, one entry on the right in (4.8.5) is zero, and for $n \geq 2$ we have

$$|v_{h,k}^{(n)}| < \tfrac{3}{4} M_{n-1} = (1 - \tfrac{1}{4}) M_{n-1} \qquad h, k \text{ on } L_1$$

On the next layer L_2, one entry on the right side of (4.8.5) comes from L_1. Therefore, if $n \geq 3$, we have for h, k on L_2,

$$|v_{h,k}^{(n)}| \leq \tfrac{3}{4} M_{n-1} + \tfrac{1}{4} \cdot \tfrac{3}{4} M_{n-2}$$

$$\leq [1 - \tfrac{1}{4} + \tfrac{1}{4}(1 - \tfrac{1}{4})] M_{n-2} = \left(1 - \frac{1}{4^2}\right) M_{n-2}$$

Continuing by induction on the number l of the layer, suppose that on L_{l-1} we have

$$|v_{h,k}^{(n)}| \leq \left(1 - \frac{1}{4^{l-1}}\right) M_{n-l+1}$$

if $n \geq l + 1$. Then on the next layer L_l, we have

$$|v_{h,k}^{(n)}| \leq \tfrac{3}{4} M_{n-1} + \tfrac{1}{4}\left(1 - \frac{1}{4^{l-1}}\right) M_{n-l}$$

$$\leq \left(1 - \frac{1}{4^l}\right) M_{n-l}$$

From this result we can also deduce that the hypothesis required for the next step of the induction is satisfied when $n \geq l + 2$. The induction can now proceed up to $l = N$, so that on L_N and all other grid points we have

$$|v_{h,k}^{(n)}| \leq \left(1 - \frac{1}{4^N}\right) M_{n-N},$$

for $n \geq +1$. By the definition of M_n we find

$$M_n \leq \left(1 - \frac{1}{4^N}\right) M_{n-N} = r M_{n-N}$$

say, with $r = 1 - 4^{-N} < 1$, when $n \geq N + 1$.
Consequently

(4.8.7) $$M_{nN+1} \leq r^n M_1$$

and the maximum error converges geometrically to zero. This completes the proof.

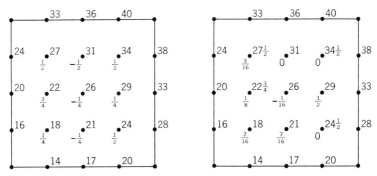

Figure 4.11 Relaxation.

In practice, since r is close to unity for large domains, the rate of convergence may be slow. Much may then depend upon an astute choice for the first approximation. To speed up the convergence, Gauss suggested that at each stage attention should be concentrated upon the points (k, l) having the largest residuals $R_{k,l}$, where

(4.8.8) $$R_{k,l} = \tfrac{1}{4}(u_{h+1,k} + u_{h,k+1} + u_{h-1,k} + u_{h,k-1}) - u_{h,k}$$

This brings us to the *method of relaxation*, as it is now known.

We shall describe the relaxation method as it may be conducted with pencil and squared paper. On the grid we enter the boundary values and the first approximation at each point, as in Figure 4.11. We then compute each residual, which measures the amount by which (4.8.2) has been "relaxed" at (k, l).

Now select the residual, or residuals, of largest magnitude, and adjust the value of the corresponding $u_{h,k}$ so as to annul that residual. That is, add $R_{h,k}$ to $u_{h,k}$. A change in $u_{h,k}$ will affect the four neighboring residuals $R_{h+1,k}$, $R_{h,k+1}$, $R_{h-1,k}$, and $R_{h,k-1}$. These we recalculate with the new values for $u_{h,k}$. Then we select the largest of the new residuals, and repeat the whole process. We continue until the largest residual has become negligibly small.

In Figure 4.11 is shown an example in which the largest residual magnitude has been reduced from $\tfrac{3}{4}$ to $\tfrac{7}{16}$ in four steps, involving adjustments at the points (1, 2), (1, 3), (3, 1) and (3, 3). At each point we have entered the value of $u_{h,k}$

178   LAPLACE'S EQUATION AND COMPLEX VARIABLES

above and to the right, and the value of the residual below and to the left. For hand calculations of this kind, it is convenient to use a pencil and eraser, or a blackboard, so that the new values can be entered directly on the diagram.

Any symmetry of the region and the boundary values should be used to reduce the size of the diagram. In Figure 4.12 is shown an elliptical region symmetric in the x- and y-axes. Only one quadrant is necessary in the relaxation diagram. Values at points immediately outside this quadrant are to be found at the neighboring mirror image points.

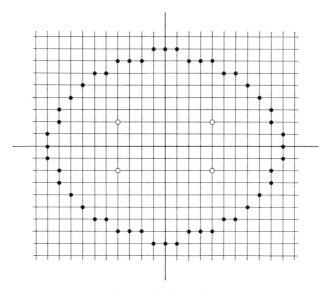

Figure 4.12 Symmetry.

The rapidity of convergence can sometimes be improved by a device of *overrelaxation*. That is, we increase $u_{h,k}$ by $(1 + \gamma)R_{h,k}$, where the overrelaxation factor γ lies between 0 and 1.

A solution found by relaxation is subject to two main causes of error. The first is the error inherent in replacing the continuous differential equation and region by a finite set of difference relations. The second is that the smallness of residuals may not imply a corresponding accuracy of the solution.

EXERCISES 4.8

1. Construct the solution of the finite difference problem corresponding to the ordinary differential equations and boundary conditions, given:

(a) $dy/dx = 2^{-x}$, $y(0) = 0$, grid interval 1, $x \geq 0$.
(b) $dy/dx = y$, $y(0) = 0$, grid interval h, $x \geq 0$.
(c) $d^2y/dx^2 = 2$, $y(0) = 0$, $y(1) = 1$, grid interval $\frac{1}{10}$.
(d) $d^2y/dx^2 = x$, $y(0) = 0$, $y(1) = \frac{1}{2}$, grid interval $\frac{1}{10}$.

2. Complete the relaxation problems shown in the figures, reducing the largest residual to 0.1 or less.

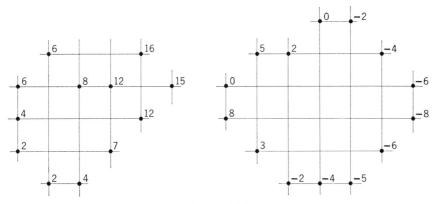

Exercise 4.8.2

3. Find a finite difference solution for $u_{xx} + u_{yy} = 1$ with boundary values zero on the grid patterns of the following figures.

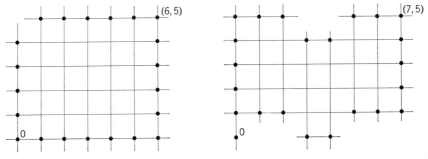

Exercise 4.8.3

4. Set up a computation scheme for the Neumann boundary condition $\partial u/\partial n = f$.

5. Discuss the possible 9-point Laplace difference templates, in which the four diagonally neighboring points appear. Show that the theorems of this section are valid for such patterns, provided the diagonal points appear with positive weights in the averaging.

6. Show how to use an approximate solution for a coarse grid to set up a first approximation for a grid of finer mesh size.

CHAPTER 5

Equations of Motion

5.1 VIBRATIONS OF A MEMBRANE

In Chapter 1 we studied the equations of motion of systems with a finite number of degrees of freedom, and showed that these ordinary differential equations could be derived from Hamilton's variational principle. We now consider the mechanics of continuous systems which involve partial differential equations. Hamilton's principle is again applicable and leads to the Lagrangian equations of motion for continuous systems.

As the derivation of the wave equation for a string in Section 2.2 shows, a continuous system may be regarded as the limit of a discrete one, and certain coordinate differences in the discrete systems will become derivatives in the continuous limit. Therefore let us suppose that the Lagrangian functions of our systems will depend on the configuration coordinates $u, v, \ldots$, and their derivatives with respect to the independent variables x, y, z, t, of space and time. For many applications it is sufficient to include first-order derivatives, and we therefore postulate a *Lagrangian density*

(5.1.1) $$\mathscr{L}(u, u_t, u_x, u_y, u_z, x, y, z, t)$$

This quantity is the difference $\mathscr{T} - \mathscr{V}$ of kinetic energy $\mathscr{T}$ and potential energy $\mathscr{V}$ per unit of volume (or area or length) occupied by the vibrating system. The total Lagrangian is therefore

$$L = \iiint \mathscr{L}\, dx\, dy\, dz, \tag{5.1.2}$$

where the integral is extended over the whole system (Goldstein, Ref. 26). The Hamiltonian integral is now

$$I = \int_{t_1}^{t_2} L\, dt = \iiiint \mathscr{L}\, dx\, dy\, dz\, dt \tag{5.1.3}$$

and the variational principle of Theorem 1.5.1 states that

$$\delta I = \iiiint \delta \mathscr{L}\, dx\, dy\, dz\, dt = 0 \tag{5.1.4}$$

The variation of $\mathscr{L}$ arises from the variations of u and its derivatives: we shall suppose that these variations vanish for $t = t_1$ and t_2, and on the boundary of the system. Now

$$\delta \mathscr{L} = \frac{\partial \mathscr{L}}{\partial u}\, \delta u + \frac{\partial \mathscr{L}}{\partial \dot{u}}\, \delta \dot{u} + \sum_{xyz} \frac{\partial \mathscr{L}}{\partial u_x}\, \delta u_x \tag{5.1.5}$$

In order to free the variations $\delta \dot{u}$, δu_x of the coordinate derivations, we must integrate by parts as was done in the discrete case of Chapter 1. A typical calculation of this type is

$$\iiint \frac{\partial \mathscr{L}}{\partial u_x}\, \delta u_x\, dx\, dy\, dz = -\iiint \frac{\partial}{\partial x}\left(\frac{\partial \mathscr{L}}{\partial u_x}\right) \delta u\, dx\, dy\, dz + \iint_S \frac{\partial \mathscr{L}}{\partial u_x}\, \delta u\, dy\, dz \tag{5.1.6}$$

We shall suppose that u is fixed on the surface S of the system, and therefore that $\delta u = 0$ there. The surface integrals then disappear. From (5.1.6) and similar formulas for derivatives with respect to y, z, and t, we obtain

$$\iiiint \left[\frac{\partial \mathscr{L}}{\partial u} - \frac{\partial}{\partial t}\left(\frac{\partial \mathscr{L}}{\partial u_t}\right) - \sum_{xyz} \frac{\partial}{\partial x}\left(\frac{\partial \mathscr{L}}{\partial u_x}\right)\right] \delta u\, dx\, dy\, dz\, dt = 0$$

Since this is to hold for arbitrary interior variations $\delta u(x, y, z, t)$, it follows that the *variational derivative* $\delta \mathscr{L}/\delta u$ defined by

$$\frac{\delta \mathscr{L}}{\delta u} \equiv \frac{\partial \mathscr{L}}{\partial u} - \sum_{xyz} \frac{\partial}{\partial x}\left(\frac{\partial \mathscr{L}}{\partial u_x}\right) \tag{5.1.7}$$

satisfies the equation

$$\frac{\partial}{\partial t}\left(\frac{\partial \mathscr{L}}{\partial u_t}\right) - \frac{\delta \mathscr{L}}{\delta u} = 0 \tag{5.1.8}$$

The variational, or functional, derivative is so called because it measures the change in $\mathscr{L}$ caused by a variation of the function $u(x, y, z, t)$ with respect to its space variables x, y, z. Since we have assumed no dependence of $\mathscr{L}$ on the space derivatives of u_t, we can write

$$\frac{\delta \mathscr{L}}{\delta u_t} = \frac{\partial \mathscr{L}}{\partial u_t}$$

When this relation is inserted in (5.1.8), we obtain the following theorem:

THEOREM 5.1.1 *The equations of motion have the Lagrangian form*

(5.1.9) $$\frac{\partial}{\partial t}\left(\frac{\delta \mathscr{L}}{\delta u_t}\right) - \frac{\delta \mathscr{L}}{\delta u} = 0$$

If the system under study is specified by several displacement functions $u, v, \ldots$, then there will be one equation of the form (5.1.9) for each such function.

Most of the Lagrangian densities $\mathscr{L} = \mathscr{T} - \mathscr{V}$ which we shall meet are quadratic functions of the displacements $u, v, \ldots$, and their derivatives. The expression for kinetic energy is always of the form $\frac{1}{2} \Sigma\, m \dot{u}^2$ in suitable coordinates. For the potential energy density $\mathscr{V}$, which we shall often calculate as the negative of work done to bring the system to its present state from equilibrium, a quadratic form is also to be expected. This is so because a linear law of force, such as Hooke's law, is a first approximation for small displacements to any force law whatever. With $F = kq$,

$$W(q) = \int_0^q F\, dq' = \frac{kq^2}{2}$$

is quadratic in the coordinate q.

In deriving these equations of motion, we often assume that the variations $\delta u, \delta u_x, \ldots$ vanish at the boundaries of the domain occupied by the continuous system. There are instances, however, where such a restriction is clearly not justified by the physical problem. In cases of this kind, the variational requirement $\delta I = 0$ for arbitrary δu then yields a natural boundary condition as well as a partial differential equation.

Let us derive by the variational method the wave equation for the string which was found in Chapter 2 as a limit of a finite system. Defining density as ρ, tension as P, and length as l, we find that the kinetic energy is

(5.1.10) $$T = \frac{1}{2} \int_0^l \rho \dot{u}^2\, dx$$

The potential energy can be found by computing the work done to lengthen the string against the tension. For any element of length dx this is

$$P(ds - dx) = P\left(\frac{ds}{dx} - 1\right) dx$$
$$= P(\sqrt{1 + u_x^2} - 1)\, dx$$
$$= \tfrac{1}{2} P u_x^2\, dx + \cdots$$

Therefore

(5.1.11) $$V = \frac{1}{2} \int_0^l P u_x^2 \, dx$$

The Lagrangian density is now

(5.1.12) $$\mathscr{L} = \tfrac{1}{2}\rho \dot{u}^2 - \tfrac{1}{2}P u_x^2$$

and the equation of motion becomes

(5.1.13) $$\frac{\partial}{\partial t}\left(\frac{\delta \mathscr{L}}{\delta \dot{u}}\right) - \frac{\delta \mathscr{L}}{\delta u} = \rho \ddot{u} - P u_{xx} = 0$$

This is the wave equation of Chapter 2.

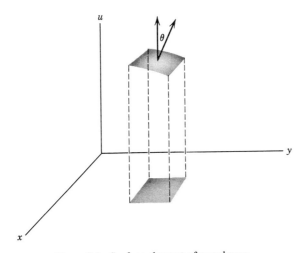

Figure 5.1 Surface element of membrane.

Consider a membrane M of density ρ under a uniform tension P. Transverse vibrations with amplitude $u(x, y, t)$ are governed by a wave equation which we can find as follows. The kinetic energy is clearly

(5.1.14) $$\frac{1}{2} \iint_M \rho \dot{u}^2 \, dx \, dy$$

To calculate the potential energy for a typical element $dx\,dy$, we compute the work done in stretching it against the tension P (Figure 5.1). For an extension δx, say in the x-direction, we perform work $(P\,\delta x)\,dy$, which is the product of tension with increase of area. Thus the extended surface element of area dS will have potential energy

$$P(dS - dx\,dy) = P(\sqrt{1 + u_x^2 + u_y^2} - 1)\,dx\,dy$$
$$= \tfrac{1}{2}P(u_x^2 + u_y^2)\,dx\,dy + \cdots$$

184   EQUATIONS OF MOTION

Then the total potential energy of the membrane is

(5.1.15) $$V = \frac{1}{2} \iint_M P(u_x^2 + u_y^2)\, dx\, dy$$

The Lagrangian density is

(5.1.16) $$\mathscr{L} = \tfrac{1}{2}\rho \dot{u}^2 - \tfrac{1}{2}P(u_x^2 + u_y^2)$$

and the equation of motion must therefore be

(5.1.17) $$\rho u_{tt} - P(u_{xx} + u_{yy}) = 0$$

This is the two-dimensional wave equation. As in the preceding case, the velocity of propagation is $c = \sqrt{P}/\sqrt{\rho}$.

A full study of wave propagation in two space dimensions is much more difficult than in one dimension, and is therefore postponed until Chapter 10. Here we shall present some examples based on the separation of variables. Let a rectangular membrane of length a and width b have one corner at the origin, and the opposite corner at (a, b). Suppose it is held fixed around the edge, that is, the displacement u is zero for $x = 0$ or a, $y = 0$ or b. Solutions of the form

(5.1.18) $$u = X(x)\,Y(y)\,T(t)$$

lead upon insertion in (5.1.17) to

(5.1.19) $$\frac{T''}{c^2 T} = \frac{X''}{X} + \frac{Y''}{Y}$$

where primes denote differentiation with respect to the arguments of each function. Since each term in (5.1.19) depends only on the single argument of the factor appearing in it, it follows that each term is a constant. Therefore,

(5.1.20) $$\begin{aligned} X'' + \alpha^2 X &= 0 \\ Y'' + \beta^2 Y &= 0 \\ T'' + c^2 \gamma^2 T &= 0 \end{aligned}$$

where

(5.1.21) $$\gamma^2 = \alpha^2 + \beta^2$$

Solutions of these equations have the forms

(5.1.22) $$\begin{aligned} X &= A \cos \alpha x + B \sin \alpha x \\ Y &= C \cos \beta y + D \sin \beta y \\ T &= E \cos \gamma c t + F \sin \gamma c t \end{aligned}$$

The fixed-edge boundary conditions in x show that $A = 0$ and

$$X(a) = B \sin \alpha a = 0$$

so that $\alpha a = m\pi$. Similarly, $C = 0$ and

$$Y(b) = D \sin \beta b = 0$$

from which follows $\beta b = n\pi$. Here m and n are independent integers. Possible modes of vibration are now given by

(5.1.23) $\quad u_{mn} = \sin \dfrac{m\pi x}{a} \sin \dfrac{n\pi y}{b} (E_{nm} \cos \gamma_{mn} ct + F_{nm} \sin \gamma_{mn} ct)$

where

(5.1.24) $\quad \gamma_{mn}^2 = \dfrac{m^2 \pi^2}{a^2} + \dfrac{n^2 \pi^2}{b^2}$

The circular frequency $\omega_{mn} = \gamma_{mn} c$ depends on the size of the membrane and on the integers m, n. The eigenfunction $u_{mn}(x, y) = \sin (m\pi x/a) \sin (n\pi y/b)$ represents a mode of oscillation with nodal (fixed) lines parallel to the edges, $m - 1$ equally spaced between the edges $x = 0$, $x = a$, and $n - 1$ between $y = 0$ and $y = b$.

To resolve various problems connected with the vibrations of the rectangular membrane, we should use a series expansion of the eigenfunctions of the domain $0 \leq x \leq a, 0 \leq y \leq b$. This leads us to double Fourier series.

THEOREM 5.1.2 *For the Fourier series expansion in two variables,*

(5.1.25) $\quad f(x, y) = \sum\limits_{n,m=1}^{\infty} c_{mn} \sin \dfrac{m\pi x}{a} \sin \dfrac{n\pi y}{b} \qquad 0 \leq x \leq a, 0 \leq y \leq b$

the coefficients c_{mn} are given by the double integrals

(5.1.26) $\quad c_{mn} = \dfrac{4}{ab} \int_0^a \int_0^b f(x, y) \sin \dfrac{m\pi x}{a} \sin \dfrac{n\pi y}{b} dx\, dy$

Proof. As in the one-dimensional theorems, we use the orthogonality of the eigenfunctions. The necessary integral is

(5.1.27) $\quad \int_0^a \int_0^b \sin \dfrac{m'\pi x}{a} \sin \dfrac{n'\pi y}{b} \sin \dfrac{m\pi x}{a} \sin \dfrac{n\pi y}{b} dx\, dy = \dfrac{ab}{4} \delta_{mm'} \delta_{nn'}$

which vanishes unless $m' = m$ and $n' = n$. We obtain (5.1.26) by multiplication of (5.1.25) by $\sin (m'\pi x/a) \sin (n'\pi y/b)$ and integration over the rectangle. We use (5.1.27) and then drop the primes.

The use of double series in the exercises will be quite similar to the use of single series in the methods of Chapters 3 and 4.

EXERCISES 5.1

1. Find the kinetic energy, the potential energy, and the equation of motion for a coiled spring with equilibrium length l, with linear density ρ, and with spring constant k extended along the x-axis. Assume the spring oscillates longitudinally and that the (small) displacement of the element at position x is $u(x, t)$.

2. Find the equation of motion if $\mathscr{L} = \mathscr{L}(u, u_t, u_x, u_{xx})$.

3. Find the natural boundary condition for vibrations of a string or membrane.

4. State and establish a double Fourier series expansion analogous to the half-range *cosine* series.

5. State and establish an eigenfunction expansion for a membrane $0 \leq x \leq a$, $0 \leq y \leq b$, if $u = 0$ for $y = 0$ and $y = b$, and $u_x = 0$ for $x = 0$ and $x = a$.

6. State and establish an eigenfunction expansion for the membrane with boundary conditions $u = 0$ for $y = 0$ and $y = b$, $u = 0$ for $x = 0$, and $u_x = 0$ for $x = a$.

7. Show that the u component of the tension across an element of the membrane of length δy parallel to the y-axis is $P(\partial u / \partial x) \delta y$. Hence show that the net u component of force due to tension on a rectangular element of sides δx, δy is $P \cdot \Delta u \cdot \delta x \, \delta y$, and so deduce the equation of motion of the membrane.

8. Show that the Fourier coefficient of f_{xx} in Theorem 5.1.2 is equal to

$$\left(\frac{4}{ab}\right)\frac{m\pi}{a}\int_0^b \sin\frac{n\pi y}{b}[f(0, y) - (-1)^m f(a, y)]\, dy - \frac{m^2\pi^2}{a^2} c_{mn}$$

9. Find the vibrations of the rectangular membrane when three sides are held fixed and $u(0, y, t) = g(y, t)$ for $0 \leq y \leq b$, $t > 0$. The initial data are zero.

10. Show how to find a series solution for membrane vibrations caused by the force function $f(x, y, t)$, where

$$u_{tt} = c^2 \Delta u + f(x, y, t)$$

11. By using Theorem 5.1.2 and Exercise 5.1.8, solve $u_{xx} + u_{yy} = u_{tt}$ in $0 \leq x \leq a$, $0 \leq y \leq b$, with the initial values $u(x, y, 0) = 0$, $u_t(x, y, 0) = f(x, y)$, and the boundary values $u = 0$ around all sides. Show that

$$u = \int_0^b\int_0^a f(x', y')G(x, y, x', y', t)\, dx'\, dy'$$

where

$$G(x, y, x', y', t) = \frac{4}{ab}\sum_{n,m=1}^{\infty} \sin\frac{m\pi x}{a} \sin\frac{n\pi y}{b} \sin\frac{m\pi x'}{a} \sin\frac{n\pi y'}{b} \frac{\sin(t\lambda_{nm})}{\lambda_{nm}}$$

with

$$\lambda_{nm} = \sqrt{\left(\frac{m\pi}{a}\right)^2 + \left(\frac{n\pi}{b}\right)^2}$$

12. Solve the initial value problem for the equation $u_{xx} = u_{tt}$ for $0 \leq x \leq a$ with $u(x, 0) = u_t(x, 0) = 0$ and $u(0, t) = 0$, $u_x(a, t) = V$ (a constant). Show by means of a

Laplace transform that

$$u = \frac{V}{2\pi i}\int_{c-i\infty}^{c+i\infty}\frac{e^{pt}\sinh px\,dp}{p^2\cosh pa}$$

Deduce that

$$u = V\left\{x - \frac{2a}{\pi^2}\sum_{n=0}^{\infty}\frac{(-1)^n}{(n+\tfrac{1}{2})^2}\cos\left[(n+\tfrac{1}{2})\frac{\pi t}{a}\right]\sin\left[(n+\tfrac{1}{2})\frac{\pi x}{a}\right]\right\}$$

5.2 LATERAL VIBRATION OF RODS AND PLATES

Consider the vibration of a stiff rod lying along the x-axis (Rayleigh, Ref. 48). We suppose that the cross section of the rod is small and also uniform for all values of x. Let ρ be the line density (mass per unit length); then, if $u(x, t)$ is the lateral displacement, the kinetic energy is

(5.2.1) $$T = \frac{1}{2}\int_0^l \rho \dot{u}^2\,dx$$

To determine the potential energy, we observe that this energy arises from internal deformation of the rod due to its curvature. Consider a rod element of length dx to be composed of parallel filaments, of thickness dh and width dw, which resist contraction or expansion like springs obeying Hooke's law. Now $dx = R\,d\theta$, where R is the radius of curvature and $d\theta$ the total angle of bending of the element. Let h denote perpendicular distance of a filament from the central or neutral surface within the rod where the expansion is zero. The extension of this filament is $h\,d\theta$. The force exerted by the filament is, by Hooke's law,

$$\frac{k}{dx}h\,d\theta\,dh\,dw$$

where k is a constant of the material. The work done to bring about the extension is half the product of this force by the extension, that is,

$$\frac{1}{2}h\,d\theta\left(\frac{k}{dx}\right)h\,d\theta\,dh\,dw = \frac{1}{2}k\left(\frac{d\theta}{dx}\right)^2 h^2\,dh\,dw\,dx$$

Therefore the work done to deform the whole element is the integral over the cross section

$$\frac{1}{2}k\left(\frac{d\theta}{dx}\right)^2\int(h^2\,dh\,dw)\,dx = \frac{1}{2}k\frac{1}{R^2}I\,dx$$

where I is the moment of inertia about the neutral surface. We therefore find a potential energy proportional to the square of the curvature

$$K = \frac{1}{R} = \frac{u_{xx}}{(1+u_x^2)^{3/2}}$$

For small vibrations $K \approx u_{xx}$, and we write

(5.2.2) $$V = \frac{B}{2}\int_0^l u_{xx}^2\,dx$$

188   EQUATIONS OF MOTION

The constant B is equal to the moment of inertia of the rod's cross section, multiplied by the material factor k, which is known as the Young's modulus of the material.

The variation of the Lagrangian integral is carried out in the following way. We suppose that the variation δu of u will vanish at the initial and final times, but not necessarily at the ends $x = 0$ and $x = l$ of the bar. Now,

(5.2.3)
$$0 = \delta \iint \mathscr{L}\, dx\, dt = \delta \iint (\tfrac{1}{2}\rho \dot{u}^2 - \tfrac{1}{2} B u_{xx}{}^2)\, dx\, dt$$
$$= \iint (\rho \dot{u}\, \delta \dot{u} - B u_{xx}\, \delta u_{xx})\, dx\, dt$$

After one partial integration with respect to t and two with respect to x, we find

(5.2.4)
$$\int_0^l \left\{ [\rho \dot{u}\, \delta u]_{t_1}^{t_2} - \int_{t_1}^{t_2} \rho \ddot{u}\, \delta u\, dt \right\} dx$$
$$- \int_{t_1}^{t_2} \left\{ [B u_{xx}\, \delta u_x - B u_{xxx}\, \delta u]_0^l + \int_0^l B u_{xxxx}\, \delta u\, dx \right\} dt$$
$$= -\int_0^l \int_{t_1}^{t_2} (\rho \ddot{u} + B u_{xxxx})\, \delta u\, dx\, dt - \int_{t_1}^{t_2} [B u_{xx}\, \delta u_x - B u_{xxx}\, \delta u]_0^l\, dt = 0$$

Since δu is arbitrary within the region, the double integral can vanish only if

(5.2.5) $$\rho \ddot{u} + B u_{xxxx} = 0$$

This is the equation of motion of the bar.

Inserting (5.2.5) into (5.2.4), we see that the double integral is zero, and therefore that the integral containing the endpoint terms must also vanish. At the endpoints $x = 0$ and $x = l$, the variations δu and δu_x can be regarded as separate and independent. If the end is *clamped*, both δu and δu_x vanish because u and u_x are fixed. If the end is *hinged*, $\delta u = 0$ but δu_x is arbitrary, so the bending moment $B u_{xx}$ must be zero. At a *free* end δu and δu_x are both arbitrary so that the bending moment $B u_{xx}$ and shear $B u_{xxx}$ must vanish. The two conditions at each end furnish a total of four boundary conditions.

Note that (5.2.5) is an equation of the fourth order in x. We shall give some examples of its solution. Separation of the two independent variables leads to solutions $u = X(x)T(t)$, where

(5.2.6)
$$T'' + \omega^2 T = 0$$
$$X^{(4)} - k^4 X = 0$$

and $k^4 B = \omega^2 \rho$.

The equation for the space factor $X(x)$ has the general solution

(5.2.7) $$X(x) = A \cosh kx + B \cos kx + C \sinh kx + D \sin kx$$

A rod clamped at its end satisfies two conditions—its position and slope are fixed. The space equation will allow four conditions: Let the rod be clamped both at the ends $x = 0$ and $x = l$ (Figure 5.2). That is, we have

$$X(0) = X'(0) = X(l) = X'(l) = 0$$

The first two of these show that $B = -A$, $D = -C$, and we then write

(5.2.8) $\quad X(x) = A(\cosh kx - \cos kx) + C(\sinh kx - \sin kx)$

We must therefore have as the two conditions at the endpoint $x = l$,

(5.2.9)
$$A(\cosh kl - \cos kl) + C(\sinh kl - \sin kl) = 0$$
$$A(\sinh kl + \sin kl) + C(\cosh kl - \cos kl) = 0$$

Figure 5.2 Clamped rod.

This is possible for nonvanishing constants A and C only if the determinant

(5.2.10) $\quad \Delta(kl) = \begin{vmatrix} \cosh kl - \cos kl, & \sinh kl - \sin kl \\ \sinh kl + \sin kl, & \cosh kl - \cos kl \end{vmatrix}$

$$= 2 - 2 \cosh kl \cos kl$$

vanishes.

The eigenvalues for k determined by this condition can be found from the roots of the transcendental equation (Figure 5.3)

(5.2.11) $\quad\quad\quad\quad \cosh \lambda = \sec \lambda$

Intersections of the graphs of these two functions occur in pairs lying on each of the positive branches of the secant curve. The apparent root $k = 0$ turns out on examination of (5.2.8) to yield no solutions.

The eigenfunctions of the doubly clamped rod have the form

(5.2.12) $\quad y_n(x) = c_n[(\sinh k_n l - \sin k_n l)(\cosh k_n x - \cos k_n x)$
$$- (\cosh k_n l - \cos k_n l)(\sinh k_n x - \sin k_n x)]$$

where c_n is a normalizing constant, and we have adopted for A and C of (5.2.8) expressions in the ratio suggested by the first equation of (5.2.9).

For the vibrations of a rod clamped at one end and free at the other, we again have (5.2.8), but now the natural conditions at $x = l$ require that bending moment X'' and shear X''' should vanish. Thus,

$$X''(l) = k^2 A(\cosh kl + \cos kl) + k^2 C(\sinh kl + \sin kl) = 0$$
$$X'''(l) = k^3 A(\sinh kl - \sin kl) + k^3 C(\cosh kl + \cos kl) = 0$$

190   EQUATIONS OF MOTION

The determinant of this system is, after cancelation of a factor k^5,

(5.2.13) $$\Delta_2(kl) = \begin{vmatrix} \cosh kl + \cos kl, & \sinh kl + \sin kl \\ \sinh kl - \sin kl, & \cosh kl + \cos kl \end{vmatrix}$$
$$= 2 + 2 \cosh kl \cos kl$$

Therefore the eigenvalues A_n satisfy

(5.2.14) $$-\cosh \lambda = +\sec \lambda$$

The intersections of these two curves are also shown in Figure 5.3. It may be noted that roots of (5.2.11) and (5.2.14) occur in pairs, and that the *pairs*

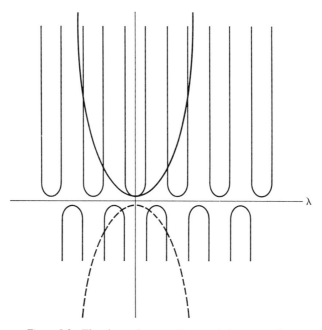

Figure 5.3 The eigenvalue equations $\cosh \lambda = \pm\sec \lambda$.

separate each other. Again, the eigenfunctions have the form (5.2.8), the ratio of A and C being determined by either of the above boundary conditions at $x = l$.

We now briefly consider the vibrations of a thin, solid plate of uniform material, constant thickness, and of areal density ρ (Love, Ref. 40; Rayleigh, Ref. 48). Then the kinetic energy for lateral motions of displacement $u(x, y, t)$ perpendicular to the plane of the plate is an integral over the area of the plate:

(5.2.15) $$T = \frac{1}{2} \rho \int_A \dot{u}^2 \, dx \, dy$$

To construct the potential energy, it is necessary to consider the curvature of the plate. We examine the curvature $1/R$ of a curve drawn through a point (x, y) of the plate as the curve rotates. The largest and smallest values of $1/R$ give us the *principal curvatures* $1/R_1$ and $1/R_2$ at (x, y), and they occur on perpendicular curves. It can be shown that the potential energy density V should be a homogeneous quadratic expression in the principal curvatures, that is,

$$(5.2.16) \qquad V = \frac{B}{2}\left(\frac{1}{R_1^2} + \frac{1}{R_2^2} + \frac{2\mu}{R_1 R_2}\right)$$

where B and μ are constants of the material. A full justification of the assumption (5.2.16) is unfortunately beyond our scope.

To express the potential energy in terms of derivatives of u, let us for the moment choose coordinate axes aligned with the directions of principal curvature at (x, y). Then to first approximation $1/R_1 \sim u_{xx}$ and $1/R_2 \sim u_{yy}$, while the mixed second derivative $u_{xy} = 0$ in these coordinates. Therefore $1/R_1 + 1/R_2 \sim u_{xx} + u_{yy} = \Delta u$, and this relation remains true in any coordinate system. The product $u_{xx}u_{yy}$ is not an invariant like the Laplacian, but the discriminant $D(u) = u_{xx}u_{yy} - u_{xy}^2$, which is equal to it in this special coordinate system, is such an invariant. Therefore we have for any coordinate system,

$$\frac{1}{R_1}\frac{1}{R_2} = u_{xx}u_{yy} - u_{xy}^2 = \frac{\partial(u_x, u_y)}{\partial(x, y)} = D(u)$$

We now write

$$V = \frac{B}{2}\left[\left(\frac{1}{R_1} + \frac{1}{R_2}\right)^2 - 2\frac{(1-\mu)}{R_1 R_2}\right]$$

$$= \frac{B}{2}[(\Delta u)^2 - 2(1-\mu)D(u)]$$

The total potential energy is the double integral

$$(5.2.17) \qquad V = \frac{B}{2}\iint [(\Delta u)^2 - 2(1-\mu)D(u)]\, dx\, dy$$

The discriminant term represents the total curvature of the plate and depends only on the configuration at the edge, since

$$\iint D(u)\, dx\, dy = \iint \frac{\partial(u_x, u_y)}{\partial(x, y)}\, dx\, dy = \iint du_x\, du_y$$

$$(5.2.18) \qquad = \frac{1}{2}\oint (u_x\, du_y - u_y\, du_x)$$

We shall consider a plate clamped at the edge, so (5.2.18) does not contribute to the potential energy. Therefore the Lagrangian is

$$(5.2.19) \qquad L = T - V = \frac{1}{2}\iint [\rho \dot{u}^2 - B(\Delta u)^2]\, dx\, dy$$

The variation of L is found after integrations by parts, and use of $\delta u = \delta u_x = \delta u_y = 0$ on the edge, to be

(5.2.20) $$\delta L = \iint (-\rho \ddot{u} - B \Delta\Delta u) \delta u \, dx \, dy = 0$$

The equation of motion is now deduced as

(5.2.21) $$\rho \ddot{u} + B \Delta\Delta u = 0$$

Let us look for separated solutions $u(x, y, t) = v(x, y)T(t)$. With

(5.2.22) $$T'' + \omega^2 T = 0$$

the equation for the space factor v is

(5.2.23) $$\Delta\Delta v - k^4 v = 0$$

where $Bk^4 = \rho\omega^2$.

For a rectangular clamped plate with boundary conditions $u = u_n = 0$, it is not possible to find the natural modes of vibration by separation of the x and y variables as was done for the membrane equation. Approximation methods, which we cannot discuss further here, are necessary. For a circular clamped plate, however, the eigenfunctions are separable in polar coordinates.

EXERCISES 5.2

1. A bar of length l is supported at an end without being clamped. One given boundary condition is $u = 0$. State the second condition for this end and give a physical interpretation.

2. A bar of length l is supported at both ends. Show that the eigenvalues are $n\pi/l$ and determine the eigenfunctions.

3. A bar of length l is supported at the origin and clamped at $x = l$. Show that the eigenvalues are roots of $\tan kl = \tanh kl$, and find the eigenfunctions.

4. A free bar slides without friction in the horizontal or "gravity-free" plane of x and u. Find the most general motion of the bar with zero potential energy and interpret.

5. Find the most general motion, without restriction on the potential energy, of the bar of Exercise 5.2.4. Explain how a moderately stiff bar held by hand can be released with a snap so that it moves rapidly.

6. Consider the kinetic and potential energy of the system consisting of a bow of length l and cross section constant B, a bowstring of length l and spring constant k, and an arrow of mass m. Find an upper limit for the velocity of the arrow, released after being drawn back a short distance a.

7. For a surface $z = z_0 + \frac{1}{2}d_{11}x^2 + d_{12}xy + \frac{1}{2}d_{22}y^2 + \cdots$, show that the principal (extreme) curvatures can be found by finding maxima or minima of $x^2 + y^2 = r^2$ with $z - z_0 = \varepsilon$ fixed.

8. Show that the principal directions are eigenvectors of $\begin{pmatrix} d_{11} & d_{12} \\ d_{21} & d_{22} \end{pmatrix}$, while the principal curvatures are reciprocals of the eigenvalues. Hence show that the product $R_1^{-1}R_2^{-1}$ is equal to the discriminant $D(u) = u_{xx}u_{yy} - u_{xy}^2$.

LATERAL VIBRATION OF RODS AND PLATES   193

9. For a plate of area A bounded by a curve C of curvature $K(s)$, where s is the arc length along C and n the normal distance from C, show that

$$\int_A \frac{dA}{R_1 R_2} = \frac{1}{2}\oint_C [K(s)(u_s^2 + u_n^2) + u_n u_{ss} - u_s u_{ns}]\, ds$$

10. Using the result of Exercise 5.2.9, show that the natural boundary conditions for the vibrations of the plate are

$$\Delta u - (1 - \mu)(K u_n + u_{ss}) = 0$$

and

$$(\Delta u)_n - (1 - \mu)[(K u_s)_s - u_{nss}] = 0$$

where subscripts n and s denote differentiation normal or parallel to C.

11. Show that the boundary conditions for a supported rectangular plate are $u = 0$, $u_{nn} = 0$. Show that the membrane eigenfunctions also satisfy these conditions, and that the plate frequencies of vibration are distributed as the squares of the membrane frequencies.

12. Deduce from Green's theorem

$$\iint_S (u\,\Delta v - v\,\Delta u)\, dx\, dy = \oint_C \left(u\frac{\partial v}{\partial n} - v\frac{\partial u}{\partial n}\right) ds$$

the identity

$$\iint_S (\Delta u\,\Delta v - v\,\Delta\Delta u)\, dx\, dy = \oint_C \left(\Delta u \frac{\partial v}{\partial n} - v\frac{\partial}{\partial n}\Delta u\right) ds$$

The function $u(x, y)$ satisfies $\Delta\Delta u = f(x, y)$ in S together with one pair of the following pairs of boundary conditions on C: either

(a) $a_1 u + a_2 \dfrac{\partial u}{\partial n} = c_1 \qquad a_1 \Delta u + a_2 \dfrac{\partial}{\partial n}\Delta u = c_2$

or

(b) $b_1 u + b_2 \Delta u = c_3 \qquad b_1 \dfrac{\partial u}{\partial n} + b_2 \dfrac{\partial}{\partial n}\Delta u = c_4$

Show that u is unique (except, in some cases, for an additive constant).

13. Show that, if $K = \theta_s$ is the curvature at a point of C,

$$\iint_S (u_{xx} v_{yy} + u_{yy} v_{xx} - 2 u_{xy} v_{xy})\, dx\, dy$$

$$= \oint_C \left[(v_{ss} + K v_n)u_n + u\frac{\partial}{\partial s}\left(\frac{\partial}{\partial s}v_n - K v_s\right)\right] ds$$

(Express the integrand on the left as a divergence and apply the divergence theorem.) By assuming the formula stated in Exercise 5.2.12, show that, if ν is any constant,

$$\iint_S [(\Delta u)^2 - u\,\Delta\Delta u - 2(1 - \nu)(u_{xx} u_{yy} - u_{xy}^2)]\, dx\, dy = \oint_C \left[M(u)\frac{\partial u}{\partial n} - uT(u)\right] ds$$

where

$$M(u) = \Delta u - (1 - v)(u_{ss} + Ku_n)$$

$$T(u) = \frac{\partial}{\partial n}\Delta u + (1 - v)\frac{\partial}{\partial s}\left(\frac{\partial}{\partial s}u_n - Ku_s\right)$$

The function u satisfies $\Delta\Delta u = f$ in S and satisfies on C one pair of the following boundary conditions: either

(a) $\quad a_1 u + a_2 \dfrac{\partial u}{\partial n} = c_1 \qquad a_1 M(u) + a_2 T(u) = c_2$

or

(b) $\quad b_1 u + b_2 M(u) = c_3 \qquad b_1 \dfrac{\partial u}{\partial n} + b_2 T(u) = c_4$

Show that u is unique (except, in some cases, for an additive constant), if $0 \leqslant v \leqslant 1$.

14. Show that the equation for transverse vibrations of a cable of density ρ and cross-section stiffness B, which is suspended under tension T, is

$$\rho u_{tt} = T u_{xx} - B u_{xxxx}$$

15. Show that sinusoidal waves of circular frequency ω are propagated along the cable of Exercise 5.2.14 with velocity v, where

$$v^2 = \frac{T}{2\rho}\left(1 + \sqrt{1 + \frac{4B\rho}{T^2}\omega^2}\right)$$

16. Let the line density of a beam be ρ, and let K be the radius of gyration of its cross-section area about an axis through its center of mass perpendicular to the plane of vibration. Show that the rotation of the material of the beam contributes a term

$$\frac{1}{2}\int K\rho\left(\frac{d^2 y}{dx\, dt}\right)^2 dx$$

to the kinetic energy. Deduce the equation of motion, taking rotational inertia into account.

17. By means of contour integration and the formulas for the heat kernel, justify the formula

$$\frac{1}{\sqrt{2\pi}}\int_{-\infty}^{\infty} \cos(bts^2) e^{-isx}\, ds = \frac{1}{2\sqrt{bt}}\left(\cos\frac{x^2}{4bt} + \sin\frac{x^2}{4bt}\right)$$

18. Show that the solution $u(x, t)$ of $u_{tt} = -b^2 u_{xxxx}$, $-\infty < x < \infty$, with $u(x, 0) = f(x)$, $u_t(x, 0) = 0$, is given by

$$u(x, t) = \frac{1}{2\sqrt{\pi bt}}\int_{-\infty}^{\infty} f(x_1) \sin\left[\frac{(x - x_1)^2}{4bt} + \frac{\pi}{4}\right] dx_1$$

5.3 INTEGRAL THEOREMS AND VECTOR CALCULUS

The equations and systems we will now encounter involve three space dimensions, and many of them will have unknown three-dimensional vectors as dependent variables. Before undertaking these more ambitious topics, we briefly review here the basic integral theorems of vector calculus, and also the properties of the vector differential operations of gradient, divergence, and curl.

Let $\mathbf{i}, \mathbf{j}, \mathbf{k}$ be an orthonormal set spanning the three-dimensional Cartesian space R^3. If $\mathbf{a}$ and $\mathbf{b}$ are vectors, inclined at an angle θ, and if

(5.3.1) $\qquad \mathbf{a} = a_1\mathbf{i} + a_2\mathbf{j} + a_3\mathbf{k} \qquad \mathbf{b} = b_1\mathbf{i} + b_2\mathbf{j} + b_3\mathbf{k}$

their scalar product is

(5.3.2) $\qquad \mathbf{a} \cdot \mathbf{b} = a_1 b_1 + a_2 b_2 + a_3 b_3 = |\mathbf{a}|\,|\mathbf{b}| \cos \theta$

Their vector product is

(5.3.3)
$$\mathbf{a} \times \mathbf{b} = (a_2 b_3 - a_3 b_2)\mathbf{i} + (a_3 b_1 - a_1 b_3)\mathbf{j} + (a_1 b_2 - a_2 b_1)\mathbf{k}$$
$$= \begin{vmatrix} \mathbf{i} & \mathbf{j} & \mathbf{k} \\ a_1 & a_2 & a_3 \\ b_1 & b_2 & b_3 \end{vmatrix} = |\mathbf{a}|\,|\mathbf{b}| \sin \theta\, \mathbf{c}$$

where $\mathbf{c}$ is a unit vector orthogonal to $\mathbf{a}$ and $\mathbf{b}$ and so oriented that $\mathbf{a}, \mathbf{b}, \mathbf{c}$ form a right-hand triad. In particular $\mathbf{a} \times \mathbf{a} \equiv 0$.

The gradient $\operatorname{grad} f$ of a function f is the vector

(5.3.4) $\qquad \operatorname{grad} f \equiv \nabla f \equiv \dfrac{\partial f}{\partial x}\mathbf{i} + \dfrac{\partial f}{\partial y}\mathbf{j} + \dfrac{\partial f}{\partial z}\mathbf{k}$

The component of $\operatorname{grad} f$ in the direction of a unit vector $\mathbf{n}$ is the directional derivative of f in that direction:

(5.3.5) $\qquad \dfrac{df}{dn} = \dfrac{\partial f}{\partial n} = \dfrac{\partial f}{\partial x} n_1 + \dfrac{\partial f}{\partial y} n_2 + \dfrac{\partial f}{\partial z} n_3$

Given a vector field $\mathbf{V} = v_1 \mathbf{i} + v_2 \mathbf{j} + v_3 \mathbf{k}$, we can construct its *divergence*, which is the scalar quantity

(5.3.6) $\qquad \operatorname{div} \mathbf{V} = \dfrac{\partial v_1}{\partial x} + \dfrac{\partial v_2}{\partial y} + \dfrac{\partial v_3}{\partial z} = \nabla \cdot \mathbf{V}$

The divergence of the gradient of a function f is its Laplacian:

$$\operatorname{div} \operatorname{grad} f = \dfrac{\partial}{\partial x}\left(\dfrac{\partial f}{\partial x}\right) + \dfrac{\partial}{\partial y}\left(\dfrac{\partial f}{\partial y}\right) + \dfrac{\partial}{\partial z}\left(\dfrac{\partial f}{\partial z}\right) = \Delta f$$

A second, this time vector-valued, differential operation is the rotation or curl:

(5.3.7)
$$\operatorname{curl} \mathbf{V} = \left(\dfrac{\partial v_3}{\partial y} - \dfrac{\partial v_2}{\partial z}\right)\mathbf{i} + \left(\dfrac{\partial v_1}{\partial z} - \dfrac{\partial v_3}{\partial x}\right)\mathbf{j} + \left(\dfrac{\partial v_2}{\partial x} - \dfrac{\partial v_1}{\partial y}\right)\mathbf{k}$$
$$= \nabla \times \mathbf{V} = \begin{vmatrix} \mathbf{i} & \mathbf{j} & \mathbf{k} \\ \partial/\partial x & \partial/\partial y & \partial/\partial z \\ v_1 & v_2 & v_3 \end{vmatrix}$$

Vectors $\mathbf{V}$ with $\operatorname{curl} \mathbf{V} = 0$ are called *irrotational*.

Although defined in Cartesian coordinates, these expressions do have an invariant significance, as will appear from their role in integral formulas. The symbolic gradient ∇ can be written in Cartesian components as

$$(5.3.8) \qquad \nabla = \mathbf{i}\frac{\partial}{\partial x} + \mathbf{j}\frac{\partial}{\partial y} + \mathbf{k}\frac{\partial}{\partial z}$$

The curl of a gradient is automatically zero:

$$(5.3.9) \qquad \operatorname{curl\ grad} f = \nabla \times \nabla f = 0$$

Therefore if a vector $\mathbf{V}$ is a gradient ∇f, its curl is necessarily zero.

Likewise the divergence of a curl is identically zero:

$$(5.3.10) \qquad \operatorname{div\ curl} \mathbf{V} = \nabla \cdot (\nabla \times \mathbf{V}) = 0$$

Hence any vector $\mathbf{W}$ which is the curl of another vector field must be *solenoidal* or divergence-free.

The Laplacian of a vector field $\mathbf{V}$ is given by the double combination

$$(5.3.11) \qquad \Delta \mathbf{V} \equiv \operatorname{grad\ div} \mathbf{V} - \operatorname{curl\ curl} \mathbf{V}$$

This identity, once verified in Cartesians, is valid in any coordinate system when the appropriate expressions for grad, div, and curl are provided.

In Section 4.6 we have used Green's theorem in two dimensions. The corresponding three-dimensional theorem gives us an interpretation of the divergence.

GAUSS' THEOREM OF THE DIVERGENCE 5.3.1 *For a vector field $\mathbf{V}$ defined in a region R bounded by the smooth surface S, the surface integral*

$$(5.3.12) \qquad \int_S \mathbf{V} \cdot d\mathbf{S} \equiv \iint_S (v_1\, dy\, dz + v_2\, dz\, dx + v_3\, dx\, dy)$$

is equal to the volume integral

$$(5.3.13) \qquad \int_R \operatorname{div} \mathbf{V}\, dV \equiv \iiint_R \left(\frac{\partial v_1}{\partial x} + \frac{\partial v_2}{\partial y} + \frac{\partial v_3}{\partial z}\right) dx\, dy\, dz$$

The surface integral can be regarded as the flow or flux of the vector field $\mathbf{V}$ through the surface S. Thus the divergence in the volume integral is to be interpreted as the total source or creation of flow within R.

To establish the theorem, we consider one of the components only, say v_1. For fixed y and z, the region R contains one segment or more. Let such a segment have endpoints x_1 and x_2. Then

$$(5.3.14) \qquad \int_{x_1}^{x_2} \frac{\partial v_1}{\partial x}\, dx = v_1(x_2) - v_1(x_1)$$

Integrate this relation over the projection of R on the y, z-coordinate plane, and insert an orienting factor -1 for the lower limit x_1 which runs over the left-hand side of S. The result is the theorem in the case $v_2 = v_3 = 0$. The other terms are supplied in a similar fashion.

There is a second integral theorem which in two dimensions also reduces to Green's theorem of Section 4.6. This theorem yields an interpretation of the curl (Figure 5.4).

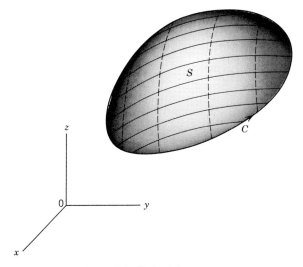

Figure 5.4 Stokes' theorem.

STOKES' THEOREM 5.3.2 *For a vector field* **V** *defined on a surface S bounded by a closed curve C, the line integral*

(5.3.15) $$\oint_C \mathbf{V} \cdot d\mathbf{s} \equiv \int_C (v_1\, dx + v_2\, dy + v_3\, dz)$$

is equal to the surface integral

(5.3.16) $$\int_S \operatorname{curl} \mathbf{V} \cdot d\mathbf{S} \equiv \iint_S \left[\left(\frac{\partial v_3}{\partial y} - \frac{\partial v_2}{\partial z} \right) dy\, dz \right.$$
$$\left. + \left(\frac{\partial v_1}{\partial z} - \frac{\partial v_3}{\partial x} \right) dz\, dx + \left(\frac{\partial v_2}{\partial x} - \frac{\partial v_1}{\partial y} \right) dx\, dy \right]$$

The line integral measures the rotation or "whirling" of the vector field around the closed curve C. Thus the curl can be interpreted as the rate of rotating or spinning of a motion following the field.

To prove the theorem, consider first only the component v_1 and fix the coordinate x. On the plane $x = \text{constant}$, the curve of intersection with S is a segment terminated by endpoints (y_1, z_1), (y_2, z_2) on C. Along this curve

(5.3.17) $$\int dv_1 = \int \left(\frac{\partial v_1}{\partial y} dy + \frac{\partial v_1}{\partial z} dz \right) = v_1(x, y_2, z_2) - v_1(x, y_1, z_1)$$

We integrate over x, and change the sign of the dy term, since for reasons of orientation, we must interpret $dy\, dx$ as $-dx\, dy$. Also the terms on the right then yield the closed curve integral $\oint v_1 \, dx$, the negative sign being consistent with the sense of traversal of the entire closed curve. By cyclic permutation we construct similar formulas for v_2 and v_3, and the theorem is established by combining them.

THEOREM 5.3.3 *For a vector field* **V** *defined in a simply-connected region R the following conditions are equivalent:*

1. *The line integral* $\int_P^Q \mathbf{V} \cdot d\mathbf{s}$ *is independent of the path of integration (in R).*
2. **V** *is irrotational:* curl **V** = 0.
3. **V** *is a gradient: there exists a potential function* φ *with* **V** = grad φ.

The proof can be indicated in cyclic order. The line integral $\oint \mathbf{V} \cdot d\mathbf{s}$ around any small surface patch S can be expressed as the difference of two paths from P to Q and, by condition 1, this difference is zero. Thus, by Stokes' theorem $\int_S \text{curl } \mathbf{V} \cdot d\mathbf{S} = 0$ and condition 2 follows. If **V** is irrotational, the conditions of integrability for the existence of φ are satisfied, so condition 3 holds. Given condition 3, the integral in condition 1 has the value $\varphi(B) - \varphi(A)$, which is independent of the path. This completes the cycle of equivalences.

We should note that this theorem can fail if R is not simply connected, as the potential function φ may then not be single-valued.

Since the divergence of a curl is identically zero, we might expect a similar theorem to the effect that a vector is a curl if and only if it is solenoidal. This can be shown by elementary means, but we shall use a method depending on Laplace's operator. For this purpose we must first construct a Green's function, or fundamental solution.

THEOREM 5.3.4 *The Green's function for the Laplace equation in three-dimensional space is*

(5.3.18) $$G(P, Q) = \frac{1}{4\pi r} \qquad r = |P - Q|$$

Proof. The required solution must satisfy

(5.3.19) $$-\Delta u(P) = \delta(P, Q)$$

where $\delta(P, Q)$ is the three-dimensional Dirac distribution with support at Q. We take Q as origin of spherical polar coordinates (see Chapter 9) and look for a solution depending only on r. We also note that

$$\delta(P, Q) = \delta(x - \xi)\delta(y - \eta)\delta(z - \zeta) \qquad P = (x, y, z), \, Q = (\xi, \eta, \zeta)$$

and in spherical coordinates

(5.3.20) $$\delta(P, Q) = \frac{\delta_+(r)}{4\pi r^2}$$

where the plus subscript on $\delta_+(r)$ indicates $\int_0^\varepsilon \delta_+(r) \, dr = 1, \, \varepsilon > 0$.

Now,

(5.3.21) $$-\Delta u(r) = -\frac{1}{r^2}\frac{\partial}{\partial r}\left(r^2\frac{\partial u}{\partial r}\right) = \frac{\delta_+(r)}{4\pi r^2}$$

so

(5.3.22) $$-r^2\frac{\partial u}{\partial r} = \frac{1}{4\pi}\int_0^r \delta_+(r')\,dr' = \frac{1}{4\pi}$$

and, therefore,

$$u(r) = \frac{1}{4\pi r} + \text{const.}$$

We choose the constant zero, to make the solution vanish at infinity. This completes the proof.

THEOREM 5.3.5 *For a vector field* **V** *which vanishes sufficiently fast at infinity, the two following conditions are equivalent:*

1. **V** *is solenoidal:* div **V** = 0.
2. *There exists a vector potential* **A** *with*

(5.3.23) $$\mathbf{V} = \operatorname{curl} \mathbf{A} \qquad \operatorname{div} \mathbf{A} = 0$$

Proof. Since div curl $\mathbf{A} \equiv 0$, condition 2 implies condition 1. We shall now establish that condition 1 implies condition 2. Given **V**, we choose for **A** the convergent integral

(5.3.24) $$\mathbf{A} = \mathbf{A}(P) = \operatorname{curl}_P \int_{R^3} \frac{\mathbf{V}(Q)}{4\pi r}\,dV \qquad dV = d\xi\,d\eta\,d\zeta$$

Where necessary, we indicate by a subscript the variable point of the differentiations. The coordinates of P are x, y, z, and those of Q are ξ, η, ζ. Since

(5.3.25) $$r = [(x-\xi)^2 + (y-\eta)^2 + (z-\zeta)^2]^{1/2}$$

any differentiation of a function of r with respect to coordinates of P and of Q will lead to equal and opposite results. Thus $\nabla_P f(r) = -\nabla_Q f(r)$.

The vector integral in (5.3.24) is solenoidal, for if S is a sphere enclosing a large volume V, then

(5.3.26) $$\operatorname{div}_P \int_V \frac{\mathbf{V}}{4\pi r}\,dV = \int_V \nabla_P\left(\frac{1}{4\pi r}\right)\cdot \mathbf{V}\,dV$$
$$= -\int_V \nabla_Q\left(\frac{1}{4\pi r}\right)\cdot \mathbf{V}\,dV = -\int_V \nabla_Q\cdot\left(\frac{1}{4\pi r}\mathbf{V}\right)dV$$
$$= -\int_S \frac{1}{4\pi r}\mathbf{V}\cdot d\mathbf{S}$$

by condition 1 and the divergence theorem. As S recedes to infinity, the surface integral will tend to zero provided **V** is small enough. By (5.3.11), this shows that

$$\operatorname{curl} \mathbf{A} = (\operatorname{curl} \operatorname{curl} - \operatorname{grad} \operatorname{div}) \int \frac{\mathbf{V} \, dV}{4\pi r}$$

(5.3.27)
$$= -\Delta \int \frac{\mathbf{V}}{4\pi r} \, dV = \int \delta(P, Q) \mathbf{V}(Q) \, dV_Q$$

$$= \mathbf{V}(P)$$

so that **A** has the required properties and the proof is complete.

By the same method, we can establish the following useful theorem of Helmholtz.

THEOREM 5.3.6 *Any vector field* **V** *which vanishes strongly at infinity can be expressed as the sum of a gradient and a curl:*

(5.3.28) $$\mathbf{V} = \operatorname{grad} \varphi + \operatorname{curl} \mathbf{A}$$

The two terms are orthogonal in the integral norm.

Proof. Given **V**, construct the vector potential

(5.3.29) $$G\mathbf{V} = \int \frac{\mathbf{V}}{4\pi r} \, dV \qquad dV = d\xi \, d\eta \, d\zeta$$

Define

(5.3.30) $$\varphi = -\operatorname{div} G\mathbf{V} \qquad \mathbf{A} = \operatorname{curl} G\mathbf{V}$$

Then

$$\operatorname{grad} \varphi + \operatorname{curl} \mathbf{A} = -\operatorname{grad} \operatorname{div} G\mathbf{V} + \operatorname{curl} \operatorname{curl} G\mathbf{V}$$

$$= -\Delta G\mathbf{V} = -\Delta_P \int \frac{\mathbf{V}}{4\pi r} \, dV$$

(5.3.31)
$$= \int \delta(P, Q) \mathbf{V}(Q) \, dV_Q$$

$$= \mathbf{V}(P)$$

and the result is established. The reader can verify that a curl and a gradient, both vanishing rapidly enough at infinity, are orthogonal under integration over three-dimensional space.

EXERCISES 5.3

1. Show that

$$(\mathbf{a} \times \mathbf{b}) \cdot \mathbf{c} = (\mathbf{b} \times \mathbf{c}) \cdot \mathbf{a} = (\mathbf{c} \times \mathbf{a}) \cdot \mathbf{b} = \begin{vmatrix} a_1 & a_2 & a_3 \\ b_1 & b_2 & b_3 \\ c_1 & c_2 & c_3 \end{vmatrix}$$

Deduce that div curl $\mathbf{V} \equiv 0$ and curl grad $\varphi \equiv 0$.

2. Show that div **V** = 0 and curl **V** = 0 imply that the Cartesian components of **V** are harmonic functions. What is the two-dimensional analogue of this statement?

3. Show that **a** × (**b** × **c**) = (**a** · **c**)**b** − (**a** · **b**)**c**, and verify that

$$\Delta \mathbf{V} = \text{grad div } \mathbf{V} - \text{curl curl } \mathbf{V}.$$

4. Verify $\nabla \cdot (f\mathbf{V}) = \nabla f \cdot \mathbf{V} + f \nabla \cdot \mathbf{V}$ and $\nabla \times (f\mathbf{V}) = \nabla f \times \mathbf{V} + f \nabla \times \mathbf{V}$.

5. Verify that $\Delta(1/r) = -4\pi\delta(P, Q)$ by evaluating $\int \mathbf{V} \cdot d\mathbf{S}$ over the surface of a sphere of radius ε, with $\mathbf{V} = \text{grad } 1/r$, and applying Gauss' theorem.

6. If S is an open surface subtending solid angle Ω at P, show

$$\int_S \nabla\left(\frac{1}{r}\right) \cdot d\mathbf{S} = -\Omega$$

7. Show that the Green's function for Laplace's equation in two dimensions is $(2\pi)^{-1} \log r^{-1}$.

8. Establish Green's first and second formulas

$$\int_R \nabla u \cdot \nabla v \, dV = -\int_R u \, \Delta v \, dV + \int_S u \frac{\partial v}{\partial n} \, dS$$

$$\int_R (u \, \Delta v - v \, \Delta u) \, dV = \int_S \left(u \frac{\partial v}{\partial n} - v \frac{\partial u}{\partial n}\right) dS$$

9. Establish the n-dimensional Fourier transforms:

$$F(\mathbf{s}) = \frac{1}{(2\pi)^{n/2}} \int_{R^n} e^{-i\mathbf{x}\cdot\mathbf{s}} f(\mathbf{x}) \, dV_x$$

$$f(\mathbf{x}) = \frac{1}{(2\pi)^{n/2}} \int_{R^n} e^{i\mathbf{x}\cdot\mathbf{s}} F(\mathbf{s}) \, dV_s$$

Hint: Employ the one-dimensional formulas repeatedly.

5.4 EQUATIONS OF MOTION OF AN ELASTIC SOLID

Elasticity is that property of a medium which enables it to regain its original position after being deformed. Thus a spring, a stiff rod, or a stiff plate are examples of elastic bodies. Here we examine the motion of an elastic solid in three dimensions. We shall introduce the concepts of strain, stress, and strain energy (Love, Ref. 40).

Strain is a geometric concept. It measures the relative motion or deformation of the points of the elastic solid. For a rigid-body displacement, all lengths are preserved, and there is no strain. To analyze strain, we consider the change in distance between two neighboring particles of the medium. Let their coordinates in the unstrained position be x_r and x'_r, $r = 1, 2, 3$. Let x_r be displaced an amount $u_r(x)$, so its strained position is $x_r + u_r(x)$. (See Figure 5.5.) Likewise the second particle has the position $x'_r + u_r(x')$ under strain. The separation distance is now

(5.4.1) $$r_1^2 = \sum_{k=1}^{3} [x'_k - x_k + u_k(x') - u_k(x)]^2$$

202 EQUATIONS OF MOTION

Here we make our first approximation, namely, that the strains u_k and their derivatives $\partial u_k/\partial x_l = u_{k,l}$ shall be small. In particular we shall write

(5.4.2) $$u_k(x') - u_k(x) = \sum_l u_{k,l}(x)(x'_l - x_l) + \cdots$$

and neglect higher order terms. Expanding the squares in (5.4.1) we have

(5.4.3)
$$r_1^2 = \sum_{k=1}^{3}\left[(x'_k - x_k)^2 + 2(x'_k - x_k)\sum_l u_{k,l}(x)(x'_l - x_l) + \cdots\right]$$
$$= r_0^2 + 2\sum_{k,l=1}^{3} u_{k,l}(x)(x'_k - x_k)(x'_l - x_l) + \cdots$$

where r_0 is the unstrained distance. We have also neglected the square of the difference (5.4.2).

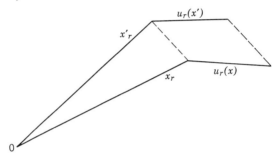

Figure 5.5 Strain.

The *extension* of the element is, by definition

(5.4.4)
$$e = \lim_{x' \to x} \frac{r_1 - r_0}{r_0} = \lim \frac{r_1 - r_0}{r_0}\left(\frac{r_1 + r_0}{2r_0}\right)$$
$$= \frac{1}{2}\lim \frac{r_1^2 - r_0^2}{r_0^2}$$
$$= \sum_{k,l=1}^{3} u_{k,l}(x)\lambda_k \lambda_l$$

where $\lambda_k = \lim (x'_k - x_k)/r_0$ is a direction cosine. The extension e is a quadratic form in which the symmetric coefficients are the *components of strain*

(5.4.5) $$e_{kl} = \tfrac{1}{2}(u_{k,l} + u_{l,k})$$

The symmetric 3 × 3 strain matrix (e_{kl}) forms a *tensor* of the second rank.

Next we introduce another tensor, the stress tensor E_{rs}. Stress has the dimensions of force per unit area, and E_{rs} is the rth component of the force per unit area exerted across a surface element perpendicular to the sth axis. It can be shown that $E_{rs} = E_{sr}$: the stress tensor is symmetric (Exercise 5.4.2).

The basic hypothesis of the classical theory of linear elasticity will now be stated: *The stress and strain tensors are linearly related*. In components, we write the stress-strain relation

(5.4.6) $$E_{rs} = \sum_{m,n} c_{rsmn} e_{mn} = c_{rsmn} e_{mn}$$

where the fourth-order tensor c_{rsmn} depends on the elastic material. In (5.4.6) we have employed the summation convention of tensor calculus, by which a repeated index implies summation over all possible values of the index. We can assume

(5.4.7) $$c_{rsmn} = c_{srmn} \qquad c_{rsmn} = c_{rsnm}$$

because stress and strain tensors are symmetric. A thermodynamic argument which we shall omit also shows that

(5.4.8) $$c_{rsmn} = c_{mnrs}$$

This reduces the total number of *elastic constants* to 21.

Next we assume the material is isotropic: its properties are invariant under rotation about any point of itself. Therefore, the coefficients c_{rsmn} must be unchanged by a rotation: $x_r = a_{rs}x_s'$, with $A' \cdot A = E$. The Kronecker delta δ_{rs} is a tensor with this property and it is possible to show that any such tensor must be composed of the delta tensor. As there are three ways to arrange the four subscripts in pairs, the most general such tensor is

(5.4.9) $$c_{rsmn} = \lambda \, \delta_{rs} \, \delta_{mn} + \mu_1 \, \delta_{rm} \, \delta_{sn} + \mu_2 \, \delta_{rn} \, \delta_{sm}$$

To satisfy (5.4.7) we must take $\mu_1 = \mu_2 = \mu$ in (5.4.9) (Exercise 5.4.1), and then the other symmetry conditions (5.4.8) are fulfilled. Thus the elasticity tensor of an isotropic body depends on the two elastic constants, the Lamé constant λ and the rigidity μ:

(5.4.10) $$c_{rsmn} = \lambda \, \delta_{rs} \, \delta_{mn} + \mu(\delta_{rm} \, \delta_{sn} + \delta_{rn} \, \delta_{sm})$$

For an isotropic medium, the stress-strain relation is now

$$E_{rs} = c_{rsmn} e_{mn} = [\lambda \, \delta_{rs} \, \delta_{mn} + \mu(\delta_{rm} \, \delta_{sn} + \delta_{rn} \, \delta_{sm})] e_{mn}$$
$$= \lambda \, \delta_{rs} \sum_m e_{mm} + 2\mu e_{rs}$$

(5.4.11) $$= \lambda \, \delta_{rs} \sum_m \frac{\partial u_m}{\partial x_m} + \mu \left(\frac{\partial u_r}{\partial x_s} + \frac{\partial u_s}{\partial x_r} \right)$$

$$= \lambda \, \delta_{rs} \, \text{div } \mathbf{u} + \mu \left(\frac{\partial u_r}{\partial x_s} + \frac{\partial u_s}{\partial x_r} \right)$$

Now we derive the equations of motion from a variational principle. The kinetic energy of strain is

(5.4.12) $$T = \frac{1}{2} \sum_r \int \rho \left(\frac{\partial u_r}{\partial t} \right)^2 dV$$

the integral being extended over the solid. The potential or strain energy is given at each point by the quadratic form

$$(5.4.13) \qquad \tfrac{1}{2} E_{rs} e_{rs} = \tfrac{1}{2} c_{rsmn} e_{rs} e_{mn}$$

This expression is equal to the work done per unit volume to strain the medium through a linear scale of intermediate states. The total strain energy W is therefore

$$(5.4.14) \qquad \begin{aligned} W &= \frac{1}{2} \int c_{rsmn} e_{rs} e_{mn} \, dV \\ &= \frac{1}{2} \int c_{rsmn} u_{r,s} u_{m,n} \, dV \end{aligned}$$

by (5.4.5) and (5.4.7). With

$$(5.4.15) \qquad \mathscr{L} = \mathscr{L}(\dot{u}_r) = \tfrac{1}{2} \rho \Sigma \dot{u}_r^2 - \tfrac{1}{2} c_{rsmn} u_{r,s} u_{m,n}$$

we find that

$$(5.4.16) \qquad \frac{\partial}{\partial t}\left(\frac{\delta \mathscr{L}}{\delta \dot{u}_r}\right) - \frac{\delta \mathscr{L}}{\delta u_r} = \rho \ddot{u}_r - c_{rsmn} u_{m,sn} = 0$$

These are the equations of motion for a general elastic medium. If a volume force F_r is present, it will appear as a nonhomogeneous term on the right side of (5.4.16).

For an isotropic medium, the equations simplify to a form we shall study later in Chapter 10. Inserting (5.4.10) for the elastic constants, we find the isotropic elastic equations

$$(5.4.17) \qquad \rho \ddot{u}_r = (\lambda + \mu) u_{s,sr} + \mu \, \Delta u_r + F_r$$

where $\Delta u_r = u_{r,ss}$ is the Laplacian operator.

The particular form of this system of wave equations for the displacement vector u_r enables us to find simpler equations for the divergence $\text{div } \mathbf{u} = u_{r,r}$ and the curl of $\mathbf{u}$.

First let us take the divergence of (5.4.17). We find

$$\rho \ddot{u}_{r,r} = (\lambda + \mu) u_{s,srr} + \mu \, \Delta u_{r,r} + F_{r,r}$$

which, if we set $\theta = \dot{u}_{r,r}$, simplifies to

$$(5.4.18) \qquad \rho \ddot{\theta} = (\lambda + 2\mu) \, \Delta \theta + F_{r,r}$$

The divergence or "expansion" θ satisfies the three-dimensional wave equation with the wave velocity c_1, where

$$(5.4.19) \qquad c_1^2 = \frac{\lambda + 2\mu}{\rho}$$

Wave motions with nonzero values of θ are called *pressure* or *voluminal* waves, and c_1 is the pressure wave velocity.

Next we take the curl of (5.4.17), observing that the curl of a gradient term such as $u_{s,sr}$ is zero. Therefore, with $\boldsymbol{\omega} \equiv \text{curl } \mathbf{u}$ we find

(5.4.20) $$\rho \dot{\boldsymbol{\omega}} = \mu \Delta \boldsymbol{\omega} + \text{curl } \mathbf{F}$$

The curl or rotation vector $\boldsymbol{\omega}$ satisfies a three-dimensional vector wave equation with the wave velocity c_2, where

(5.4.21) $$c_2^2 = \frac{\mu}{\rho}$$

Wave motions involving $\boldsymbol{\omega} = \text{curl } \mathbf{u}$ are called *rotational* or *shear* waves, and c_2 is the shear wave velocity. Since λ and μ are positive constants, shear waves travel less rapidly than pressure waves.

EXERCISES 5.4

1. Verify that the symmetry $c_{rsmn} = c_{srmn}$ implies that $\mu_1 = \mu_2$ in (5.4.9).

2. By considering the forces tending to rotate an element of the medium, show that $E_{rs} = E_{sr}$.

3. By considering a small rectangular element, show that the force vector F_r arising from a system of stresses E_{rs} is $F_r = E_{rs,s}$.

4. *Airy's stress function.* For a stationary plane stress system with only E_{11}, $E_{12} = E_{21}$ and E_{22} different from zero, and if there are no body forces F_r, show that there exists a function χ with $E_{11} = \chi_{yy}$, $E_{12} = -\chi_{xy}$, $E_{22} = \chi_{xx}$.

5. For a system in equilibrium, show that div $\mathbf{u}$ and curl $\mathbf{u}$ are harmonic functions. Show also that the function χ of Exercise 5.4.4 is biharmonic: $\Delta\Delta\chi = 0$.

6. An elastic body is in equilibrium under the action of external forces P_i applied at points of its bounding surface S. By considering the equilibrium of a surface element Q, show that $P_i = E_{ij}l_j$, where (l_1, l_2, l_3) denotes the unit normal to S at Q.

7. Elastic material in a state of plane stress occupies the half-plane $-\infty < x < \infty$, $0 \leq y < \infty$, and is subject to a normal stress $q(x)$ on the face $y = 0$. Show that the boundary conditions on the face $y = 0$ for the Airy stress function are $\chi_{xy} = 0$ and $\chi_{xx} = -q(x)$. With Fourier integrals show that

$$\chi(x, y) = +\frac{1}{\pi} \int_{-\infty}^{\infty} (x' - x) \tan^{-1}\left(\frac{y}{x' - x}\right) q(x')\, dx$$

apart from terms linear in x and y.

By setting $q(x) = \delta(x)$, obtain the stress field due to an isolated normal point force applied at the origin.

8. (a) Let E_{rs} denote the components of stress with respect to axes x, y, z and let E'_{rs} be the stress components corresponding to axes x', y', z obtained by a positive rotation through an angle θ about the z-axis. Show that

$$E'_{11} + E'_{22} = E_{11} + E_{22}$$

$$E'_{11} - E'_{22} + 2iE'_{12} = (E_{11} - E_{22} + 2iE_{12})e^{-2i\theta} \qquad E'_{13} + iE'_{23} = (E_{13} + iE_{23})e^{-i\theta}$$

(b) Deduce that in polar coordinates the polar stress components are expressible in terms of the Airy stress function $\chi(r, \theta)$ in the form

$$E'_{11} = \frac{1}{r^2}\chi_{\theta\theta} + \frac{1}{r}\chi_r \qquad E'_{22} = \chi_{rr}$$

$$E'_{12} = -\frac{\partial}{\partial r}\left(\frac{1}{r}\chi_\theta\right)$$

(c) Show that the biharmonic equation now becomes

$$\left[\frac{1}{r}\frac{\partial}{\partial r}\left(r\frac{\partial}{\partial r}\right) + \frac{1}{r^2}\frac{\partial^2}{\partial \theta^2}\right]^2 \chi = 0$$

(d) If $\bar{\chi}(s, \theta)$ denotes the Mellin transform $\int_0^\infty \chi(r, \theta) r^{s-1}\, dr$ show that

$$\left[\frac{d^2}{d\theta^2} + (s+2)^2\right]\left[\frac{d^2}{d\theta^2} + s^2\right]\bar{\chi} = 0$$

Hence discuss the stress problem for the region $0 \leq r < \infty$, $0 \leq \theta \leq \alpha < 2\pi$, when the normal stress distribution $E'_{22} = q(r)$, $E'_{12} = 0$, is applied on the faces $\theta = 0$, $\theta = \alpha$.

9. (a) Deduce from (5.4.11) and (5.4.17) the body stress equations

$$\rho \ddot{u}_r = E_{rs,s} + F_r$$

(b) In rectangular Cartesian coordinates x, y, z, show that the displacement vector $(-kyz, kxz, k\varphi(x, y))$, where k is a constant, corresponds to the stress field

$$E_{11} = E_{22} = E_{33} = E_{12} = 0$$

$$E_{13} = \mu k(\varphi_x - y) \qquad E_{23} = \mu k(\varphi_y + x)$$

Show that $\Delta\varphi = 0$.

(c) A cylinder with generators parallel to the z-axis is twisted through a small angle so that the displacement is of the above form. If the outward unit normal at any point of the curved surface S has components $(l, m, 0)$, and this surface is stress-free, obtain the boundary condition $\partial\varphi/\partial n = ly - mx$. Writing $\varphi_x = \psi_y$, $\varphi_y = -\psi_x$, show that $\Delta\psi = 0$ and that $\psi - \frac{1}{2}(x^2 + y^2)$ reduces to a constant on S.

(d) If S is the circle $0 \leq r \leq a$, $0 \leq \theta \leq 2\pi$, show that the boundary condition is $\partial\varphi/\partial n = 0$ and deduce that the only nonvanishing stresses are $E_{13} = -\mu ky$, $E_{23} = \mu kx$.

(e) If S is the rectangle $-a \leq x \leq a$, $-b \leq y \leq b$, show that

$$\psi = b^2 + \frac{x^2 - y^2}{2} - \frac{4}{b}\sum_{n=0}^{\infty}\frac{(-1)^n}{\lambda_n^3}\frac{\cosh \lambda_n x}{\cosh \lambda_n a}\cos \lambda_n y$$

where $\lambda_n = (n + \frac{1}{2})\pi/b$.

5.5 MOTION OF A FLUID

We shall now study the equations which govern the motion of a fluid (Courant-Friedrichs, Ref. 12; Howarth, Ref. 33; Lamb, Ref. 38; Temple, Ref. 57). Here it is necessary to employ a vector $\mathbf{v} = (v_1, v_2, v_3)$ representing the *velocity* of the fluid at any point. In general, the relative motion between adjacent elements of fluid is resisted by tangential, frictional, or *viscous* forces. To analyse these forces we must, as in the preceding section, introduce a stress tensor τ_{ij} such that the stress across a small element of area with unit normal n_j is $\tau_{ij}n_j$.

To express the components of stress as functions of the components of velocity we consider the parallel shear flow for which $\mathbf{v} = (v_1(x_2), 0, 0)$. The shear stress between adjacent layers is assumed to be given by the law $\tau_{12} = \mu v_1'(x_2)$, where μ is known as the first coefficient of viscosity. Thus the viscous force depends on *relative* motion and is assumed to be proportional to the space derivative of the velocity component. To analyze the relative *motion* of moving fluid elements, we consider not space derivatives of displacements as in the theory of elasticity but space derivatives of velocities. We therefore introduce, by analogy with (5.4.5), the *rate of strain* tensor,

(5.5.1) $$e_{ij} = \tfrac{1}{2}(v_{i,j} + v_{j,i})$$

We shall generalize the above simple law for the shear force by assuming that there exists a linear relationship between the tensors τ_{ij} and e_{ij} so that

$$\tau_{ij} = c_{ij} + \sigma_{ijrs}e_{rs}$$

where c_{ij} denotes the static values of the stress tensor. If the fluid properties are independent of direction, then the tensors c_{ij}, σ_{ijrs} must be isotropic. Thus we may write $c_{ij} = -p_0 \delta_{ij}$, where p_0 is independent of the e_{rs}, and obtain σ_{ijrs} by analogy with (5.4.10), giving

(5.5.2) $$\tau_{ij} = -p_0 \delta_{ij} + \lambda e_{rr} \delta_{ij} + 2\mu e_{ij}$$

The quantities λ, μ introduced here are known as the coefficients of viscosity. If these coefficients vanish, then the stress vector across an element with normal n_i reduces to $\tau_{ij}n_j = -p_0 n_i$. This stress is entirely normal and of magnitude independent of the direction $\mathbf{n}$. The viscous (tangential) component is now zero. We may then refer to p_0 as the *pressure* exerted at a point of the fluid. If the fluid is viscous, the above simple concept of pressure is no longer possible. We shall *define* the pressure p to be the negative mean of the three principal stresses, or eigenvalues, of τ_{ij}. Thus

(5.5.3) $$p = -\tfrac{1}{3}\tau_{ii} = p_0 - (\lambda + \tfrac{2}{3}\mu)\nabla \cdot \mathbf{v}$$

By means of this relation equation (5.5.2) becomes

(5.5.4) $$\tau_{ij} = -[p + \tfrac{2}{3}\mu \nabla \cdot \mathbf{v}]\delta_{ij} + 2\mu e_{ij}$$

It is necessary now to consider more fully the implications of (5.5.3). If the fluid is in equilibrium, then $\mathbf{v} = 0$, there are no viscous forces, and the quantity p reduces to the all round pressure p_0. This pressure p_0 is, by the equation of state [see (5.5.15) below] a function of the density and temperature. However, for a

208   EQUATIONS OF MOTION

viscous compressible gas in motion, the pressure p as defined by (5.5.3) will depend on the rate-of-stress components, and is no longer a true thermodynamic variable. This difficulty is avoided only if $\lambda + \tfrac{2}{3}\mu = 0$, an assumption we shall now make and which is supported by the kinetic theory of gases.

We now obtain the equations of flow. The first of these is known as the equation of continuity and is a statement of the fact that mass is conserved. Let S

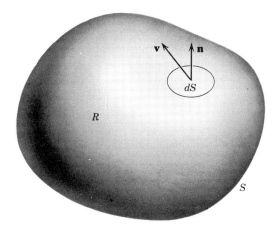

Figure 5.6   Flow across a surface.

be a fixed closed surface containing a volume V of fluid. We assume that there are no sources or sinks through which fluid is artificially injected or withdrawn. Then the rate of increase of the mass of fluid in V must equal the net rate at which fluid enters V across S (Figure 5.6). Thus, if ρ is the fluid density,

$$\frac{\partial}{\partial t}\int_V \rho\, dV = -\int_S \rho \mathbf{v}\cdot\mathbf{n}\, dS$$

Applying the divergence theorem to the surface integral, we find that

$$\int_V [\rho_t + \nabla\cdot(\rho\mathbf{v})]\, dV = 0$$

Since the volume V may be chosen arbitrarily we deduce that

(5.5.5) $$\rho_t + \operatorname{div}(\rho\mathbf{v}) = 0$$

This is the equation of continuity of mass.

Next we construct the equations of motion. We now take a closed surface S which moves with the fluid in such a way that it always contains the same fluid elements. The total mass M contained in this surface remains unaltered and moves under the action of the stress $\tau_{ij}n_j$ exerted across S by the surrounding fluid. If F_i refers to possible body forces, then application of the Newtonian equations of motion leads to the equation

$$\int_V \rho\frac{dv_i}{dt}\, dV = \int_V \rho F_i\, dV + \int_S \tau_{ij}n_j\, dS$$

Transforming the surface integral by means of the divergence theorem gives the equation

$$\int_V \rho \left(\frac{dv_i}{dt} - \frac{1}{\rho} \tau_{ij,j} - F_i \right) dV = 0$$

so that, as the initial choice of V is quite arbitrary, we find

(5.5.6) $$\frac{dv_i}{dt} = \frac{1}{\rho} \tau_{ij,j} + F_i$$

The time derivative occurring here is known as the *Eulerian derivative* following the motion of the fluid element, and is defined as

(5.5.7) $$\frac{dv_i}{dt} = \frac{\partial v_i}{\partial t} + v_j v_{i,j}$$

This follows on finding the change of $v(x, y, z; t)$ due to simultaneous increments in each of the variables x, y, z, t subject to the equations $dx_i = v_i \, dt$.

If we insert the expression (5.5.4) for the components of stress into (5.5.6), we obtain after some reduction, the Navier-Stokes equations of motion:

(5.5.8) $$\mathbf{v}_t + (\mathbf{v} \cdot \nabla)\mathbf{v} = \mathbf{F} - \frac{1}{\rho}\nabla p + \frac{\nu}{3}\nabla \nabla \cdot \mathbf{v} + \nu \Delta \mathbf{v}$$

Here we have taken the viscosity μ to be constant and have introduced the quantity $\nu = \mu/\rho$, which is known as the kinematic viscosity.

The equations (5.5.5) and (5.5.8) together provide four scalar equations connecting the five variables p, ρ, v_1, v_2, v_3. To make the problem determinate, it is necessary to obtain one further equation relating these variables. This may be obtained from the principle of conservation of energy. It is now necessary to consider more closely what is meant by the energy of a gas and, in fact, we shall call on results of a thermodynamic nature. A mass of gas at rest under pressure contains energy which may be released on expansion. This energy we call internal, and a gas in motion possesses internal energy which is additive to its kinetic energy. Let us denote by E the internal energy per unit mass, so that the total energy density $\mathscr{L}$ is given by

(5.5.9) $$\mathscr{L} = E + \frac{v^2}{2}$$

We may construct the required equation of energy by recalling the first law of thermodynamics which asserts the equivalence of heat and mechanical energy. Let S be a fixed closed surface enclosing a volume V. Then the rate of change of the total energy of the gas in V is

(5.5.10) $$\int_S \rho \mathscr{L} v_i n_i \, dS + \frac{\partial}{\partial t} \int_V \rho \mathscr{L} \, dV$$

210   EQUATIONS OF MOTION

This expression must equal the net rate of working of all the forces acting on the gas in V and is therefore equal to

(5.5.11) $$\int_S \tau_{ij} v_i n_j \, dS + \int_V \rho F_i v_i \, dV$$

We equate (5.5.10) to (5.5.11) and apply the divergence theorem to the surface integrals. Then, since V is arbitrary, we find the following equation:

$$\text{div}\,(\rho \mathscr{L} \mathbf{v}) + \frac{\partial}{\partial t}(\rho \mathscr{L}) = \rho F_i v_i + \frac{\partial}{\partial x_j}(\tau_{ij} v_i)$$

By expanding the derivatives, this equation can be set in the following form:

(5.5.12) $[\rho_t + \text{div}\,(\rho \mathbf{v})]\mathscr{L} + \rho(\mathscr{L}_t + \mathbf{v} \cdot \nabla \mathscr{L}) = \tau_{ij} v_{i,j} + v_i(\tau_{ij,j} + \rho F_i)$

By means of the continuity equation (5.5.5) and the equations of motion (5.5.6), we find that (5.5.12) becomes

(5.5.13) $$\rho(\mathscr{L}_t + \mathbf{v} \cdot \nabla \mathscr{L}) = \tau_{ij} v_{i,j} + \rho v_i \frac{dv_i}{dt}$$

If we now insert the expression (5.5.9) for $\mathscr{L}$, then (5.5.13) reduces to the equation

(5.5.14) $$\rho(E_t + \mathbf{v} \cdot \nabla E) = \tau_{ij} v_{i,j}$$

To proceed further, it is necessary to introduce certain thermodynamic notions. First there is assumed a relationship known as the equation of state connecting the variables p, ρ, and the temperature T. We shall confine attention to perfect gases for which this relationship takes the form

(5.5.15) $$p = \rho R T$$

Here R is a constant whose value depends on the molecular weight of the gas.

We shall furthermore assume the existence of a function S, the *entropy*, with the property that

(5.5.16) $$dS = \frac{1}{T}(dE + p\, d\tau) \qquad \tau = \frac{1}{\rho}$$

This equation is a consequence of the second law of thermodynamics and is assumed valid for gases in motion. It implies that the expression on the right of (5.5.16) is an exact differential and that, in fact, T is an integrating factor of the combination $dE + p\, d\tau$. A full explanation of (5.5.16) would carry us too far afield. For a detailed discussion, see Howarth, Ref. 33; Chapman and Cowling, Ref. 8. An important consequence of the assumption (5.5.16) is that for a perfect gas the internal energy E is reducible to a function of the single variable T (see Exercise 5.5.11). In particular, we shall assume the linear relationship $E = cT$, where c is a constant known as the specific heat at constant volume. We may then integrate (5.5.16) directly and find, after using the equation of state $p = \rho RT$, that

(5.5.17) $\qquad S = c \log(p\rho^{-\gamma}) + \text{const.} \qquad \gamma = 1 + R/c$

By means of (5.5.16) we may transform the energy equation (5.5.14). If d/dt denotes the Eulerian derivative, then (5.5.16) and (5.5.5) become

$$\rho T \frac{dS}{dt} = \rho \frac{dE}{dt} - \frac{p}{\rho} \frac{d\rho}{dt}$$

and

$$\frac{1}{\rho} \frac{d\rho}{dt} + \nabla \cdot \mathbf{v} = 0$$

From these, by eliminating the term $d\rho/dt$, we obtain

$$\rho T \frac{dS}{dt} = \rho \frac{dE}{dt} + p(\nabla \cdot \mathbf{v})$$

The equation (5.5.14) can now be transformed into

(5.5.18) $\qquad \rho T \dfrac{dS}{dt} = \tau_{ij} v_{i,j} + p(\nabla \cdot \mathbf{v})$

If we insert (5.5.4) into (5.5.18), we obtain finally the energy equation

(5.5.19) $\qquad \rho T \dfrac{dS}{dt} = -\tfrac{2}{3}\mu(\nabla \cdot \mathbf{v})^2 + 2\mu e_{ij} e_{ij}$

From this relation we deduce immediately by setting $\mu = 0$ that the entropy S of each particle of an inviscid fluid remains constant. In this case, we see from (5.5.17), that for any given particle p is proportional to ρ^γ. The changes of state are then referred to as *adiabatic*.

It should be noted that the entropy S may vary from particle to particle, however, this possibility is often ruled out by a consideration of initial conditions. Suppose, for example, that the flow commences in a region where p and ρ are constant. The entropy there will likewise be constant initially and, therefore, constant subsequently in the region occupied by the same fluid elements. Such flows are termed *isentropic*.

The reader should note that equations (5.5.5), (5.5.8), (5.5.19), in contrast to all others so far encountered, are nonlinear, even in the nonviscous case $\mu = 0$. Therefore, in general, a superposition theorem for fluid motions is not admissible. However, if the velocity components v_i are sufficiently small, we can neglect "second order" terms and thus find linear equations for small or slow motions.

We illustrate such approximations by deriving the sound-wave equation for which we can confine attention to small disturbances in air otherwise at rest

(Friedlander, Ref. 21; Rayleigh, Ref. 48). Let p_0, ρ_0 denote the pressure and density in still air. For small fluctuations, we may write $\rho = \rho_0(1 + s)$, where the numerically small quantity s is called the *condensation*. Equations (5.5.5), (5.5.8) are now replaced by the approximate equations in which $v = 0$:

(5.5.20) $$s_t + \nabla \cdot \mathbf{v} = 0$$

(5.5.21) $$\mathbf{v}_t + \frac{1}{\rho_0} \nabla p = 0$$

Also we see from (5.5.19) that the motion is adiabatic within the approximation accepted, and by virtue of the initial conditions and the above remarks, isentropic also. Therefore p reduces to a function of ρ, so that $\nabla p = (dp/d\rho)\nabla \rho = \rho_0 c_0^2 \nabla s$, approximately, where

(5.5.22) $$c_0^2 = \left(\frac{dp}{d\rho}\right)_0$$

The subscript zero indicates that the derivative is evaluated at the values p_0 and ρ_0. Equation (5.5.21) now becomes

(5.5.23) $$\mathbf{v}_t + c_0^2 \nabla s = 0$$

Therefore

$$\frac{\partial}{\partial t} \nabla \times \mathbf{v} = 0$$

so that $\nabla \times \mathbf{v}$ is independent of the time. Since the air is originally at rest, $\nabla \times \mathbf{v}$ is zero initially and, therefore, $\nabla \times \mathbf{v} = 0$ for all subsequent times. Thus we may infer the existence of a *potential* φ such that $\mathbf{v} = \nabla \varphi$. Equations (5.5.20), (5.5.23) now become

(5.5.24)
$$s_t + \Delta \varphi = 0$$
$$\nabla(\varphi_t + c_0^2 s) = 0$$

The second of these equations implies that the combination $(\varphi_t + c_0^2 s)$ is a function of time only, which function we may incorporate in φ without affecting the requirement that $\mathbf{v} = \nabla \varphi$. Thus we take $\varphi_t = -c_0^2 s$ and, by substitution into (5.5.24), obtain the acoustic wave equation

(5.5.25) $$c_0^2 \Delta \varphi = \varphi_{tt}$$

Since the small disturbances result in approximately isentropic flow, we may use (5.5.17) to find the pressure. Clearly $p/p_0 = (\rho/\rho_0)^\gamma$, so $dp/d\rho = \gamma p/\rho$ and, therefore, $c_0^2 = \gamma p_0/\rho_0$. Also, approximately, $p = p_0(1 + \gamma s)$ and, therefore,

(5.5.26) $$p = p_0 - \rho_0 \varphi_t$$

When air vibrates in a fixed region or room, the normal component of velocity at the boundary will be zero. Thus

(5.5.27) $$v_n = \mathbf{n} \cdot \nabla \varphi = \frac{\partial \varphi}{\partial n} = 0$$

is the proper boundary condition for sound waves in a closed room.

Separated solutions of the form $X(x)Y(y)Z(z)T(t)$ can be found for the normal modes in a rectangular room. If the dimensions are a, b, c, then the oscillations in the region $0 \leq x \leq a, 0 \leq y \leq b, 0 \leq z \leq c$ are

$$\cos \frac{l\pi x}{a} \cos \frac{m\pi y}{b} \cos \frac{n\pi z}{c} (A \cos kc_0 t + B \sin kc_0 t)$$

with

(5.5.28) $$k_{lmn}^2 = \pi^2 \left(\frac{l^2}{a^2} + \frac{m^2}{b^2} + \frac{n^2}{c^2} \right)$$

where l, m, n are integers. The spacing of these natural or resonant frequencies k_{lmn} greatly influences the acoustic properties of a room.

EXERCISES 5.5

1. Find the normalized eigenfunctions for sound vibrations in a room of dimensions a, b, c.

2. For a room of length a, width b, and height c, which has no ceiling and is open above, find the eigenvalues and normalized eigenfunctions for sound vibrations. As boundary condition at the top set $\varphi = 0$.

3. State theorems for triple Fourier series analogous to those of Sections 5.1 for double series. Show how to establish the formulae by three successive applications of the one-dimensional series.

4. Given that a tube or pipe has at an open end the boundary condition $\varphi = 0$, and at a closed end $\partial \varphi / \partial n = 0$, show that the natural period of a pipe of length l is $2l/c$ if it is open at both ends, and $4l/c$ if it is open at one end and closed at the other.

5. For a long, rectangular tunnel of length a, width b, and height c ($a \gg b, a \gg c$) show that the eigenvalues for sound vibrations fall into closely spaced sequences determined by m and n in (5.5.28). Find the minimum circular frequency $\omega = kc$ possible for sound propagation in a tunnel of unlimited length.

6. Let ζ be the elevation above the undisturbed level of a small surface wave on water of depth h. Show that the pressure at depth z is $p = g\rho(\zeta + z)$ and that the x and y components of the equations of motion can be written

$$\frac{\partial v_x}{\partial t} = -g \frac{\partial \zeta}{\partial x} \qquad \frac{\partial v_y}{\partial t} = -g \frac{\partial \zeta}{\partial y}$$

7. Assuming incompressibility, show that

$$\frac{\partial \zeta}{\partial t} = v_z = -h \left(\frac{\partial v_x}{\partial x} + \frac{\partial v_y}{\partial y} \right)$$

and deduce the wave equation governing ζ. Assume that h is constant.

8. *Bernoulli's theorem.* If **v** is a gradient, $\mathbf{v} = \operatorname{grad} \varphi$, and F has a potential, $\mathbf{F} = -\operatorname{grad} V$, show that (5.5.8) can be integrated for an inviscid fluid with $\nu = 0$ to give

$$\int \frac{dp}{\rho} + \frac{\partial \varphi}{\partial t} + \frac{(\nabla \varphi)^2}{2} + V = \text{const.}$$

Deduce that in a steady flow with $V = 0$, low pressure indicates high velocity.

9. Show that a steady flow **v** of an incompressible fluid, which is also irrotational, is harmonic.

10. Show that a flow (5.5.8), where $\nu = 0$, implies that the circulation $\int \mathbf{v} \cdot d\mathbf{s}$ on any closed curve moving with the fluid is constant in time, provided the force **F** is the gradient of a potential. Assume that the entropy is constant.

11. Regarding E as defined by (5.5.16) as a function of S and T, show that

$$R \frac{\partial E}{\partial S} + T \frac{\partial E}{\partial T} = 0$$

Deduce that $E = g(\sigma)$, where g is some function of the single variable $\sigma = Te^{-S/R}$. Show also that $RT = -\sigma g'(\sigma)$, and deduce that E is a function of T alone.

5.6 EQUATIONS OF THE ELECTROMAGNETIC FIELD

In electromagnetic theory we have to consider primarily two field vectors—**E** the electric field strength and **H** the magnetic field strength (Morse-Feshbach, Ref. 45; Stratton, Ref. 55). Together with these are associated the electric displacement vector

(5.6.1) $$\mathbf{D} = \varepsilon \mathbf{E}$$

and the magnetic induction vector

(5.6.2) $$\mathbf{B} = \mu \mathbf{H}$$

Here the electric and magnetic inductive capacities ε and μ depend upon the material and in empty space are both equal to unity. The density ρ of electric charge is, in mks units,

(5.6.3) $$\rho = \operatorname{div} \mathbf{D}$$

Since free magnetic poles do not exist, there is a corresponding magnetic charge density of zero:

(5.6.4) $$\operatorname{div} \mathbf{B} = 0$$

The rate of change with time of the magnetic field **H** is determined by an observation of Faraday. Such a change produces in any closed electric circuit an electromotive force defined by the line integral $\int \mathbf{E} \cdot d\mathbf{s}$ around the circuit. The value of the electromotive force is

(5.6.5) $$\int \mathbf{E} \cdot d\mathbf{s} = -\frac{\partial}{\partial t} \int B_n \, dS$$

That is, the electromotive force is the rate of decrease with time of the flux or amount of magnetic induction which passes through the circuit. We apply Stokes' theorem 5.3.2 to (5.6.5) and write

(5.6.6) $$\int \left(\frac{\partial \mathbf{B}}{\partial t} + \operatorname{curl} \mathbf{E}\right) \cdot d\mathbf{S} = 0$$

Since the surface is arbitrary, we must have

(5.6.7) $$\frac{\partial \mathbf{B}}{\partial t} + \operatorname{curl} \mathbf{E} = 0$$

throughout space and time. This is the first of the two Maxwell vector equations.

The second equation is connected with Oersted's observation that a current of electricity gives rise to a magnetic field. Again, the line integral $\int \mathbf{H} \cdot d\mathbf{s}$, or magnetomotive force around a circuit, is equal to the total current flow enclosed by the curve. We denote the total current vector by $\mathbf{J}_1$ and thus

(5.6.8) $$\int \mathbf{H} \cdot d\mathbf{s} = \int \mathbf{J}_1 \cdot d\mathbf{S}.$$

From Stokes' theorem we have

$$\int (\operatorname{curl} \mathbf{H} - \mathbf{J}_1) \cdot d\mathbf{S} = 0$$

and there follows the vector equation

(5.6.9) $$\operatorname{curl} \mathbf{H} = \mathbf{J}_1$$

From this relation we see that the total current vector must be solenoidal:

(5.6.10) $$\operatorname{div} \mathbf{J}_1 = \operatorname{div} \operatorname{curl} \mathbf{H} \equiv 0$$

However the conduction current $\mathbf{J}$ satisfies the equation of continuity, or charge conservation,

(5.6.11) $$\frac{\partial \rho}{\partial t} + \operatorname{div} \mathbf{J} = 0$$

Hence a supplementary term must be adjoined to $\mathbf{J}$ in order to satisfy (5.6.10). It was suggested by Maxwell that the total current should therefore include a *displacement current* contribution. Maxwell's hypothesis that a changing electric displacement acts like a current completed the foundations of the electromagnetic theory of light.

To determine the displacement contribution, we observe the equation of continuity (5.6.11), and the charge-density relation (5.6.3). Differentiating the latter with respect to t, and combining, we can write

$$\text{div}\left(\mathbf{J} + \frac{\partial \mathbf{D}}{\partial t}\right) = 0$$

Thus the vector

(5.6.12) $$\mathbf{J}_1 = \mathbf{J} + \frac{\partial \mathbf{D}}{\partial t}$$

is solenoidal, as befits the total current. Maxwell placed this vector field in (5.6.9) to obtain the second vector equation

(5.6.13) $$\frac{\partial \mathbf{D}}{\partial t} + \mathbf{J} = \text{curl } \mathbf{H}$$

Equations (5.6.7) and (5.6.13) together with the subsidiary relations (5.6.3) and (5.6.4) constitute Maxwell's equations.

We now envisage a medium with constant values for ε and μ. For simplicity, suppose also that $\mathbf{J}$ and ρ are zero. Since conductivity σ is defined by the general relation

(5.6.14) $$\mathbf{J} = \sigma \mathbf{E}$$

we are, in effect, supposing that the medium is nonconducting. Then by (5.6.7) and (5.6.13),

(5.6.15)
$$\frac{\partial^2 \mathbf{B}}{\partial t^2} = \mu \frac{\partial^2 \mathbf{H}}{\partial t^2} = -\frac{\partial}{\partial t} \text{curl } \mathbf{E} = -\text{curl } \frac{\partial \mathbf{E}}{\partial t}$$
$$= -\frac{1}{\varepsilon} \text{curl } \frac{\partial \mathbf{D}}{\partial t} = -\frac{1}{\varepsilon} \text{curl curl } \mathbf{H}$$

We again use the vector identity

(5.6.16) $$\Delta \mathbf{H} = \text{grad div } \mathbf{H} - \text{curl curl } \mathbf{H}$$

and observe from (5.6.4) that div $\mathbf{H}$ is zero. Since $\mathbf{B} = \mu\mathbf{H}$, we find from (5.6.15) the wave equation

(5.6.17) $$\mu\varepsilon \frac{\partial^2 \mathbf{H}}{\partial t^2} = \Delta \mathbf{H}$$

Likewise we can show that

(5.6.18) $$\mu\varepsilon \frac{\partial^2 \mathbf{E}}{\partial t^2} = \Delta \mathbf{E}$$

THEOREM 5.6.1 *In a nonconducting medium, the Cartesian components of $\mathbf{E}$ and $\mathbf{H}$ satisfy the three-dimensional wave equation with a propagation velocity $c = (\varepsilon\mu)^{-1/2}$.*

For empty space this velocity c is the velocity of light. In a conducting medium the equation is modified by the presence of a first derivative with respect to time (Exercise 5.6.5).

Maxwell's equations can also be treated by means of a scalar and a vector potential. The solenoidal field **B** (div **B** ≡ 0) can be written in the form

(5.6.19) $$\mathbf{B} = \operatorname{curl} \mathbf{A}$$

where **A** is the vector-potential field. Inserting this in (5.6.7), we have

$$\frac{\partial \mathbf{B}}{\partial t} = \operatorname{curl} \frac{\partial \mathbf{A}}{\partial t} = -\operatorname{curl} \mathbf{E}$$

so that

(5.6.20) $$\operatorname{curl}\left(\frac{\partial \mathbf{A}}{\partial t} + \mathbf{E}\right) = 0$$

Therefore the irrotational field $\partial \mathbf{A}/\partial t + \mathbf{E}$ is the gradient of a scalar potential function φ:

(5.6.21) $$\frac{\partial \mathbf{A}}{\partial t} + \mathbf{E} = -\operatorname{grad} \varphi$$

The divergence of this relation is

(5.6.22) $$\frac{\partial}{\partial t} \operatorname{div} \mathbf{A} + \frac{\rho}{\varepsilon} = -\operatorname{div} \operatorname{grad} \varphi = -\Delta \varphi$$

The assignment of div **A** is at our disposal since only curl **A** has yet been determined. Let us choose

(5.6.23) $$\operatorname{div} \mathbf{A} = -\varepsilon\mu \frac{\partial \varphi}{\partial t}$$

Then (5.6.22) becomes a wave equation for φ:

(5.6.24) $$\frac{\partial^2 \varphi}{\partial t^2} - \frac{1}{\varepsilon\mu} \Delta \varphi = \frac{\rho}{\mu\varepsilon^2}$$

Likewise the vector potential **A** satisfies the wave equation

(5.6.25) $$\frac{\partial^2 \mathbf{A}}{\partial t^2} = \frac{1}{\varepsilon\mu} \Delta \mathbf{A} + \frac{\mathbf{J}}{\varepsilon}$$

as the reader may verify from (5.6.13), (5.6.21), and (5.6.23).

The electromagnetic fields generated by a given distribution of charge and current can now be determined by solution of (5.6.24) and (5.6.25).

As an example of the use of vector potentials in the construction of electromagnetic fields, let us consider the transverse magnetic fields within a wave guide parallel to the z-axis. For these fields only the z component of **A** is different from zero: $\mathbf{A} = A_z \mathbf{k}$. From (5.6.23) we have

(5.6.26) $$\operatorname{div} \mathbf{A} = \frac{\partial A_z}{\partial z} = -\varepsilon\mu \frac{\partial \varphi}{\partial t}$$

and this is the condition for the existence of a potential function ψ such that

(5.6.27) $$\varphi = -\frac{\partial \psi}{\partial z} \qquad A_z = \varepsilon\mu \frac{\partial \psi}{\partial t}$$

We impose on the scalar Herz potential ψ the wave equation

(5.6.28) $$\psi_{tt} = \frac{1}{\varepsilon\mu} \Delta \psi$$

and then both φ and $\mathbf{A} = A_z\mathbf{k}$ will satisfy homogeneous forms of the wave equations (5.6.24) and (5.6.25). From (5.6.21) we have

$$\mathbf{E} = -\operatorname{grad} \varphi - \frac{\partial \mathbf{A}}{\partial t}$$

$$= \operatorname{grad} \frac{\partial \psi}{\partial z} - \varepsilon\mu \frac{\partial^2 \psi}{\partial t^2} \mathbf{k}$$

$$= \operatorname{grad} \frac{\partial \psi}{\partial z} - \Delta\psi \mathbf{k}$$

Thus,

(5.6.29) $$E_x = \frac{\partial^2 \psi}{\partial x\, \partial z} \qquad E_y = \frac{\partial^2 \psi}{\partial y\, \partial z} \qquad E_z = -\left(\frac{\partial^2 \psi}{\partial x^2} + \frac{\partial^2 \psi}{\partial y^2}\right)$$

Likewise from (5.6.19),

$$\mu\mathbf{H} = \operatorname{curl}(A_z\mathbf{k}) = \varepsilon\mu \operatorname{curl}\left(\frac{\partial \psi}{\partial t}\mathbf{k}\right)$$

so that

(5.6.30) $$H_x = \varepsilon \frac{\partial^2 \psi}{\partial y\, \partial t} \qquad H_y = -\varepsilon \frac{\partial^2 \psi}{\partial x\, \partial t} \qquad H_z = 0$$

Let us study the transmission along the wave guide of transverse magnetic waves of impressed circular frequency ω. We denote the attenuation factor in the z direction by ih: real values of h will appear in the *absence* of exponential damping along the guide. We set

(5.6.31) $$\psi = e^{i(\omega t - hz)}\psi_0(x, y)$$

Then (5.6.28) becomes

(5.6.32) $$\frac{\partial^2 \psi_0}{\partial x^2} + \frac{\partial^2 \psi_0}{\partial y^2} + (k^2 - h^2)\psi_0 = 0$$

where $\omega = kc = k/\sqrt{\varepsilon\mu}$.

For infinitely conducting (metallic) walls, an appropriate boundary condition is $\psi = 0$. Suppose, for instance, that the cross section is a rectangle of width a, length b. Then

$$\psi_0 = \sin\frac{m\pi x}{a} \sin\frac{n\pi y}{b}$$

is a possible solution, with

$$k^2 - h^2 = \pi^2 \left(\frac{m^2}{a^2} + \frac{n^2}{b^2} \right)$$

Since $k = \omega/c$ is given, and m, n are integers, there will exist an undamped mode of propagation, provided

$$h^2 = \frac{\omega^2}{c^2} - \pi^2 \left(\frac{m^2}{a^2} + \frac{n^2}{b^2} \right)$$

is positive for $m = n = 1$. For a given frequency (i.e., given wavelength), this expression will be negative if m, n are large enough, so that, in general, only a finite number of undamped modes exist. The dimensions a and b of the guide are usually chosen so that only one mode of the given frequency will propagate undamped.

EXERCISES 5.6

1. Show from (5.6.7) that $\mathbf{E}$ is a gradient field $\mathbf{E} = -\operatorname{grad} \varphi$, when $\mathbf{B}$ is constant in time.

2. Show that (5.6.7) implies that $\operatorname{div} \mathbf{B} = 0$ for all time, provided $\operatorname{div} \mathbf{B} = 0$ at some initial time.

3. Derive the wave equation (5.6.18) from Maxwell's equations, given $\sigma = 0$.

4. From (5.6.13) and (5.6.14), deduce that in a medium of conductivity σ,

$$\operatorname{div} \mathbf{E}(t) = e^{-(\sigma/\varepsilon)t} \operatorname{div} \mathbf{E}(0)$$

5. Show that in a medium of conductivity σ, $\mathbf{E}$ and $\mathbf{H}$ satisfy wave equations

$$\mu\varepsilon \frac{\partial^2 \mathbf{E}}{\partial t^2} + \sigma\mu \frac{\partial \mathbf{E}}{\partial t} = \Delta \mathbf{E} - \frac{1}{\varepsilon} \operatorname{grad} \rho$$

$$\mu\varepsilon \frac{\partial^2 \mathbf{H}}{\partial t^2} + \sigma\mu \frac{\partial \mathbf{H}}{\partial t} = \Delta \mathbf{H}$$

where Δ is the vector Laplacian grad div $-$curl curl.

6. Applying the divergence theorem to $\operatorname{div} \mathbf{B} = 0$ in a small coin-shaped region parallel to and enclosing a patch of a boundary or interface, show that B_n is continuous across the boundary. *Hint:* Let the thickness of the region tend to zero.

7. Apply Stokes' theorem to $\partial \mathbf{B}/\partial t + \operatorname{curl} \mathbf{E} = 0$ on a small rectangle intersecting a boundary or interface lengthwise, and let the width of the rectangle tend to zero. Thus show that any tangential component E_t of $\mathbf{E}$ is continuous across the boundary. Deduce that $E_t = 0$ at a perfectly conducting surface.

8. *The Hertz vector.* Show that every vector solution $\mathbf{\Pi}$ of

$$\mu\varepsilon \frac{\partial^2 \mathbf{\Pi}}{\partial t^2} = \Delta \mathbf{\Pi}$$

where $\Delta = \text{grad div} - \text{curl curl}$ defines an electromagnetic field

$$\mathbf{B} = \mu\varepsilon\,\text{curl}\,\frac{\partial \mathbf{\Pi}}{\partial t} \qquad \mathbf{E} = \text{grad div}\,\mathbf{\Pi} - \mu\varepsilon\,\frac{\partial^2 \mathbf{\Pi}}{\partial t^2}$$

9. Show that every vector solution $\mathbf{\Pi}^*$ of the above vector wave equation determines an electromagnetic field with

$$\mathbf{D} = -\mu\varepsilon\,\text{curl}\,\frac{\partial \mathbf{\Pi}^*}{\partial t} \qquad \mathbf{H} = \text{grad div}\,\mathbf{\Pi}^* - \mu\varepsilon\,\frac{\partial^2 \mathbf{\Pi}^*}{\partial t^2}$$

10. Show that the wave guide problem has transverse electric field solutions with

$$D_1 = -\mu\varepsilon\psi^*_{yt} \qquad D_2 = \mu\varepsilon\psi^*_{xt} \qquad D_3 = 0$$
$$H_1 = \psi^*_{xz} \qquad H_2 = \psi^*_{yz} \qquad H_3 = -\psi^*_{xx} - \psi^*_{yy}$$

Find the differential equation satisfied by $\psi^*_0(x, y)$, where $\psi = e^{i(\omega t - hz)}\psi^*_0(x, y)$, and find the lowest modes for a rectangular wave guide of width a, height b.

11. Discuss the propagation of electromagnetic waves of frequency ω between two parallel, perfectly conducting planes separated by distance a. Show that for propagation in a given tangential direction there is an infinity of modes for transverse electric and magnetic waves, and that each mode has a maximum wavelength in the direction of propagation.

12. *The cavity resonator.* Show that, if $K_1 e_1 + K_2 e_2 + K_3 e_3 = 0$ and $K_1^2 + K_2^2 + K_3^2 = \omega^2 \varepsilon \mu$, electromagnetic oscillations with

$$E_1 = e_1 \cos K_1 x \sin K_2 y \sin K_3 z\, e^{-i\omega t}$$
$$E_2 = e_2 \sin K_1 x \cos K_2 y \sin K_3 z\, e^{-i\omega t}$$
$$E_3 = e_3 \sin K_1 x \sin K_2 y \cos K_3 z\, e^{-i\omega t}$$

are possible for a rectangular cavity of dimensions a, b, and c, enclosed by perfectly conducting walls. Here

$$K_1 = \frac{l\pi}{a} \qquad K_2 = \frac{m\pi}{b} \qquad K_3 = \frac{n\pi}{c}$$

where l, m, n are integers.

5.7 EQUATIONS OF QUANTUM MECHANICS

The advent of quantum mechanics has been accompanied by profound changes in the foundations of physical science, as the existence of quanta implies an absolute meaning of size. Thus experiments on atomic systems cannot be performed without affecting the systems themselves, and Newtonian concepts of prediction, determinism, and causality have had to be modified. The mathematical structure of quantum mechanics is therefore basically different from classical mechanics; however, it reduces to the latter when Planck's constant is considered negligible.

In the following brief account we shall omit all historical considerations and instead present the leading concepts axiomatically (see also Morse-Feshbach, Ref. 45; Schiff, Ref. 52). An atomic or nuclear system is described in quantum mechanics by its configuration or *state*. To every possible condition or motion of the system there corresponds a different state, and the mathematical representation of a state is as a vector of a suitable linear space.

POSTULATE I *The states of a quantum system are elements ψ of a linear vector space carrying an inner product.*

From this postulate follows the principle of superposition in quantum mechanics. States ψ_1, ψ_2 of a quantum system can be superposed, giving rise to a new state

$$(5.7.1) \qquad c_1\psi_1 + c_2\psi_2$$

which may be regarded as consisting partly of state ψ_1 and partly of state ψ_2 in a manner that is not possible in classical mechanics. The superposition of a state with itself does not lead to a new state: we regard the vector $c\psi$ as representing the same state for $c \neq 0$, and no state at all if $c = 0$.

For a quantum system enclosed in a region R of space, the vector space of states ψ will be the complex Hilbert space of functions on the domain R (Exercise 1.1.11).

The observations or measurements made on the system are mathematically represented by operations on the state vectors ψ. Physical quantities such as energy, position, or momentum are called *observables*.

POSTULATE II *To every observable there corresponds a Hermitian operator.*

In the complex Hilbert space with inner product $(\varphi, \psi) = \int \varphi\bar\psi \, dV$, an operator P is Hermitian when

$$(5.7.2) \qquad (\varphi, P\psi) = (P\varphi, \psi)$$

for all φ, ψ. The measurement of coordinates x_k is represented by the operation of multiplication by x_k. The measurement of momenta p_k, on the other hand, is represented by the operator $-i\hbar \, \partial/\partial x_k$, where $\hbar$ is Planck's constant divided by 2π; this operator is also Hermitian.

The eigenvalues of a Hermitian operator are real, for, if $P\psi = \lambda\psi$, then

$$(5.7.3) \qquad \begin{aligned} \lambda \, |\psi|^2 &= \lambda(\psi, \psi) = (\lambda\psi, \psi) = (P\psi, \psi) \\ &= (\psi, P\psi) = (\psi, \lambda\psi) = \bar\lambda(\psi, \psi) \\ &= \bar\lambda \, |\psi|^2 \end{aligned}$$

where $\bar\lambda$ is the complex conjugate and $|\psi| > 0$. The existence of eigenvalues of self-adjoint operators is studied in Chapter 6, Sections 6.1 and 6.2, where it is shown that such operators have an infinity of eigenvalues and that their eigenvectors or eigenfunctions form an orthogonal basis for the inner product space. Similar results are true for Hermitian operators, and we shall assume here that every such operator has infinitely many eigenvalues, together with eigenfunctions which form a basis for the vector space of states. For observables represented by differential operators, suitable boundary conditions are to be included.

For example, the observable x has the eigenvalues ξ, where ξ is any real number, and its corresponding eigenfunction is $\delta(x - \xi)$, since

(5.7.4) $$x\delta(x - \xi) = \xi\delta(x - \xi)$$

On the interval $-\infty < x < \infty$ the momentum operator $p = -i\hbar\, \partial/\partial x$ has eigenvalues λ, where $-\infty < \lambda < \infty$, and eigenfunctions $e^{i\lambda x/\hbar}$:

(5.7.5) $$-i\hbar \frac{\partial}{\partial x} e^{i\lambda x/\hbar} = \lambda e^{i\lambda x/\hbar}$$

POSTULATE III *The numerical values possible for any measurement of an observable P are the eigenvalues λ_n of P.*

If the state ψ of the system is a "pure" state, or eigenstate of the operator P,

(5.7.6) $$P\psi_n = \lambda_n \psi_n$$

then measurement of the observable P will yield the numerical value λ_n with certainty. In general, however, the state ψ is expressible by an eigenfunction expansion

(5.7.7) $$\psi = \Sigma\, c_n \psi_n$$

which is a superposition of the pure states ψ_n. A measurement of P will then yield one or other of the eigenvalues λ_n, but we cannot say with certainty which one. Instead, we can only predict the average value of a sequence of many measurements.

POSTULATE IV *The expected mean value of a sequence of measurements of P is*

(5.7.8) $$(\psi, P\psi) = \int \bar{\psi} P\psi \, dV = \Sigma\, \lambda_n |c_n|^2$$

It is desirable to normalize the state vector ψ:

(5.7.9) $$|\psi|^2 = \int \bar{\psi}\psi \, dV = 1$$

Then, if the eigenfunctions ψ_n are also normalized, we have [see (6.4.19)]

$$\Sigma |c_n|^2 = 1$$

and we can refer to the squared moduli of the Fourier coefficients c_n as *probability amplitudes*. For instance, the eigenfunction expansion of the state ψ in eigenfunctions of the position operator x is

(5.7.10) $$\psi(x) = \int \psi(\xi)\, \delta(x - \xi)\, d\xi$$

Thus the Fourier coefficient of $\psi(x)$ with respect to this system is $\psi(\xi)$. The probability that the particle or system is located between ξ and $\xi + d\xi$ is therefore $|\psi(\xi)|^2 \, d\xi$. This interpretation of the squared modulus of ψ as the position probability density suggests the term *wave function* for the state vector $\psi(x)$.

Now consider two operators P and Q, which have the same eigenfunctions ψ_n, corresponding to the eigenvalues λ_n and μ_n respectively. Then

(5.7.11) $$PQ\psi_n = P(\mu_n\psi_n) = \mu_n P\psi_n = \mu_n \lambda_n \psi_n$$

and, similarly,

(5.7.12) $$QP\psi_n = Q(\lambda_n\psi_n) = \lambda_n Q\psi_n = \lambda_n \mu_n \psi_n = PQ\psi_n$$

Since the ψ_n form a basis for the space of state vectors, the operator $PQ - QP$ is identically zero. Thus the operators P and Q commute: $PQ = QP$.

Conversely, if $PQ\psi = QP\psi$ for every ψ, and if ψ_n is an eigenfunction of P, then

(5.7.13) $$PQ\psi_n = QP\psi_n = Q(\lambda_n\psi_n) = \lambda_n Q\psi_n$$

That is, $Q\psi_n$ is also an eigenfunction of P with the same eigenvalue λ_n. If for simplicity we assume that P has no multiple eigenvalues, then $Q\psi_n$ must be equal to a scalar multiple $\mu_n\psi_n$ of ψ_n. Therefore ψ_n is also an eigenfunction of Q. That is, commuting operators have the same pure states. They are compatible in the sense that both can have definite numerical values simultaneously. For example, the operator x of numerical multiplication by x commutes with y and z, and with $p_y = -i\hbar\, \partial/\partial y$ and $p_z = -i\hbar\, \partial/\partial z$.

However x and p_x do not commute, for

(5.7.14) $$(xp_x - p_x x)\psi = -i\hbar x \frac{\partial \psi}{\partial x} + i\hbar \frac{\partial}{\partial x}(x\psi) = i\hbar \psi$$

Thus x and p_x satisfy the symbolic operator equation

(5.7.15) $$xp_x - p_x x = i\hbar$$

known as a *commutation relation*. Observables related in this way are said to be *conjugate*. As these operators have different eigenstates, it is impossible to measure simultaneously both the exact position and the exact momentum of a quantum system. In general, if P and Q do not commute, and the state ψ_n is an eigenfunction of P but not of Q, then measurements of Q will lead to a statistical distribution of values. Thus a measurement of Q, which transforms ψ_n into $Q\psi_n$, will prevent any accurate measurement of P, so that neither measurement can be made with precision.

The physical interpretation of the noncommutation of operators is the famous *uncertainty principle* of Heisenberg, to which we can give a quantitative form by means of the Schwarz inequality. Let P and Q be Hermitian operators, with expected means

(5.7.16) $$\bar{p} = (\psi, P\psi) \qquad \bar{q} = (\psi, Q\psi)$$

respectively. If we write $P = \bar{p} + P_1$ and $Q = \bar{q} + Q_1$, then the means of P_1 and Q_1 are zero. Also $PQ - QP = P_1Q_1 - Q_1P_1$, as the reader can easily verify.

Now consider the expected mean of the expression $(1/i\hbar)(PQ - QP)$, namely,

(5.7.17) $$\left(\psi, \frac{1}{i\hbar}(PQ - QP)\psi\right) = \left(\psi, \frac{1}{i\hbar}(P_1Q_1 - Q_1P_1)\psi\right)$$

$$= \frac{1}{\hbar}(iP_1\psi, Q_1\psi) + \frac{1}{\hbar}(Q_1\psi, iP_1\psi)$$

$$= \frac{1}{\hbar}[(iP_1\psi, Q_1\psi) + \overline{(iP_1\psi, Q_1\psi)}]$$

$$= \frac{2}{\hbar}\,\text{Re}\,(iP_1\psi, Q_1\psi)$$

By the Schwarz inequality for the complex Hilbert space (Exercise 1.1.12), this expression is less in magnitude than

$$\frac{2}{\hbar}|P_1\psi||Q_1\psi|$$

But the norm $|P_1\psi| = |(P - \bar{p})\psi|$ is the standard deviation δp of the measurements of P about their mean $\bar{p}$. Therefore we may write

(5.7.18) $$\left|\left(\psi, \frac{1}{i\hbar}(PQ - QP)\psi\right)\right| \leq \frac{2}{\hbar}\delta p\,\delta q$$

For conjugate operators P and Q, with $PQ - QP = i\hbar$, the left side of (5.7.18) is unity since ψ is assumed normalized. The Heisenberg uncertainty relation

(5.7.19) $$\delta p \cdot \delta q \geq \tfrac{1}{2}\hbar$$

is thus established.

A quantum system is defined when expressions for its kinetic energy T and potential energy V are given. The sum of these is $T + V$, the total energy. Now energy is an observable which is conjugate to time in the same way that momentum is conjugate to space coordinates. As in classical mechanics, the observable corresponding to total energy is called the Hamiltonian H, and we have

(5.7.20) $$H = T + V$$

By the Heisenberg relation $\delta E \cdot \delta t \geq \frac{1}{2}\hbar$, where E is the expected energy. For the eigenstates ψ_n of H, which satisfy

(5.7.21) $$H\psi_n = E_n\psi_n$$

we can measure the energy with precision and, therefore, the time must be completely indeterminate. The energy eigenstates represent the stationary states of the system, the eigenvalues E_n being the energy levels.

The simplest possible system is a free particle with mass m and zero potential energy. Then,

(5.7.22) $$H\psi = \frac{1}{2m}(p_1^2 + p_2^2 + p_3^2)\psi = \frac{-\hbar^2}{2m}\Delta\psi$$

The equation for the energy levels is

(5.7.23) $$H\psi = \frac{-\hbar^2}{2m}\Delta\psi = E\psi$$

If the particle is enclosed within a rectangular solid of side lengths a, b, c, with one corner at the origin, we can find the stationary states at once. The wave amplitude must be zero outside the region and, by continuity, it vanishes on the sides. The eigenfunctions are

(5.7.24) $$\psi_{lmn} = c \sin\frac{l\pi x}{a} \sin\frac{m\pi y}{b} \sin\frac{n\pi z}{c}$$

while the eigenvalues must be

(5.7.25) $$E_{lmn} = \frac{\pi^2 \hbar^2}{2m}\left[\left(\frac{l}{a}\right)^2 + \left(\frac{m}{b}\right)^2 + \left(\frac{n}{c}\right)^2\right]$$

where l, m, and n are integers.

For the equation of motion of a quantum system, we have

POSTULATE V *The equation of motion is*

(5.7.26) $$i\hbar \frac{\partial \psi}{\partial t} = H\psi$$

For a particle of mass m in a potential field of force $V(x, y, z)$, this equation takes the form

(5.7.27) $$i\hbar \frac{\partial \psi}{\partial t} = \frac{-\hbar^2}{2m}\Delta\psi + V\psi$$

which is known as Schrödinger's equation.

If the stationary states of the system are ψ_n, with energy levels E_n, this time-dependent equation has separated solutions $\psi_n e^{-iE_n t/\hbar}$, and its general solution, representing a superposition of energy states, is

(5.7.28) $$\psi = \sum_n c_n \psi_n e^{-iE_n t/\hbar}$$

The probability density of such a state is

$$\bar{\psi}\psi = \sum_{m,n} c_m c_n \bar{\psi}_m \psi_n e^{i(E_m - E_n)t/\hbar}$$

which is an oscillating function of time, the characteristic frequencies ω_{mn} being given by $\hbar\omega_{mn} = E_m - E_n$.

For further study of this fascinating subject, we refer the reader to the bibliography (Morse-Feshbach, 45, and Schiff, 52).

EXERCISES 5.7

1. Show that $i\hbar\, \partial/\partial x$ is Hermitian, and find its eigenvalues and eigenfunctions on $-l \leq x \leq l$, with the boundary condition $\psi(-l) = \psi(l)$.

2. Show how to expand an arbitrary $\psi(x)$ in a series of eigenfunctions of Exercise 5.7.1. Find a matrix which transforms the vector of coefficients of $\psi(x)$ in this expansion into the vector of coefficients of $i\hbar\psi'(x)$. Similarly, find a matrix representing for the given basis of eigenfunctions the operation of multiplication of $\psi(x)$ by x.

3. For any operator H independent of t, show that (5.7.26) has the solution $\psi = e^{-iHt/\hbar}\psi_0$.

4. Show that the general solution of (5.7.26) can be expressed as $\psi = \Sigma e^{-iE_n t/\hbar}\psi_n$, where $H\psi_n = E_n\psi_n$. Show how to solve an initial value problem for Schrödinger's equation, supposing that the eigenfunctions of H are known.

5. Show that the Hamiltonian of the harmonic oscillator (1.5.23) is $H = (1/2m)(p^2 + m^2\omega^2 x^2)$, where ω is the circular frequency. Show that the Schrödinger equation (in x only) is

$$\frac{d^2\psi}{dx^2} + \left(\frac{2mE}{\hbar^2} - \frac{m^2\omega^2}{\hbar^2}x^2\right)\psi = 0$$

6. Show that Hermite's differential equation

$$y'' - 2xy' + 2ny = 0 \qquad -\infty < x < \infty,$$

has polynomial solutions $H_n(x)$ of degree n when n is an integer. Show that $\psi_n(x) = ce^{-(\beta/2)x^2} H_n(\sqrt{\beta}x)$, where $\beta\hbar = m\omega$, is an eigenfunction for Exercise 5.7.5, with energy level $E_n = (n + \tfrac{1}{2})\hbar\omega$.

7. Show that if $P_k = i\hbar\, \partial/\partial x_k$, $Q_k = x_k$, where multiplication by x_k is indicated, then

$$P_k Q_l - Q_l P_k = i\hbar\, \delta_{kl}$$

8. Show that if two Hermitian matrices P, Q commute, there is a similarity transformation which renders both P and Q diagonal.

9. Show that solutions ψ of (5.7.26), where H is a Hermitian operator, have norms which are constant in time.

10. Show that $\partial\rho/\partial t + \text{div}\,\mathbf{J} = 0$, where $\rho = \bar{\psi}\psi$ is the probability density and $\mathbf{J} = -\dfrac{i\hbar}{2m}(\bar{\psi}\,\nabla\psi - \psi\,\nabla\bar{\psi})$ is the probability "current." Interpret and characterize the current-free states ψ.

11. By integrating e^{-z^2} around the sector $0 < \theta < \pi/4$, $0 < r < R$, and letting $R \to \infty$, show that

$$\int_0^\infty e^{ix^2} \, dx = \sqrt{\frac{\pi}{2}} \frac{1+i}{2}$$

12. Show that the Green's function of the one-dimensional Schrödinger equation for a free particle of mass m is

$$K(x, t) = \frac{\sqrt{m}}{2\sqrt{\pi \hbar t}} (1 - i) e^{-imx^2/2\hbar t}$$

Calculate at time t the position probability density of the particle, assuming it was located at the origin when $t = 0$.

CHAPTER 6

General Theory of Eigenvalues and Eigenfunctions

6.1 THE MINIMUM PROBLEM

We have already given numerous examples of the solution of partial differential equations by the method of separation of variables. Actually most of the explicit solutions known in applied mathematics have been found this way. Moreover, it is through this approach that eigenvalues and eigenfunctions appear. In this chapter we shall study the general properties of eigenvalues and eigenfunctions, and will indicate the wide scope and also the limitations inherent in their use (Courant-Hilbert, Ref. 13, vol. 1).

By separation of variables we are able to reduce the number of independent variables in an equation, and in many cases to reduce a partial differential equation to ordinary differential equations. It is comparatively easy to compile information about the solutions of the ordinary differential equations encountered in this way. However, we should not regard a problem as "solved" if it is reduced to ordinary differential equations, and we should not regard an ordinary differential equation as solved if we have only invented a notation for its solutions. The process of solving a problem is gradual, involving the gathering of information and an increase of understanding of the system. Even if an explicit Fourier series solution is known, there may be much work needed to make numerical predictions in a specific instance.

The reader will therefore appreciate that the "solution" of a problem in applied mathematics neither begins nor ends with the finding of a formula. Conversely, much can be discovered about a system even if it is too complicated to admit of an explicit solution. It is from this viewpoint that we shall study eigenvalues and eigenfunctions—we wish to find their qualitative properties, and this is possible in situations of great generality.

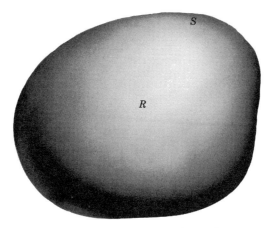

Figure 6.1 Region and boundary surface.

If in the wave equation $U_{tt} = \Delta U$, we set $U = e^{ikt}u$, or if in the heat flow equation $U_t = \Delta U$, we set $U = e^{-k^2 t}u$, then we find that the space factor $u = u(x, y, z)$ satisfies the Helmholtz equation

(6.1.1) $$\Delta u + k^2 u = 0$$

With this equation will be prescribed a single boundary condition, which may have one of the following forms:

1. $u = 0$: first, or Dirichlet, condition.
2. $\dfrac{\partial u}{\partial n} = 0$: second, or Neumann, condition.
3. $\dfrac{\partial u}{\partial n} + hu = 0$: third, or mixed, condition.

The boundary condition will apply on a surface S enclosing a region of space R in three dimensions, or a closed curve C bounding a plane area A in two dimensions, or at the two endpoints (x_1, x_2) of a segment $x_1 \leq x \leq x_2$ in one space dimension. We shall follow the three-dimensional terminology in this chapter (Figure 6.1).

For regions bounded by coordinate surfaces of certain special types, which we call *separable*, it is possible to find eigenvalues and eigenfunctions by further separations of the space coordinates. For these regions it is sufficient to analyze certain particular ordinary differential equations obtained after separation of the variables has taken place. In Chapters 8 and 9 we shall discuss in detail the theory associated with the Bessel and Legendre equations, which arise in this way for cylinders and spheres.

However the separable coordinate systems are not numerous, being in practice restricted to ellipsoidal systems or particular degenerate cases of such systems. For regions of all other types, it is therefore necessary to treat (6.1.1) directly. Let us suppose, then, that all possible separations have already been performed. We can accommodate some separated problems with no great difficulty by considering the slightly more general Sturm-Liouville equation

$$(6.1.2) \qquad \Delta u - q(P)u + \lambda \rho(P)u = 0$$

where $\rho = \rho(P) > 0$ is a positive function of position. Here $q(P)$ is a bounded function which we can take as positive by addition of a constant to λ.

Let us reverse the standpoint of Chapter 5 and find an integral which is minimized by the solutions of the given differential equation. We shall then use this integral to show that eigenvalues and eigenfunctions exist and have certain useful properties. Such an integral, which represents the "total energy" of a given function, is the Dirichlet integral

$$(6.1.3) \qquad E(u) = \int_R [(\nabla u)^2 + q(P)u^2] \, dV$$

The Dirichlet integral is a positive definite quadratic functional defined over the space of testing functions $u(P)$, P in R. Indeed $E(u) = 0$ implies that $u \equiv 0$ in R, since the second term is positive otherwise.

If we require $u = 0$ on S, and ask for the function minimizing $E(u)$, then the answer is clearly that $u = 0$. If our problem is to be significant, a further restriction is required. Let us ask for the normalized function u such that

$$(6.1.4) \qquad H(u) = \int_R \rho u^2 \, dV = 1$$

which minimizes $E(u)$. The theory of the Lagrange multiplier for constrained extremum problems is applicable to our infinite-dimensional problem and it indicates that we should look for the free minimum of

$$(6.1.5) \qquad E(u) - \lambda H(u) = \int_R [(\nabla u)^2 + (q - \lambda \rho)u^2] \, dV$$

The mathematical problem of minimizing this quadratic functional is both difficult and historic. The method for minimizing this functional was first proposed by Riemann in connection with Dirichlet's problem, which we discussed in Section 4.2, and it was subsequently criticized by Weierstrass. The difficulty is to show that a minimizing function *exists*. It is not always true that such minimum problems have solutions, and examples of unusual domains for which Dirichlet's problem may have no solution have been given by Lebesgue. Thus the actual attainment of a minimum of the quadratic integral (6.1.3) cannot be taken for granted. However, we shall not enter into the searching analysis required for such existence theorems, as it is beyond our scope. (Garabedian, Ref. 24, Chapter 8). Instead, we assume that a minimum value for $E(u)$ is attained by a minimizing function $u_1(P)$. This assumption, which we may call "Dirichlet's principle," is true for smooth, bounded domains.

We shall also assume that the minimizing function $u_1(P)$ is twice continuously differentiable: this can be proved by advanced methods which we omit. We wish to show here that $u_1(P)$ is an eigenfunction which satisfies (6.1.2). Suppose that $v(P)$ is any other function which vanishes on the boundary S. Then $u = u_1 + \varepsilon v$ also vanishes on S, and since u_1 is the minimizing function, then $E(u) - \lambda H(u)$ cannot be less than the minimum value. However, if we expand the quadratic integrals, we find

(6.1.6) $\qquad E(u) = E(u_1 + \varepsilon v) = E(u_1) + 2\varepsilon E(u_1, v) + \varepsilon^2 E(v)$

and

(6.1.7) $\qquad H(u) = H(u_1 + \varepsilon v) = H(u_1) + 2\varepsilon H(u_1, v) + \varepsilon^2 H(v)$

Here the bilinear terms are

(6.1.8) $\qquad E(u_1, v) = \int_R [\nabla u_1 \cdot \nabla v + q u_1 v] \, dV$

and

(6.1.9) $\qquad H(u_1, v) = \int_R \rho u_1 v \, dV$

Both of these expressions can be regarded as scalar products.

From our inequality

(6.1.10) $\qquad E(u) - \lambda H(u) \geq E(u_1) - \lambda H(u_1)$

we now find the further inequality

(6.1.11) $\qquad 2\varepsilon[E(u_1, v) - \lambda H(u_1, v)] + \varepsilon^2[E(v) - \lambda H(v)] \geq 0$

LEMMA 6.1.1 *For every smooth function v vanishing on S,*

(6.1.12) $\qquad E(u_1, v) - \lambda H(u_1, v) = 0$

Proof. To establish this result, let us suppose that the left side of (6.1.12) is not zero. With no loss of generality, we can suppose it is positive. Then choose ε negative, and so small that the magnitude of the second term on the left in (6.1.11), the term with the factor ε^2, is less than the magnitude of the first term. If this is true, we can delete the second term from (6.1.11) without altering the inequality sign. But now the left side is negative, which contradicts the inequality. Therefore our assumption that (6.1.12) does not hold must be false. This completes the proof of the lemma.

Since (6.1.12) holds for arbitrary functions v vanishing on S, it will imply a restriction on the function u_1. To show what this is, we apply Green's first formula (Exercise 5.3.8).

$$(6.1.13) \qquad \int_R \nabla u_1 \cdot \nabla v \, dV = -\int_R v \Delta u_1 \, dV + \int_S v \frac{\partial u_1}{\partial n} \, dS$$

The surface integral vanishes since $v = 0$ on S, and we find

$$(6.1.14) \qquad \int_R v(\Delta u_1 - qu_1 + \lambda \rho u_1) \, dV = 0$$

THEOREM 6.1.1 *The minimizing function $u_1(P)$ is an eigenfunction of (6.1.2) and the corresponding eigenvalue λ_1 is the minimum value $E(u_1)$.*

We must show that the differential equation $\Delta u_1 + (\lambda \rho - q)u_1 = 0$ holds throughout R. Suppose, to the contrary, that it does not. Then the expression on the left side is a continuous function different from zero at some point P_0 of R. Suppose it is positive. Then there is a neighborhood N of P_0 in which this quantity is positive. The neighborhood N can be chosen as a sphere of radius ε. Now choose for v a testing function with support N, which is positive within N and zero outside N. But in this case the integral (6.1.14) must be positive, which is a contradiction. This shows that the differential equation (6.1.2) holds throughout R.

To prove the second part of the theorem we calculate as follows:

$$E(u_1) = E(u_1, u_1) = \int_R [(\nabla u_1)^2 + qu_1^2] \, dV$$

$$= -\int_R u_1(\Delta u_1 - qu_1) \, dV$$

$$= \int_R u_1 \lambda \rho u_1 \, dV = \lambda H(u_1) = \lambda$$

Therefore the eigenvalue, which we now denote by λ_1, is the minimum energy itself.

For concreteness we quote explicitly the simplest example of this theorem. All functions $u(x)$, $0 \leq x \leq a$, with $u(0) = u(a) = 0$, satisfy

$$\int_0^a \left(\frac{du}{dx}\right)^2 dx \geq \frac{\pi^2}{a^2} \int_0^a u^2 \, dx$$

with equality only when $u = c \sin(\pi x/a)$. This general inequality is a consequence of the fact that the lowest eigenvalue of the operator $-(d^2/dx^2)$ on the interval $0 \leq x \leq a$ with Dirichlet end conditions is π^2/a^2.

EXERCISES 6.1

1. Find the minimum of $\int_0^a (du/dx)^2 \, dx$, given $\int_0^a u^2 \, dx = 1$.

2. Find the minimum of $\int_0^a \int_0^b (u_x^2 + u_y^2) \, dx \, dy$, given $u(0, y) = u(a, y) = u(x, 0) = u(x, b) = 0$ and $\int_0^a \int_0^b u^2 \, dx \, dy = 1$. Also find the minimizing function $u(x, y)$ for $0 \leq x \leq a, 0 \leq y \leq b$.

3. Find the minimum in Exercise 6.1.2 if the conditions $u(x, 0) = u(x, b) = 0$ are omitted. Find the new minimizing function.

4. Given a sequence $y_0, y_1, \ldots, y_n$, find the minimum of $\sum_{k=1}^n y_k y_{k+1}$, when $\sum_{k=1}^n y_k^2 = 1$ and $y_0 = y_{n+1} = 0$.

5. Find the lowest (and all other) eigenvalues of the operator $d^2/d\theta^2$ on the circle with parameter angle θ, where $\theta \pm 2m\pi$, m integral, represent the same point. Why is it not necessary to specify homogeneous boundary conditions?

6. Show that the function $y(x)$, $a \leq x \leq b$, which gives the minimum of

$$\int_a^b \left[p(x) \left(\frac{dy}{dx} \right)^2 + q(x) y^2 \right] dx + \beta y^2(b) - \alpha y^2(a)$$

with $\int_a^b p(x) y^2(x) \, dx = 1$ is the lowest eigenfunction of

$$\frac{d}{dx}\left[p(x) \frac{dy}{dx} \right] + [\lambda \rho(x) - q(x)] y = 0$$

with

$$p(a) y'(a) + \alpha y(a) = 0 \qquad p(b) y'(b) + \beta y(b) = 0$$

6.2 SEQUENCES OF EIGENVALUES AND EIGENFUNCTIONS

In Section 1.2 we showed how the eigenvalues and eigenfunctions of a symmetric finite matrix can be found by successive minimum methods. We shall now extend this scheme to the infinite-dimensional differential operator $\Delta - q$ of (6.1.2). The formal properties of the matrix eigenvalues and eigenvectors will certainly carry over to the more general case. A most significant example is the following analogue of Theorem 1.2.1.

THEOREM 6.2.1 *The eigenvalues of the self-adjoint operator $\Delta - q$ are real, and eigenfunctions corresponding to distinct eigenvalues are orthogonal with respect to the weight function ρ.*

Proof. Again we first establish the second statement. Let u_k, u_l be eigenfunctions with distinct eigenvalues λ_k, λ_l. That is,

(6.2.1) $$\Delta u_k - q u_k + \lambda_k \rho u_k = 0$$

and

(6.2.2) $$\Delta u_l - q u_l + \lambda_l \rho u_l = 0$$

We multiply these equations by u_l and u_k respectively and subtract. We find

(6.2.3) $$u_l \Delta u_k - u_k \Delta u_l = (\lambda_l - \lambda_k)\rho u_k u_l$$

To the left-hand side we can apply Green's second formula by integrating over R. This gives

(6.2.4) $$(\lambda_l - \lambda_k)\int_R \rho u_k u_l \, dV = \int_R (u_l \Delta u_k - u_k \Delta u_l) \, dV$$
$$= \int_S \left(u_l \frac{\partial u_k}{\partial n} - u_k \frac{\partial u_l}{\partial n}\right) dS$$

But the eigenfunctions u_k, u_l satisfy one of the boundary conditions of Section 6.1. For the first or second type of boundary condition, the surface integral on the right of (6.2.4) clearly vanishes. We leave as an exercise the same result for the third or mixed boundary condition.

As the right side of (6.2.4) is zero, and $\lambda_k - \lambda_l \neq 0$, it follows that

(6.2.5) $$\int_R \rho u_k u_l \, dV = 0$$

as stated.

To prove the reality of the eigenvalues, we now suppose that $\lambda = \mu + i\nu$, with $\nu \neq 0$, is a complex eigenvalue with corresponding eigenfunction u. Then, since all other quantities in (6.1.2) and the boundary condition are real, the complex conjugate value $\bar\lambda = \mu - i\nu$ is an eigenvalue with eigenfunction $\bar u$. But now, by (6.2.5), u and $\bar u$ are orthogonal, so

(6.2.6) $$\int_R \rho u \bar u \, dV = \int_R \rho |u|^2 \, dV = 0$$

which forces $u = 0$ and shows that $\lambda = \mu + i\nu$ was not a true eigenvalue. This completes the proof of Theorem 6.2.1.

Before turning to our next theorem, we make one observation. If u_k is an eigenfunction with eigenvalue λ_k, and u any function that vanishes on S, then

(6.2.7) $$E(u_k, u) = \int_R [\nabla u_k \cdot \nabla u + q u_k u] \, dV$$
$$= -\int_R (\Delta u_k - q u_k) u \, dV$$
$$= \lambda_k \int_R \rho u_k u \, dV = \lambda_k H(u_k, u)$$

Thus if $H(u_k, u) = 0$, then also $E(u_k, u) = 0$, so the functions u_k, u are orthogonal with respect to both the E and H inner products.

Any other eigenfunctions $u_2, u_3, \ldots$ must be sought among functions orthogonal to u_1. Therefore we adjoin this condition to the minimum problem, which now reads: Find the minimum of $E(u)$ for all functions u satisfying

(6.2.8) $\qquad H(u_1, u) = 0 \qquad H(u) = 1 \qquad u = 0 \text{ on } S$

Let the minimizing function, the existence and smoothness of which we assume, be denoted by u_2. As in Section 6.1, we can show that the minimum property implies

(6.2.9) $\qquad E(u_2, v) - \lambda_2 H(u_2, v) = 0$

for every v orthogonal to u_1 and which vanishes on S.

But now we encounter a complication. To deduce from (6.2.9) after integration by parts that u_2 is an eigenfunction, we need to have (6.2.9) valid for *all* functions that vanish on S, not just those functions v of the subspace orthogonal to u_1. Here, however, we can use (6.2.7). Let w be any function which vanishes on S, and resolve w into components parallel and perpendicular to u_1. Thus

(6.2.10) $\qquad w = c_1 u_1 + v$

where we shall require $H(u_1, v) = 0$. Then $H(w, u_1) = H(v, u_1) + c_1 H(u_1) = c_1$, so that the coefficient c_1 is determined as the Fourier coefficient of w with respect to u_1. Now v was defined as the component of w orthogonal to u_1 and so v satisfies (6.2.9). In (6.2.7) we set $k = 2$ and $u = u_1$, multiply by c_1, and add to (6.2.9). The result is

(6.2.11) $\quad E(u_2, w) - \lambda H(u_2, w) = -\int_R (\Delta u_2 - q u_2 + \lambda \rho u_2) w \, dV = 0$

Now we can follow the method of (6.1.14) and Theorem 6.1.1 and show that u_2 is also an eigenfunction.

If in (6.2.11) we set $w = u_2$, we find $E(u_2) = \lambda_2 H(u_2) = \lambda_2$. Since λ_2 is the value of the constrained minimum of $E(u)$, λ_2 is at least as large as λ_1, the minimum without the constraint $H(u_1, u) = 0$.

This process of defining eigenvalues and eigenfunctions now continues by induction. To find any higher eigenvalue and eigenfunction, we minimize the energy integral over the normalized functions orthogonal to all the known eigenfunctions.

THEOREM 6.2.2 *The minimizing function u_n of the variational problem*

(6.2.12) $\qquad E(u) = minimum$

over all functions u vanishing on the boundary and satisfying

(6.2.13) $\qquad H(u) = 1, H(u, u_1) = 0, \ldots, H(u, u_{n-1}) = 0$

is the nth eigenfunction of (6.2.1), and the nth successive minimum value of $E(u)$ attained in this problem is the corresponding eigenvalue λ_n, where $\lambda_1 \leq \lambda_2 \leq \lambda_3 \leq \cdots \leq \lambda_n \leq \cdots$.

Proof. We again suppose that the minimizing function exists, and denote it by u_n. Let v be any function vanishing on S and orthogonal to $u_1, \ldots, u_{n-1}$

in the H inner product. Then the sum $u_n + \varepsilon v$ has these same properties and is admissible in the minimizing problem. Therefore, since u_n gives a minimum of $E - \lambda H$, we have

$$E(u_n + \varepsilon v) - \lambda H(u_n + \varepsilon v)$$
$$= E(u_n) - \lambda H(u_n) + 2\varepsilon[E(u_n, v) - \lambda H(u_n, v)] + \varepsilon^2[E(v) - \lambda H(v)]$$
$$\geq E(u_n) - \lambda H(u_n)$$

Canceling certain terms and choosing ε appropriately, we see that this inequality implies

(6.2.14) $$E(u_n, v) - \lambda H(u_n, v) = 0$$

To extend this relation to functions w which vanish on S but are arbitrary within R, we write

(6.2.15) $$w = v + \sum_{k=1}^{n-1} c_k u_k$$

and choose the coefficients as

(6.2.16) $$c_k = H(w, u_k)$$

Then v is orthogonal to u_k, as required for (6.2.14). Also, from (6.2.7) we see that, since u_n is orthogonal to the eigenfunction u_k in the H inner product, it is also orthogonal in the E inner product. Therefore

(6.2.17) $$E(u_n, w) - \lambda H(u_n, w) = E(u_n, v) - \lambda H(u_n, v)$$
$$+ \sum_{k=1}^{n-1} c_k[E(u_n, u_k) - \lambda H(u_n, u_k)] = 0$$

by (6.2.7) and (6.2.14). We are now able to integrate by parts in (6.2.17) and deduce that

(6.2.18) $$\int_R (\Delta u_n - q u_n + \lambda \rho u_n) w \, dV = 0$$

Since this relation holds for functions w which can be chosen arbitrarily within R, the factor in parentheses of the integrand must vanish:

(6.2.19) $$\Delta u_n - q u_n + \lambda \rho u_n = 0$$

Therefore u_n is an eigenfunction.

The eigenvalue we identify by setting $u = u_k = u_n$ in (6.2.7). Then,

(6.2.20) $$E(u_n) = \lambda H(u_n) = \lambda$$

so the eigenvalue λ, which we henceforth denote by λ_n, is the nth successive minimum value of the energy integral. Clearly $\lambda_n \geq \lambda_{n-1}$, since the minimum value of $E(u)$ is maintained or increased by the additional constraint $H(u_{n-1}, u) = 0$.

Continuing the induction, we can find an infinite sequence of eigenvalues λ_n, ranged in increasing order, and corresponding eigenfunctions u_n. These are eigenfunctions of the Dirichlet problem.

By a minor modification we can adapt this process to the boundary conditions of the second or third kind. For the Neumann or normal derivative condition, we simply drop the constraint that u vanishes on S for our class of admissible functions in the minimum problem. Then the vanishing of the normal derivative of the eigenfunction on the boundary appears as a "natural" boundary condition. We will demonstrate this as a particular case of the mixed type of boundary condition.

To consider the third, or mixed, boundary condition, we shall add a boundary term to the energy integral. Let the mixed condition be

$$(6.2.21) \qquad \frac{\partial u}{\partial n} + hu = 0$$

Then we construct the energy integral

$$(6.2.22) \qquad E_h(u) = \int_R [(\nabla u)^2 + qu^2] \, dV + \int_S hu^2 \, dS$$

Again, this is a positive definite quadratic functional, and it reduces to $E(u)$ when the function h on the boundary is identically zero.

We set up the minimum problem for $E_h(u)$ with the normalizing condition $H(u) = 1$ and the side conditions $H(u_1, u) = 0, \ldots, H(u_{n-1}, u) = 0$, as before. Here $u_1, \ldots, u_{n-1}$ are the eigenfunctions of this new problem. The requirement of vanishing on the surface S is omitted.

We find now the analogue of (6.2.17), which is

$$(6.2.23) \qquad E_h(u_n, w) - \lambda H(u_n, w) = 0$$

where u_n is the new minimizing function. When we integrate by parts, a surface integral term is present, and we obtain

$$(6.2.24) \qquad -\int_R (\Delta u_n - qu_n + \lambda \rho u_n) w \, dV + \int_S \left(\frac{\partial u_n}{\partial n} + hu_n \right) w \, dS = 0$$

First we can choose w to vanish on S, and then the arbitrary choice of w in the interior of R leads us to the eigenfunction equation (6.2.19). Thus the volume integral in (6.2.24) vanishes altogether. This leaves the surface integral, in which we have an arbitrary choice of w on S. But this is possible only when

$$(6.2.25) \qquad \frac{\partial u_n}{\partial n} + hu_n \equiv 0 \qquad \text{on } S$$

This shows that u_n is an eigenfunction of the differential equation with eigenvalue λ_n, for the third boundary condition.

EXERCISES 6.2

1. Find the maximum and minimum values of $x_1^2 + 2x_2^2 + 3x_3^2$ on the sphere $x_1^2 + x_2^2 + x_3^2 = 1$. Find the maximum and minimum values under the additional constraints

   (a) $x_1 = 0$ or (b) $x_1 = 0, x_2 = 0$ or (c) $x_1 = 2x_2$

2. Discuss the analogy of Theorem 6.2.2 and the standard method of finding the principal axes of a quadric $\sum_{kl} a_{kl} x_k x_l = 1$.

3. If B is a positive definite matrix and the eigenvectors of a matrix A with respect to B satisfy $Ax = \lambda Bx$, and if A and B are symmetric, show:
   (a) The eigenvalues are real.
   (b) The eigenfunctions are orthogonal with respect to B in a suitable sense.

4. Find the sequence of constrained minimum values of the quadratic form $\sum_{k=1} y_k y_{k+1}$, with $\Sigma y_k^2 = 1$. *Hint:* see Chapter 2, Section 2.3.

5. By referring to the formulas stated in Exercises 5.2.12 and 5.2.13, show that

$$\oint_C \left[(u_{ss} + Ku_n)v_n - (v_{ss} + Kv_n)u_n + v\frac{\partial}{\partial s}\left(\frac{\partial u_n}{\partial s} - Ku_s\right) - u\frac{\partial}{\partial s}\left(\frac{\partial v_n}{\partial s} - Kv_s\right) \right] ds = 0$$

and that

$$\iint_S (u \,\Delta\Delta v - v \,\Delta\Delta u)\, dx\, dy = \oint_C \left[uT(v) - vT(u) - \frac{\partial u}{\partial n} M(v) + \frac{\partial v}{\partial n} M(u) \right] ds$$

$$= \oint_C \left[u\frac{\partial}{\partial n} \Delta v - v\frac{\partial}{\partial n} \Delta u - \frac{\partial u}{\partial n} \Delta v + \frac{\partial v}{\partial n} \Delta u \right] ds$$

6. Let u_n, u_m be two different eigenfunctions satisfying the respective equations $\Delta\Delta u_n = \lambda_n u_n$ and $\Delta\Delta u_m = \lambda_m u_m$ together with one of the following pairs of boundary conditions: either

   (a) $a_1 u + a_2 \dfrac{\partial u}{\partial n} = 0 \qquad a_1 M(u) + a_2 T(u) = 0$

or

   (b) $b_1 u + b_2 \dfrac{\partial u}{\partial n} = 0 \qquad b_1 \Delta u + b_2 \dfrac{\partial}{\partial n} \Delta u = 0$

or

   (c) $c_1 u + c_2 T(u) = 0 \qquad d_1 \dfrac{\partial u}{\partial n} + d_2 M(u) = 0$

If $\lambda_n \neq \lambda_m$, obtain the orthogonality property

$$\iint_S u_n u_m\, dx\, dy = 0$$

6.3 VARIATIONAL PROPERTIES OF EIGENVALUES AND EIGENFUNCTIONS

The minimum problem method presented in the two preceding sections shows that a boundary value problem for a finite region R does in general have a sequence of eigenvalues $\lambda_1 \leq \lambda_2, \ldots$, and a set of orthonormal eigenfunctions u_n. However, we can pursue this method further and find many specific properties, both qualitative and quantitative, of these sequences.

The minimum problem for $E(u)$ with the constraint $H(u) = 1$ was discussed above by the Lagrange multiplier device. However, the same result is also obtained if we minimize the "Rayleigh quotient"

$$(6.3.1) \qquad \frac{E(u)}{H(u)}$$

without the normalizing constraint. The minimum value is λ_1, the minimum when $H(u_1, u) = 0$ is λ_2, and so on, so that the constrained minimum with $H(u, u_k) = 0$ for $k = 1, \ldots, n-1$ is λ_n. The eigenvalues λ_n are the successive minimum values of the Rayleigh quotient, where the successive constraints are the orthogonal conditions $H(u_k, u) = 0$, $k = 1, \ldots, n-1$.

We shall now establish a maximum-minimum principle, which states, in effect, that any other constraints will lead to lower successive minimum values. In other words, the orthogonal constraints most effective in raising the numerical values of the higher eigenvalues are the orthogonal conditions $H(u_k, u) = 0$ naturally associated with the problem (Courant-Hilbert, Ref. 13, vol. 1).

THEOREM 6.3.1 (*Maximum-Minimum Principle*). *Let* $\mu_1, \mu_2, \ldots, \mu_n, \ldots$ *denote the successive minimum values of the Rayleigh quotient* $E(u)/H(u)$, *subject for $n > 1$ to the $n - 1$ successive orthogonal constraints*

$$(6.3.2) \qquad H(v_1, u) = 0, H(v_2, u) = 0, \ldots, H(v_{n-1}, u) = 0$$

defined by a given sequence of functions $v_1, v_2, \ldots, v_{n-1}, \ldots$ *Then*

$$(6.3.3) \qquad \mu_1 = \lambda_1, \mu_2 \leq \lambda_2, \ldots, \mu_n \leq \lambda_n, \ldots$$

with equality in these relations up to λ_n only if $v_1 = u_1, v_2 = u_2, \ldots, v_{n-1} = u_{n-1}$, where the u_k are the eigenfunctions, and the λ_k the corresponding eigenvalues.

Proof. Suppose that $v_1, \ldots, v_{n-1}$ are given, and let $H(v_j, u) = 0$, $j = 1, \ldots, n-1$. Then let us choose a linear combination

$$(6.3.4) \qquad u = \sum_{k=1}^{n} c_k u_k$$

of the eigenfunctions u_k, which satisfies

$$(6.3.5) \qquad H(v_j, u) = \sum_{k=1}^{n} c_k H(v_j, u_k) = 0 \qquad j = 1, \ldots, n-1$$

We can regard these $n - 1$ conditions as linear homogeneous equations for the n coefficients $c_1, \ldots, c_n$. Such a system of n linear equations for $n - 1$ unknowns always has a solution with $c_1, \ldots, c_n$ not all zero (see p. 3), and since the conditions are homogeneous, the solution is undetermined by a common multiplicative factor. We impose one further condition of normalization to determine this factor, namely,

$$(6.3.6) \quad H(u) = H\left(\sum_{k=1}^{n} c_k u_k\right) = \sum_{k,l=1}^{n} H(u_k, u_l) c_k c_l$$

$$= \sum_{k,l=1}^{n} \delta_{kl} c_k c_l = \sum_{k=1}^{n} c_k^2 = 1$$

Now let us show that the Rayleigh quotient of this particular function u is at most λ_n. Since the minimum μ_n cannot exceed this particular value of the Rayleigh quotient, we shall then have $\mu_n \leq \lambda_n$. The Rayleigh quotient of (6.3.4) is

$$(6.3.7) \quad \frac{E(u)}{H(u)} = E(u) = \sum_{k,l=1}^{n} c_k c_l E(u_k, u_l)$$

$$= \sum_{k,l=1}^{n} c_k c_l \lambda_k \delta_{kl} = \sum_{k=1}^{n} \lambda_k c_k^2$$

since $E(u_k, u_l) = \lambda_k \delta_{kl}$ by (6.2.7). This Rayleigh quotient is at most equal to λ_n, since $\lambda_k \leq \lambda_n$ for $k \leq n$, so that

$$(6.3.8) \quad \frac{E(u)}{H(u)} \leq \lambda_n \sum_{k=1}^{n} c_k^2 = \lambda_n$$

This establishes the general result. For completeness, however, we must examine the possible cases of equality in (6.3.8). There are two such cases, the first being if the n leading eigenvalues $\lambda_1, \ldots, \lambda_n$ are all equal. In this case we also have $\mu_1 = \lambda_1 = \cdots = \mu_n = \lambda_n$, while the minimizing functions are the eigenfunctions. Since this case is trivially true for $n = 1$, we shall always have equality in the first relation $\mu_1 \equiv \lambda_1$ for which no constraint of orthogonality is present.

The second case of equality in (6.3.8) arises if $c_1 = c_2 = \cdots = c_{n-1} = 0$ and $c_n = 1$ in (6.3.4); then $u \equiv u_n$. If there is also equality in the preceding relations $\mu_k = \lambda_k$ for $k = 1, 2, \ldots, n - 1$, then we must have $v_1 = u_1, v_2 = u_2, \ldots, v_{n-1} = u_{n-1}$, so that the given constraints are exactly those of the variational problem of Section 6.2. That is, we have demonstrated the maximum-minimum property that the eigenfunctions themselves provide the most effective orthogonal constraints.

As a first application of this theorem, let us prove the following result.

THEOREM 6.3.2 *All the eigenvalues λ_n are increased (or not decreased) if an additional constraint is imposed on the system.*

Proof. Let λ_k, u_k be the successive eigenvalues and eigenfunctions of the system. Let an extra constraint be imposed which reduces the class of functions admissible in the variational problem, and let λ'_k, u'_k be the new constrained sequences. Since λ_1 is an absolute minimum, it is clear that $\lambda'_1 \geq \lambda_1$. However, to compare the higher eigenvalues, we must set up an artificial intermediate problem. Let λ''_k denote the minimum of the Rayleigh quotient for functions u' of the restricted class which are subject to the *original* orthogonality conditions $H(u_k, u') = 0$. Then, by the maximum-minimum principle, $\lambda'_k \geq \lambda''_k$, since the u_k are not the natural eigenfunctions of the restricted problem. However, we also have $\lambda''_k \geq \lambda_k$, since the intermediate problem defining the λ''_k has the same orthogonality conditions as the original unconstrained problem, and a smaller class of admissible functions. Therefore any function u' admissible in the minimizing problem defining λ''_k is also admissible in the problem defining λ_k. Thus $\lambda'_k \geq \lambda''_k \geq \lambda_k$, which establishes the theorem.

For example, the eigenvalues μ_k of the equation $\Delta u - qu + \mu \rho u = 0$, with the Neumann condition $\partial u/\partial n = 0$ on S, are found by minimizing the Rayleigh quotient without any boundary restriction on the functions. Suppose now that on some part S_1 of S we constrain all admissible functions to be zero. That is, we impose a Dirichlet condition on S_1. Then the new eigenvalues μ'_k will be larger: $\mu'_k \geq \mu_k$. If the Dirichlet condition applies to the entire boundary and the eigenvalues are denoted by λ_k, we have

$$\lambda_k \geq \mu'_k \geq \mu_k$$

Actually these inequalities will be strict.

THEOREM 6.3.3 *If the eigenvalue problem with Dirichlet condition $u = 0$ on S_1 and Neumann condition $\partial u/\partial n = 0$ on $S - S_1$ has eigenvalues μ_k, then μ_k is monotonic increasing with S_1.*

Another type of comparison theorem can be derived for the mixed boundary condition, for which we minimize $E_h(u)$ of (6.2.22). Suppose the boundary function h is increased to h'. Then values $E_h(u)$ are also increased and, in particular, the lowest eigenvalue increases: $\lambda_1 \leq \lambda'_1$. To show a similar relation for the higher eigenvalues λ'_k of the new problem, we shall consider the successive minimum values λ''_k of $E'_h(u)$ subject to the orthogonal constraints $H(u_k, u) = 0$, $k = 1, \ldots, n - 1$. These are less than the values λ'_k by the maximum-minimum principle, since the constraints $H(u_k, u) = 0$ are not natural to the increased value of h. But also $\lambda_k \leq \lambda''_k$, simply because $E_h(u) \leq E'_h(u)$ and the admissible functions for the kth successive minimum problem satisfy the same constraints. Finally, $\lambda_k \leq \lambda''_k \leq \lambda'_k$.

THEOREM 6.3.4 *The eigenvalues λ_k of the third boundary problem are monotonic increasing with the coefficient function h, for $h \geq 0$.*

Next we shall use comparison methods to determine the asymptotic behavior of the eigenvalues as n tends to infinity. It will turn out that expressions involving the size of the region R appear in a quite simple way in the leading terms of these estimates.

Consider a region R and let us divide it into a number of partial regions $R_1, \ldots, R_m$ by inserting certain internal dividing surfaces S_j. The inequalities we can find will depend on the boundary condition, so we first consider the Dirichlet problem for R. Let us impose further constraints, namely, that the admissible functions should vanish on the internal boundaries S_j. By Theorem 6.3.2 this will raise the eigenvalues.

But the constrained problem includes among its eigenvalues all the eigenvalues of the partial regions $R_1, \ldots, R_m$. Indeed, if we consider all the functions which vanish identically in all these regions except one, we have a class identical with the admissible functions of that part region.

To express the relationships now before us, we require *enumeration functions* for the sequences of eigenvalues. Let us denote by $N_R^1(\lambda)$ the number of Dirichlet eigenvalues λ_k of the given problem on R, which satisfy $\lambda_k \leq \lambda$. For the partial domains $R_1, \ldots, R_m$ we have the corresponding enumeration functions $N_{R_1}^1(\lambda), \ldots, N_{R_m}^1(\lambda)$. Let us assemble in increasing order all the eigenvalues of all the part domains and denote the nth one by λ_n^*. We have shown that $\lambda_n^* \geq \lambda_n$. The enumeration function for the combined sequence is clearly the sum of the separate enumeration functions, and since the sequence of larger eigenvalues has the smaller enumeration function, we find

(6.3.9) $$N_{R_1}^1(\lambda) + \cdots + N_{R_m}^1(\lambda) \leq N_R^1(\lambda)$$

Secondly, consider the Neumann boundary condition which requires no restriction on the admissible functions of the variation problem. When R is divided into the part regions $R_1, \ldots, R_m$, we shall relax the restriction of continuity for our admissible functions on the internal boundaries S_j. By Theorem 6.3.2 we shall obtain lower values for the combined sequence of eigenvalues of the part regions. Let us denote the enumeration function for the Neumann problem on R by $N_R^2(\lambda)$. Reasoning parallel to that which led to (6.3.9) now yields

(6.3.10) $$N_R^2(\lambda) \leq N_{R_1}^2(\lambda) + \cdots + N_{R_m}^2(\lambda)$$

If we also denote by $N_R^{(h)}(\lambda)$ the enumeration function for the mixed boundary condition, then the limit $h \to \infty$ corresponds to the Dirichlet condition. By Theorem 6.3.4, (6.3.9), and (6.3.10), we find the following theorem.

THEOREM 6.3.5 *The enumeration functions satisfy the chain of inequalities*

(6.3.11) $$\sum_{k=1} N_{R_k}^1(\lambda) \leq N_R^1(\lambda) \leq N_R^{(h)}(\lambda) \leq N_R^2(\lambda) \leq \sum_{k=1} N_{R_k}^2(\lambda)$$

where $R = R_1 + \cdots + R_m$.

We must now turn to the simplest example—the enumeration function for a rectangle. The Helmholtz equation

(6.3.12) $$\Delta u + \lambda u = 0$$

in the rectangle $0 \le x \le a$, $0 \le y \le b$, has the eigenfunctions and eigenvalues

(6.3.13) $$u_{lm} = \sin\frac{l\pi x}{a} \sin\frac{m\pi y}{b} \qquad \lambda_{lm} = \pi^2\left(\frac{l^2}{a^2} + \frac{m^2}{b^2}\right)$$

where $l, m = 1, 2, \ldots$, for the Dirichlet condition. For the Neumann condition the eigenfunctions are cosines and the values $l, m = 0, 1, 2, \ldots$ are included

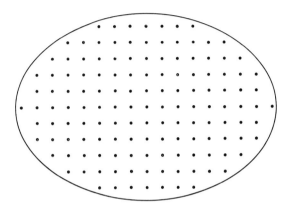

Figure 6.2 Lattice points within ellipse.

in the count. The respective enumeration functions are just the number of integer lattice points within one quadrant of the ellipse (Figure 6.2)

(6.3.14) $$\frac{x^2}{a^2} + \frac{y^2}{b^2} = \frac{\lambda}{\pi^2}$$

For large values of λ this number is asymptotic to the area

(6.3.15) $$\frac{ab\lambda}{4\pi}$$

of one quadrant. The error is of the order of magnitude of the length of the quarter ellipse curve, that is, of order $\sqrt{\lambda}$. The Dirichlet and Neumann enumerations also differ by a term of order $\sqrt{\lambda}$, the number of integer points on the positive semiaxes of the ellipse.

Referring now to (6.3.11), we see that the enumeration functions are bounded above and below by sums which in their leading asymptotic terms contain the sums of areas $A = ab$ of rectangular partial domains. From the theory of the Riemann integral we know that the area of the given region R can be expressed as a limit of such "rectangular" sums of partial areas. Thus we conclude that (6.3.15) holds now for the Helmholtz equation (6.3.12) on a general plane domain of finite area A. That is, we have for all the enumeration functions

(6.3.16) $$N(\lambda) = \frac{A\lambda}{4\pi} + O(\lambda^{1/2})$$

Inverting this relationship, we find since $N(\lambda_n) = n$ by definition that

(6.3.17) $$\lambda_n \sim \frac{4\pi}{A} N(\lambda_n) = \frac{4\pi n}{A}$$

To apply these considerations to the more general equation (6.1.2), we observe that, since $q(P)$ is bounded, it will have no effect on the leading term of order λ or on any term which grows large with λ. It represents a finite displacement, at every point (x, y), along the λ-axis. In a small rectangle at (x, y) we can allow for the values of the coefficient function $\rho(x, y)$ by taking the eigenvalue parameter to be $\lambda \rho(x, y)$. When we now sum the rectangles over the full domain R, we are led to the Riemann integral

(6.3.18) $$N_R(\lambda) \sim \frac{\lambda}{4\pi} \int_R \rho \, dx \, dy$$

which generalizes (6.3.16). The finiteness of this integral ensures a discrete spectrum, and we have established the following theorem.

THEOREM 6.3.6 *The nth eigenvalue of* (6.1.2) *for a two-dimensional region R is*

(6.3.19) $$\lambda_n \sim \frac{4\pi n}{\int_R \rho \, dx \, dy} \qquad n \to \infty$$

EXERCISES 6.3

1. Show that the nth eigenvalue λ_n of the first boundary problem decreases if R is enlarged.

2. Show that λ_n decreases if $\rho(P)$ is increased.

3. Show that (6.3.11) remains true if the summation on the left is over a set of part regions contained in R, and on the right over part regions the sum of which includes R.

4. For the eigenvalue problems on a rectangle of length a and height b, show that $N^2(\lambda) - N^1(\lambda) \sim (a + b)\sqrt{\lambda}/\pi$.

5. Show that for the ordinary differential equation $y'' + \lambda \rho(x) y = 0$, $y(0) = y(a) = 0$,
$$\lambda_n \sim \pi^2 n^2 \Big/ \left(\int_0^a \sqrt{\rho} \, dx \right)^2.$$

6. Show that for the region R in three dimensions,
$$\lambda_n \sim 6^{2/3} \pi^{4/3} n^{2/3} \Big/ \left(\int_R \rho^{3/2} \, dx \, dy \, dz \right)^{2/3}$$

7. From (6.3.9) deduce the *nodal line theorem* for the Dirichlet eigenfunctions: the nodes (zeros) of u_n divide R into at most n connected part regions. *Hint:* Show that (6.3.9) yields a contradiction otherwise.

6.4 EIGENFUNCTION EXPANSIONS

So far in this book we have encountered Fourier series of several different kinds and Fourier integral transforms. A considerable variety of the problems in mathematical physics arising in connection with line intervals, rectangles, or rectangular solids can be treated by Fourier series in one or more variables. The sines and cosines appearing in those series are, as we have already indicated, the eigenfunctions of the eigenvalue problems that arise in these cases. In this chapter we have shown that regions of arbitrary shape also have eigenvalues and eigenfunctions. We have seen some properties of these when the differential operator of the problem is the Laplacian, and there are similar eigenfunction and eigenvalue sequences for any elliptic differential operator. However, the problems of applied mathematics lead most often to the Laplace operator, for reasons of symmetry discussed in Chapter 4. In this book we shall consider only the eigenvalues and eigenfunctions of the Laplace operator for certain regions. Most of the special functions of mathematical physics are introduced as eigenfunctions for regions of specific shapes. In Chapters 8 and 9 we shall study in detail the eigenfunctions of cylinders and spheres.

Here, however, we are going to use the eigenfunctions $u_n(P)$ of a general region R to generalize the Fourier series expansion theorem. That is, we will show that an arbitrary function on the region R can be expanded in a Fourier series.

If an arbitrary function f of a certain vector space can be expressed as the sum of a linear combination of given functions u_n,

$$(6.4.1) \qquad f = \sum c_n u_n$$

then we can say that the u_n span the vector space. However, for vector spaces of infinite dimension, in contrast to finite dimensional spaces, the notion of spanning a vector space depends on the kind of convergence specified for the series (6.4.1).

We shall actually use a new kind of convergence—convergence in *mean square*—which is especially suited to inner product spaces. The reader should understand that convergence of the series at each point of the region is not implied by mean square convergence. The advantage of this type of convergence is that simpler statements and theorems are now possible.

DEFINITION 6.4.1 *A sequence of functions $f_n(P)$ defined on a region R converge in mean square to a function f if*

$$(6.4.2) \qquad \lim_{n \to \infty} \int_R \rho |f_n - f|^2 \, dV = 0$$

Likewise, a series $\sum_{n=1}^{\infty} c_n u_n$ converges in mean square to f when its partial sums $f_n = \sum_{k=1}^{n} c_k u_k$ converge as in (6.4.2) to f. That is, given $\varepsilon > 0$, we can find an $n = n(\varepsilon)$ such that

$$(6.4.3) \qquad \int_R \rho \left| f - \sum_{k=1}^{n} c_k u_k \right|^2 dV < \varepsilon$$

whenever $n > n(\varepsilon)$.

DEFINITION 6.4.2 *If a sequence of functions u_n spans a vector space V, with convergence in mean square understood, then the sequence u_n is called complete in V.*

For example, the functions

$$\frac{1}{\sqrt{2l}} \qquad \frac{1}{\sqrt{l}}\cos\frac{n\pi x}{l} \qquad \frac{1}{\sqrt{l}}\sin\frac{n\pi x}{l} \qquad n = 1, 2, \ldots$$

form a complete orthonormal set on the interval $(-l, l)$. If we remove the function $1/\sqrt{2l}$, the sequence is no longer complete because, for instance, the arbitrarily chosen function 1 cannot be expressed as a linear combination of the remaining functions.

Let V be the vector space (its dimension is infinite) of functions of integrable square on the region R.

THEOREM 6.4.1 *The eigenfunctions of the Helmholtz equation in R and any one of the boundary conditions of Sections 6.1 and 6.2 on S form a complete orthonormal sequence in V.*

The proof of this theorem will occupy the remainder of this section. We first have some formal preliminaries. If the formal expansion (6.4.1) holds, then the *Fourier coefficients* c_k are given by

$$(6.4.4) \qquad c_k = \int_R \rho f u_k \, dV = H(f, u_k)$$

To demonstrate this, we multiply (6.4.1) by ρu_k and integrate over R. All terms of the series, with one exception, yield zero because the eigenfunctions u_k are orthogonal. Since u_k is normalized, we now find (6.4.4).

THEOREM 6.4.2 *A linear combination*

$$(6.4.5) \qquad \sum_{k=1}^{n} a_k u_k$$

yields the best mean square approximation to a given function f, when the coefficients a_i are the same as the Fourier coefficients c_k.

Proof. The best mean square approximation is that for which the norm of the remainder $r_n = f - \sum_{k=1}^{n} a_k u_k$ is least. This norm is

$$(6.4.6) \qquad \begin{aligned} H(r_n) &= H\left(f - \sum_{k=1}^{n} a_k u_k\right) \\ &= H(f) - 2\sum_{k=1}^{n} a_k H(f, u_k) + \sum_{k,l=1}^{n} a_k a_l H(u_k, u_l) \\ &= H(f) - 2\sum_{k=1}^{n} a_k c_k + \sum_{k=1}^{n} a_k^2 \end{aligned}$$

where we have used (6.4.4). We shall minimize this quadratic form in the coefficients a_k by completing the square in every term. Thus we find

(6.4.7) $$H\left(f - \sum_{k=1}^{n} a_k u_k\right) = H(f) + \sum_{k=1}^{n} (a_k - c_k)^2 - \sum_{k=1}^{n} c_k^2$$

where the last sum has been added and again subtracted on the right. Since the first term and the last sum do not depend on the a_k, we shall get the minimum value by choosing $a_k = c_k$ as stated. This completes the proof of Theorem 6.4.2.

For this most efficient mean square approximation of Theorem 6.4.2 we have a geometrical interpretation, namely, that the shortest distance from the point f to the hyperplane W_n spanned by $u_1, \ldots, u_n$ is the *perpendicular* distance. Also the projection of f on W_n is

(6.4.8) $$P_{W_n} f = \sum_{k=1}^{n} c_k u_k$$

To prove this, we must show that the remainder

(6.4.9) $$r_n = f - P_{W_n} f = f - \sum_{k=1}^{n} c_k u_k$$

is orthogonal to W_n. But if $1 \leq l \leq n$, then

(6.4.10) $$H(r_n, u_l) = H(f, u_l) - \sum_{k=1}^{n} c_k H(u_k, u_l)$$
$$= c_l - \sum_{k=1}^{n} c_k \delta_{kl} = 0$$

and this proves the orthogonality.

Now let us refer to the variational problem of Sections 6.1 and 6.2. Suppose that f is an admissible function for the variational problem. This implies only that, for the Dirichlet problem, f is a function vanishing on S. By (6.4.9), the remainder r_n is also an admissible function, and by (6.4.10) it satisfies the auxiliary conditions of the variational problem for the $(n+1)$st eigenvalue λ_{n+1}. Therefore the Rayleigh quotient of r_n is at least as great as λ_{n+1}:

(6.4.11) $$\frac{E(r_n)}{H(r_n)} \geq \lambda_{n+1}$$

We rearrange this relation to read

(6.4.12) $$H(r_n) \leq \frac{E(r_n)}{\lambda_{n+1}}$$

Let us now show that $E(r_n)$ is bounded above independently of n. To calculate this quadratic integral we shall need to evaluate [see (6.2.22)]

$$E(f, u_n) = \int_R (\nabla f \cdot \nabla u_n + qfu_n) \, dV + \int_S hfu_n \, dS$$

(6.4.13)
$$= -\int_R f(\Delta u_n - qu_n) \, dV + \int_S f\left(\frac{\partial u_n}{\partial n} + hu_n\right) dS$$

$$= \lambda_n \int_R \rho f u_n \, dV + 0 = \lambda_n c_n$$

Here the surface integral vanishes for the Dirichlet problem because, in that case, we have assumed f vanishes on S. For the second and third boundary conditions, the factor containing u_n is zero on S, so the conclusion is the same.

With the help of (6.4.13) we compute the energy integral of the remainder r_n:

(6.4.14)
$$E(r_n) = E\left(f - \sum_{k=1}^n c_k u_k\right)$$
$$= E(f) - 2\sum_{k=1}^n c_k E(f, u_k) + \sum_{k,l=1}^n c_k c_l E(u_k, u_l)$$

By (6.4.13) with $f = u_k$ and l replacing n we find $E(u_k, u_l) = \lambda_l H(u_k, u_l) = \lambda_l \delta_{kl}$. Thus

(6.4.15)
$$E(r_n) = E(f) - 2\sum_{k=1}^n c_k \lambda_k c_k + \sum_{k=1}^n c_k^2 \lambda_k$$
$$= E(f) - \sum_{k=1}^n \lambda_k c_k^2 \leq E(f)$$

From (6.4.12) we now conclude

(6.4.16)
$$H(r_n) \leq \frac{E(f)}{\lambda_{n+1}}$$

Since $\lambda_{n+1} \to \infty$ as $n \to \infty$, we can make $H(r_n)$ as small as we please by choosing n sufficiently large. This completes the proof of Theorem 6.4.1.

We conclude this section with some remarks on the sum of the squares of the coefficients. If in (6.4.7) we take $a_k = c_k$, then, since the term on the left is not negative, we find

(6.4.17)
$$\sum_{k=1}^n c_k^2 = H(f) - H(r_n) \leq H(f)$$

But the norm now on the right-hand side does not depend on n, so if we allow n to tend to infinity in the sum, we find in the limit

(6.4.18)
$$\sum_{k=1}^\infty c_k^2 \leq H(f)$$

This statement is known as *Bessel's inequality* and it holds for any orthonormal system whether complete or not.

However, if the orthonormal set u_n is complete, as are the eigenfunctions of R, the remainder term $H(r_n)$ will always tend to zero, and the equality sign will hold in (6.4.18). This yields an alternative criterion, known as *Parseval's theorem*, for the completeness of the set $\{u_n\}$.

THEOREM 6.4.3 *The orthonormal set u_n of V is complete if and only if, for every function f of V, the relation*

$$(6.4.19) \qquad \sum_{k=1}^{\infty} c_k^2 = H(f)$$

holds for the Fourier coefficients c_k of f.

We can interpret the Fourier coefficients c_k as the components, or coordinates, in a coordinate system with axes defined by the mutually orthogonal unit vectors u_k. Then Parseval's relation (6.4.19) is the infinite dimensional analogue of Pythagoras' theorem.

EXERCISES 6.4

1. Is the set $\sin n\pi x/l$, $n = 1, 2, \ldots$, complete on the interval $(0, l)$? On the interval $(0, 2l)$? (Let the weight function be unity.)

2. Is the set $\exp(in\pi x/l)$, $n = 0, \pm 1, \pm 2, \ldots$, with weight function unity complete on the intervals $(0, l)$, $(-l, l)$, $(-2l, 2l)$?

3. Indicate the changes necessary in Theorems 6.4.1, 6.4.2, and 6.4.3 if the inner product is Hermitian and the eigenfunctions u_n are complex.

4. Show that, if $\{u_n\}$ is a complete set, and the Fourier coefficients of f are c_n while those of g are d_n then

$$\int_R \rho f g \, dV = \sum_{k=1}^{\infty} c_k d_k$$

5. Show that a continuous function which is orthogonal to every function of a complete set must be identically zero.

6. Using the fact that a uniformly convergent series can be integrated term by term, show that, if the partial sums of a Fourier series converge uniformly, they converge to the value of the function.

7. From Exercise 6.4.4, deduce that $E(f) = \Sigma \lambda_n c_n^2$ if f is a twice differentiable function. *Hint:* Let $\Delta f - qf = g$.

8. The Dirac distribution δ_Q with the weight function $\rho(P)$ is defined by

$$H(\delta_Q, \varphi) = \int_R \rho(Q)\delta(P, Q)\varphi(Q) \, dV_Q = \varphi(P)$$

Show that the Fourier coefficients of $\delta(P, Q)$ with respect to the basis $u_n(P)$ are $u_n(Q)$, and deduce that $\delta(P, Q) = \sum_{k=1}^{\infty} u_n(P)u_n(Q)$.

6.5 THE RAYLEIGH–RITZ APPROXIMATION METHOD

One advantage of the variational outlook we have adopted in this chapter is that it leads naturally to certain useful methods of approximation. The numerical determination of eigenvalues, which is very necessary in engineering applications, cannot be carried out exactly except for the simplest problems for regions of very special shapes. Therefore, a good method of numerical approximation, particularly for the lowest eigenvalue, is valuable.

The Rayleigh-Ritz minimum method to be presented here is, in effect, a replacement of the continuous mechanical system by a system of a finite number of degrees of freedom, the eigenvalues of which can be found by solving a polynomial equation (Rayleigh, Ref. 48; Sagan, Ref. 50). The choice of this finite approximating system, and even of the number of degrees of freedom in it, is at our disposal, and we can use any prior knowledge to make the approximation efficient. For example, the continuous vibrating string can be approximated by a chain of beads.

Though physical insight should be used whenever possible to gain a good first approximation, it is not absolutely necessary as the finite-dimensional approximating system can be chosen in a purely mathematical way.

We wish to find, approximately, the leading (lowest) eigenvalues and eigenfunctions of a continuous system. Let us select n functions $f_1, \ldots, f_n$, defined over the domain of the system, and which are admissible in the variational problem. That is, the f_k satisfy the boundary condition only in the case of the Dirichlet problem. These *trial functions* shall be chosen by inspection, or otherwise, to be as good approximations to the eigenfunctions as we can find. Now we form the linear combination

(6.5.1) $$u_0 = \sum_{k=1}^{n} a_k f_k$$

with undetermined coefficients a_k, and insert it into the Rayleigh quotient $E(u_0)/H(u_0)$. This quotient now becomes a function of the n numbers a_k.

By analogy with the full variational problem, we shall try to find the minimum value of the Rayleigh quotient, regarded as a function of the n variables $a_1, \ldots, a_n$. This minimum value we take as the approximate eigenvalue. As we shall see, the minimizing process will also yield approximate eigenfunctions.

The problem of minimizing the quotient $E(u_0)/H(u_0)$ is equivalent, mathematically, to the Lagrange-multiplier problem of finding the minimum of $E(u_0) - \Lambda H(u_0)$. This was remarked in Section 6.3 for the infinite-dimensional problem and is also true for the present finite case.

From (6.5.1) we have

(6.5.2) $$E(u_0) - \Lambda H(u_0) = \sum_{k,l=1}^{n} a_k a_l [E(f_k, f_l) - \Lambda H(f_k, f_l)]$$

Let us define two $n \times n$ matrices E and H by their matrix elements

(6.5.3) $$e_{kl} = E(f_k, f_l) = e_{lk}$$

and

(6.5.4) $$h_{kl} = H(f_k, f_l) = h_{lk}$$

Then we have the quadratic form

(6.5.5) $$E(u_0) - \Lambda H(u_0) = \sum_{k,l=1}^{n} (e_{kl} - \Lambda h_{kl}) a_k a_l$$

to minimize. We differentiate with respect to a typical variable a_k and equate the derivative to zero. The result is

(6.5.6) $$\sum_l (e_{kl} - \Lambda h_{kl}) a_l = 0 \qquad k = 1, \ldots, n$$

Let us denote by **a** the column vector with components $a_1, \ldots, a_n$. Then we can express (6.5.6) in matrix form:

(6.5.7) $$E\mathbf{a} = \Lambda H \mathbf{a}$$

If the original set of functions $f_1, \ldots, f_n$ had been orthonormal with respect to the H functional, then the H matrix would be the unit matrix. In that case the vectors **a** satisfying (6.5.7) would be the eigenvectors and the values of Λ eigenvalues of the matrix E. When H is not a unit matrix, we shall still call the vectors **a** of (6.5.7) *eigenvectors of E with respect to H* and, similarly, for eigenvalues. From (6.5.6) we find that the eigenvalue equation or condition for the existence of eigenvectors is

(6.5.8) $$\det(e_{kl} - \Lambda h_{kl}) = 0$$

This is a polynomial equation of degree n in Λ. We shall suppose that its n roots $\Lambda_1, \ldots, \Lambda_n$ can be found, perhaps approximately. These roots are real and can be ranged in increasing order, $\Lambda_1 \leq \Lambda_2 \leq \cdots \leq \Lambda_n$, as approximate values for the lowest n roots $\lambda_1, \ldots, \lambda_n$ of the full variational problem.

Let $\mathbf{a}^{(j)}$ be the eigenvector of E with respect to H, corresponding to the root Λ_j. Then (6.5.1) yields an approximate eigenfunction

(6.5.9) $$U^{(j)} = \sum_{k=1}^{n} a_k^{(j)} f_k \qquad j = 1, \ldots, n$$

We can easily verify that $\mathbf{a}^{(j)}$ and $\mathbf{a}^{(h)}$ are orthogonal with respect to the H matrix. Multiplying

$$\sum_{l=1}^{n} (e_{kl} - \Lambda_j h_{kl}) a_l^{(j)} = 0$$

by $a_k^{(h)}$ and summing over k, and multiplying

$$\sum_{l=1}^{n} (e_{kl} - \Lambda_h h_{kl}) a_l^{(h)} = 0$$

by $a_k^{(j)}$ and summing over k, we find after subtraction that, since the matrices e_{kl} and h_{kl} are symmetric,

$$(6.5.10) \qquad (\Lambda_j - \Lambda_h)\left(\sum_{k,l=1}^{n} h_{kl}a_k^{(j)}a_l^{(h)}\right) = 0$$

When $\Lambda_j \neq \Lambda_h$, this forces the second factor to vanish.

THEOREM 6.5.1 *The eigenvectors of E with respect to H are orthogonal with respect to H in the sense that for $\Lambda_j \neq \Lambda_h$,*

$$(6.5.11) \qquad \sum_{k,l=1}^{n} h_{kl}a_k^{(j)}a_l^{(h)} = 0$$

We can now deduce the following property of the approximate eigenfunctions (6.5.9).

COROLLARY 6.5.1 *The approximate eigenfunctions $U^{(j)}$ are orthogonal with respect to the H functional:*
Proof:

$$(6.5.12) \qquad \begin{aligned} H(U^{(j)}, U^{(h)}) &= \sum_{k,l} a_k^{(j)}a_l^{(h)}H(f_k, f_l) \\ &= \sum_{k,l} a_k^{(j)}a_l^{(h)}h_{kl} = 0 \end{aligned}$$

by Theorem 6.5.1.

We can subsequently normalize these eigenfunctions, and we now suppose this accomplished, so that

$$(6.5.13) \qquad H(U^{(j)}, U^{(h)}) = \delta_{jh} \qquad j, h = 1, \ldots, n$$

Since all the f_j are admissible in the variational problem of Sections 6.1 and 6.2, it is seen that the lowest approximate eigenvalue Λ_1 is not less than the true minimum λ_1. We can compare the remaining eigenvalues if we regard the choice of the trial functions f_k as a constraint upon the system. Thus the original infinite-dimensional system has been constrained to fall within the finite-dimensional vector space spanned by $f_1, \ldots, f_n$. But now, by Theorem 6.3.2, we see that every eigenvalue of the original system has been raised.

THEOREM 6.5.2 *The approximate eigenvalues Λ_j are upper bounds for the n lowest eigenvalues:*

$$(6.5.14) \qquad \lambda_j \leq \Lambda_j \qquad j = 1, \ldots, n$$

To find numerical lower bounds for the eigenvalues is much more difficult, and we are not able to discuss it here. However, the accuracy of the approximation Λ_1 is often very close, as we now shall see in some examples.

Example 6.5.1 Find, approximately, the lowest eigenvalue of $y'' + \lambda y = 0$, $0 \leq x \leq 1$, with $y(0) = y(1) = 0$.

We shall not suppose that the exact lowest eigenfunction $\sin \pi x$ is known, but will use one trial function. This is the simplest possible function that satisfies the boundary conditions, namely, $f = x(1 - x)$. The matrices are now scalars, and $\Lambda = E(f)/H(f)$. An elementary calculation yields $\Lambda_1 = 10$. The true value $\lambda_1 = \pi^2 = 9.8696$ is $1\frac{1}{2}\%$ less.

Example 6.5.2 The radially symmetric vibrations of a circular membrane satisfy the ordinary differential equation

$$u_{rr} + \frac{1}{r} u_r + \lambda u = 0 \qquad\qquad 0 \le r \le 1$$

Find the two lowest frequencies.

The Rayleigh quotient appropriate here is

$$\frac{E(u)}{H(u)} = \frac{\int_0^1 \left(\frac{du}{dr}\right)^2 r\, dr}{\int_0^1 u^2 r\, dr}$$

At $r = 1$ we have $u = 0$, and at $r = 0$ we find $u_r = 0$ since u is not permitted to have a cusp or other singularity at the origin. The simplest trial function satisfying these conditions is $f_1 = 1 - r^2$. We also choose, as a second trial function, $f_2 = (1 - r^2)^2$. The matrix elements h_{kl} and e_{kl} are elementary integrals which can be evaluated more rapidly if we use $x = r^2$ as variable of integration. The resulting eigenvalue equation for Λ is

$$\begin{vmatrix} 1 - \dfrac{\Lambda}{6} & \dfrac{2}{3} - \dfrac{\Lambda}{8} \\ \dfrac{2}{3} - \dfrac{\Lambda}{8} & \dfrac{2}{3} - \dfrac{\Lambda}{10} \end{vmatrix} = \frac{\Lambda^2}{960} - \frac{2\Lambda}{45} + \frac{2}{9} = 0$$

The roots are $\Lambda_1 = 5.784$ and $\Lambda_2 = 36.9$. Accurate values can be calculated from the zeros of Bessel functions (Chapter 8), and are 5.783 and 27.3 respectively. Thus the accuracy of the second eigenvalue is comparatively poor.

EXERCISES 6.5

1. Show that $\sum_{k,l=1}^{n} e_{kl}\alpha_k\alpha_l > 0$ and $\sum_{k,l=1}^{} h_{kl}\alpha_k\alpha_l > 0$ unless $\alpha_k \equiv 0$. (This proves that (e_{kl}) and (h_{kl}) are positive definite matrices.)

2. Using Theorem 6.5.1, show that the eigenvalues Λ_j are real.

3. Show that

$$\sum_{k,l=1}^{n} e_{kl} a_k^{(j)} a_l^{(h)} = 0 \qquad\qquad j \ne h$$

where $a_h^{(j)}$ are the eigenvectors of (6.5.6).

4. From (6.5.13) show that $\Lambda_j = E(U^{(j)})$, provided the $U^{(j)}$ are normalized.

5. Find an approximate value for the second eigenvalue of Example 6.5.1, using $f_1 = x(1 - x), f_2 = x^2(1 - x)$.

6. Find approximately the first two eigenvalues of $y'' + \lambda y = 0$, $0 \le x \le 1$, $y(0) = y'(1) = 0$.

7. Show that the ordinary differential equation of the vibrating beam, $u_{xxxx} - \lambda u = 0$, will have eigenvalues given by minima of $\int u_{xx}^2\, dx / \int u^2\, dx$. For the clamped beam of length 1, show that the lowest eigenvalue is approximately 25.4.

6.6 ON THE SEPARATION OF VARIABLES

The reader has now seen numerous examples of the separation of variables method and can appreciate the power and convenience of the procedure. Of all the methods known for solving the partial differential equations of mathematical physics, the separation of variables technique does seem the most

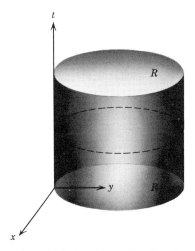

Figure 6.3 Product region $R \times I_+$.

useful, and it has certainly been the most widely used. And yet there is an accidental or fortuitous appearance to the whole scheme. One may well ask, why should an apparently arbitrary hypothesis, namely, that certain solutions are products of separated factors, be so successful? Perhaps our study of the vibrating string in Chapter 2 has provided one answer to this. The separation into normal modes is very natural for such vibrations. Indeed, by perception of the fundamental frequency and its overtones, the human ear itself acts as a kind of spectral analyzer. In many other systems the hypothesis of separated factors of a solution also corresponds to a genuine physical property of the system, namely, the existence of standing waves and normal modes.

We present in this section some mathematical considerations relevant to the separation of variables method, which should provide a mathematical justification and clarification of the question raised above.

First, let us observe that we use this device in two different circumstances. The first is in the separation of the time variable from space variables in a wave or heat equation. This gives an eigenvalue problem in the space variables. The second circumstance arises when the eigenfunctions of this spatial problem can be found by further separations.

In the first of these cases, we will show that if a certain type of problem has a solution, then that solution can be found by the separation method. For a fixed region R of space, we consider the *product region* $R \times I_+$, where I_+ is the interval $t > 0$ of time (Figure 6.3). Let the wave equation

(6.6.1) $$u_{tt} = \Delta u$$

hold on $R \times I_+$, with $u = 0$ on the boundary $S \times I_+$, S being the boundary of R. Let initial data on R be

(6.6.2) $$u(P, 0) = f(P) \qquad u_t(P, 0) = g(P)$$

THEOREM 6.6.1 *Let $u(P, t)$ be a smooth solution of the initial problem (6.6.1) and (6.6.2) on $R \times I_+$, where*

(6.6.3) $$H(f) \qquad E(f) \qquad H(g) \qquad E(g)$$

are finite. Then the eigenfunction solution

(6.6.4) $$\sum_n c_n(t) u_n(P)$$

obtained by separation of the time variable converges in mean square to $u(P, t)$ for each value of t.

Proof. We define the Fourier coefficient

(6.6.5) $$c_n(t) = \int_R u(P, t) u_n(P) \, dV_P$$

where $u_n(P)$ is the nth Dirichlet eigenfunction of R. Thus

$$\ddot{c}_n(t) = \int_R \ddot{u}(P, t) u_n(P) \, dV_P$$
$$= \int_R \Delta u(P, t) u_n(P) \, dV_P$$
(6.6.6) $$= \int_R u(P, t) \Delta u_n(P) \, dV_P$$
$$= -\lambda_n \int_R u(P, t) u_n(P) \, dV_P$$
$$= -\lambda_n c_n$$

Here we have used Green's second formula, the boundary conditions for u and u_n, and the equation $\Delta u_n + \lambda_n u_n = 0$. Let us write $\lambda_n = k_n^2$, where k_n is real because λ_n is positive. The Fourier coefficients of $f(P)$ and $g(P)$ being f_n and g_n respectively, we should solve the ordinary differential equation (6.6.6)

256   GENERAL THEORY OF EIGENVALUES AND EIGENFUNCTIONS

with $c_n(0) = f_n$, $\dot{c}_n(0) = g_n$. Thus

(6.6.7) $$c_n(t) = f_n \cos k_n t + g_n \frac{\sin k_n t}{k_n}$$

Now let us define a remainder solution

(6.6.8) $$r_n(P, t) = u(P, t) - \sum_{j=1}^{n} c_j(t) u_j(P)$$

Clearly $r_n(P, t)$ is a solution of the wave equation and the boundary condition, and satisfies the initial conditions

(6.6.9)
$$r_n(P, 0) = f(P) - \sum_{j=1}^{n} f_n u_n(P)$$
$$\frac{\partial}{\partial t} r_n(P, 0) = g(P) - \sum_{j=1}^{n} g_n u_n(P)$$

Since $f(P)$ and $g(P)$ have Fourier expansions which converge in mean square, we can render the square integrals $H(r_n(0))$ and $H(\dot{r}_n(0))$ as small as we please by choosing n large enough. From (6.4.15) and Exercise 6.4.7 we see that

$$E(r_n) = E(f) - \sum_{k=1}^{n} \lambda_k f_k^2 = \sum_{k=n+1}^{\infty} \lambda_k f_k^2$$

can also be made small.

LEMMA 6.6.1   *Let $u(P, t)$ be a solution of the wave equation* (6.6.1) *on $R \times I_+$, which vanishes on the boundary $S \times I_+$. Then the energy integral*

(6.6.10) $$\mathscr{E}(t) = \int_R [u_t^2 + (\nabla u)^2] \, dV$$

is constant in time.

Proof. We compute the time derivative

(6.6.11)
$$\frac{d\mathscr{E}(t)}{dt} = 2 \int_R (u_t u_{tt} + \nabla u \cdot \nabla u_t) \, dV$$
$$= 2 \int_R u_t (u_{tt} - \Delta u) \, dV + 2 \int_S u_t \frac{\partial u}{\partial n} \, dS$$

by Green's first formula. The surface integral vanishes, since $u = 0$ on $S \times I_+$ implies $u_t = 0$ there. The volume integral vanishes, since u is a solution of (6.6.1). Therefore $E(t)$ is a constant, as asserted.

To complete the proof of Theorem 6.6.1, we observe that $\mathscr{E}_n(t) \equiv \mathscr{E}(r_n(t))$ is equal to its initial value, which is

(6.6.12)
$$\mathscr{E}(r_n(0)) = \int_R \{[\dot{r}_n(0)]^2 + [\nabla r_n(0)]^2\} \, dV$$
$$= H(\dot{r}_n(0)) + E(r_n(0))$$

By choosing n sufficiently large, we shall make each item on the right of (6.6.12) less than ε. From (6.4.12) we know that

$$H(r_n(t)) \leq \frac{E(r_n(t))}{\lambda_{n+1}} < E(r_n(t))$$

when $\lambda_{n+1} > 1$, so that

(6.6.13) $H(r_n(t)) + H(\dot{r}_n(t)) \leq H(\dot{r}_n(t)) + E(r_n(t)) = \mathscr{E}(r_n(t)) < 2\varepsilon$

Therefore $H(r_n(t)) < 2\varepsilon$, and this proves Theorem 6.6.1. The proof also shows that the series for $u_t(P, t)$ converges in mean square for each value of t.

This theorem shows that the separation of variables procedure will yield the correct solution of the problem if there is one. This is an independent

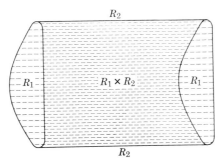

Figure 6.4 Product region $R_1 \times R_2$.

justification of the formal operations on series and integrals, such as interchange of summation and integration, that appear in this method. A direct analytical justification of all these steps would be very tedious. Therefore, whether we are interested in the mathematical properties of the solution or its physical applications, we should first find the formal solution.

Now consider the second situation where separation of variables can appear. When are the eigenfunctions of a space region R products of separate factor eigenfunctions? Clearly R must be a product region such as a rectangle. Also the examples so far considered concern a differential operator with separate differentiation terms in each of the variables.

Let L_i, for $i = 1, 2$, be a differential operator in one (or more) variables x_i, on the domain R_i, and let u_n^i denote the complete set of eigenfunctions and λ_n^i the corresponding eigenvalues. Suppose, for simplicity, Dirichlet boundary conditions for the product domain $R_1 \times R_2$ (Figure 6.4).

THEOREM 6.6.2 *Let $L = L_1 + L_2$, and $R = R_1 \times R_2$. Then the eigenvalues of L are $\lambda_m^1 + \lambda_n^2$ and the eigenfunctions are $u_m^1(x_1)u_n^2(x_2)$, where $m, n = 1, 2, \ldots$.*

Proof. Clearly the products $u_m^1(x_1)u_n^2(x_2)$ are eigenfunctions with the eigenvalues $\lambda_m^1 + \lambda_n^2$ for

(6.6.14) $\begin{aligned}Lu_m^1(x_1)u_n^2(x_2) &= L_1 u_m^1(x_1)u_n^2(x_2) + L_2 u_m^1(x_1)u_n^2(x_2) \\ &= \lambda_m^1 u_m^1(x_1)u_n^2(x_2) + \lambda_n^2 u_m^1(x_1)u_n^2(x_2)\end{aligned}$

258   GENERAL THEORY OF EIGENVALUES AND EIGENFUNCTIONS

What we have to show is, that there are no other eigenfunctions. First we will show that there are no other eigenvalues. The orthogonality of Theorem 6.2.1 applies to eigenfunctions on the product domain. Suppose then that $u(x_1, x_2)$ is an eigenfunction with eigenvalue λ not equal to any combination $\lambda_m^1 + \lambda_n^2$. Then $u(x_1, x_2)$ is orthogonal to every product eigenfunction:

$$(6.6.15) \qquad \int_{R_1} \int_{R_2} u(x_1, x_2) u_m^1(x_1) u_n^2(x_2) \, dx_1 \, dx_2 = 0$$

That is,

$$(6.6.16) \qquad \int_{R_1} u_m^1(x_1) v_n(x_1) \, dx_1 = 0 \qquad m = 1, 2, \ldots$$

where

$$(6.6.17) \qquad v_n(x_1) = \int_{R_2} u(x_1, x_2) u_n^2(x_2) \, dx_2$$

Since $u_m^1(x_1)$ form a complete set on R_1, and $v_n(x_1)$ is a continuous function there, we can use Exercise 6.4.5. There it was shown that a continuous function orthogonal to every function of a complete set is identically zero. Therefore the integrals (6.6.17) all vanish for $n = 1, 2, \ldots$. Again, since the $u_n^2(x_2)$ form a complete set and $u(x_1, x_2)$ is continuous, we find that $u(x_1, x_2)$ must be identically zero. Therefore λ is not an eigenvalue.

Suppose next that $u(x_1, x_2)$ is an eigenfunction with eigenvalue λ equal to one or more of the combinations $\lambda_m^1 + \lambda_n^2$. Let c_{mn} be the Fourier coefficient of $u(x_1, x_2)$ with respect to the product eigenfunctions $u_m^1(x_1) u_n^2(x_2)$ having the same eigenvalue λ.

Then the difference

$$(6.6.18) \qquad u(x_1, x_2) - \sum_{\lambda_m^1 + \lambda_n^2 = \lambda} c_{mn} u_m^1(x_1) u_n^2(x_2)$$

is orthogonal to all the product eigenfunctions with eigenvalue λ, and also to all those with other eigenvalues. By the reasoning used above, the function (6.6.18) must be identically zero, and so $u(x_1, x_2)$ is a linear combination of the product eigenfunctions with the same eigenvalue. This completes the proof of Theorem 6.6.2.

This theorem justifies the hypothesis of separate factors when the differential operator is a sum and the domain a product. Observe that a domain need not be a product in Cartesian coordinates; for example, the circle $r^2 = x^2 + y^2 \leq a^2$ is a product domain $0 \leq r \leq a, 0 \leq \theta \leq 2\pi$, in polar coordinates. The Helmholtz equation in polar coordinates can be written

$$r^2 u_{rr} + r u_r + u_{\theta\theta} + \lambda r^2 u = 0$$

and we observe that the eigenvalue parameter appears multiplied by a function of one of the coordinates. This situation, which is also very common, means that the two "factors" must be treated on a slightly different footing.

Suppose the eigenvalue equation is

$$(6.6.19) \qquad L_1 u + L_2 u = \lambda f(x_1) u$$

where L_1 and L_2 contain differentiations and functions only of x_1 and x_2 respectively. The separation hypothesis $u(x_1, x_2) = u_1(x_1)u_2(x_2)$ now leads to the pair of equations

(6.6.20)
$$L_1 u_1 = [\lambda f(x_1) + \mu]u_1$$
$$L_2 u_2 + \mu u_2 = 0$$

where μ is the new separation constant. Consider μ as an eigenvalue parameter for L_2 and suppose that the boundary or auxiliary conditions applied to u_2 yield a complete orthogonal set of eigenfunctions $u_{2m}(x_2)$ with eigenvalues μ_m. By inserting μ_m in the first of (6.6.20), we are led to eigenfunctions $u_{1mk}(x_1)$ with eigenvalues λ_{mk}, $k = 1, 2, \ldots$, and for each m these will form a complete set in the x_1 variable. Orthogonality of the $u_{1mk}(x_1)$ is with respect to $f(x_1)$ as weight function. The following result can be established.

THEOREM 6.6.3 *The product eigenfunctions*

(6.6.21) $\qquad\qquad u_{1mk}(x_1)u_{2m}(x_2) \qquad\qquad m, k = 1, 2, \ldots$

form a complete orthogonal set on the product domain $R_1 \times R_2$, *with weight function* $f(x_1)$.

In particular, every eigenfunction has this product form. The proof is included in the exercises.

EXERCISES 6.6

1. Separate the partial differential equation

$$\frac{\partial}{\partial x}\left[T(x)\frac{\partial u}{\partial x}\right] = \rho(x)\frac{\partial^2 u}{\partial t^2}$$

for a string of variable density $\rho(x)$ and tension $T(x)$. Show that the normal modes of oscillation are orthogonal with "weight" function $\rho(x)$.

2. Determine the eigenvalues and normal modes for the equation of Exercise 6.6.1 on $1 \leq x \leq 2$, with $u(1, t) = u(2, t) = 0$ if $T(x) = x^2$, $\rho(x) = 1$.

3. Show that the equation

$$u_{xx} + u_{yy} + [a + b(x^2 + y^2)]u = 0$$

has separated solutions of the form $X(x)Y(y)$, and also of the form $R(r)\Theta(\theta)$, where $x = r\cos\theta$, $y = r\sin\theta$. Exhibit the differential equations satisfied by the factors.

4. In spheroidal coordinates (α, β, φ), Laplace's equation takes the form

$$\frac{\partial}{\partial \alpha}\left[(\alpha^2 - 1)\frac{\partial u}{\partial \alpha}\right] + \frac{\partial}{\partial \beta}\left[(1 - \beta^2)\frac{\partial u}{\partial \beta}\right] + \left(\frac{1}{\alpha^2 - 1} + \frac{1}{1 - \beta^2}\right)\frac{\partial^2 u}{\partial \varphi^2} = 0$$

Find separated solutions and give the differential equations satisfied by each factor.

5. For the damped wave equation $u_{tt} - \Delta u + cu = 0$, show that Lemma 6.6.1 holds for the energy integral

$$\int_R (u_t^2 + (\nabla u)^2 + cu^2)\,dV$$

6. Show that the problem (6.6.1)–(6.6.2) has at most one continuous solution. *Hint:* Consider the proof of Theorem 6.6.1.

7. Show that all eigenvalues in Theorems 6.6.2 and 6.6.3 correspond to a finite number of linearly independent eigenfunctions.

8. Adapt the theorems of this section to the second and third boundary conditions.

9. Show that the eigenfunctions (6.6.21) are orthogonal on $R_1 \times R_2$ with weight function $f(x_1)$.

10. Show that any continuous function on $R_1 \times R_2$ orthogonal to all product eigenfunctions (6.6.21) is identically zero.

6.7 SERIES EXPANSIONS AND INTEGRAL TRANSFORMS

The separation of variables method, the eigenfunction expansions including Fourier series, the Fourier integral and other transforms, all form part of a single body of knowledge. This topic can be described by one word—the *spectrum*—which has a special significance in pure mathematics, in physics, and in engineering. Here we wish to emphasize the unity of this group of methods by describing various cases which commonly arise, and relating them to one another. We should see the forest as well as the trees.

Also we discuss an aspect often found difficult by students: when a specific problem is given, how do we decide whether any eigenfunction expansion or integral transform is appropriate, and if so, which one? To continue our metaphor, we need to learn the systematic identification of various species of trees. In Chapters 8 and 9 we shall extend our field of knowledge by numerous examples in cylindrical and spherical functions. The reader will find it a revealing exercise to relate those calculations to the general theory of this chapter (see also Morse-Feshbach, Ref. 45; Tranter, Ref. 59).

The principal stages in the solution of a boundary or initial value problem by the eigenvalue method are:

1. *Formulation.* Derivation of the partial differential equation and the auxiliary conditions. This somewhat separate art, which involves the construction of an abstract model of the problem and so requires physical insight, was discussed in Chapter 5.

2. *Separation of variables.* When the domain is a product in suitable coordinates, and the differential equation is separable, we form the two or more separated equations and specify the auxiliary conditions for each of them.

3. *Determination of eigenvalues and eigenfunctions.* For each separate equation we must choose appropriate solutions. For an equation in the time variable, we have an initial problem of the type of Chapter 1. For an equation in space coordinates, we have to find eigenvalues and eigenfunctions by a sequence of separations as in Theorem 6.6.3, if necessary. To do this, we must call on our knowledge of the special functions defined for this purpose, and set up the required series expansions or integral transforms.

4. *Construction and evaluation of the solution.* By assembling the series or integrals provided by these transforms, we construct a formal solution. If a

general formula is needed, the formal solution provides a Green's function formula to be evaluated. For a particular solution, we try to evaluate the various transforms as they appear, and so emerge with an explicit solution. Where this is too difficult, approximate or asymptotic methods of evaluation will be employed.

Consider now the passage from the second stage to the third. Suppose the equation

(6.7.1) $$L_1 u = \rho(x) L_2 u$$

is given for a function $u(x, t)$, where L_1 is the ordinary differential operator

(6.7.2) $$L_1 y(x) = \frac{d}{dx}\left[p(x)\frac{dy}{dx}\right] - q(x)y$$

$\rho(x)$ is a positive function of x, and L_2 is an operator not involving x. Suppose the range for x is $a \leq x \leq b$ and the boundary conditions for u include the nonhomogeneous conditions

(6.7.3) $$u_x + h_1 u = a_1(t) \quad \text{at } x = a$$
$$u_x + h_2 u = a_2(t) \quad \text{at } x = b$$

where h_1 and h_2 are independent of t.

For product solutions

(6.7.4) $$u(x, t) = y(x)v(t)$$

we find the separated equations

(6.7.5) $$L_1 y(x) + \lambda \rho(x) y(x) = 0$$
$$L_2 v(t) + \lambda v(t) = 0$$

with eigenvalue parameter λ.

Now consider the ordinary differential equation for $y(x)$:

(6.7.6) $$L_1 y(x) + \lambda \rho(x) y(x) = \frac{d}{dx}\left[p(x)\frac{dy}{dx}\right] + [\lambda \rho(x) - q(x)]y(x) = 0$$

together with the corresponding *homogeneous* boundary conditions.

(6.7.7) $$y'(a) + h_1 y(a) = 0$$
$$y'(b) + h_2 y(b) = 0$$

Together (6.7.6) and (6.7.7) are a Sturm-Liouville problem of the type considered in Sections 6.1 and 6.2. Let the eigenvalues be λ_n and normalized eigenfunctions $y_n(x)$. The Fourier coefficients of $u(x, t)$ are now

(6.7.8) $$c_n(t) = \int_a^b \rho(x) u(x, t) y_n(x)\, dx$$

We regard these coefficients as components of a vector which itself is a transform of $u(x, t)$. The dual of the variable x is the discrete index n. Observe that for an integral transform, in a problem with a continuous spectrum, the transform of x becomes a continuous variable.

Since the boundary conditions (6.7.3) are nonhomogeneous, we expect the transform $c_n(t)$ to satisfy a nonhomogeneous equation in t. This is found by integrations by parts or, in other words, by Green's formula for the operator L_1. For functions $u(x, t)$ and $y_n(x)$, this formula is

$$(6.7.9) \quad \int_a^b u(x, t) L_1 y_n(x)\, dx - \int_a^b y_n(x)\, L_1 u(x, t)\, dx$$
$$= \left[p(x)\left(u(x, t) \frac{dy_n(x)}{dx} - y_n(x) \frac{du(x, t)}{dx} \right) \right]_a^b$$

From (6.7.5), $L_1 y_n(x) = -\lambda_n \rho(x) y_n(x)$, so the first integral on the left is equal by (6.7.8) to $-\lambda_n c_n(t)$. In the second integral we use (6.7.1) and withdraw the operator L_2 through the integration over x; this is possible since L_2 does not involve x. The integral becomes $L_2 c_n(t)$, and we have

$$(6.7.10) \quad L_2 c_n(t) + \lambda_n c_n(t) = [p(x)(u_x y_n - u y_n')]_a^b$$

By (6.7.7) the right-hand side becomes

$$(6.7.11) \quad p(b)(u_x + h_2 u) y_n(b) - p(a)(u_x + h_1 u) y_n(a)$$

and we see that this combination of u and its derivative at each endpoint is exactly that required in (6.7.3). This is the justification of the homogeneous boundary conditions (6.7.7).

Finally, then, the nonhomogeneous equation in t for the transform $c_n(t)$ is

$$(6.7.12) \quad L_2 c_n(t) + \lambda_n c_n(t) = p(b) y_n(b) a_2(t) - p(a) y_n(a) a_1(t)$$

From this equation all dependence on x has been removed and replaced by the index n. After solving (6.7.12) we have still to invert the "transform" (6.7.8); this is done by the series expansion

$$(6.7.13) \quad u(x, t) = \sum_{n=1}^{\infty} c_n(t) y_n(x)$$

Note that this summation over the spectrum will be replaced by an integral if the Sturm-Liouville problem (6.7.6)–(6.7.7) is of the singular or infinite type with a continuous spectrum. In most cases, but not all, a problem on a finite interval, with $p(x)$ and $\rho(x)$ bounded away from zero, has a discrete spectrum. An infinite interval, on the other hand, usually leads to a continuous spectrum. For example, the Fourier and Hankel transforms (Chapter 8) behave in this way.

Many important problems in quantum mechanics lead to singular problems having both discrete and continuous spectra. For example, the Schrödinger equation for the hydrogen atom,

$$-\frac{\hbar^2}{2m} \Delta \psi + \frac{e^2}{r} \psi = E \psi$$

has the discrete energy eigenvalues $E_n = -\frac{1}{2}(me^4/n^2\hbar^2)$, and a continuous spectrum for $E > 0$. In this equation the coefficient q is unbounded at the origin $r = 0$. Such singular expansion theorems are not treated in this book, as a general theory including them is beyond our scope (Birkhoff-Rota, Ref. 6).

For the "finite" transform or series expansion formulas (6.7.8) and (6.7.13), we prefer the eigenfunctions to be normalized. It is interesting to note that this can sometimes be achieved without direct evaluation of the integrals. Let $y(x, \lambda)$ denote a solution of (6.7.6) which satisfies the first boundary condition (6.7.7) at $x = a$, and consider λ as a parameter. Clearly any such solution is a constant multiple of this one. Then the equation to determine the eigenvalues is

(6.7.14) $$y'(b, \lambda) + h_2 y(b, \lambda) = 0.$$

We shall differentiate (6.7.6), with $y(x, \lambda)$ as solution, with respect to λ. Then

(6.7.15) $$(py'_\lambda)' + (\lambda \rho - q) y_\lambda + \rho y = 0 \qquad y_\lambda \equiv \frac{\partial y}{\partial \lambda}$$

Now set up Green's formula for y and y_λ. That is, multiply (6.7.6) by y_λ and (6.7.15) by y and then subtract. The result is, after integration,

(6.7.16) $$[p(y' y_\lambda - y'_\lambda y)]_a^b = \int_a^b \rho y^2 \, dx$$

The integral on the right is exactly the normalizing integral required. Since y and y_λ both satisfy the boundary condition at $x = a$, the contribution on the left from the lower limit disappears. Therefore,

(6.7.17) $$y_n(x) = \frac{y(x, \lambda_n)}{\sqrt{p(b)(y' y_\lambda - y'_\lambda y)(b)}}$$

is a normalized eigenfunction.

EXERCISES 6.7

1. Invert the formulas

$$\frac{1}{n^2} = \int_0^\pi f(x) \cos nx \, dx \qquad n = 0, 1, 2, \ldots .$$

$$x = \sum_{n=1}^\infty b_n \sin nx \qquad 0 \leq x \leq \pi$$

2. Find the series expansion formulas for an arbitrary function $f(x)$ in the eigenfunctions of the Sturm-Liouville problem $y''(x) + \lambda y(x) = 0$, $y'(0) + hy(0) = 0$, $y(l) = 0$, where h is a constant.

3. Find the integral transform formulas which appear as $l \to \infty$ in Exercise 2.6.7.

4. Recalculate (6.7.12) for Dirichlet boundary conditions in one or both of (6.7.3).

5. Recalculate formula (6.7.12) if (6.7.1) has on the right side a forcing term $f(x, t)$.

6. In (6.7.17) why can we not use (6.7.14) and its derivative with respect to λ to simplify?

7. Show that for a Dirichlet condition $y(b) = 0$ (6.7.16) yields

$$p(b) y'(b, \lambda) y_\lambda(b, \lambda) = \int_a^b \rho(x) y^2(x, \lambda) \, dx$$

Deduce that any zero of $y(x, \lambda)$ to the right of $x = a$ is a decreasing function of λ. Deduce further that $y_n(x)$ has exactly $n - 1$ zeros in $a \leq x \leq b$.

CHAPTER 7

Green's Functions

7.1 INVERSES OF DIFFERENTIAL OPERATORS

We have already met with a variety of Green's functions for ordinary and partial differential equations of different types with a variety of boundary conditions. Now we shall examine these functions more systematically and study their relation to the linear differential operators which appear in the equations. As stated in Chapter 1, we intend to represent certain functions by vectors, and it is equally natural to represent operators by matrices. We shall see that Green's functions, which depend on two points, can be regarded as matrices, and that they represent operators which are inverse to the differential operator. The various analytical properties of all the Green's functions we encounter are consequences of this single underlying relation (Bergman-Schiffer, Ref. 4; Lanczos, Ref. 39).

Consider a linear differential operator $L = -\Delta + q(P)$, where Δ is the Laplacian and P a typical point with coordinates x, y, z. A region R and a boundary condition are also given. For simplicity we shall assume that the weight function is unity. The Green's function $G(P, Q)$ of this three-dimensional

Sturm-Liouville problem is the solution of the partial differential equation

(7.1.1) $$L_P G(P, Q) = \delta(P, Q)$$

which satisfies the homogeneous boundary condition. For simplicity we shall here use the Dirichlet condition, so that

(7.1.2) $$G(P, Q) = 0 \qquad P \text{ on } S$$

The fixed point Q is sometimes called a *source* point, from the hydrodynamical interpretation of the equation div $V \equiv$ div grad $\varphi = \Delta \varphi = -\rho$. The Green's function itself is also known as a source function, response function, or influence function.

THEOREM 7.1.1 *The Green's function is symmetric:*

(7.1.3) $$G(Q_1, Q_2) = G(Q_2, Q_1)$$

Proof. Apply Green's second formula to the functions $G(P, Q_1)$ and $G(P, Q_2)$, where Q_1 and Q_2 are any two points in R. Thus

$$\int_R [G(P, Q_1) L_P G(P, Q_2) - G(P, Q_2) L_P G(P, Q_1)] \, dV$$

$$= -\int_S \left[G(P, Q_1) \frac{\partial G(P, Q_2)}{\partial n} - G(P, Q_2) \frac{\partial G(P, Q_1)}{\partial n} \right] d\varSigma = 0$$

since G satisfies the boundary condition (7.1.2). We use (7.1.1) with $Q = Q_1$ and $Q = Q_2$ and the substitution property of $\delta(P, Q)$ yields

$$0 = \int_R [G(P, Q_1) \delta(P, Q_2) - G(P, Q_2) \delta(P, Q_1)] \, dV$$

$$= G(Q_2, Q_1) - G(Q_1, Q_2)$$

If the Green's function is known, then a solution for the boundary value problem can be written down at sight.

THEOREM 7.1.2 *The nonhomogeneous problem*

(7.1.4) $$Lu = f(P)$$

with boundary condition

(7.1.5) $$u(P) = g(P)$$

on S has the formal solution

(7.1.6) $$u(P) = \int_R G(P, Q) f(Q) \, dV_Q - \int_S g(Q) \frac{\partial G(P, Q)}{\partial n_Q} \, dS_Q$$

Proof. Apply Green's second formula to the functions $u(Q)$ and $G(P, Q)$ of the point Q. Then, after integration over Q, we have

$$\int_R [u(Q)L_Q G(P, Q) - G(P, Q)Lu(Q)]\, dV$$
(7.1.7)
$$= -\int_S \left[u(Q) \frac{\partial G(P, Q)}{\partial n_Q} - G(P, Q) \frac{\partial u(Q)}{\partial n_Q} \right] dS$$

By the symmetry of $G(P, Q)$, we have

(7.1.8) $\qquad L_Q G(P, Q) = L_Q G(Q, P) = \delta(Q, P) = \delta(P, Q)$

Also from (7.1.2) the second term in the surface integral disappears. The solution formula (7.1.6) now follows from the substitution property of the delta function when inserted in the first term on the left in (7.1.7).

In Chapter 4 we have already given several examples of this formula (7.1.6) containing the surface integral term. We shall calculate explicitly a number of Green's functions in Section 7.2, but now let us examine their Fourier expansions.

THEOREM 7.1.3 *The Green's function has the bilinear Fourier expansion*

(7.1.9) $\qquad G(P, Q) = \sum_{n=1}^{\infty} \frac{u_n(P)u_n(Q)}{\lambda_n}$

Proof. If we hold Q fixed, the Fourier coefficients of $G(P, Q)$ with respect to P are

(7.1.10) $\qquad g_n(Q) = \int_R u_n(P)G(P, Q)\, dV_P$

where $u_n(P)$ are the orthonormal eigenfunctions of Chapter 6. These eigenfunctions satisfy the boundary condition $u_n(P) = 0$ on S as well as the differential equation

(7.1.11) $\qquad Lu_n = -\Delta u_n + qu_n = \lambda_n u_n$

where the eigenvalues are denoted by λ_n as in Chapter 6. From Theorem 7.1.2 we observe that $u_n(P)$, which is the solution of $Lu_n = f(P)$, $f(P) = \lambda_n u_n$, and which vanishes on S, must be equal to

(7.1.12) $\qquad \lambda_n \int_R u_n(Q)G(P, Q)\, dV = \lambda_n g_n(P)$

Therefore $g_n(P) = \lambda_n^{-1} u_n(P)$, and this gives (7.1.9) immediately, since $G(P, Q) = \Sigma g_n(Q) u_n(P)$.

It should be remarked that this bilinear series does not always converge in the conventional pointwise sense. Even in such cases, however, (7.1.9) has a correct meaning if convergence is interpreted in the distribution sense. We return to this question in Section 7.6.

We shall find it useful to be able to shift the eigenvalues.

COROLLARY 7.1.1 *The Green's function $G(P, Q, \lambda)$ of the differential equation*

$$LG - \lambda G = \delta(P, Q) \qquad \lambda \neq \lambda_n$$

has the bilinear expansion

(7.1.13) $$G(P, Q, \lambda) = \sum_{n=1}^{\infty} \frac{u_n(P)u_n(Q)}{\lambda_n - \lambda}$$

Proof. Since L has eigenvalues λ_n, with $Lu_n = \lambda_n u_n$, we find $(L - \lambda)u_n = (\lambda_n - \lambda)u_n$, that is, $L - \lambda$ has eigenvalues $\lambda_n - \lambda$. Therefore Theorem 7.1.3 yields the result.

Example 7.1.1 The Green's function for $\Delta u + \lambda u = 0$ on the rectangle $0 \leq x \leq a, 0 \leq y \leq b$ is

$$G(x, y; x_1, y_1; \lambda) = \frac{4}{ab} \sum_{m,n=1}^{\infty} \frac{\sin \frac{m\pi x}{a} \sin \frac{m\pi x_1}{a} \sin \frac{n\pi y}{b} \sin \frac{n\pi y_1}{b}}{\pi^2 \left(\frac{m^2}{a^2} + \frac{n^2}{b^2}\right) - \lambda}$$

The differential operator L transforms a function $u(P)$ on R into the function $Lu(P) = [q(P) - \Delta]u(P)$. In Chapter 1 we saw that operators can be represented by matrices. To represent L, let us consider formal matrices with a continuous number of rows and of columns, one for each point P of R. Such a formal matrix A would be specified by a function $a(P, Q)$ of two points P, Q in R, P being, say, the row index and Q the column index. It is customary to call $a(P, Q)$ the *kernel* of the operator A. Multiplication of operators involves integration over adjacent row and column indices of the corresponding matrices. Thus $A_1 A_2$ has the kernel

$$a_{12}(P_1, P_2) = \int_R a_1(P_1, Q) a_2(Q, P_2) \, dV_Q$$

The unit operator E has matrix elements $\delta(P, Q)$ because the substitution property of the Dirac distribution corresponds to the unit property $AE = EA = A$ for all operators A:

(7.1.14) $$\int a(P_1, Q)\delta(Q, P_2) \, dV_Q = \int \delta(P_1, Q)a(Q, P_2) \, dV_Q = a(P_1, P_2)$$

Multiplication by a scalar function $q(P)$ is represented by the kernel $q(P)\delta(P, Q)$, and the operation L should be represented by the kernel

$$L\delta(P, Q) = [q(P) - \Delta_P]\delta(P, Q).$$

Let the Green's function $G(P, Q)$ define the operator G which transforms functions $f(P)$ into functions

(7.1.15) $$Gf(P) = \int_R G(P, Q)f(Q) \, dV_Q$$

268 GREEN'S FUNCTIONS

The Green's function is the kernel of the Green's operator G. Then the operator LG is represented by the kernel

(7.1.16)
$$[q(P) - \Delta_P] \int_R \delta(P, Q_1) G(Q_1, Q) \, dV_{Q_1} = [q(P) - \Delta_P] G(P, Q)$$
$$= L_P G(P, Q) = \delta(P, Q)$$

according to (7.1.1). That is, the product LG is equal to E. Likewise we can show that $GL = E$. For GL is represented by the kernel

(7.1.17) $\quad \int_R G(P, Q_1) L_{Q_1} \delta(Q_1, Q) \, dV_{Q_1} = \int_R L_{Q_1} G(P, Q_1) \delta(Q_1, Q) \, dV_{Q_1}$

by Green's second formula, and since $G(P, Q) = 0$ for Q on S and $\delta(Q_1, Q) = 0$ for Q_1 on S, Q in $R - S$. By the substitution property this integral becomes $L_Q G(P, Q) = \delta(P, Q)$ by (7.1.1), and this completes the proof.

THEOREM 7.1.4 *The Green's function is the kernel of the integral operator G inverse to the differential operator L; thus $GL = LG = E$.*

In Chapter 1, Section 1.2, we saw that a symmetric matrix has real eigenvalues and orthogonal eigenvectors. By Theorem 6.2.1 the same result holds for the symmetric kernel of the operator L, and therefore also for the kernel $G(P, Q)$ of its inverse G. To solve a matrix equation $A\mathbf{x} = \mathbf{f}$, we can diagonalize A by an orthogonal transformation, thus resolving the set of equations into separate components.

Likewise, to solve $Lu(P) = f(P)$, we can use the eigenfunctions $u_n(P)$ for a Fourier expansion of $u(P)$ and $f(P)$. Let

$$u(P) = \sum_{n=1}^{\infty} c_n u_n(P) \qquad f(P) = \sum_{n=1}^{\infty} f_n u_n(P)$$

Then

$$Lu(P) = \sum_{n=1}^{\infty} c_n L u_n(P) = \sum_{n=1}^{\infty} \lambda_n c_n u_n(P)$$

and the equation $Lu = f$ transforms into a fully separated algebraic system

$$\lambda_n c_n = f_n \qquad\qquad n = 1, 2, \ldots$$

with the immediate solution $c_n = \lambda_n^{-1} f_n$. Thus the formal solution is

(7.1.18)
$$u(P) = \sum_{n=1}^{\infty} \frac{1}{\lambda_n} f_n u_n(P)$$

The transformation of L and G to principal axes involves a change to a discrete set of rows and columns of the formal matrices, one for each eigenvalue λ_n. The unit matrix is now (δ_{mn}). Relative to the basis $\{u_n(P)\}$ of eigenfunctions the operator L is represented by the infinite diagonal matrix $(\lambda_n \delta_{mn})$ and the inverse operator G by the infinite matrix $(\lambda_n^{-1} \delta_{mn})$.

The formal transforming matrix X by which L and G are diagonalized has continuous rows labeled by P and discrete columns labeled by the eigenvalue index n. The elements of X are the numerical values $u_n(P)$ of the eigenfunctions: thus $X = (u_n(P))$. Comparing with Section 1.2, we see that X is indeed orthogonal since

(7.1.19) $$X'X = \left(\int u_m(P) u_n(P) \, dV_P \right) = (\delta_{mn})$$

which is the discrete countably infinite unit matrix. Also

(7.1.20) $$XX' = \left(\sum_{n=1}^{\infty} u_n(P) u_n(Q) \right) = (\delta(P, Q))$$

which is the continuous unit matrix or unit kernel.

In Chapter 2 and Chapter 3, Section 3.6, we have given examples of the bilinear formula

(7.1.21) $$\delta(P, Q) = \sum_{n=1}^{\infty} u_n(P) u_n(Q)$$

for the Dirac distribution. A formal proof of this relation, which we suggest as an exercise, can be based upon the substitution property of the Dirac distribution. The series on the right, though not strongly convergent, may, in some cases, be evaluated by summation methods for divergent series. For $P \neq Q$, the value is, of course, zero. As an equation in distributions, (7.1.21) is equivalent to the substitution property.

EXERCISES 7.1

1. Verify that (7.1.1) and (7.1.2) determine $G(P, Q)$ uniquely, provided $q(P) \geq 0$.
2. Verify that

$$\frac{2}{\pi^2} \sum_{n=1}^{\infty} \frac{\sin n\pi x \sin n\pi x_1}{n^2} = \begin{cases} x(1 - x_1) & 0 \leq x \leq x_1 \leq 1 \\ x_1(1 - x) & 0 \leq x_1 \leq x \leq 1 \end{cases}$$

and

$$\frac{2}{\pi^2} \sum_{n=1}^{\infty} \frac{\sin (n + \tfrac{1}{2})\pi x \sin (n + \tfrac{1}{2})\pi x_1}{(n + \tfrac{1}{2})^2} = \begin{cases} x & 0 \leq x \leq x_1 \leq 1 \\ x_1 & 0 \leq x_1 \leq x \leq 1 \end{cases}$$

Hint: See Chapter 1, Section 1.8.

3. The gravitational potential of a mass m at a point Q is $-Gm/r$, where r is the distance from the field point P to Q and G the gravitational constant. Write down the potential V of a mass distribution in a region R, and show that:
   (a) V is harmonic outside R.
   (b) $\Delta V = -4\pi \rho G$, where ρ is mass density.

4. Verify the bilinear formula (7.1.9) from the matrix analogy for the Green's function. *Hint:* See (1.2.34).

5. Establish the formal relation

$$[q(P) - \Delta_P] \delta(P, Q) = \sum_{n=1}^{\infty} \lambda_n u_n(P) u_n(Q)$$

In what sense is this equation to be understood?

6. State in detail the two-dimensional and one-dimensional versions of Theorem 7.1.2.

7. Show that (7.1.9) remains valid for any choice of the weight function $\rho(P) \geq 0$. Restate the matrix analogy (7.1.14)–(7.1.20) for a general weight function. *Hint:* $\sqrt{\rho(P)}\, u_n(P)$ form a complete orthonormal set with unit weighting.

8. Show that the formal solution (7.1.18) is found by substitution of the bilinear series for the Green's function in Theorem 7.1.1.

7.2 EXAMPLES OF GREEN'S FUNCTIONS

For every linear initial or boundary value problem it is possible to define a source function or Green's function. In this section we will consider elliptic equations. If we can find the Green's function explicitly, then the formal solution of our problem is immediate, by Theorem 7.1.2. The determination of formulas for Green's functions is therefore an important part of our knowledge of boundary value problems. Here we shall present a number of leading examples (Courant-Hilbert, Ref. 13, vol. 1; Morse-Feshbach, Ref. 45).

For one-dimensional problems the construction of Green's functions for two-point boundary problems was treated in Chapter 1, Section 1.8, which the reader will recall.

The foremost example in two dimensions is the Green's function of the Laplace operator:

$$(7.2.1) \qquad -\left(\frac{\partial^2}{\partial x^2} + \frac{\partial^2}{\partial y^2}\right) G(x, y, x_1, y_1) = \delta(x - x_1)\delta(y - y_1)$$

To be consistent with our usage in Section 7.1, we have placed a negative sign on the left side of (7.2.1). This is because the Laplacian itself is an essentially negative operator. By the Helmholtz equation $-\Delta u_n = \lambda_n u_n$, where $\lambda_n \geq 0$, the eigenvalues of $-\Delta$ are positive. Thus the operator $L = -\Delta$ is inherently positive and so is its inverse kernel $G(P, Q)$.

For (7.2.1) we first consider the whole plane as region R. By symmetry, the Green's function will depend only on r, the distance of source and field points. Taking the source point as origin, and writing (7.2.1) in polar coordinates r, θ, we find, as in Section 5.3,

$$(7.2.2) \qquad G_{rr} + \frac{1}{r} G_r = -\frac{1}{2\pi r} \delta_+(r)$$

Integration leads to the particular solution

$$(7.2.3) \qquad G = -\frac{1}{2\pi} \log r$$

where r is the distance from source to field point. The reader will remember that the logarithm of r is one of the basic single-valued harmonic functions employed in Chapter 4. In two dimensions it does not seem possible to fulfill the condition of vanishing as $r \to \infty$; instead (7.2.3) tends to $-\infty$.

At this point we shall utilize a theorem of Riemann on the conformal mapping of a domain D of the complex plane onto the unit circle $|w| \leq 1$ of the complex w-plane (Figure 7.1). The map of points (x, y) to points $w = w_1 + iw_2$ defined by $w = f(z)$ is conformal, that is, locally angle-preserving wherever $f'(z) \neq 0$.

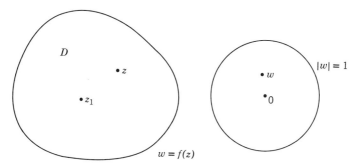

Figure 7.1 Conformal mapping onto unit circle.

Moreover, the real part of any analytic function of z is harmonic. Therefore let $w = f(z)$ map D onto the unit circle $|w| \leq 1$ with the given "source" point z_1 being mapped into $w = 0$: $f(z_1) = 0$, and $f'(z_1) \neq 0$. By Riemann's mapping theorem (Copson, Ref. 11, p. 185) this function $f(z)$ exists for every simply connected bounded domain with piecewise smooth boundary curves.

THEOREM 7.2.1 *The Green's function of D is*

$$(7.2.4) \qquad G = -\frac{1}{2\pi} \log |f(x + iy)|$$

Proof. This expression is the real part of $-(1/2\pi) \log f(z)$ and is therefore harmonic. At $z = z_1$ it has a logarithmic singularity and it satisfies $\Delta G = -\delta(x - x_1)\delta(y - y_1)$. Also when (x, y) is a point of the boundary of D, then $f(z) = f(x + iy)$ is a point of the unit circle $|w| = 1$, so the logarithm vanishes as required for the boundary condition.

Example 7.2.1 If D is the unit circle, z_1 the source point, then the linear mapping

$$(7.2.5) \qquad w = \frac{z - z_1}{\bar{z}_1 z - 1} \qquad \bar{z}_1 = x_1 - iy_1$$

carries $|z| \leq 1$ onto $|w| \leq 1$, with z_1 being mapped on the origin. The Green's function is therefore

$$(7.2.6) \qquad G(x, y; x_1, y_1) = -\frac{1}{2\pi} \log \left| \frac{z - z_1}{\bar{z}_1 z - 1} \right|$$

272   GREEN'S FUNCTIONS

In Chapter 8 we shall encounter further two-dimensional Green's functions involving Bessel functions. We now turn to some examples in three dimensions. For the harmonic equation in three dimensions, there is no analogue of Theorem 7.2.1 since the only nontrivial conformal transformations are inversions in a sphere. We can study these to obtain the Green's function for a sphere.

The three-dimensional harmonic Green's function for the full space was found in Chapter 5. Here we write $\Delta G = -\delta(x - x_1)\delta(y - y_1)\delta(z - z_1)$ and obtain

(7.2.7) $$G(P, Q) = G(x, y, z; x_1, y_1, z_1) = \frac{1}{4\pi r}$$

where

(7.2.8) $$r^2 = (x - x_1)^2 + (y - y_1)^2 + (z - z_1)^2$$

This Green's function or "fundamental solution" appears in potential theory as the potential of a point mass or point charge. As $r \to \infty$ its limiting value is zero.

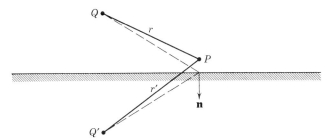

Figure 7.2   Source and image for a half-space.

In Chapters 2, 3, and 4 we made frequent use of the method of reflection involving even or odd functions. Here we deduce the Green's function for a half-space $z \geq 0$ by the same method, using an *image* point. (See Figure 7.2.) If we let r' be the distance from field to image point,

(7.2.9) $$r'^2 = (x - x_1)^2 + (y - y_1)^2 + (z + z_1)^2$$

then

(7.2.10) $$G(P, Q) = \frac{1}{4\pi r} - \frac{1}{4\pi r'}$$

clearly satisfies the boundary condition of vanishing for $z = 0$. This expression is the potential of a positive unit charge at the source point Q and of a negative unit charge at the image point Q_1.

The source function for the second boundary condition $\partial u/\partial n = 0$ is often called a Neumann function. The harmonic Neumann function for the half-plane $z \geq 0$ is found by taking a positive charge at the image point:

(7.2.11) $$N(P, Q) = \frac{1}{4\pi r} + \frac{1}{4\pi r'}$$

Since this is an even function of z, analytic for $z = 0$, the normal or z derivative will be zero on the boundary plane $z = 0$.

The Green's function for a sphere of radius R can also be expressed as a source and image point combination (Figure 7.3). For any point Q at radial distance a within the sphere, we have an inverse point Q' outside, on the same

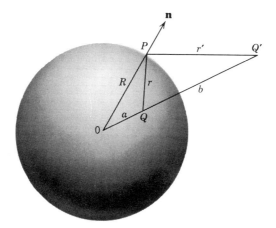

Figure 7.3 Inverse image point in a sphere.

line through the center and at a radial distance b, where $ab = R^2$. If P is any point of the surface of the sphere, the triangles OPQ, OPQ' have a common angle at the origin, and proportional sides including that angle, for

$$\frac{a}{R} = \frac{R}{b}$$

Therefore the triangles are similar, and so

(7.2.12) $$\frac{r'}{r} = \frac{R}{a} \quad \text{i.e.} \quad \frac{1}{r} = \frac{R}{a} \cdot \frac{1}{r'}$$

for a point P on the surface. Therefore an additional negative charge $-R/a$ situated at the inverse point will yield the Green's function for a sphere, namely,

(7.2.13) $$G(P, Q) = \frac{1}{4\pi}\left(\frac{1}{r} - \frac{R}{a}\frac{1}{r'}\right)$$

Let us employ Theorem 7.1.2 in conjunction with this expression to compute the formula for a harmonic function with given values $u(R, \theta, \varphi) = f(\theta, \varphi)$ on the surface of the sphere. From the triangles OPQ, OPQ' we have

$$a^2 = R^2 + r^2 + 2Rr \cos \mathbf{nr}$$
$$b^2 = R^2 + r'^2 + 2Rr' \cos \mathbf{nr'}$$

where **n** is the outward normal to the sphere. The required kernel in the surface integral expression becomes

$$\frac{\partial G}{\partial n} = \frac{1}{4\pi}\left(-\frac{1}{r^2}\frac{\partial r}{\partial n} + \frac{R}{a}\cdot\frac{1}{r'^2}\frac{\partial r'}{\partial n}\right)$$

$$= \frac{1}{4\pi}\left(+\frac{\cos \mathbf{nr}}{r^2} - \frac{R}{a}\frac{\cos \mathbf{nr'}}{r'^2}\right)$$

$$= \frac{1}{4\pi}\left(\frac{a^2 - R^2 - r^2}{2Rr^3} - \frac{R}{a}\frac{b^2 - R^2 - r'^2}{2Rr'^3}\right)$$

To simplify, we may use (7.2.12) and the inverse relation $ab = R^2$, and so find

(7.2.14) $$\frac{\partial G}{\partial n} = -\frac{R^2 - a^2}{4\pi Rr^3}$$

From Theorem 7.1.2 the representation formula is

(7.2.15) $$u(P) = \frac{R(R^2 - a^2)}{4\pi}\int_0^\pi\int_0^{2\pi}\frac{f(\theta, \varphi)\sin\theta\, d\theta\, d\varphi}{(a^2 + R^2 - 2aR\cos\theta)^{3/2}}$$

This is the three-dimensional Poisson formula.

The harmonic Neumann function for a sphere or other bounded region is subject to a complication that was discussed in Section 4.2 and will be further considered in Section 7.3. Here we continue with the Green's function of the three-dimensional Helmholtz equation

(7.2.16) $$\Delta u + k^2 u = 0$$

The point-source solution is found by asking for solutions dependent only on the source distance r. This gives

$$u_{rr} + \frac{2}{r}u_r + k^2 u = 0 \qquad r > 0$$

or

$$(ru)_{rr} + k^2(ru) = 0 \qquad r > 0$$

an equation with the easily recognizable solutions

(7.2.17) $$ru = Ae^{ikr} + Be^{-ikr} = C\cos kr + D\sin kr$$

One real-valued point-source solution is therefore

(7.2.18) $$\frac{\cos kr}{4\pi r}$$

However in physical applications it is usual to select one of the exponential forms. The reason is that (7.2.18) represents physically a radial standing wave, which is never realized in practice. If we adjoin an exponential time factor e^{ikct}, then the solutions

$$\frac{e^{ik(r+ct)}}{r} \qquad \frac{e^{-ik(r-ct)}}{r}$$

represent convergent and divergent traveling waves respectively. But the convergent radial waves, representing an implosion, are rare in practice. Therefore it is customary to use the exponential Green's function

(7.2.19)
$$\frac{e^{-ikr}}{4\pi r}$$

This solution satisfies the emission or radiation condition of Sommerfeld:

(7.2.20)
$$\lim_{r \to \infty} r\left(\frac{\partial u}{\partial r} + iku\right) = 0$$

which excludes solutions that represent incoming or convergent waves with the time-dependent factor e^{ikct}.

EXERCISES 7.2

1. Show that the harmonic Green's function of the half-space $y \geq 0$ is

$$G(x, y; x_1, y_1) = -\frac{1}{2\pi} \log \frac{r}{r_1}$$

where $r^2 = (x - x_1)^2 + (y - y_1)^2$, $r_1^2 = (x - x_1)^2 + (y + y_1)^2$. Verify (4.5.12).

2. Show that the harmonic Neumann function of the half-space $y \geq 0$ is

$$N(x, y; x_1, y_1) = -\frac{1}{2\pi} \log rr_1$$

and verify (4.5.22).

3. Show that the harmonic Green's function for a circle $|z| \leq R$ is

$$G(x, y; x_1, y_1) = -\frac{1}{2\pi} \log \frac{rR}{r'a}$$

with the notation of Figure 7.3, and verify that this agrees with (7.2.6) for $R = 1$.

4. Show that the biharmonic equation $\Delta\Delta u = 0$ has the fundamental solution $-r^2 \log r/8\pi$ in two dimensions.

5. Find harmonic source functions in the quarter plane $x \geq 0$, $y \geq 0$ which satisfy the following conditions:
   (a) $u = 0$ on $x = 0$ and on $y = 0$.
   (b) $u = 0$ on $x = 0$ and $\partial u/\partial y = 0$ on $y = 0$.
   (c) $\partial u/\partial x = 0$ on $x = 0$ and $\partial u/\partial y = 0$ on $y = 0$.
   (d) $\partial u/\partial x + hu = 0$ on $x = 0$ and $u = 0$ on $y = 0$.

6. Show how to find the Green's function for the sector $0 \leq \theta \leq \pi/3$, $0 \leq r \leq 1$, in polar coordinates in two ways:
   (a) By finding the mapping function of Theorem 7.2.1.
   (b) By the method of images.

7. Find by the image method the Green's function of the sector $0 \le \theta \le \pi/3$, $0 \le r \le 1$, with the following boundary conditions:
   (a) $u = 0$ for $\theta = 0$ and $\theta = \pi/3$; $\partial u / \partial n = 0$ for $r = 1$.
   (b) $u = 0$ for $\theta = 0$; $\partial u / \partial n = 0$ for $\theta = \pi/3$ and $r = 1$.
   (c) $\partial u / \partial n = 0$ for $\theta = 0$ and $\theta = \pi/3$; $u = 0$ for $r = 1$.

8. For what sector angles α can the Green's function be constructed as in Exercise 7.2.7 for the sector $0 \le \theta \le \alpha$, $0 \le r \le 1$?

9. Show that the function harmonic for $z \ge 0$ with given values $f(x, y)$ for $z = 0$, and well behaved at infinity, is

$$u(x, y, z) = \frac{z}{2\pi} \int_{-\infty}^{\infty} \int_{-\infty}^{\infty} \frac{f(x_1, y_1)\, dx_1\, dy_1}{[(x - x_1)^2 + (y - y_1)^2 + z^2]^{3/2}}$$

10. Show that the harmonic Green's function for the exterior of the sphere of radius R is also given by (7.2.13), where now $R < a$. Deduce the representation formula analogous to (7.2.15) for the exterior region.

11. Show how to find the Green's function for the octant $x > 0, y > 0, z > 0$ of the sphere of radius R centred at the origin. Discuss various possible boundary conditions amenable to the image method.

12. Find the Green's and Neumann's functions of the half-space $z > 0$ for the equation of damped waves $\Delta u - k^2 u = 0$, k real.

13. Show that the Green's function of the equation $\Delta u - k^2 u = 0$ for the infinite strip $0 \le x \le a$, $-\infty < y < \infty$, subject to the condition of vanishing on both sides, is given by

$$G(x, x', y, y') = \frac{-1}{\pi} \int_0^\infty \frac{\cos s(y - y') \sinh \lambda x' \sinh \lambda(x - a)}{\lambda \sinh \lambda a}\, ds$$

where $0 < x' < x < a$ and $\lambda = \sqrt{k^2 + s^2}$. Show also that

$$G = \frac{1}{a} \sum_{n=1}^{\infty} \frac{1}{\sigma} e^{-\sigma|y - y'|} \sin \frac{n\pi x}{a} \sin \frac{n\pi x'}{a}$$

where $\sigma = \sqrt{k^2 + n^2 \pi^2 / a^2}$.

14. Show that the Green's function of the equation $\Delta u - k^2 u = 0$ for the rectangle $0 \le x \le a$, $0 \le y \le b$ vanishing around all sides is given by

$$G = -\frac{2}{a} \sum_{n=1}^{\infty} \frac{\sinh \sigma y' \sinh \sigma(y - b)}{\sigma \sinh \sigma b} \sin \frac{n\pi x}{a} \sin \frac{n\pi x'}{a}$$

where $0 < y' < y < b$ and $\sigma = \sqrt{k^2 + n^2 \pi^2 / a^2}$. Obtain also the eigenfunction expansion of this Green's function.

15. Show that, in polar coordinates, the harmonic Green's function for the infinite sector $0 \le \theta \le \alpha$, $0 \le r < \infty$ vanishing on both faces is given by

$$G(r, \theta, r', \theta') = \frac{1}{4\pi} \log \left[\frac{\cosh\left(\frac{\pi}{\alpha} \log \frac{r}{r'}\right) - \cos \frac{\pi}{\alpha}(\theta + \theta')}{\cosh\left(\frac{\pi}{\alpha} \log \frac{r}{r'}\right) - \cos \frac{\pi}{\alpha}(\theta - \theta')} \right]$$

Hence construct the harmonic function $u(r, \theta)$ in the stated region subject to the boundary values $u(r, 0) = u(r, \alpha) = H(a - r)$, where $H(r)$ denotes the Heaviside unit function.

16. Show that the conformal mapping $\zeta = -\cos \pi z/a$, $z = x + iy$, $\zeta = \xi + i\eta$ maps the semi-infinite strip $0 \leq x \leq a$, $0 \leq y < \infty$ onto the half-plane $\eta \geq 0$. Deduce that the harmonic Green's function vanishing on all three sides of the strip is given by

$$G(x, y, x', y') = \frac{1}{2\pi} \log \left| \frac{\cos \dfrac{\pi z}{a} - \cos \dfrac{\pi \bar{z}'}{a}}{\cos \dfrac{\pi z}{a} - \cos \dfrac{\pi z'}{a}} \right|$$

17. Show that the transformation $\zeta = -e^{-i\pi z/a}$ maps the infinite strip $0 \leq x \leq a$, $-\infty < y < \infty$ into the half plane $I(\zeta) \geq 0$. Deduce that the harmonic Green's function vanishing on both sides is

$$G(x, y, x', y') = \frac{1}{2\pi} \log \left| \frac{\sin \dfrac{\pi(z + \bar{z}')}{2a}}{\sin \dfrac{\pi(z - z')}{2a}} \right|$$

Interpret this result as an infinite series of images by recalling the infinite product

$$\sin t = t \prod_{n=1}^{\infty} \left(1 - \frac{t^2}{n^2 \pi^2}\right)$$

18. By referring to Exercises 5.2.13, 6.2.5 and 6.2.6, show that, if $\Delta \Delta u - \lambda u = f$, $\Delta \Delta G - \lambda G = \delta(x - x')\delta(y - y')$, then

$$u(x, y) = \iint_S fG \, dx \, dy + \oint_C \left[uT(G) - GT(u) - \frac{\partial u}{\partial n} M(G) + \frac{\partial G}{\partial n} M(u) \right] ds$$

Show how to choose the Green's function G for the following cases:
(a) u and $\partial u/\partial n$ given on C.
(b) u and $M(u)$ given on C.
(c) $M(u)$ and $T(u)$ given on C.
(d) $\partial u/\partial n$ and $T(u)$ given on C.

If u_n, λ_n are the corresponding eigenfunctions and eigenvalues, show that

$$G = \sum_{n=1}^{\infty} \frac{u_n(x, y)u_n(x', y')}{\lambda_n - \lambda}$$

Show that the biharmonic Green's function H such that $\Delta \Delta H = \delta(x - x')\delta(y - y')$ is given by

$$H(x, x'; y, y') = \sum_{n=1}^{\infty} \frac{1}{\lambda_n} u_n(x, y) u_n(x', y')$$

7.3 THE NEUMANN AND ROBIN FUNCTIONS

The source functions we have been studying have a variety of physical interpretations. For instance, the Green's function $G(P, Q)$ of the harmonic equation on a domain R can be interpreted as the electrostatic potential (of positive sign) of a point charge at Q when the boundary of the region is grounded so that the potential is zero there. A somewhat different interpretation will help us to visualize the source functions appropriate to the various types of boundary conditions. Suppose that R is a region, or body, immersed in a circulating fluid maintained at constant temperature zero. If the conduction of heat across the boundary S is perfect and without resistance, the temperature on the "inner" side of the surface will also be zero. Suppose also that a unit source of heat is present at the interior point Q. Then the stationary temperature distribution throughout region R is also given by the Green's function $G(P, Q)$.

Now consider the third boundary condition

(7.3.1) $$\frac{\partial u}{\partial n} + hu = 0 \qquad h > 0$$

which relates the flow of heat $-\partial u/\partial n$ over the boundary to the temperature drop u across it. The constant of proportionality h is infinite in the case of perfect conductivity just discussed. For finite positive values of h, which may vary from point to point on S, the temperature distribution due to a point source at Q is known as the Robin function, and is denoted by $R(P, Q)$. Thus $R(P, Q)$ is the solution of

(7.3.2) $$LR(P, Q) = \delta(P, Q) \qquad \text{in } R$$

where in the present instance $L = -\Delta$, with

(7.3.3) $$\frac{\partial R(P, Q)}{\partial n} + h(P)R(P, Q) = 0 \qquad \text{on } S$$

This function is uniquely determined by (7.3.2) and (7.3.3) for $h > 0$; we suggest the proof of this as an exercise. The actual existence of $R(P, Q)$ we have not proved mathematically. We shall accept the argument of physical plausibility for the consistency of these conditions and the existence of the Robin function.

The same argument of a physical nature suggests, as is indeed true, that $R(P, Q)$ depends monotonically upon the boundary function $h(P)$. If $h(P)$ is decreased, the resistance of the boundary to the passage of heat is increased, and the temperature distribution throughout R also increases. Thus $R(P, Q)$ is a *monotonically decreasing functional* of the function $h(P)$ (Bergman-Schiffer, Ref. 4).

The Robin function is, like the Green's function, symmetric. As the proof is nearly the same as in Theorem 7.1.1, we propose it also as an exercise, and quote the result:

(7.3.4) $$R(P, Q) = R(Q, P)$$

The proofs of the bilinear formulas of Theorem 7.1.3 and Corollary 7.1.1 are also valid here without significant alteration. However, the new boundary condition leads to a slightly different representation formula for solutions of the third boundary value problem.

THEOREM 7.3.1 *The nonhomogeneous problem*

(7.3.5) $$Lu = f(P) \quad \text{in } R$$

with boundary condition

(7.3.6) $$\frac{\partial u}{\partial n} + h(P)u = g(P) \quad \text{on } S$$

has the formal solution

(7.3.7) $$u(P) = \int_R R(P, Q) f(Q) \, dV_Q + \int_S R(P, Q) g(Q) \, dS_Q$$

To establish this formula, we apply Green's second theorem to $u(Q)$ and $R(P, Q)$, regarding P as a parameter point. Since L is $-\Delta$, we find

$$\int_R (u \, LR - R \, Lu) \, dV = -\int_S \left(u \frac{\partial R}{\partial n} - R \frac{\partial u}{\partial n} \right) dS$$

On the left, we use (7.3.2) and (7.3.5) while on the right (7.3.3) gives

$$\int_R u \delta \, dV - \int_R Rf \, dV = -\int_S \left[u(-hR) - R \frac{\partial u}{\partial n} \right] dS$$

$$= \int_S R \left(\frac{\partial u}{\partial n} + hu \right) dS$$

Therefore, by the substitution property of $\delta(P, Q)$ and the boundary condition (7.3.6), we find the result stated.

The reader will notice that this formal calculation succeeds for any values of the coefficient function $h(P)$. We might expect, therefore, that the result will hold when $h(P)$ is everywhere zero. However, the physical interpretation as a temperature problem should warn us that in this case the source function will not exist. For the heat released by the steady source at Q cannot escape through a completely insulated boundary and, in the long run the temperature must rise to infinity. In Section 4.2, when describing the Neumann problem for harmonic functions, we observed that a solution cannot exist unless a certain integral or average condition is satisfied by the data. If $\Delta u = 0$ in R and

(7.3.8) $$\frac{\partial u}{\partial n} = g(P) \quad \text{on } S$$

then from the Gauss divergence theorem we find

(7.3.9) $$\int_S g(P) \, dS = \int_S \frac{\partial u}{\partial n} \, dS = \int_R \Delta u \, dV = 0$$

Thus (7.3.9) is a *necessary* condition for the harmonic Neumann problem.

On the other hand, the solution, if it exists, is not unique, for any constant can be added to it. In fact, this constant is an eigenfunction of the Neumann eigenvalue problem

(7.3.10) $$\Delta u + \lambda u = 0 \qquad \frac{\partial u}{\partial n} = 0 \quad \text{on} \quad S$$

and its corresponding eigenvalue is zero.

The presence of this eigenvalue also affects the construction of a Neumann function (Duff, Ref. 15, Chapter 8). The bilinear formula (7.1.9) of Theorem 7.1.3 fails for a Neumann function. The presence of a zero eigenvalue means that one term has a vanishing denominator and the series is not defined. From the series of Corollary 7.1.1 we observe that the function has a pole at $\lambda = \lambda_n$ for every eigenvalue λ_n, and, in particular, if $\lambda_n = 0$.

To remove this difficulty from the construction of Neumann's function, we shall project our functions on the subspace orthogonal to the critical eigenfunction. Consider the nonhomogeneous problem

(7.3.11) $$\Delta u = -f(P)$$

with $\partial u/\partial n = 0$ on S. If there is a solution

(7.3.12) $$\int_V f(P) \, dV = -\int_V \Delta u \, dV = -\int_S \frac{\partial u}{\partial n} \, dS = 0$$

and if this necessary condition is not satisfied, no solution u can possibly exist.

The orthogonal projection of $f(P)$ on the constant function is the average value

(7.3.13) $$\mathbf{P}_c f(P) = \frac{1}{V} \int_V f(P) \, dV_P = \bar{f}$$

The difference $f(P) - \bar{f}$ has zero average value and its projection is zero. Therefore, for any function $f(P)$, the problem

(7.3.14) $$Lu = -\Delta u = f(P) - \bar{f}$$

with $\partial u/\partial n = 0$ on S satisfies the necessary condition (7.3.12). Let us construct a Neumann function for the problem (7.3.14). That is, let us choose $f(P) = \delta(P, Q)$. In place of the delta function on the right-hand side, we should write $\delta(P, Q) - \bar{\delta}$. We observe that the average value $\bar{\delta}$ is the reciprocal of the volume V. Therefore we set

(7.3.15) $$-\Delta N(P, Q) = \delta(P, Q) - \frac{1}{V}$$

while

(7.3.16) $$\frac{\partial N(P, Q)}{\partial n_P} = 0 \qquad\qquad P \quad \text{on} \quad S$$

The Fourier expansion of this function can be found as in Theorem 7.1.3. We shall require $N(P, Q)$ to be orthogonal to the lowest eigenfunction $1/\sqrt{V}$.

In other words, we set

$$N(P, Q) = \sum_{n=2}^{\infty} n_n(Q) u_n(P)$$

Then

$$-\Delta_P N(P, Q) = \sum_{n=2}^{\infty} n_n(Q) \lambda_n u_n(P)$$

and, according to (7.3.15), this is equal to

$$\delta(P, Q) - \frac{1}{\sqrt{V}} \frac{1}{\sqrt{V}} = \sum_{n=2}^{\infty} u_n(P) u_n(Q)$$

This last equation is an instance of the bilinear expansion (7.1.21).

Equating coefficients of $u_n(P)$, we can determine the functions $n_n(Q)$ as $\lambda_n^{-1} u_n(Q)$. The result is that

(7.3.17) $$N(P, Q) = \sum_{n=2}^{\infty} \frac{u_n(P) u_n(Q)}{\lambda_n}$$

Example 7.3.1 The Neumann function for the rectangle $0 \leq x \leq a$, $0 \leq y \leq b$ is

(7.3.18) $$N(x, y; x_1, y_1) = \frac{4}{ab} \sum_{m,n=0}^{\infty} \gamma_{mn} \frac{\cos \frac{m \pi x}{a} \cos \frac{m \pi x_1}{a} \cos \frac{n \pi y}{b} \cos \frac{n \pi y_1}{b}}{\pi^2 \left(\frac{m^2}{a^2} + \frac{n^2}{b^2} \right)}$$

where $\gamma_{00} = 0$, $\gamma_{m0} = \gamma_{0n} = \frac{1}{2}$, and $\gamma_{mn} = 1$ for $m > 0, n > 0$.

Let us now consider the most general nonhomogeneous Neumann problem for a region R for which (7.3.11) is the differential equation and (7.3.8) the boundary condition. The necessary constraint is

(7.3.19) $$\int_R f(P) \, dV_P = -\int_R \Delta u \, dV = -\int_S \frac{\partial u}{\partial n} \, dS = -\int_S g(P) \, dS$$

Let us suppose that it is satisfied.

Suppose now that $u(P)$ is a solution and apply Green's second formula to $u(Q)$ and $N(P, Q)$. We find

(7.3.20) $$\int_R [u(Q) \Delta N(P, Q) - N(P, Q) \Delta u(Q)] \, dV_Q$$
$$= \int_S \left[u(Q) \frac{\partial N(P, Q)}{\partial n_Q} - N(P, Q) \frac{\partial u(Q)}{\partial n_Q} \right] dS_Q$$

The volume integral on the left becomes

$$\int_R \left\{ u(Q) \left[-\delta(P, Q) + \frac{1}{V} \right] + N(P, Q) f(Q) \right\} dV_Q$$
$$= -u(P) + \frac{1}{V} \int_R u(Q) \, dV_Q + \int_R N(P, Q) f(Q) \, dV_Q$$

The integral over $u(Q)$ is simply an additive constant, representing the average value $\bar{u}$ of the solution. On the right side of (7.3.20) the first term disappears, by (7.3.16). We insert the boundary values (7.3.8) and so establish

THEOREM 7.3.2 *The nonhomogeneous Neumann problem*

(7.3.21) $$-\Delta u = f(P)$$

with boundary condition

(7.3.22) $$\frac{\partial u}{\partial n} = g(P) \qquad P \text{ on } S$$

satisfying (7.3.19) *has the representation formula*

(7.3.23) $$u(P) = \int_R N(P, Q) f(Q) \, dV_Q + \int_S N(P, Q) g(Q) \, dS_Q + \bar{u}$$

where $\bar{u}$ is an arbitrary constant equal to the average value of u.

EXERCISES 7.3

1. Show that $R(P, Q)$ is unique and symmetric. Assume $h > 0$.

2. Find the one-dimensional Robin function for the equation $u_{xx} = 0$ on $0 \leq x \leq l$, and verify that it becomes infinite as $h \to 0$ in the conditions $-u_x + hu = 0$ at $x = 0$, $u_x + hu = 0$ at $x = l$.

3. Construct the bilinear series for the Robin function of the condition $u_n + hu = 0$, h a constant, for:
   (a) The interval $0 \leq x \leq a$.
   (b) The rectangle $0 \leq x \leq a$, $0 \leq y \leq b$.
   (c) The rectangular solid $0 \leq x \leq a$, $0 \leq y \leq b$, $0 \leq z \leq c$.
The differential operator is in each case the Laplacian.

4. If $u(x, y, z)$ is harmonic and bounded for $z > 0$ and satisfies $-u_z + hu = g(x, y)$ on $z = 0$, where $h > 0$, show that
$$u(x, y, z) = \int_z^\infty v(x, y, z_1) e^{-h(z_1 - z)} \, dz_1$$
where $v(x, y, z)$ is harmonic and bounded for $z > 0$ and $v(x, y, 0) = g(x, y)$. Assume h is constant.

5. Verify, as in (7.3.13), that the projection of a function (vector) $f(P)$ on the constant function is its average value $\bar{f}$. See Chapter 1, Section 1.1. Verify also that its projection orthogonal to the constant function is $f(P) - \bar{f}$.

6. Find the Green's function G of the equation $\Delta u - k^2 u = 0$ in the quarter-infinite region $0 \leq x < \infty$, $0 \leq y < \infty$ subject to the mixed boundary conditions $G = 0$ on $y = 0$ and $G_x = 0$ on $x = 0$. Show that
$$G(x, y, x', y') = \frac{1}{\pi} \int_0^\infty \frac{1}{\sigma} (e^{-\sigma|x-x'|} + e^{-\sigma(x+x')}) \sin sy \sin sy' \, ds$$

where $\sigma = \sqrt{k^2 + s^2}$. Show also that

$$G(x, y, x', y') = \frac{1}{\pi} \int_0^\infty \frac{1}{\sigma} (e^{-\sigma|y-y'|} - e^{-\sigma(y+y')}) \cos sx \cos sx' \, ds$$

7. Show that the Helmholtz equation Green's function vanishing on both sides of the strip $0 \leq x \leq a$, $-\infty < y < \infty$ and representing outgoing waves at infinity is given by

$$G = \sum_{n=1}^{n_0} \frac{e^{-i\mu|y-y'|}}{ia\mu} \sin\frac{n\pi x}{a} \sin\frac{n\pi x'}{a} + \sum_{n=n_0+1}^\infty \frac{e^{-\lambda|y-y'|}}{a\lambda} \sin\frac{n\pi x}{a} \sin\frac{n\pi x'}{a}$$

where

$$\mu = \sqrt{k^2 - \frac{n^2\pi^2}{a^2}}, \qquad \lambda = \sqrt{\frac{n^2\pi^2}{a^2} - k^2}$$

and n_0 is that integer such that $n_0 < ak/\pi < n_0 + 1$. Explain how to treat the case when ak/π is an integer.

7.4 DIFFERENTIAL AND INTEGRAL EQUATIONS

In Section 7.1 we discussed the Green's function of a boundary value problem and its interpretation as an inverse matrix. Here we shall pursue this analogy which now leads us to integral equations. We shall also study the inverse of the translated operator $L - \lambda$ as a function of the spectral parameter λ. The Green's function for this operator is known as the *resolvent* of the original operator L.

For vectors in a finite-dimensional space an operator equation of the "first kind"

(7.4.1) $$T\mathbf{x} = \mathbf{f}$$

signifies a system of a finite number of linear algebraic equations

(7.4.2) $$\sum_{l=1}^n T_{kl} x_l = f_k \qquad k = 1, \ldots, n$$

We shall suppose that the operator T and also the matrix (T_{kl}) are symmetric. If the determinant $\det T$ is not zero, that is, if $\lambda = 0$ is not an eigenvalue of T, then for every vector $\mathbf{f}$ there exists a unique solution vector $\mathbf{x}$ of (7.4.1). However, if zero is an eigenvalue of T, then one or more corresponding eigenvectors $\mathbf{x}_k$ exist. In this case a solution $\mathbf{x}$ of (7.4.1) exists if and only if the nonhomogeneous term $\mathbf{f}$ is orthogonal to $\mathbf{x}_k$: $\mathbf{f} \cdot \mathbf{x}_k = 0$. Moreover, the solution vector $\mathbf{x}$ contains additive eigenvector terms $\mathbf{x}_k$.

Similarly, an operator equation of the second kind

(7.4.3) $$\mathbf{x} - \lambda T\mathbf{x} = \mathbf{f}$$

corresponds to a system

(7.4.4) $$x_k - \lambda \sum_{l=1}^n T_{kl} x_l = f_k \qquad k = 1, \ldots, n$$

284   GREEN'S FUNCTIONS

This system has a unique solution if and only if $1/\lambda$ is not an eigenvalue of T, that is, $\det(E - \lambda T) \neq 0$. If $1/\lambda$ is an eigenvalue, there exists a solution vector $\mathbf{x}$ if and only if f is orthogonal to the corresponding eigenvector or eigenvectors.

The symbolic solution of (7.4.3) is

$$(7.4.5) \qquad \mathbf{x} = \frac{1}{E - \lambda T}\mathbf{f} = \mathbf{f} + \lambda T\mathbf{f} + \lambda^2 T^2\mathbf{f} + \cdots + \lambda^n T^n\mathbf{f} + \cdots$$

This *resolvent series* converges for sufficiently small λ.

We shall now discuss the boundary value problems and the matrix analogy of Section 7.1 in the light of these standard matrix theorems. Perhaps the simplest way to do this would be to use eigenfunction expansions and express the problem by means of infinite matrices. However, this would raise further questions of convergence and existence of solutions of such systems of infinite order. These we can avoid by working directly with the Green's functions already introduced.

We shall now examine the equation

$$(7.4.6) \qquad Lu(P) - \lambda u(P) = f(P)$$

where λ is a parameter. This equation would arise, for instance, from a steady-state solution of a wave equation $u_{tt} + Lu = f$ with a periodic time factor e^{ikt}, where $k^2 = \lambda$. The Dirichlet boundary condition

$$(7.4.7) \qquad u(P) = 0$$

will also be assumed.

We shall express this problem by means of an *integral equation*. Suppose that the Green's function $G(P, Q)$ of L is known. Transpose $\lambda u(P)$ to the right side of (7.4.6) and consider it as a nonhomogeneous term.

THEOREM 7.4.1   *A solution $u(P)$ of (7.4.6) and (7.4.7) satisfies the integral equation of the second kind*

$$(7.4.8) \qquad u(P) = \lambda \int_R G(P, Q)u(Q)\, dV_Q + F(P)$$

where

$$(7.4.9) \qquad F(P) = Gf(P) = \int_R G(P, Q)f(Q)\, dV_Q$$

Conversely, a solution of the integral equation satisfies both the differential equation and the boundary condition.

Proof. To establish (7.4.8), we apply Theorem 7.1.2 to the differential equation, noting that by (7.4.7) the surface integral term will disappear. That is, we take the "inner product" of the inverse matrix $G(P, Q)$ with the terms $\lambda u(P) + f(P)$ and so obtain (7.4.8).

If $u(P)$ is a solution of the integral equation, we compute

$$Lu(P) = \lambda \int_R L_P G(P, Q) u(Q) \, dV_Q + LF(P)$$
$$= \lambda \int_R \delta(P, Q) u(Q) \, dV_Q + f(P)$$
$$= \lambda u(P) + f(P)$$

again using Theorem 7.1.2 to evaluate $LF(P)$. Therefore the differential equation is satisfied. As for the boundary condition, we note by Theorem 7.1.2 that $F(P) = 0$ on S. The same is true of the integral over $u(Q)$, since $G(P, Q) = 0$ for P on S, Q in R. Some caution is necessary in drawing this conclusion because the singularity of $G(P, Q)$ when $P \to Q$ prevents the boundary values of G being attained uniformly with respect to Q. However, it is shown in Courant-Hilbert, Ref. 13 (vol. II, 2nd Ed., p. 263) that this nonuniformity does not impair the formal result.

Finally, therefore, we see that the integral equation is equivalent to the differential equation together with the boundary condition.

To invert the differential operator $L - \lambda$ directly, we should consider a Green's function $G(P, Q, \lambda)$ satisfying

(7.4.10) $\qquad L_P G(P, Q, \lambda) = \lambda G(P, Q, \lambda) + \delta(P, Q)$

The bilinear series for this "resolvent" function is given in Corollary 7.1.1. For the present we shall regard the right side of (7.4.10) as nonhomogeneous, and "solve" the equation by using $G(P, Q)$. We note that

(7.4.11) $\qquad\qquad G(P, Q, \lambda) = 0 \qquad\qquad P$ on S

and apply Theorem 7.1.2. This yields the *resolvent formula*

(7.4.12) $\qquad G(P, Q, \lambda) = \lambda \int_R G(P, Q_1, \lambda) G(Q_1, Q) \, dV_{Q_1} + G(P, Q)$

This is an integral equation connecting the Green's functions of L and $L - \lambda$.

By means of the resolvent relation we can verify that the solution of the integral equation (7.4.8) is

(7.4.13) $\qquad\qquad u(P) = \int_R G(P, Q, \lambda) f(Q) \, dV_Q$

provided that $G(P, Q, \lambda)$ is defined for the given value of λ. Substitution of (7.4.13) into (7.4.8) gives

(7.4.14)
$$u(P) - \lambda \int_R G(P, Q_1) u(Q_1) \, dV_{Q_1}$$
$$= \int_R G(P, Q, \lambda) f(Q) \, dV_Q - \lambda \int_R G(P, Q_1) \int_R G(Q_1, Q, \lambda) f(Q) \, dV_Q \, dV_{Q_1}$$
$$= \int_R \left[G(P, Q, \lambda) - \lambda \int_R G(P, Q_1) G(Q_1, Q, \lambda) \, dV_{Q_1} \right] f(Q) \, dV_Q$$

By (7.4.12) and the symmetry of the two Green's functions, the first factor in the integrand is $G(P, Q)$, so the whole expression becomes the integral $F(P)$ of (7.4.9) as required.

THEOREM 7.4.2 *If λ is not an eigenvalue the differential equation (7.4.6) with the boundary condition (7.4.7) has a unique solution given by (7.4.13).*

Proof. If there were two distinct solutions, their difference would be an eigenfunction, contrary to hypothesis. Now apply the operator L to (7.4.13) and use (7.4.10) and (7.4.13) again. We find that the differential equation is formally satisfied by $u(P)$. Likewise, the boundary condition is satisfied by $G(P, Q, \lambda)$ for all Q within the region, and so holds formally for the integral (7.4.13). This completes the proof.

Now let us consider the exceptional case where λ is equal to a simple eigenvalue λ_m with the eigenfunction $u_m(P)$. The bilinear series

$$(7.4.15) \qquad G(P, Q, \lambda) = \sum_n \frac{u_n(P) u_n(Q)}{\lambda_n - \lambda}$$

shows that the resolvent $G(P, Q, \lambda)$ is not defined for this value of λ. However, the summation with the term for $n = m$ omitted is defined. A direct way to define the truncated function is to observe that this term is the *projection* of $G(P, Q, \lambda)$ on the *eigenspace* of λ_m. Therefore we set

$$(7.4.16) \qquad \begin{aligned} G^m(P, Q, \lambda_m) &= [G(P, Q, \lambda) - P_m G(P, Q, \lambda)]_{\lambda = \lambda_m} \\ &= \sum_{n \neq m} \frac{u_n(P) u_n(Q)}{\lambda_n - \lambda_m} \end{aligned}$$

By construction this modified Green's function is orthogonal to $u_m(P)$.

The differential equation satisfied by $G^m(P, Q, \lambda_m)$ is

$$(7.4.17) \qquad LG^m(P, Q, \lambda_m) = \lambda_m G^m(P, Q, \lambda_m) + \delta(P, Q) - u_m(P) u_m(Q)$$

This can be verified if we recall the series

$$(7.4.18) \qquad \delta(P, Q) = \sum_{n=1}^{\infty} u_n(P) u_n(Q)$$

for the delta function and observe that its projection on the eigenspace of λ_m is the product $u_m(P) u_m(Q)$ which has been subtracted in (7.4.17).

From this differential equation we deduce by Theorem 7.1.2 the modified resolvent formula

$$(7.4.19)$$

$$G^m(P, Q, \lambda_m) = \lambda_m \int_R G(P, Q_1) G^m(Q_1, Q, \lambda_m) \, dV_{Q_1} + G(P, Q) - \frac{u_m(P) u_m(Q)}{\lambda_m}$$

THEOREM 7.4.3 *If λ is equal to an eigenvalue λ_m of L, the equation*

$$(7.4.20) \qquad Lu(P) - \lambda u(P) = f(P)$$

with $u(P) = 0$ on S, has a solution if and only if

(7.4.21) $$\int_R f(P)u_m(P)\,dV_P = 0$$

that is, $f(P)$ is orthogonal to the eigenfunction u_m. The most general solution contains an additive multiple of u_m.

Proof. Condition (7.4.21) is certainly necessary, for if a solution $u(P)$ exists, then

$$\int_R f(P)u_m(P)\,dV_P = \int_R [Lu(P) - \lambda_m u(P)]u_m(P)\,dV_P$$
$$= \int_R u(P)[Lu_m(P) - \lambda_m u_m(P)]\,dV_P - \int_S \left(u_m \frac{\partial u}{\partial n} - u \frac{\partial u_m}{\partial n}\right) dS_P$$
$$= 0$$

The volume term vanishes because u_m is an eigenfunction with eigenvalue λ_m, the surface term because u and u_m satisfy the boundary conditions.

Suppose the condition is satisfied. Then

(7.4.22) $$U^m(P) = \int_R G^m(P, Q, \lambda_m) f(Q)\,dV_Q$$

is a solution. We can establish this result either with the differential equation (7.4.17) or the resolvent formula (7.4.19). As the latter includes the boundary condition, we shall use it. Inserting the expression for $G^m(P, Q, \lambda_m)$ into (7.4.22), we obtain

$$U^m(P) = \lambda_m \int_R \int_R G(P, Q_1)G^m(Q_1, Q, \lambda_m)\,dV_{Q_1} f(Q)\,dV_Q$$
$$+ \int_R G(P, Q)f(Q)\,dV_Q - \frac{u_m(P)}{\lambda_m} \int_R f(Q)u_m(Q)\,dV_Q$$

The last integral is zero by hypothesis. Using (7.4.22) again, we find

(7.4.23) $$U^m(P) = \lambda_m \int_R G(P, Q)U^m(Q)\,dV_Q + F(P)$$

where $F(P)$ is given by (7.4.9). This shows that $U^m(P)$ satisfies the integral equation (7.4.8) and is therefore a solution of the boundary value problem.

These theorems can be extended in various ways some of which are suggested as exercises. The subject of integral equations and their relation to algebraic and differential equations is a fascinating one. More detailed accounts can be found in Courant-Hilbert, Ref. 13 (vol. 1); Epstein, Ref. 20; and Lovitt, Ref. 41.

EXERCISES 7.4

1. If a finite symmetric matrix T has a null vector $\mathbf{x}_0$, and $T\mathbf{x} = \mathbf{f}$ has a solution $\mathbf{x}$, show that $\mathbf{f} \cdot \mathbf{x}_0 = 0$.

2. Show that, if $\mathbf{x}_0$ is the only null vector of T, the condition $\mathbf{f} \cdot \mathbf{x}_0 = 0$ is sufficient as well as necessary for the existence of a solution vector $\mathbf{x}$ of $T\mathbf{x} = \mathbf{f}$.

3. Prove the statement in the text that $\mathbf{x} - \lambda T\mathbf{x} = \mathbf{f}$ has a solution if and only if $\mathbf{f}$ is orthogonal to every eigenvector with eigenvalue $1/\lambda$.

4. Show that the problem $Lu - \lambda u = f$, with $\partial u/\partial n + hu = g$ on S, has a solution if and only if $\int_R fu_m \, dV + \int_S gu_m \, dS = 0$ for every eigenfunction u_m of L with eigenvalue λ.

5. Show that u satisfies the integral equation $u(P) = \lambda \int_R G(P, Q)u(Q) \, dV_Q$ of the first kind if and only if u is an eigenfunction of L with eigenvalue λ.

6. Establish the resolvent formula
$$G(P, Q, \lambda) - G(P, Q, \mu) = (\lambda - \mu) \int_R G(P, Z, \lambda)G(Z, Q, \mu) \, dV_Z$$

7.5 SOURCE FUNCTIONS FOR PARABOLIC EQUATIONS

Consider the heat flow equation

(7.5.1) $$\frac{\partial u}{\partial t} = \Delta u$$

on a domain R. The boundary condition can be taken as

(7.5.2) $\qquad\qquad u(P, t) = 0 \qquad\qquad P$ on S

and an initial distribution of temperature

(7.5.3) $\qquad\qquad u(P, 0) = f(P)$

will be assigned.

The eigenfunction expansion

(7.5.4) $$u(P, t) = \sum_{n=1}^{\infty} c_n(t)u_n(P)$$

with eigenfunctions $u_n(P)$ satisfying $\Delta u_n + \lambda_n u_n = 0$ will be assumed for the solution. Thus,

(7.5.5) $$c_n(t) = \int_R u_n(P)u(P, t) \, dV$$

where the eigenfunctions $u_n(P)$ are normalized, while

(7.5.6) $$\begin{aligned}\dot{c}_n(t) &= \int_R u_n(P)\dot{u}(P, t) \, dV \\ &= \int_R u_n(P) \Delta u(P, t) \, dV \\ &= \int_R \Delta u_n(P) u(P, t) \, dV\end{aligned}$$

by Green's formula and the vanishing of $u_n(P)$ and of $u(P, t)$ on S. Therefore

(7.5.7)
$$\dot{c}_n(t) = -\lambda_n \int_R u_n(P)u(P, t)\,dV$$
$$= -\lambda_n c_n(t)$$

The solution of this first order equation is

(7.5.8)
$$c_n(t) = c_n(0)e^{-\lambda_n t}$$

Since (7.5.3) holds, the initial values of the Fourier coefficients (7.5.5) will be

(7.5.9)
$$c_n(0) = f_n = \int_R f(P)u_n(P)\,dV$$

Insertion of (7.5.8) and (7.5.9) in (7.5.4) gives

(7.5.10)
$$u(P, t) = \sum_{n=1}^{\infty} e^{-\lambda_n t} u_n(P) \int_R f(Q)u_n(Q)\,dV_Q$$
$$= \int_R f(Q) \left[\sum_{n=1}^{\infty} e^{-\lambda_n t} u_n(P)u_n(Q) \right] dV_Q$$

upon inversion of the order of summation and integration. Therefore, we shall define

(7.5.11)
$$K(P, Q, t) = \sum_{n=1}^{\infty} e^{-\lambda_n t} u_n(P)u_n(Q)H(t)$$

as the *Green's function for the parabolic equation*. The Heaviside factor $H(t)$ is inserted as in Chapter 3, Section 3.4, because the solution is identically zero for all negative times.

When t is positive, $H(t) \equiv 1$, and so

(7.5.12)
$$\lim_{t \to 0+} K(P, Q, t) = \sum_{n=1}^{\infty} u_n(P)u_n(Q) = \delta(P, Q)$$

This shows that the integral in (7.5.10) formally satisfies the initial condition.

By direct differentiation, we have

$$\left(\frac{\partial}{\partial t} - \Delta_P \right) K(P, Q, t) = \sum_{n=1}^{\infty} (-\lambda_n e^{-\lambda_n t}) u_n(P)u_n(Q)H(t)$$
$$- \sum_{n=1}^{\infty} e^{-\lambda_n t}[-\lambda_n u_n(P)]u_n(Q)H(t) + \sum_{n=1}^{\infty} e^{-\lambda_n t} u_n(P)u_n(Q)\,\delta(t)$$

The first two series cancel, since K is a solution of the homogeneous parabolic equation away from the source point. In the third series we can set $e^{-\lambda_n t}\delta(t) = \delta(t)$, and so find the bilinear series for the delta function. Therefore

(7.5.13)
$$\frac{\partial K}{\partial t} - \Delta K = \delta(P, Q)\,\delta(t)$$

Example 7.5.1 For the entire three-dimensional space the Laplace operator has a continuous spectrum and the normalized eigenfunctions [see (3.6.18)] are

$$\frac{e^{i(s_1 x_1 + s_2 x_2 + s_3 x_3)}}{(2\pi)^{3/2}}$$

The Green's function is then the integral

$$K(x, t) = \frac{H(t)}{(2\pi)^3} \iiint e^{is\cdot x - s^2 t} \, ds_1 \, ds_2 \, ds_3$$

where $s^2 = s_1^2 + s_2^2 + s_3^2$ is the squared length of $\mathbf{s} = (s_1, s_2, s_3)$. This expression, which would also be found by a direct Fourier analysis, factors into the product of three one-dimensional heat source solutions. Therefore, by Exercise 3.4.2,

$$(7.5.14) \qquad K(x, t) = \frac{H(t)}{(2\sqrt{\pi t})^3} e^{-r^2/4t}$$

where r is the Euclidean length of x, that is, $r^2 = x_1^2 + x_2^2 + x_3^2$.

Let us use the function $K(P, Q, t)$, which is the temperature at P due to the release at Q of a unit of heat t moments earlier, to represent the solution of the most general boundary and initial conditions. Suppose now that heat sources

$$(7.5.15) \qquad \frac{\partial u}{\partial t} - \Delta u = f(P, t)$$

boundary temperatures

$$(7.5.16) \qquad u(P, t) = g(P, t) \qquad\qquad P \text{ on } S$$

and initial temperatures

$$(7.5.17) \qquad u(P, 0) = u_0(P)$$

are given.

Consider a product domain $R \times (0, t - \varepsilon)$ of the region R and a time interval $0 \leq \tau \leq t - \varepsilon$ (Figure 7.4). To set up a suitable integral formula, we shall integrate $K(P, Q, t - \tau)(u_\tau - \Delta u)$ over the product domain and subsequently integrate by parts. Thus

$$\int_{0+}^{t-\varepsilon} \int_R K(P, Q, t - \tau)(u_\tau - \Delta u) \, dV_Q \, d\tau$$

$$(7.5.18) \quad = \int_R \left\{ [K(P, Q, t - \tau) u(Q, \tau)]_0^{t-\varepsilon} + \int_0^{t-\varepsilon} K_t(P, Q, t - \tau) u(Q, \tau) \, d\tau \right\} dV_Q$$

$$- \int_0^{t-\varepsilon} \left[\int_S \left(K \frac{\partial u}{\partial n} - u \frac{\partial K}{\partial n} \right) dS + \int_R u \Delta K \, dV \right] d\tau$$

SOURCE FUNCTIONS FOR PARABOLIC EQUATIONS

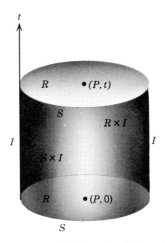

Figure 7.4 Parabolic domain of dependence.

We can simplify the various terms as follows. On the left side we substitute (7.5.15). We use $K(P, Q, 0 + \varepsilon) \to \delta(P, Q)$ in the first term on the right, and we let ε tend to zero. The boundary conditions $K = 0$ and (7.5.16) apply to the surface integrals. Also $K_t - \Delta K = 0$ for $t - \tau > 0$. Thus we find

(7.5.19)
$$\int_0^t \int_R K(P, Q, t - \tau) f(Q, \tau) \, dV_Q \, d\tau$$
$$= \int_R K(P, Q, 0+) u(Q, t) \, dV_Q - \int_R K(P, Q, t) u_0(Q) \, dV_Q$$
$$+ \int_0^t d\tau \int_S g(Q, \tau) \frac{\partial K(P, Q, t - \tau)}{\partial n_Q} \, dS_Q$$

Therefore, finally, we have by (7.5.12) the following theorem.

THEOREM 7.5.1 *The representation of the solution $u(P, t)$ of (7.5.15), (7.5.16), and (7.5.17) is*

(7.5.20)
$$u(P, t) = \int_R K(P, Q, t) u_0(Q) \, dV_Q + \int_0^t \int_R K(P, Q, t - \tau) f(Q, \tau) \, dV_Q \, d\tau$$
$$- \int_0^t \int_S \frac{\partial K}{\partial n_Q}(P, Q, t - \tau) g(Q, \tau) \, dS_Q \, d\tau$$

By detailed analysis into which we shall not enter here, it can be shown that each of these integrals or *thermal potentials* has the properties required by the conditions of this heat flow problem (Tychonov-Samarskii, Ref. 60, p. 533).

EXERCISES 7.5

1. Evaluate the function $K(x, x_1, t)$ for the interval $(0, \infty)$, with temperatures vanishing at the origin, in two ways:
   (a) By source and image.
   (b) By bilinear expansion.
Show the equivalence of the two methods.

2. Show that the function $K_h(P, Q, t)$ for the radiation equation $u_t = \Delta u - hu$ is given by $K_h(P, Q, t) = e^{-ht} K(P, Q, t)$.

3. Determine the function $K(P, Q, t)$ for the Schrödinger equation of a free particle of mass m in three dimensions. If the particle is located at the origin at time zero, find the probability that it is located in an interval dx at x at time t.

4. Show that $\int_0^\infty K(P, Q, t)\, dt = G(P, Q)$ and interpret physically.

5. Show that the solution of a heat flow problem with given boundary temperatures tends to the harmonic equilibrium distribution as $t \to \infty$, and estimate its rate of approach to the limit, supposing the lowest eigenvalue λ_1 of R is known.

6. The density of neutrons diffusing through an atomic reactor occupying a region R satisfies the equation

$$u_t = a^2 \Delta u + \beta u$$

where $\beta > 0$ determines the source density due to chain reaction multiplication. Determine a criterion for the *critical size* of the region R.

7. Show that the Green's function of the equation $\Delta u - u_t = 0$ for the region $0 \leq x \leq a$, $-\infty < y < \infty$, subject to the condition of vanishing on both sides, is given by

$$K(x, y, x', y', t) = \frac{e^{-(y-y')^2/4t}}{a\sqrt{\pi t}} \sum_{n=1}^\infty e^{-n^2\pi^2 t/a^2} \sin\frac{n\pi x}{a} \sin\frac{n\pi x'}{a} H(t)$$

8. By letting $a \to \infty$ in the preceding result, show that the Green's function for the half-plane $0 \leq x < \infty$, $-\infty < y < \infty$ vanishing for $x = 0$ is given by

$$K(x, y, x', y', t) = \frac{1}{4\pi t} \left\{ \exp\left[-\frac{(x-x')^2 + (y-y')^2}{4t}\right] - \exp\left[-\frac{(x+x')^2 + (y-y')^2}{4t}\right] \right\} H(t)$$

9. For the equation $G_{xx} - G_t = -\delta(x - x')\,\delta(t)$ and the interval $0 \leq x \leq a$ with $G = 0$ at both extremities, show that

$$G(x, x', t) = \frac{2}{a} \sum_{n=1}^\infty e^{-n^2\pi^2 t/a^2} \sin\frac{n\pi x}{a} \sin\frac{n\pi x'}{a} H(t)$$

10. A heat source of strength $f(t)H(t)$ appears at the origin at time zero and moves along the positive x-axis with constant speed U. Show that

$$\varphi_{xx} - \varphi_t = -\delta(x - Ut)H(t)f(t)$$

and that

$$\varphi = \frac{1}{2\sqrt{\pi}} \int_0^t (t - t')^{-1/2} e^{-(x-Ut')^2/4(t-t')} f(t')\, dt'$$

7.6 CONVERGENCE OF SERIES OF DISTRIBUTIONS

We have seen that the Green's function for the diffusion equation can be evaluated by means of an appropriate bilinear series. It would be useful to have this method available for other types of differential equations as well. The formal series thus encountered is often divergent, in the ordinary sense, so we shall discuss its convergence in a sense appropriate for distributions.

A sequence T_n of distributions is said to *converge* if the numerical sequence

(7.6.1) $$T_n(\varphi)$$

converges for every testing function φ. The limit, which we denote by $T(\varphi)$, defines a new linear functional. It can be shown (Schwartz, Ref. 51, p. 71; Gelfand-Shilov, Ref. 25, p. 368) that $T(\varphi)$ is continuous and defines a distribution T. That is, the linear space of distributions is *closed* under limits taken in this distribution sense.

If a sequence $T_n(\varphi)$ of distributions converge to a distribution $T = T(\varphi)$, then we shall say that the corresponding symbolic functions *converge weakly*. That is, if

(7.6.2) $$T_n(\varphi) = \int f_n(x)\varphi(x)\,dx$$

and

(7.6.3) $$T(\varphi) = \int f(x)\varphi(x)\,dx$$

where $\varphi(x)$ is any testing function, we shall say that $f_n(x)$ tends weakly to $f(x)$, as $n \to \infty$.

We give two examples. First, let $f(x)$ be any positive function of compact support with

(7.6.4) $$\int f(x)\,dx = 1$$

Then, as the reader can easily verify, we have the weak limit

(7.6.5) $$\lim_{\varepsilon \to 0} \frac{1}{\varepsilon} f\left(\frac{x}{\varepsilon}\right) = \delta(x)$$

Second, let ε_n be a sequence of real numbers with $\lim_{n \to \infty} \varepsilon_n = 0$. Then the weak limit

$$\lim_{n \to \infty} \delta(x + \varepsilon_n)$$

is equal to $\delta(x)$. Note that the functions $f_n(x)$ of the sequence need not be classical functions but can be symbolic functions.

Let us show how weak convergence can be established in a particular case. The Fourier sine series on $(0, l)$ gives rise to the expansion

(7.6.6) $$\delta(x - y) = \frac{2}{l} \sum_{n=1}^{\infty} \sin \frac{n\pi x}{l} \sin \frac{n\pi y}{l}$$

of the Dirac distribution, where $0 < x < l$, $0 < y < l$. The partial sums of this bilinear series are the functions

(7.6.7) $$\sigma_m(x, y) = \frac{2}{l} \sum_{n=1}^{m} \sin \frac{n\pi x}{l} \sin \frac{n\pi y}{l}$$

294   GREEN'S FUNCTIONS

The value of the distribution $\sigma_m = \sigma_m(\varphi)$ is

$$\sigma_m(\varphi)(x) = \int_0^l \sigma_m(x, y)\varphi(y)\, dy$$

(7.6.8)
$$= \sum_{n=1}^m \frac{2}{l} \int_0^l \sin\frac{n\pi y}{l}\, \varphi(y)\, dy\, \sin\frac{n\pi x}{l}$$

$$= \sum_{n=1}^m b_n \sin\frac{n\pi x}{l} = s_m(x)$$

say, where b_n is the Fourier sine coefficient of φ and $s_m(x)$ the partial sum of the Fourier sine series. Since $\varphi(x)$ is a smooth function, we know from Section 2.7 that the Fourier sine series is convergent to the value $\varphi(x)$. Therefore

(7.6.9) $$\lim_{m\to\infty} \sigma_m(\varphi)(x) = \lim_{m\to\infty} s_m(x) = \varphi(x) = \delta_x(\varphi)$$

That is, the weak limit of the sequence $\sigma_m(x, y)$ is the delta distribution. Equivalently, we say that the bilinear series converges weakly to the delta distribution.

Here we have used the ordinary convergence of Fourier series to establish weak convergence of the bilinear series. There are many functions $f(x)$ not as smooth as testing functions for which the Fourier series converges, so we would expect our conclusion to be less strong than the convergence theorem we have utilized. Weak convergence, therefore, is truly a weaker concept, which applies to a wider class of series than ordinary convergence.

Another example is the series

$$\sum_{n=1}^\infty n \sin nx$$

which converges to $-\delta'(x)$ in the weak sense, for $-\pi \leq x \leq \pi$. This can be deduced by differentiation of the preceding example. Indeed, any Fourier series with coefficients that are polynomials in n converges in the weak sense to a polynomial expression in the derivatives of the Dirac delta.

We can extend the notion of weak convergence to integrals, as well as series, by considering a continuous family of distributions T_ν depending upon a continuous parameter ν. Thus $\lim_{\nu\to\infty} T_\nu = T$ means that $\lim_{\nu\to\infty} T_\nu(\varphi) = T(\varphi)$ for every testing function φ. Then the weak convergence of an integral such as

(7.6.10) $$T_\nu(x) = \int_0^\nu t(x, y)\, dy$$

defines the weak limit

(7.6.11) $$T(x) = \int_0^\infty t(x, y)\, dy$$

Let us indicate the proof of weak convergence for the integral

(7.6.12) $$\delta(x - y) = \frac{1}{2\pi} \int_{-\infty}^\infty e^{i(x-y)s}\, ds$$

Setting

(7.6.13) $$T_R(x-y) = \frac{1}{2\pi}\int_{-R}^{R} e^{i(x-y)s}\, ds = \frac{1}{\pi}\frac{\sin R(x-y)}{x-y}$$

we see that

(7.6.14) $$T_R(\varphi)(x) = \frac{1}{2\pi}\int_{-\infty}^{\infty} \varphi(y)\, dy \int_{-R}^{R} e^{i(x-y)s}\, ds$$

The integration over y is actually finite, since all testing functions $\varphi(y)$ have bounded support. We can therefore interchange the order of integration and write

(7.6.15) $$T_R(\varphi)(x) = \frac{1}{2\pi}\int_{-R}^{R} e^{ixs}\, ds \int_{-\infty}^{\infty} \varphi(y) e^{-iys}\, dy$$

By the classical proofs of Fourier's integral theorem, the integral on the right converges to $\varphi(x)$ in the limit $R \to \infty$. That is,

(7.6.16) $$\lim_{R \to \infty} T_R(\varphi)(x) = \varphi(x) = \delta_x(\varphi)$$

and this proves the result stated in (7.6.12).

As an example of an integral with a polynomially divergent integrand, we show that

(7.6.17) $$\delta'(x-y) = \frac{i}{2\pi}\int_{-\infty}^{\infty} s e^{i(x-y)s}\, ds$$

is valid in the weak sense. Let us define the symbolic function

(7.6.18) $$T'_R(x-y) = \frac{i}{2\pi}\int_{-R}^{R} s e^{i(x-y)s}\, ds$$

Then

(7.6.19)
$$T'_R(\varphi)(x) = \frac{i}{2\pi}\int_{-\infty}^{\infty} \varphi(y)\, dy \int_{-R}^{R} s e^{i(x-y)s}\, ds$$
$$= \frac{i}{2\pi}\int_{-R}^{R} s e^{ixs}\, ds \int_{-\infty}^{\infty} \varphi(y) e^{-iys}\, dy$$
$$= \frac{-1}{2\pi}\int_{-R}^{R} e^{ixs}\, ds \int_{-\infty}^{\infty} \varphi(y)\, d_y(e^{-iys})$$
$$= \frac{1}{2\pi}\int_{-R}^{R} e^{ixs}\, ds \int_{-\infty}^{\infty} e^{-iys}\varphi'(y)\, dy$$

by integration by parts. Since $\varphi'(y)$ is also a very smooth function, the classical Fourier theorem shows that

(7.6.20) $$\lim_{R \to \infty} T'_R(\varphi)(x) = \varphi'(x) = \delta'_x(\varphi)$$

this being an ordinary numerical limit. By the definition of weak convergence, therefore, (7.6.17) now holds.

Let us calculate the Green's function for Laplace's equation in three-dimensional space:

(7.6.21) $$-\Delta G(\mathbf{x}, \mathbf{y}) = \delta(\mathbf{x}, \mathbf{y})$$

The eigenfunctions are exponentials $e^{i\mathbf{k}\cdot\mathbf{x}}$, the eigenvalues are $\lambda = |\mathbf{k}|^2 = k_1^2 + k_2^2 + k_3^2$. The bilinear formula (7.1.9) yields the continuous "sum"

(7.6.22) $$G(\mathbf{x}, \mathbf{y}) = \frac{1}{(2\pi)^3} \int\int\int \frac{e^{i\mathbf{k}\cdot(\mathbf{x}-\mathbf{y})}}{|\mathbf{k}|^2} \, dk_1 \, dk_2 \, dk_3$$

which we can interpret as a weakly convergent integral. For spherical polar coordinates $|\mathbf{k}|, \theta, \varphi$ with polar axis parallel to $\mathbf{x} - \mathbf{y}$, we have

(7.6.23)
$$G(\mathbf{x}, \mathbf{y}) = \frac{1}{(2\pi)^3} \int_0^\infty \int_0^\pi \int_0^{2\pi} \frac{e^{i|\mathbf{k}||\mathbf{x}-\mathbf{y}|\cos\theta}}{|\mathbf{k}|^2} |\mathbf{k}|^2 \, d|\mathbf{k}| \sin\theta \, d\theta \, d\varphi$$

$$= \frac{1}{(2\pi)^2} \int_0^\infty \left(\int_0^\pi e^{i|\mathbf{k}||\mathbf{x}-\mathbf{y}|\cos\theta} \sin\theta \, d\theta \right) d|\mathbf{k}|$$

$$= \frac{2}{(2\pi)^2} \int_0^\infty \frac{\sin|\mathbf{k}||\mathbf{x}-\mathbf{y}|}{|\mathbf{k}||\mathbf{x}-\mathbf{y}|} \, d|\mathbf{k}|$$

$$= \frac{1}{4\pi|\mathbf{x}-\mathbf{y}|} = \frac{1}{4\pi r}$$

EXERCISES 7.6

1. Evaluate as distributions the series

$$\sum_{n=1}^\infty n^2 \cos nx \qquad \sum_{n=1}^\infty n^3 \sin \frac{n\pi x}{l}$$

2. Determine the distributions represented by the weakly convergent integrals

$$\int_0^\infty s \sin sx \, ds \qquad \int_0^\infty s^2 \cos sx \, ds$$

3. If a sequence of distributions T_n has limit T, show that $\lim_{n\to\infty} T_n' = T'$. *Hint:* Use the definition of limit given for the sequence (7.6.1).

4. If $f(s)$ and $g(s)$ are symbolic functions with $sf(s) = g(s)$, show that $f(s) = g(s)/s + C\delta(s)$. Find the general solution $f(s)$ if $s^2 f(s) = h(s)$.

5. Show that

$$x^{-1}(\varphi) = \int_0^\infty \frac{\varphi(x) - \varphi(-x)}{x} \, dx$$

defines a distribution which is equal to the function $1/x$ for $x \neq 0$. Show that

$$x^{-2}(\varphi) = \int_0^\infty \frac{\varphi(x) + \varphi(-x) - 2\varphi(0)}{x^2} \, dx$$

defines a distribution which is equal to the function $1/x^2$ for $x \neq 0$.

6. Show that
$$\lim_{R\to\infty} \frac{\cos Rx}{x} = 0$$
where the limit is understood in the weak sense. Show that, in the weak sense,
$$\int_0^\infty \sin xs \, ds = \frac{1}{x}$$
Evaluate the Fourier transform of $H(x)$, and check by comparing with the Fourier transform of $\delta(x)$.

7. Show that
$$\lim_{y\to 0+} \frac{d}{dx} \log(x + iy) = \frac{1}{x} - i\pi\, \delta(x)$$

8. Show that, if $a < 0 < b$, then
$$x^{-2}(\varphi) - \int_b^\infty \frac{\varphi(x)\,dx}{x^2} - \int_{-\infty}^a \frac{\varphi(x)\,dx}{x^2}$$
is equal to the integral
$$\int_a^b \frac{\varphi(x)\,dx}{x^2}$$
whenever the latter converges. Hence show how a finite value can be assigned to a formally divergent integral.

9. By summation of the bilinear series, construct the Green's function of $\Delta u = 0$ in three dimensions for the half-space $z \geq 0$ with the boundary conditions
(a) $u = 0$
(b) $\dfrac{\partial u}{\partial n} = 0$
(c) $\dfrac{\partial u}{\partial n} + hu = 0$

CHAPTER 8

Cylindrical Eigenfunctions

8.1 BESSEL FUNCTIONS

One of the most interesting aspects of applied mathematics is the relationship between the general theorems of analysis and the variety of particular cases which arise in the applications. For general theories we strip away and discard all but the essential relationships, which are therefore more clearly exposed to view. In a particular application we can fully work out the detailed consequences of a hypothesis and put it to the test of observation and comparison with the real world. Also the particular cases will serve as illustration and justification of the general theory which they exemplify.

For particular problems on any domain we require special functions adapted to the geometry of that domain. The Bessel functions which we study in this chapter are next in simplicity and frequency of appearance to the elementary functions used so far. They arise in problems on circles, cylinders, and spheres, and are also known as cylindrical functions. Particular cases of Bessel functions were encountered during the eighteenth century by Bernoulli, Euler, Lagrange, and later Fourier and Poisson. However, their name derives from a famous memoir written by the astronomer F. W. Bessel in 1824. Since then an immense amount of information has been gathered about Bessel functions of various kinds, and we can present only a few of the more prominent applications.

We shall introduce the theory of Bessel functions in the following way. Let us consider the wave equation in two space dimensions

(8.1.1) $$v_{xx} + v_{yy} = \frac{1}{c^2} v_{tt}$$

If we extract a harmonic time dependence by writing

(8.1.2) $$v(x, y, t) = e^{-ikct} u(x, y)$$

then from (8.1.1) we obtain the equation of Helmholtz

(8.1.3) $$u_{xx} + u_{yy} + k^2 u = 0$$

A particular solution of this equation is e^{iky}. This corresponds to the wave function $v = e^{ik(y-ct)}$ which represents plane progressive waves traveling in the y-direction. If we transform to plane polar coordinates $x = r \cos \theta$, $y = r \sin \theta$, we find that

(8.1.4) $$u = e^{ikr \sin \theta}$$

is a solution of the equation

(8.1.5) $$u_{rr} + \frac{1}{r} u_r + \frac{1}{r^2} u_{\theta\theta} + k^2 u = 0$$

Now let us expand the function $u(r, \theta)$ as an exponential Fourier series on the interval $-\pi \leq \theta \leq \pi$ by writing

(8.1.6) $$u(r, \theta) = \sum_{n=-\infty}^{\infty} e^{in\theta} u_n(r)$$

so that

(8.1.7) $$u_n(r) = \frac{1}{2\pi} \int_{-\pi}^{\pi} e^{-in\theta} u(r, \theta) \, d\theta$$

By virtue of the cyclic properties $u(r, -\pi) = u(r, \pi)$, $u_\theta(r, -\pi) = u_\theta(r, \pi)$, we find that (8.1.5) converts into the ordinary differential equation

(8.1.8) $$\frac{d^2 u_n}{dr^2} + \frac{1}{r} \frac{du_n}{dr} + \left(k^2 - \frac{n^2}{r^2}\right) u_n = 0$$

A solution of this equation is given by inserting (8.1.4) into (8.1.7). We denote this solution by $J_n(kr)$ and write (Watson, Ref. 61)

(8.1.9) $$J_n(kr) = \frac{1}{2\pi} \int_{-\pi}^{\pi} e^{-in\theta + ikr \sin \theta} \, d\theta$$

Equation (8.1.8) is *Bessel's differential equation*, and the function $J_n(kr)$ is the *Bessel function of the first kind of order n*. The function $J_n(kr)$ so defined satisfies Bessel's equation (8.1.8) only when n is an integer or zero. If we insert (8.1.4) into (8.1.6) and recall that $u_n(r) = J_n(kr)$, then we find the Fourier series

$$e^{ikr \sin \theta} = \sum_{n=-\infty}^{\infty} e^{in\theta} J_n(kr) \tag{8.1.10}$$

If in this result and in (8.1.9) we set $t = e^{i\theta}$ and $z = kr$, then we obtain the *generating function*

$$e^{(z/2)(t-1/t)} = \sum_{n=-\infty}^{\infty} t^n J_n(z) \tag{8.1.11}$$

The Cauchy integral equivalent to (8.1.9) is

$$J_n(z) = \frac{1}{2\pi i} \int_C t^{-n-1} e^{(z/2)(t-1/t)} \, dt \tag{8.1.12}$$

where C is a simple closed contour enclosing the origin in the complex t-plane. By using the relation (8.1.11) it is possible to obtain recurrence relations

Figure 8.1 Contour of integration L.

connecting the Bessel functions, and their derivatives, of integral order. Some such relations are given in the exercises at the end of this section.

For future applications it is desirable to obtain solutions of Bessel's equation (8.1.8) for any value of the index n. Let s be unrestricted and introduce the function $J_s(z)$ by generalizing the expression (8.1.12). We write

$$J_s(z) = \frac{1}{2\pi i} \int_L t^{-s-1} e^{(z/2)(t-1/t)} \, dt \tag{8.1.13}$$

This is known as Schläfli's integral. Because the integrand is no longer single-valued, it is necessary to insert a barrier or cut which we place along the negative real axis from $-\infty$ to the origin. The amplitude θ of any point t not on the barrier is made definite by adopting the convention that $-\pi < \theta < \pi$. In this way the integrand appearing in (8.1.13) is made single-valued in the plane cut along the negative real axis. The contour L illustrated in Figure 8.1 consists of an open loop round the origin commencing at $-\infty - 0i$ and ending at $-\infty + 0i$. With this contour the definition (8.1.13) is applicable when $R(z) > 0$ (otherwise the integral is divergent), and it is readily verified by direct substitution that the function $J_s(z)$ defined as above satisfies Bessel's equation (8.1.8) with n set equal to s therein.

If s is an integer or zero, the above contour L may be replaced by a circuit enclosing the origin, as in (8.1.12).

We may obtain a series expansion for $J_s(z)$ in the following manner. In the integral (8.1.13) we write $tz = 2w$ and obtain

$$(8.1.14) \qquad J_s(z) = \frac{1}{2\pi i}\left(\frac{z}{2}\right)^s \int_L w^{-s-1} e^{w-(z^2/4w)}\, dw$$

Here the many-valued function w^{-s-1} is defined as before by means of a barrier along the negative real axis in the w-plane. The path of integration in (8.1.14) is then taken as the same as that in (8.1.13). We now substitute the expression

$$e^{-z^2/4w} = \sum_{m=0}^{\infty} \frac{1}{m!}\left(\frac{-z^2}{4w}\right)^m$$

and obtain

$$(8.1.15) \qquad J_s(z) = \frac{1}{2\pi i}\left(\frac{z}{2}\right)^s \sum_{m=0}^{\infty} \frac{1}{m!}\left(\frac{-z^2}{4}\right)^m \int_L e^w w^{-m-s-1}\, dw$$

To express this result in a convenient form, we introduce the gamma function which we define for $R(z) > 0$ by means of the integral

$$(8.1.16) \qquad \Gamma(z) = \int_0^\infty e^{-t} t^{z-1}\, dt$$

Certain properties of this function are collected in the exercises at the end of this section together with hints for their solution. Here we need only the alternative formula (Exercise 8.1.6)

$$(8.1.17) \qquad \frac{1}{\Gamma(z)} = \frac{1}{2\pi i}\int_L e^w w^{-z}\, dw$$

where L is a loop contour similar to that used in (8.1.13). This enables us to write (8.1.15) as the series

$$(8.1.18) \qquad J_s(z) = \left(\frac{z}{2}\right)^s \sum_{m=0}^{\infty} \frac{1}{m!\,\Gamma(m+s+1)}\left(\frac{-z^2}{4}\right)^m$$

So far we have constructed only one solution of Bessel's equation and a second solution linearly independent of the first is required. If s is not an integer or zero, we will show that the function $J_{-s}(z)$ may be used as the second solution. We write the equation of Bessel in the Sturm-Liouville form

$$\frac{d}{dz}[zJ'_s(z)] + \left(z - \frac{s^2}{z}\right)J_s(z) = 0$$

Similarly,

$$\frac{d}{dz}[zJ'_{-s}(z)] + \left(z - \frac{s^2}{z}\right)J_{-s}(z) = 0$$

Multiply these equations by $J_{-s}(z)$, $J_s(z)$ respectively and subtract. Then, after slight rearrangement, we obtain the identity

$$\frac{d}{dz}\{z[J_s(z)J'_{-s}(z) - J_{-s}(z)J'_s(z)]\} = 0$$

so that, for some constant A,

$$z[J_s(z)J'_{-s}(z) - J_{-s}(z)J'_s(z)] = A$$

To determine A, we take the limiting value of the expression on the left as z tends to zero. When z is small, we obtain from (8.1.18) the approximate form

(8.1.19) $$J_s(z) = \frac{1}{\Gamma(s+1)}\left(\frac{z}{2}\right)^s + \cdots$$

The derivative $J'_s(z)$ may be evaluated similarly and we find, on using the identity $\Gamma(s)\Gamma(1-s) = \pi \operatorname{cosec} s\pi$ developed in Exercise 8.1.5 that $A = -(2/\pi)\sin s\pi$. Therefore,

(8.1.20) $$J_s(z)J'_{-s}(z) - J'_s(z)J_{-s}(z) = \frac{-2\sin s\pi}{\pi z}$$

The expression on the left is the Wronskian of the pair of functions $J_s(z)$, $J_{-s}(z)$. If s is not an integer or zero, this Wronskian does not vanish and so J_s and J_{-s} are linearly independent. However if $s = n$, where $n = 0, \pm 1, \pm 2, \ldots$, then these solutions are no longer independent. In fact, we can deduce directly from (8.1.9), on replacing θ by $\pi - \theta$ therein, that for such values of n,

(8.1.21) $$J_{-n}(z) = (-1)^n J_n(z)$$

It is customary to introduce for general values of s the function

(8.1.22) $$Y_s(z) = \frac{\cos s\pi J_s(z) - J_{-s}(z)}{\sin s\pi}$$

This provides another solution of Bessel's equation independent of $J_s(z)$ and of $J_{-s}(z)$. If $s = n$, an integer or zero, the above expression becomes indeterminate in view of (8.1.21) and it is necessary to take its limiting value as $s \to n$. By means of (8.1.20) and (8.1.22) we can easily obtain the Wronskian

(8.1.23) $$J_s(z)Y'_s(z) - J'_s(z)Y_s(z) = \frac{2}{\pi z}$$

which verifies that $J_s(z)$ and $Y_s(z)$ do form a basic pair of solutions of Bessel's equation whatever the value of s.

We conclude this section with a discussion of the behavior of the Bessel functions near the origin. We see from (8.1.19) that the function $z^{-s}J_s(z)$ remains finite as $z \to 0$. This means that $J_s(z)$ vanishes at the origin like z^s if $s > 0$, but is singular there if $s < 0$. An exception arises when $s = n$, an integer or zero, for then (8.1.21) holds and both $J_n(z)$ and $J_{-n}(z)$ possess zeros of order $|n|$ at the origin. The function $J_0(z)$ reduces to unity when $z = 0$.

The function $Y_s(z)$ is singular at the origin for every real value of s. If s is not an integer or zero, it follows from (8.1.22) and the above remarks that $Y_s(z)$ behaves like $z^{-|\operatorname{Re} s|}$. If s is an integer n or zero, then we must evaluate the limit of (8.1.22) by differentiation. After some reduction we find the formula

$$(8.1.24) \qquad Y_n(z) = \frac{1}{\pi} \lim_{s \to n} \frac{\partial}{\partial s} [J_s(z) - (-1)^n J_{-s}(z)]$$

We shall confine attention to the series for $Y_0(z)$ only. The series for $Y_n(z)$ is more difficult to develop and we shall only state the final result in this case. From the series (8.1.18) we have

$$\lim_{s \to 0} \frac{\partial}{\partial s} J_s(z) = \sum_{m=0}^{\infty} \frac{1}{(m!)^2} \left(\frac{-z^2}{4} \right)^m \left[\log \frac{z}{2} - \frac{\Gamma'(m+1)}{\Gamma(m+1)} \right]$$

$$= J_0(z) \log \frac{z}{2} - \sum_{m=0}^{\infty} \frac{1}{(m!)^2} \left(\frac{-z^2}{4} \right)^m \frac{\Gamma'(m+1)}{\Gamma(m+1)}$$

We find the negative of this expression as the limiting value of $(\partial/\partial s) J_{-s}(z)$ when $s \to 0$. Using these results in (8.1.24), we deduce that

$$(8.1.25) \qquad Y_0(z) = \frac{2}{\pi} J_0(z) \log \frac{z}{2} - \frac{2}{\pi} \sum_{m=0}^{\infty} \frac{1}{(m!)^2} \left(\frac{-z^2}{4} \right)^m \frac{\Gamma'(m+1)}{\Gamma(m+1)}$$

This formula shows that $Y_0(z)$ possesses a logarithmic singularity at the origin. We can establish similarly that, when n is a positive integer, the function $Y_n(z)$ behaves like z^{-n} near the origin. In fact,

$$(8.1.26) \qquad \begin{aligned} Y_n(z) = {} & \frac{2}{\pi} J_n(z) \log \frac{z}{2} - \frac{1}{\pi} \sum_{m=0}^{n-1} \frac{(n-m-1)!}{m!} \left(\frac{2}{z} \right)^{n-2m} \\ & - \frac{1}{\pi} \left(\frac{z}{2} \right)^n \sum_{m=0}^{\infty} \frac{1}{m!(m+n)!} \left(\frac{-z^2}{4} \right)^m \left[\frac{\Gamma'(m+1)}{\Gamma(m+1)} + \frac{\Gamma'(m+n+1)}{\Gamma(m+n+1)} \right] \end{aligned}$$

We also note the following formula which is a consequence of (8.1.22):

$$(8.1.27) \qquad Y_{-s}(z) = Y_s(z) \cos s\pi + J_s(z) \sin s\pi$$

If s is an integer, this relation reduces to

$$(8.1.28) \qquad Y_{-n}(z) = (-1)^n Y_n(z)$$

EXERCISES 8.1

1. By differentiating the relation (8.1.11) with respect to z and comparing coefficients of t show that

$$J_{n-1}(z) - J_{n+1}(z) = 2 J_n'(z)$$

By differentiating (8.1.11) with respect to t show that

$$J_{n-1}(z) + J_{n+1}(z) = \frac{2n}{z} J_n(z)$$

Deduce that

$$\int z^{n+1} J_n(z) \, dz = z^{n+1} J_{n+1}(z)$$

2. By replacing t by $-t^{-1}$ in (8.1.11) show that

$$J_n(z) = (-1)^n J_{-n}(z)$$

By replacing t by $-t$ in (8.1.11) show also that

$$J_n(-z) = (-1)^n J_n(z)$$

3. Show by integration by parts that $\Gamma(z+1) = z\Gamma(z)$. Show also that $\Gamma(1) = 1$ and deduce that $\Gamma(n+1) = n!$ when n is a positive integer.

4. Show how the relation $\Gamma(z+1) = z\Gamma(z)$ may be used to define the Γ function in the whole plane except for the points $z = 0, -1, -2, -3, \ldots$.

5. Show directly from (8.1.16) that if $0 < R(z) < 1$,

$$\Gamma(z)\Gamma(1-z) = 2 \int_0^\infty re^{-r^2} \, dr \int_0^\infty \frac{t^{z-1} \, dt}{1+t}$$

Deduce that

$$\Gamma(z)\Gamma(1-z) = \frac{\pi}{\sin z\pi}$$

6. Let L be the contour of Figure 8.1. By deforming this contour onto both sides of the branch cut along the negative real axis, connected by a small circle round the origin, show that

$$\int_L e^w w^{-z} \, dw = 2i \sin \pi z \, \Gamma(1-z)$$

Deduce *Hankel's formula*

$$\frac{1}{\Gamma(z)} = \frac{1}{2\pi i} \int_L e^w w^{-z} \, dw$$

7. Show that $\Gamma(z)$ has a simple pole at the point $z = -n$ of residue $(-1)^n/n!$, where $n = 0, 1, 2, 3, \ldots$.

8. Show that

$$\Gamma(n) = 2 \int_0^\infty e^{-t^2} t^{2n-1} \, dt$$

Deduce that $\Gamma(\tfrac{1}{2}) = \sqrt{\pi}$.

By considering the product $\Gamma(n)\Gamma(m)$ as a repeated integral, show that if n and m are positive

$$\int_0^{\pi/2} (\cos\theta)^{2m-1} (\sin\theta)^{2n-1} \, d\theta = \frac{\Gamma(n)\Gamma(m)}{2\Gamma(n+m)}$$

By setting $m = n$ obtain the *Legendre duplication formula*

$$\Gamma(2n) = \frac{2^{2n-1}}{\sqrt{\pi}} \Gamma(n)\Gamma(n + \tfrac{1}{2})$$

9. Show that
$$\int_0^{\pi/2} \frac{a\, d\theta}{a^2 + b^2 \cos^2\theta} = \frac{\pi}{2\sqrt{a^2 + b^2}} \qquad a > 0$$

Deduce that
$$\int_0^\infty e^{-pz} J_0(\lambda z)\, dz = \frac{1}{\sqrt{p^2 + \lambda^2}} \qquad p > 0$$

10. By means of (8.1.9) show that
$$J_0(x) = \frac{2}{\pi} \int_0^1 \frac{\cos sx\, ds}{\sqrt{1 - s^2}}$$

Deduce that
$$\int_0^\infty \cos sx J_0(x)\, dx = \frac{H(1 - s^2)}{\sqrt{1 - s^2}}$$

11. By means of Fourier integrals show that the solution for $0 < r < \infty$, $-\infty < t < \infty$ of the wave equation
$$u_{rr} + \frac{1}{r} u_r = u_{tt}$$
with the condition $u(0, t) = f(t)$ is
$$u(r, t) = \frac{1}{\pi} \int_{t-r}^{t+r} \frac{f(t')\, dt'}{\sqrt{r^2 - (t - t')^2}}$$

12. Using the results of Exercise 8.1.8 show that
$$\int_{-1}^1 (\mu^2 - 1)^n\, d\mu = \frac{(-1)^n 2^{2n}(n!)^2}{(n + \tfrac{1}{2})(2n)!}$$

13. The surface elevation $\eta(x, t)$ of long waves in a canal satisfies the equation
$$\frac{\partial^2 \eta}{\partial t^2} = \frac{g}{b} \frac{\partial}{\partial x}\left(A \frac{\partial \eta}{\partial x}\right)$$
where $A = A(x)$ and $b = b(x)$ denote the cross-sectional area and surface breadth at the place x. The mean depth h at any section is defined by the relation $A = bh$. The canal which is of length l opens into a sea, and the surface oscillations at the mouth are expressible in the form $\eta = a \cos \omega t$.

(a) If the canal is of constant mean depth but of varying breadth $b = cx$, where c is a constant, obtain the elevation of waves in the canal as
$$\eta = a \cos \omega t\, \frac{J_0(\lambda x)}{J_0(\lambda l)}$$
where $\lambda = \omega/\sqrt{gh}$.

(b) If the canal is of constant surface breadth but of varying depth $h = cx$, where c is a constant, show that
$$\eta = a \cos \omega t\, \frac{J_0(k\sqrt{x})}{J_0(k\sqrt{l})}$$
with $k = 2\omega/\sqrt{gc}$.

8.2 EIGENFUNCTIONS FOR FINITE REGIONS

In Section 6.4 we discussed the theory of eigenfunctions and the associated expansion theorems for the equation of Helmholtz. Important applications of that theory occur when we consider circles, cylinders, and other regions bounded by the natural coordinate surfaces of a cylindrical coordinate system. Let us for the present confine attention to plane regions for which the equation in question is (8.1.5). As in previous instances, we may construct these eigenfunctions explicitly by separation of the variables. We shall illustrate the

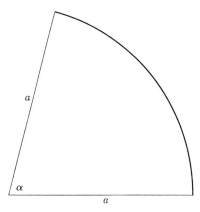

Figure 8.2 Sector.

procedure in detail for the sector $0 \leq \theta \leq \alpha < 2\pi, 0 \leq r \leq a$ (Figure 8.2) with the boundary conditions

(8.2.1)
$$
\begin{array}{lllll}
(a) & u = 0 & & \theta = 0 & 0 < r < a \\
(b) & u = 0 & & \theta = \alpha & 0 < r < a \\
(c) & \dfrac{\partial u}{\partial n} + hu = 0 & & r = a & 0 < \theta < \alpha, h > 0
\end{array}
$$

If we set $u = v(r)g(\theta)$ in (8.1.5), we find the separate equations

(8.2.2)
$$v_{rr} + \frac{1}{r} v_r + \left(k^2 - \frac{s^2}{r^2}\right)v = 0$$

and

(8.2.3)
$$g_{\theta\theta} + s^2 g = 0$$

where s is the separation constant.

The solution of (8.2.3) is $g = A \cos s\theta + B \sin s\theta$, and it is evident from the first two of the boundary conditions (8.2.1) that $A = 0$ and $s = n\pi/\alpha$, where n is a positive integer. Equation (8.2.2) is Bessel's equation with general solution

$CJ_s(kr) + DY_s(kr)$, whatever the value of s. However, the condition of boundedness on the closed interval $0 \le r \le a$ requires that the function $Y_s(kr)$ be omitted. The boundary condition (8.2.1c) implies that k be a root of the equation

(8.2.4) $$hJ_s(ka) + kJ'_s(ka) = 0 \qquad h > 0$$

We know already by Theorem 6.2.2 that the eigenvalues k^2 are real and positive. Therefore the values of k itself are real. Since $s > 0$, (8.2.4) possesses the apparent root $k = 0$. However, if $h > 0$, this root is not a true eigenvalue.

We note also that to every zero k of (8.2.4) there corresponds another zero $-k$. This follows by virtue of the identities $J_s(-z) = e^{is\pi}J_s(z)$ and $J'_s(-z) = -e^{is\pi}J'_s(z)$ which follow from the series (8.1.18). However, because of the first of these identities, it is apparent that the corresponding normalized eigenfunctions are identical.

We also mention that (8.2.4) possesses an infinite number of real roots. This can be inferred from an examination of the asymptotic expressions for the Bessel functions. These are developed in Section 8.6 and we refer the reader to Exercise 8.6.5.

On collecting these results, we find the eigenfunctions $J_{n\pi/\alpha}(kr) \sin(n\pi\theta/\alpha)$, where n is any integer and k is a positive root of (8.2.4). The orthogonality property of two different eigenfunctions follows at once as a special case of the general result developed in Section 6.2. Suppose that

$$J_{n\pi/\alpha}(kr) \sin \frac{n\pi\theta}{\alpha} \quad \text{and} \quad J_{n'\pi/\alpha}(k'r) \sin \frac{n'\pi\theta}{\alpha}$$

are two different eigenfunctions. Then the orthogonality property is

$$\int_0^\alpha \sin \frac{n\pi\theta}{\alpha} \sin \frac{n'\pi\theta}{\alpha} \, d\theta \int_0^a rJ_{n\pi/\alpha}(kr) J_{n'\pi/\alpha}(k'r) \, dr = 0$$

The θ integral which occurs here vanishes for all unequal integers n and n', so that the above equation reduces to

(8.2.5) $$\int_0^a rJ_{n\pi/\alpha}(kr) J_{n\pi/\alpha}(k'r) \, dr = 0$$

where k, k' are two unequal roots of (8.2.4).

It is instructive to verify this result independently by direct reference to Bessel's equation. Let $v = J_s(kr)$ and $w = J_s(k'r)$, so that

$$(rv_r)_r + \left(k^2 r - \frac{s^2}{r}\right) v = 0$$

and

$$(rw_r)_r + \left(k'^2 r - \frac{s^2}{r}\right) w = 0$$

308   CYLINDRICAL EIGENFUNCTIONS

Multiply the first of these equations by w, the second by v, subtract and integrate; then we obtain

(8.2.6) $$(k^2 - k'^2)\int_0^a rvw\,dr = [r(vw_r - wv_r)]_0^a.$$

Now if $v(r)$ and $w(r)$ are both eigenfunctions satisfying the same boundary condition (8.2.4), then we see that $v(a)w'(a) - w(a)v'(a) = 0$. Furthermore, by virtue of the fact that v and w both behave like r^s near the origin, it follows, since $s \geq 0$, that

$$\lim_{r \to 0} r[v(r)w'(r) - w(r)v'(r)] = 0$$

Thus the terms on the right-hand side of (8.2.6) vanish, so that, if k, k' are unequal positive roots of (8.2.4), then

(8.2.7) $$\int_0^a rJ_s(kr)J_s(k'r)\,dr = 0$$

To construct the normalized eigenfunctions, it is necessary to evaluate the integral

(8.2.8) $$\int_0^\alpha \sin^2\frac{n\pi\theta}{\alpha}\,d\theta \int_0^a rJ_{n\pi/\alpha}(kr)^2\,dr$$

The θ integral occurring here is equal to $\alpha/2$, and the r integral is given by

(8.2.9) $$\int_0^a rJ_s(kr)^2\,dr = \frac{1}{2k^2}[k^2a^2J'_s(ka)^2 + (k^2a^2 - s^2)J_s(ka)^2]$$

This integral can be found directly from the differential equation (8.2.2). Actually we shall obtain a slightly more general result, namely, the indefinite integral

(8.2.10) $$\int rvw\,dr = \frac{1}{2k^2}[r^2v_rw_r + (k^2r^2 - s^2)vw]$$

where v and w are any solutions of (8.2.2). Thus we assume that

(8.2.11) $$(rw_r)_r + \left(k^2r - \frac{s^2}{r}\right)w = 0$$

The procedure is now to multiply (8.2.2) by rw_r, (8.2.11) by rv_r, and add. After some rearrangement we find the equation

$$\frac{d}{dr}[r^2v_rw_r + (k^2r^2 - s^2)vw] = 2k^2rvw$$

from which (8.2.10) follows at once. The important case occurs when we set $w = v$ and thus obtain the normalizing integral

(8.2.12) $$\int rv^2\,dr = \frac{1}{2k^2}[r^2v_r^2 + (k^2r^2 - s^2)v^2]$$

In particular, if we set $v = J_s(kr)$, we find that the terms on the above right vanish at $r = 0$ and we obtain (8.2.9).

If the boundary condition (8.2.1c) is $u = 0$ on $r = a$, then the eigenvalues are the positive zeros of the functions $J_{n\pi/\alpha}(ka)$, where n is any integer or zero. Equation (8.2.9) now gives the formula

(8.2.13) $$\int_0^a r J_{n\pi/\alpha}(kr)^2 \, dr = \frac{a^2}{2} J'_{n\pi/\alpha}(ka)^2 \qquad J_{n\pi/\alpha}(ka) = 0$$

THEOREM 8.2.1 *The normalized eigenfunctions for the sector $0 \leq r \leq a$, $0 \leq \theta \leq \alpha < 2\pi$, with the condition of vanishing on the whole boundary are*

(8.2.14) $$\frac{2}{a\sqrt{\alpha}} \frac{J_{n\pi/\alpha}(k_{nm}r)}{J'_{n\pi/\alpha}(k_{nm}a)} \sin \frac{n\pi\theta}{\alpha}$$

Here the k_{nm} are the positive zeros of $J_{n\pi/\alpha}(ka)$.

For the complete circle $0 \leq r \leq a$, $0 \leq \theta \leq 2\pi$, with the condition of vanishing on $r = a$, we find by separation of variables the basic solutions

$$(A \cos s\theta + B \sin s\theta) J_s(kr)$$

These functions will be periodic in θ of period 2π only if s is an integer n or zero, and will vanish on $r = a$ only if $J_s(ka) = 0$. Thus we find the eigenfunctions

$$J_n(k_{nm}r) \cos n\theta \qquad J_n(k_{nm}r) \sin n\theta$$

where $n = 0, 1, 2, \ldots$, and the k_{nm} are the positive zeros of $J_n(ka)$. To normalize these eigenfunctions, we note from (8.2.9) that

(8.2.15) $$\int_0^a r J_n(kr)^2 \, dr = \frac{a^2}{2} J'_n(ka)^2 \qquad J_n(ka) = 0$$

Furthermore,

$$\int_0^{2\pi} \cos^2 n\theta \, d\theta = \int_0^{2\pi} \sin^2 n\theta \, d\theta = \pi \qquad n = 1, 2, 3, \ldots$$

THEOREM 8.2.2 *The normalized eigenfunctions for the circle (vanishing on the boundary) are*

(8.2.16) $$\frac{1}{a\sqrt{\pi}} \frac{J_0(k_{0m}r)}{J'_0(k_{0m}a)} \qquad \frac{1}{a}\sqrt{\frac{2}{\pi}} \frac{J_n(k_{nm}r)}{J'_n(k_{nm}a)} \cos n\theta \qquad \frac{1}{a}\sqrt{\frac{2}{\pi}} \frac{J_n(k_{nm}r)}{J'_n(k_{nm}a)} \sin n\theta$$

where the k_{nm} are the positive zeros of $J_n(ka)$.

It is interesting to compare the eigenfunctions (8.2.16) with those obtained from (8.2.14) by setting $\alpha = 2\pi$. The sector $0 \leq \theta \leq \alpha$ now becomes the whole circle and we obtain the eigenfunctions which vanish on the radius $\theta = 0$ as well as on the boundary circle $r = a$. These eigenfunctions are

$$\frac{1}{a}\sqrt{\frac{2}{\pi}} \frac{J_{n/2}(k'_{nm}r)}{J'_{n/2}(k'_{nm}a)} \sin \frac{n\theta}{2} \qquad n = 1, 2, 3, \ldots$$

The eigenvalues k'_{nm} are now the positive zeros of $J_{n/2}(k'a)$.

EXERCISES 8.2

1. Verify the eigenvalues and eigenfunctions for each of the following regions and boundary conditions, where r, θ, z denote cylindrical polar coordinates. Evaluate the normalizing factor in each case.

(a) The cylinder $0 \leq r \leq a$, $0 \leq \theta \leq 2\pi$, $0 \leq z \leq h$, with $u = 0$ around the entire surface:

$$u_{nmp} = e^{\pm in\theta} J_n(k_{nm}r) \sin \frac{p\pi z}{h} \qquad \lambda_{nmp} = k_{nm}^2 + \left(\frac{p\pi}{h}\right)^2$$

where n, p are integers and k_{nm} is the mth zero of $J_n(ka)$.

(b) The wedge $0 \leq r \leq a$, $0 \leq \theta \leq \alpha < 2\pi$, $0 \leq z \leq h$, with $u = 0$ on $r = a$ and $\partial u / \partial n = 0$ elsewhere:

$$u_{nmp} = \cos \frac{n\pi \theta}{\alpha} \cos \frac{p\pi z}{h} J_{n\pi/\alpha}(k_{nm}r) \qquad \lambda_{nmp} = k_{nm}^2 + \left(\frac{p\pi}{h}\right)^2$$

where n, p are integers (or zero) and $J_{n\pi/\alpha}(k_{nm}a) = 0$.

(c) The semicircle $0 \leq r \leq a$, $0 \leq \theta \leq \pi$ in plane polar coordinates, with $u = 0$ on $\theta = 0$ and $\partial u / \partial n = 0$ elsewhere:

$$u_{nm} = \sin(n + \tfrac{1}{2})\theta J_{n+\frac{1}{2}}(k_{nm}r)$$

where n is an integer (or zero) and the eigenvalues k_{nm} are the zeros of $J'_{n+\frac{1}{2}}(ka)$.

8.3 THE FOURIER–BESSEL SERIES

We now illustrate a particular case of the eigenfunction expansion theorem developed in Section 6.4. Suppose the function $f(r)$ is defined for $0 \leq r \leq a$ and let us propose the formal expansion

$$(8.3.1) \qquad f(r) = \sum_{m=1}^{\infty} a_m J_s(k_m r) \qquad 0 \leq r \leq a$$

Here the summation is extended over all the positive zeros $k_1, k_2, \ldots$ of the function $hJ_s(ka) + kJ'_s(ka)$. By (8.2.7) the functions $J_s(k_m r)$ form an orthogonal set with weight function r. Multiplying (8.3.1) by $rJ_s(k_m r)$, integrating, and using (8.2.9), we obtain the equation

$$(8.3.2) \qquad \int_0^a rf(r) J_s(k_m r)\, dr = a_m \int_0^a rJ_s(k_m r)^2\, dr$$

$$= \frac{a_m}{2k_n^2} [a^2 k_m^2 J'_s(k_m a)^2 + (a^2 k_m^2 - s^2) J_s(k_m a)^2]$$

THEOREM 8.3.1 *If*

$$(8.3.3) \qquad c_m = \int_0^a rf(r) J_s(k_m r)\, dr$$

then

$$(8.3.4) \qquad f(r) = \sum_{m=1}^{\infty} \frac{2k_m^2 c_m J_s(k_m r)}{a^2 k_m^2 J'_s(k_m a)^2 + (a^2 k_m^{\prime 2} - s^2) J_s(k_m a)^2}$$

where k_m is the mth *positive zero of* $hJ_s(ka) + kJ'_s(ka)$.

THE FOURIER-BESSEL SERIES

It should be noted that in expansions of this type the order s of the Bessel functions is held fixed, the summation running over the positive eigenvalues k_m. The expansion (8.3.4) is known as a *Fourier-Bessel series*. Several instances of such expansions are illustrated in the exercises at the end of this section. These expansions are generated by Sturm-Liouville problems involving Bessel's differential equation, and they arise from problems associated with partial differential equations.

The actual expansion (8.3.4) could occur in a variety of circumstances. If we consider the circular domain $0 \le r \le a$, $0 \le \theta \le 2\pi$, and take for simplicity the condition of vanishing on the boundary $r = a$, then the eigenfunctions are

$$(8.3.5) \qquad J_n(k_{nm}r)e^{in\theta} \qquad n = 0, \pm 1, \pm 2, \ldots$$

where k_{nm} is the mth positive zero of $J_n(ka)$. If $f(r, \theta)$ is a function defined on the circle $r \le a$, then we may propose the expansion

$$(8.3.6) \qquad f(r, \theta) = \sum_{n=-\infty}^{\infty} \sum_{m=1}^{\infty} a_{mn} J_n(k_{mn}r)e^{in\theta}$$

The coefficients a_{mn} are found after multiplication by a typical eigenfunction and integration over the whole circle. Let $J_n(k_{nm}r)e^{-in\theta}$ be such an eigenfunction. We find that

$$\int_0^{2\pi} \int_0^a f(r, \theta) r J_n(k_{nm}r) e^{-in\theta} \, dr \, d\theta = \sum_{n'=-\infty}^{\infty} \sum_{m'=1}^{\infty} a_{m'n'} \int_0^{2\pi} e^{i(n'-n)\theta} \, d\theta$$
$$\times \int_0^a r J_n(k_{nm}r) J_{n'}(k_{n'm'}r) \, dr$$

The θ integral appearing on the right vanishes unless $n = n'$ when it equals 2π. Also when $n = n'$, the r integral vanishes if k_{nm} and $k_{n'm'}$ are unequal positive zeros of $J_n(ka)$, by virtue of the orthogonality property (8.2.7). Therefore,

$$\int_0^{2\pi} \int_0^a f(r, \theta) r J_n(k_{nm}r) e^{-in\theta} \, dr \, d\theta = 2\pi a_{mn} \int_0^a r J_n(k_{nm}r)^2 \, dr$$

The surviving Bessel function integral on the right is given by (8.2.13). Inserting the value given there we find the explicit formula

$$(8.3.7) \qquad a_{mn} = \frac{1}{\pi a^2 J_n'(k_{nm}a)^2} \int_0^{2\pi} \int_0^a r J_n(k_{nm}r) e^{-in\theta} f(r, \theta) \, dr \, d\theta$$

If the function $f(r, \theta)$ is a product $e^{ip\theta}g(r)$, where $p = 0, \pm 1, \pm 2, \ldots$, then the expansion (8.3.6) itself becomes the Fourier-Bessel expansion of order p of the function $g(r)$. In particular, if the function $f(r, \theta)$ is a function of r only, expansion (8.3.6) reduces to a Fourier-Bessel series of order zero. We shall leave the reader to verify these results for himself.

In applying Theorem 8.3.1 to the solution of specific boundary value problems, it is useful to employ the formula

$$(8.3.8) \quad \int_0^a r J_s(k_m r)\left(f_{rr} + \frac{1}{r}f_r - \frac{s^2}{r^2}f\right) dr$$

$$= -k_m^2 c_m + a f'(a) J_s(k_m a) - a k_m f(a) J_s'(k_m a)$$

This result is established by repeated integration by parts. It is a particular case of the general relation (6.7.9) and is valid provided that

$$\lim_{r \to 0} [rf'(r)J_s(kr) - krf(r)J_s'(kr)] = 0$$

Since k_m satisfies $hJ_s(ka) + kJ_s'(ka) = 0$, the right-hand side of (8.3.8) will actually contain only the combination $hf(a) + f'(a)$. We may regard (8.3.3) as defining an integral operator which will transform the group of terms $f_{rr} + f_r/r - s^2 f/r^2$ on the interval $0 \le r \le a$ subject to the boundary condition that $hf(a) + f'(a)$ is prescribed.

Such a transform is called a finite Hankel transform and its inversion formula is the series (8.3.4). Application of these transforms follows the same procedure used for the various Fourier transforms. The following exercises can be solved by means of finite Hankel transforms.

EXERCISES 8.3

1. Show that the Green's function of the diffusion equation $\Delta v - v_t = 0$ for the circular domain $0 \le r \le a$, $0 \le \theta \le 2\pi$, vanishing on $r = a$ is given by

$$v = \frac{1}{\pi a^2} \sum_{n=-\infty}^{\infty} \sum_{m=1}^{\infty} \frac{e^{-k^2_{nm} t} J_n(k_{nm} r) J_n(k_{nm} r') e^{in(\theta - \theta')}}{J_n'(k_{nm} a)^2} H(t)$$

where k_{nm} is the mth positive zero of $J_n(ka)$.

2. Show that the Green's function of the wave equation $\Delta v - v_{tt} = 0$ for the region $0 \le r \le a$, $0 \le \theta \le \pi$, with the boundary conditions $v = 0$ on $\theta = 0$, $v = 0$ on $\theta = \pi$, and $\partial v/\partial n = 0$ on $r = a$, is given by (see Chapter 10, Section 10.2)

$$v = \frac{4}{\pi} \sum_{n=1}^{\infty} \sum_{m=1}^{\infty} \frac{k'_{nm} J_n(k'_{nm} r) J_n(k'_{nm} r') \sin n\theta \sin n\theta' \sin(k'_{nm} t)}{(a^2 k'^2_{nm} - n^2) J_n(k'_{nm} a)^2}$$

where k'_{nm} is the mth positive zero of $J_n'(ka)$.

3. By writing $kxJ_0(kx) = -J_0'(kx) - kxJ_0''(kx)$ show that

$$\int x J_0(kx)\, dx = -\frac{x}{k} J_0'(kx)$$

$$\int x \log x J_0(kx)\, dx = \frac{1}{k^2} J_0(kx) - \frac{x}{k} \log x J_0'(kx)$$

4. Assuming the results of the preceding question, show that the Fourier-Bessel series of order zero on the interval $(0, a)$ of the function

$$G(r, r') = \begin{cases} \log r'/a & 0 < r < r' < a \\ \log r/a & 0 < r' < r < a \end{cases}$$

is

$$G(r, r') = \frac{-2}{a^2} \sum_{n=1}^{\infty} \frac{J_0(k_n r) J_0(k_n r')}{k_n^2 J_0'(k_n a)^2}$$

where the summation is extended over all the positive zeros k_n of $J_0(ka)$. Show also that the Fourier-Bessel series of order n on the interval $(0, a)$ of the function

$$H(r, r') = \begin{cases} \dfrac{1}{2n}\left(\dfrac{r}{a}\right)^n \left[\left(\dfrac{r'}{a}\right)^n - \left(\dfrac{a}{r'}\right)^n\right] & 0 < r < r' < a \\ \dfrac{1}{2n}\left(\dfrac{r'}{a}\right)^n \left[\left(\dfrac{r}{a}\right)^n - \left(\dfrac{a}{r}\right)^n\right] & 0 < r' < r < a \end{cases}$$

when $n > 0$ is

$$H(r, r') = \frac{-2}{a^2} \sum_{m=1}^{\infty} \frac{J_n(k_{nm}r) J_n(k_{nm}r')}{k_{nm}^2 J_n'(k_{nm}a)^2}$$

where k_{nm} is the mth zero of $J_n(ka)$.

5. Show that the harmonic Green's function for the circle $0 \le r \le a$, $0 \le \theta \le 2\pi$, vanishing on the boundary is

$$G(r, \theta, r', \theta') = \frac{2}{\pi a^2} \sum_{n,m} \varepsilon_n \frac{J_n(k_{nm}r) J_n(k_{nm}r') \cos n(\theta - \theta')}{k_{nm}^2 J_n'(k_{nm}a)^2}$$

where $\varepsilon_n = 1$ for $n \ge 1$, $\varepsilon_0 = \frac{1}{2}$, and the summation is extended for $m = 1, 2, 3, \ldots$ over all the zeros λ_{nm} of the function $J_n(ka)$, and $n = 0, 1, 2, 3, \ldots$. Show that for $0 < r < r' < a$

$$G(r, \theta, r', \theta') = \frac{-1}{2\pi} \log\left(\frac{r'}{a}\right) - \frac{1}{2\pi}\sum_{n=1}^{\infty}\frac{1}{n}\left(\frac{r}{a}\right)^n\left[\left(\frac{r'}{a}\right)^n - \left(\frac{a}{r'}\right)^n\right]\cos n(\theta' - \theta)$$

(Interchange r, r' when $0 < r' < r < a$.) Verify (7.2.6).

6. The function $f(r, \theta, z)$ is defined in the region $0 \le r \le a$, $0 \le \theta \le 2\pi$, $0 \le z \le h$. Obtain the following eigenfunction expansions:

(a) $$f(r, \theta, z) = \sum_{n=-\infty}^{\infty} \sum_{p=1}^{\infty} \sum_{m=1}^{\infty} a_{mnp} e^{in\theta} J_n(k_{nm}r) \sin\frac{p\pi z}{h}$$

$$a_{mnp} = \frac{2}{\pi h a^2 J_n'(k_{nm}a)^2} \int_0^a \int_0^{2\pi} \int_0^h r J_n(k_{nm}r) e^{-in\theta} \sin\frac{p\pi z}{h}\, dr\, d\theta\, dz$$

where $J_n(k_{nm}a) = 0$.

(b) $$f(r, \theta, z) = \sum_{n=-\infty}^{\infty} \sum_{p=0}^{\infty} \sum_{m=1}^{\infty} a_{mnp} e^{in\theta} J_n(k_{nm}r) \sin\left(p + \tfrac{1}{2}\right)\frac{\pi z}{h}$$

$$a_{mnp} = \frac{2k_{nm}^2}{\pi h(a^2 k_{nm}^2 - n^2) J_n(k_{nm}a)^2} \int_0^a \int_0^{2\pi} \int_0^h r J_n(k_{nm}r) e^{-in\theta}$$

$$\times \sin\left(p + \tfrac{1}{2}\right)\frac{\pi z}{h}\, dr\, d\theta\, dz$$

where $J_n'(k_{nm}a) = 0$.

314 CYLINDRICAL EIGENFUNCTIONS

For what initial and boundary value problems would the above expansions be appropriate?

7. Solve the Helmholtz equation $\Delta u + k^2 u = 0$ in the cylinder $0 \leq r \leq a$, $0 \leq z \leq h$, $0 \leq \theta \leq 2\pi$, with the boundary conditions that u vanishes on the curved surface and on the end $z = h$ while $u = f(r)$ on the base $z = 0$. Show that

$$u = \frac{2}{a^2} \sum_{m=1}^{\infty} \frac{J_0(k_n r)}{J_0'(k_n a)^2} \frac{\sinh \sigma(h-z)}{\sinh \sigma h} \int_0^a r' f(r') J_0(k_n r') \, dr'$$

where

$$\sigma = \sqrt{k_n^2 - k^2} \quad \text{and} \quad J_0(k_n a) = 0$$

8. Show that the Green's function of $\Delta u - k^2 u = 0$ for the interior of the infinite cylinder $0 \leq r \leq a$, $0 \leq \theta \leq 2\pi$, $-\infty < z < \infty$, with the condition of vanishing on the surface, is given by

$$u = \frac{1}{2\pi a^2} \sum_{n=-\infty}^{+\infty} e^{in(\theta-\theta')} \sum_{m=1}^{\infty} \frac{J_n(k_{nm} r) J_n(k_{nm} r') e^{-\sigma|z-z'|}}{\sigma J_n'(k_{nm} a)^2}$$

where

$$\sigma = \sqrt{k^2 + k_{nm}^2} \quad \text{and} \quad J_n(k_{nm} a) = 0$$

9. Show that the harmonic function u for the cylinder $0 \leq r \leq a$, $0 \leq z \leq h$, subject to the boundary conditions $u = 0$ on $z = 0$, $u = 0$ on $z = h$, and $u = V$ (a constant) on $r = a$, is given by

$$u = \frac{4V}{\pi} \sum_{m=0}^{\infty} \frac{I_0(\lambda_m r)}{I_0(\lambda_m a)} \frac{\sin \lambda_m z}{(2m+1)}$$

where $\lambda_m = (2m+1)(\pi/h)$. The function $I_0(x)$ is defined in Section 8.7. Show also that

$$u = \frac{2V}{a} \sum_{n=1}^{\infty} \frac{\sinh [k_n(h-z)] + \sinh (k_n z) - \sinh (k_n h)}{k_n \sinh (k_n h)} \frac{J_0(k_n r)}{J_0'(k_n a)}$$

where $J_0(k_n a) = 0$. Reconcile these two results.

8.4 THE GREEN'S FUNCTION

We shall calculate the explicit form of a Fourier-Bessel series in just one case, that of expanding a Green's function in a series of the associated eigenfunctions. We consider again the equation of Helmholtz (8.1.5) written in plane polar coordinates and we now construct the Green's function for the circular domain $0 \leq r \leq a$, $-\pi \leq \theta \leq \pi$, vanishing on the boundary $r = a$. In accordance with the procedure explained in Section 7.1, we write for $G = G(r, \theta; r', \theta')$ the differential equation

(8.4.1) $$G_{rr} + \frac{1}{r} G_r + \frac{1}{r^2} G_{\theta\theta} + k^2 G = -\frac{\delta(r-r') \, \delta(\theta-\theta')}{r}$$

Since the Green's function is to be periodic in θ of period 2π, we may extract the θ dependence in a complex Fourier series by writing

(8.4.2) $$\tilde{G}_n = \tilde{G}_n(r; r', \theta') = \int_{-\pi}^{\pi} e^{in\theta} G(r, \theta; r', \theta') \, d\theta$$

(8.4.3) $$G(r, \theta; r', \theta') = \frac{1}{2\pi} \sum_{n=-\infty}^{\infty} e^{-in\theta} \tilde{G}_n(r; r', \theta')$$

If we apply the integral transform (8.4.2) to (8.4.1) and invoke the periodic property of G, we obtain

(8.4.4) $$\tilde{G}_{nrr} + \frac{1}{r}\tilde{G}_{nr} + \left(k^2 - \frac{n^2}{r^2}\right)\tilde{G}_n = -\frac{\delta(r-r')e^{in\theta'}}{r}$$

We now require the Green's function of this ordinary differential equation bounded in the interval $0 \leq r \leq a$ and vanishing on $r = a$. Let

$$\tilde{\alpha}_n = \begin{cases} A_1 J_n(kr) & 0 \leq r < r' \leq a \\ A_2[J_n(kr)Y_n(ka) - J_n(ka)Y_n(kr)] & 0 \leq r' < r \leq a \end{cases}$$

In the first of these expressions the Bessel function $Y_n(kr)$ is omitted since it is singular at the origin. The second expression is constructed to vanish at $r = a$ as desired. The constants A_1, A_2 are found as described in Section 1.8 by requiring the Green's function to be continuous, and its derivative to have a negative discontinuity of $(1/r')e^{in\theta'}$ at $r = r'$. We find after using the Wronskian relation (8.1.23) that

(8.4.5) $$\tilde{G}_n = \frac{\pi}{2}\frac{J_n(kr)}{J_n(ka)}[J_n(kr')Y_n(ka) - J_n(ka)Y_n(kr')]e^{in\theta'}$$

for $r < r'$, with a similar but interchanged expression for $r > r'$. Inserting this expression into (8.4.3), we obtain

(8.4.6) $$G(r, \theta; r', \theta') = \frac{1}{4}\sum_{n=-\infty}^{\infty} e^{in(\theta'-\theta)}\frac{J_n(kr')}{J_n(ka)}$$
$$\times [J_n(kr)Y_n(ka) - J_n(ka)Y_n(kr)]$$

This form is valid for $0 \leq r' < r \leq a$, while for $0 \leq r \leq r' \leq a$ we interchange r and r'.

From the preceding expression we may deduce the form of the Green's function singularity of the Helmholtz equation for *finite* plane regions. To do this, we shall place the singularity at the origin by taking the limit as $r' \to 0$ in (8.4.6). As noted in Section 8.1, all the Bessel functions $J_n(kr')$ vanish at the origin except $J_0(kr')$, which reduces to unity. Therefore (8.4.6) yields in the limit the formula

(8.4.7) $$G_0 = -\frac{1}{4}Y_0(kr) + \frac{1}{4}\frac{J_0(kr)Y_0(ka)}{J_0(ka)}$$

The only singular part on the right-hand side of this expression is the leading term $-\frac{1}{4}Y_0(kr)$ which, as we saw in (8.1.25), behaves like $(-1/2\pi) \log r$ near the origin. The remaining terms appearing on the right-hand side of (8.4.7) are nonsingular and are inserted to meet the condition of vanishing on $r = a$. Therefore we may deduce that the Helmholtz equation Green's function appropriate to finite domains must in the neighborhood of the singularity behave like the function $-\frac{1}{4}Y_0(kr)$. Here r is the distance of a general point from the singularity.

In general, when the singularity is located at the point (x', y'), the expression for the singularity becomes

(8.4.8)
$$-\tfrac{1}{4} Y_0(kR)$$

where

(8.4.9) $\quad R = \sqrt{(x-x')^2 + (y-y')^2} = \sqrt{r^2 + r'^2 - 2rr' \cos(\theta - \theta')}$

To obtain the eigenfunction expansions of the Green's function defined by the Fourier series (8.4.6), we shall make use of the bilinear expansion of Section 7.1. This enables us to avoid the direct use of (8.3.7). The normalized eigenfunctions for the circular domain vanishing on the boundary are the functions (8.2.16). The bilinear formula (7.1.13) now yields the expression

(8.4.10) $\quad G = \dfrac{-1}{\pi a^2} \sum_{n=0}^{\infty} \sum_{m=1}^{\infty} \dfrac{\varepsilon_n J_n(k_{nm}r) J_n(k_{nm}r') \cos n(\theta - \theta')}{(k^2 - k_{nm}^2) J_n'(k_{nm}a)^2}.$

Here $\varepsilon_0 = 1$, $\varepsilon_n = 2$ for $n \geq 1$, and the k_{nm} are the positive zeros of $J_n(ka)$. If we recall that $J_{-n} = (-1)^n J_n$, then it is obvious that we may write (8.4.10) in the complex form

(8.4.11) $\quad G(r, \theta) = \dfrac{-1}{\pi a^2} \sum_{n=-\infty}^{\infty} \sum_{m=1}^{\infty} \dfrac{J_n(k_{nm}r) J_n(k_{nm}r') e^{in(\theta-\theta')}}{(k^2 - k_{nm}^2) J_n'(k_{nm}a)^2}$

Comparing this expression with that of (8.4.6), we deduce that

(8.4.12) $\quad \dfrac{J_n(kr')}{J_n(ka)} [J_n(kr) Y_n(ka) - J_n(ka) Y_n(kr)]$

$$= -\dfrac{4}{\pi a^2} \sum_{m=1}^{\infty} \dfrac{J_n(k_{nm}r) J_n(k_{nm}r')}{(k^2 - k_{nm}^2) J_n'(k_{nm}a)^2}$$

for $0 < r' < r < a$. The expression on the right-hand side of (8.4.12) is the Fourier-Bessel series on the interval $0 \leq r \leq a$ of the composite function $H(r, r')$ defined by

(8.4.13) $\quad H(r, r') = \begin{cases} \dfrac{J_n(kr)}{J_n(ka)} [J_n(kr') Y_n(ka) - J_n(ka) Y_n(kr')] & r < r' \\[2mm] \dfrac{J_n(kr')}{J_n(ka)} [J_n(kr) Y_n(ka) - J_n(ka) Y_n(kr)] & r > r' \end{cases}$

EXERCISES 8.4

1. Show that the Green's function of the equation

$$r^2 u_{rr} + 2r u_r + [k^2 r^2 - n(n+1)] u = 0$$

bounded on the interval $0 \leq r \leq a$ and vanishing for $r = a$ is given for $0 < r < r' < a$ by

$$u(r, r') = \dfrac{-\pi J_s(kr)}{2\sqrt{rr'} \sin s\pi J_s(ka)} [J_s(kr') J_{-s}(ka) - J_s(ka) J_{-s}(kr')]$$

where $s = n + \frac{1}{2}$. Show also that

$$u(r, r') = -\frac{2}{a^2} \sum_{m=1}^{\infty} \frac{J_s(k_{nm}r)J_s(k_{nm}r')}{(k^2 - k_{nm}^2)J_s'(k_{nm}a)^2 \sqrt{rr'}}$$

where the k_{nm} are the positive zeros of $J_s(ka)$.

2. By letting $k \to 0$ in the results of Exercise 8.4.1, show that the Green's function of the equation $r^2 v_{rr} + 2r v_r - n(n+1)v = 0$ bounded on the interval $0 \le r \le a$ and vanishing at $r = a$ is given for $0 < r < r' < a$ by

$$v(r, r') = \frac{-1}{(2n+1)a} \left(\frac{r}{a}\right)^n \left[\left(\frac{r'}{a}\right)^n - \left(\frac{a}{r'}\right)^{n+1}\right]$$

$$= \frac{2}{a^2 \sqrt{rr'}} \sum_{m=1}^{\infty} \frac{J_s(k_{nm}r)J_s(k_{nm}r')}{k_{nm}^2 J_s'(k_{nm}a)^2}$$

where $s = n + \frac{1}{2}$ and $J_s(k_{nm}a) = 0$.

3. Show that the harmonic Green's function for the cylinder $0 \le r \le a$, $0 \le \theta \le 2\pi$, $0 \le z \le h$ vanishing on the boundary is given by

$$G(r, \theta, z; r', \theta', z') = \frac{-1}{\pi a^2} \sum_{n=-\infty}^{+\infty} \sum_{m=1}^{\infty}$$

$$\times \frac{\sinh(k_{nm}z')\sinh[k_{nm}(z-h)]J_n(k_{nm}r)J_n(k_{nm}r')}{k_{nm}\sinh(k_{nm}h)J_n'(k_{nm}a)^2} e^{in(\theta-\theta')}$$

for $0 < z' < z < h$. Show also that

$$G(r, \theta, z; r', \theta', z') = \frac{2}{\pi a^2 h} \sum_{n=-\infty}^{\infty} \sum_{m=1}^{\infty} \sum_{p=1}^{\infty}$$

$$\times \frac{\sin\frac{p\pi z}{h}\sin\frac{p\pi z'}{h} J_n(k_{nm}r)J_n(k_{nm}r')}{\left[k_{nm}^2 + \left(\frac{p\pi}{h}\right)^2\right]J_n'(k_{nm}a)^2} e^{in(\theta-\theta')}$$

where $J_n(k_{nm}a) = 0$.

8.5 FUNCTIONS OF LARGE ARGUMENT

Hitherto in this chapter we have dealt entirely with finite domains. In many physical situations, however, the region of interest extends to infinity. In the applications that follow we require to know the behavior of certain Bessel functions when the argument is large. It is evident that power series like (8.1.18) are then unsuitable and that series in descending powers of the argument are required. We can find such a series, but it is of a new kind called an *asymptotic series*.

We shall define asymptotic series in the following way. Suppose that

(8.5.1) $$z^n \left[f(z) - \sum_{m=0}^{n} a_m z^{-m}\right] \to 0$$

for each n as $|z| \to \infty$. Then we call the series $\Sigma a_m z^{-m}$ an asymptotic expansion of the function $f(z)$ and write

(8.5.2) $$f(z) \sim a_0 + \frac{a_1}{z} + \frac{a_2}{z^2} + \cdots$$

The asymptotic series itself may or may not converge. The essential feature of (8.5.2) is that the error incurred in approximating $f(z)$ by the first n terms of the series is $O(z^{-n})$, and *this error can be made as small as we desire by choosing $|z|$ large enough.*

A slightly more general situation is often encountered, for the given function $f(z)$ may possess no asymptotic expansion of the type (8.5.2). In such a case it may be possible to extract a factor, say $g(z)$, such that the new function $f(z)/g(z)$ does possess an asymptotic expansion. We should then write

$$\frac{f(z)}{g(z)} \sim b_0 + \frac{b_1}{z} + \frac{b_2}{z^2} + \cdots$$

or

(8.5.3) $$f(z) \sim g(z)\left(b_0 + \frac{b_1}{z} + \frac{b_2}{z^2} + \cdots\right)$$

In the applications we shall need to retain only the first or dominant term in the series (8.5.3). To find this term we shall use the *method of steepest descents*, which we now describe.

Consider the contour integral

(8.5.4) $$F(z) = \int_L f(t) e^{zg(t)}\, dt$$

where L is a given path joining points t_1 and t_2 of the complex t-plane. We wish to find the leading term in the asymptotic expansion of $F(z)$ when z is large and positive. If $z > 0$ is large enough, the dominant contribution will arise from those points, if any, where Re $g(t)$ is greatest. The original path of integration may not pass through these points, but we shall deform the path so that it does. This will not alter the value of $F(z)$ provided the integrand is regular in the region traversed by the deformation. If the integrand possesses poles in this region, then the method is still applicable provided the residues at these poles are taken into account. Similarly, we must not cross any branch cuts in the process of deforming the contour. The points at which Re $g(t)$ is greatest are termed *saddle points*, and the object is to find a path L' running through these points (Figure 8.3).

If $g(t) = u(\xi, \eta) + iv(\xi, \eta)$, where $t = \xi + i\eta$, then the function u will be stationary at points where $u_\xi = u_\eta = 0$. By the Cauchy-Riemann equations this implies that $v_\xi = v_\eta = 0$ also. Thus $g'(t) = 0$ at a saddle point. Now since $u(\xi, \eta)$ is harmonic, it cannot attain a true maximum (or a true minimum). The value which u attains at a saddle point will be a maximum for some paths through it and a minimum for others. If we project onto the ξ, η-plane the curves of constant u, that is, the geographical contour lines, then u will change most rapidly along the orthogonal trajectories of these contour lines. Along any such orthogonal trajectories the conjugate quantity v is constant. We shall refer to these paths as the *steepest paths*.

Through any saddle point there will, in general, be two steepest paths. On one of these paths the function u will attain a minimum value at the saddle point. Along the other path $v = $ constant, through the saddle point, the function u will attain a maximum value so that this path will be termed a path of *steepest descent*. Having formed a contour L' to pass through the saddle point, the procedure now is to orient it locally so that it passes through this saddle point in the direction of the path of steepest descent.

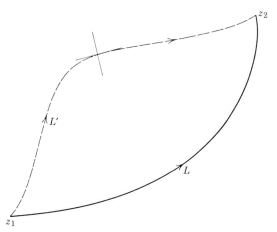

Figure 8.3 Deformed contour through saddlepoint.

When z is large enough, we expect that the greatest contribution to the value of the integral along L' will arise from that part l of L' in the immediate neighborhood of the saddle point. By taking L' as a steepest path, we ensure that the part l of L' where u is greatest will be as short as possible. We may then obtain an estimate of the value of the integral on replacing the original contour by the segment l.

Let $t_1 = \xi_1 + i\eta_1$ be a saddle point. Then the direction of the path of steepest descent may be determined as follows. On this path and near t_1 we shall write $t = t_1 + re^{i\alpha}$, and by Taylor's theorem expand the function $g(t)$ as

$$g(t) = g(t_1) + (t - t_1)g'(t_1) + \tfrac{1}{2}(t - t_1)^2 g''(t_1) + \cdots$$

Since $g'(t_1) = 0$, this reduces to

$$g(t) = g(t_1) + \frac{r^2}{2} e^{2i\alpha} g''(t_1) + \cdots$$

Along a steepest path the function $v = \operatorname{Im} g(t)$ remains constant and equal to its value at the saddle point t_1:

$$\operatorname{Im} g(t) = \operatorname{Im} g(t_1)$$

Furthermore, if the steepest path is one of descent, the value of u at any point on it will be less than the value $u(\xi_1, \eta_1)$. Thus we see that $g(t) - g(t_1)$ must be real and negative, so $e^{2i\alpha}g''(t_1)$ must be real and negative. This condition determines the angle α up to an additive multiple of π. This ambiguity of α affects only the *sense* in which the steepest path must be described. Reference to the original contour L should enable us to determine this sense.

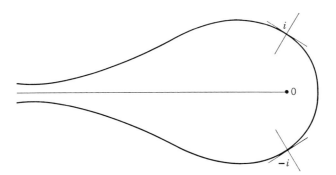

Figure 8.4 Steepest descent contour for $J_s(z)$.

We shall now apply this method to the Bessel function integral (8.1.13) which we now rewrite as

$$(8.5.5) \qquad J_s(z) = \frac{1}{2\pi i} \int_L t^{-s-1} e^{(z/2)(t-1/t)} \, dt$$

Here L is the loop round the negative real axis crossing the positive real axis. By comparison with (8.5.4), we see that

$$g(t) = \frac{1}{2}\left(t - \frac{1}{t}\right)$$

from which it follows that $g'(t)$ vanishes when $t = \pm i$ (Figure 8.4). There are thus two saddle points and we may deform the path L to pass through them. Considering first the value $t = i$, we write $t = i + re^{i\alpha}$ for points on a steepest path near the saddle point. By Taylor's theorem,

$$g(t) - i = \frac{r^2}{2} \sin 2\alpha - \frac{ir^2}{2} \cos 2\alpha + \cdots$$

For a path of steepest descent, the expression on the right is required to be real and negative. This means that $\cos 2\alpha = 0$ and $\sin 2\alpha = -1$, so that $\alpha = 3\pi/4$. Therefore, along this path, we have

$$t = i + re^{3i\pi/4} \qquad g(t) = i - \frac{r^2}{2} + \cdots$$

FUNCTIONS OF LARGE ARGUMENT   321

Inserting these values in (8.5.5), we find a contribution from the first saddle point of

$$(8.5.6) \quad \frac{1}{2\pi i} \int_{l} (i + \cdots)^{-s-1} \exp\left[z\left(i - \frac{r^2}{2} + \cdots\right)\right] dr e^{3i\pi/4}$$

$$= \frac{1}{2\pi} \exp\left\{i\left[z - \left(s + \frac{1}{2}\right)\frac{\pi}{2}\right]\right\} \int_{-\gamma}^{\gamma} e^{-zr^2/2} \, dr$$

where γ is a suitable limit whose exact value becomes immaterial if z is large enough. By an obvious change of variable, $\sigma = (z/2)^{1/2} r$, we see that

$$\int_{-\gamma}^{\gamma} e^{-zr^2/2} \, dr = \sqrt{\frac{2}{z}} \int_{-\gamma\sqrt{z/2}}^{\gamma\sqrt{z/2}} e^{-\sigma^2} \, d\sigma$$

As z increases, the integral on the right rapidly approaches the value

$$\int_{-\infty}^{\infty} e^{-\sigma^2} \, d\sigma = \sqrt{\pi}.$$

Thus we find that the contribution (8.5.6) is

$$(8.5.7) \quad \frac{1}{\sqrt{2\pi z}} \exp\left\{i\left[z - \left(s + \frac{1}{2}\right)\frac{\pi}{2}\right]\right\} \qquad z \to \infty$$

For the second saddle point where $t = -i$, we find the path of steepest descent is $t = -i + re^{i\pi/4}$. We obtain the conjugate contribution

$$(8.5.8) \quad \frac{1}{\sqrt{2\pi z}} \exp\left\{-i\left[z - \left(s + \frac{1}{2}\right)\frac{\pi}{2}\right]\right\}, \qquad z \to \infty$$

On adding the expressions (8.5.7) and (8.5.8), we reach the final result that

$$(8.5.9) \quad J_s(z) \sim \sqrt{\frac{2}{\pi z}} \cos\left[z - \left(s + \frac{1}{2}\right)\frac{\pi}{2}\right] \qquad z \to \infty$$

Changing the sign of s in this expression to obtain the asymptotic behavior of the function $J_{-s}(z)$, and then inserting these results in the definition (8.1.22) of $Y_s(z)$, we find that

$$(8.5.10) \quad Y_s(z) \sim \sqrt{\frac{2}{\pi z}} \sin\left[z - \left(s + \frac{1}{2}\right)\frac{\pi}{2}\right] \qquad z \to \infty$$

From the two preceding expressions we see that the Bessel functions $J_s(z)$, $Y_s(z)$ behave when z is large like damped trigonometric functions. In this section we have supposed that z is real and positive, but it can be shown that the same expressions are valid for $|\arg z| < \pi$. When z is real, both functions $J_s(z)$, $Y_s(z)$ tend to zero as $z \to \pm\infty$. If z is not real, these functions diverge as $|z| \to \infty$.

It is opportune here to introduce the Hankel functions defined by the equations

$$(8.5.11) \quad \begin{aligned} H_s^{(1)}(z) &= J_s(z) + iY_s(z) \\ H_s^{(2)}(z) &= J_s(z) - iY_s(z) \end{aligned}$$

CYLINDRICAL EIGENFUNCTIONS

These functions together form a basis of solutions for Bessel's equation. From (8.5.9) and (8.5.10) we note the asymptotic relations

(8.5.12)
$$H_s^{(1)}(z) \sim \sqrt{\frac{2}{\pi z}} \exp\left\{i\left[z - \left(s + \frac{1}{2}\right)\frac{\pi}{2}\right]\right\} \qquad z \to \infty$$

$$H_s^{(2)}(z) \sim \sqrt{\frac{2}{z\pi}} \exp\left\{-i\left[z - \left(s + \frac{1}{2}\right)\frac{\pi}{2}\right]\right\} \qquad z \to \infty$$

Because of their asymptotic behavior the Hankel functions are used extensively in wave propagation problems.

For future use we record here the relation

(8.5.13)
$$J_s(z) H_s^{(1)\prime}(z) - H_s^{(1)}(z) J_s'(z) = \frac{2i}{\pi z}$$

which follows at once from the Wronskian (8.1.23).

EXERCISES 8.5

1. By writing $t = z(1 + x)$ in (8.1.16), show that
$$\Gamma(z + 1) = e^{-z} z^{z+1} \int_{-1}^{\infty} e^{zg(x)}\, dx$$
where $g(x) = -x + \ln(1 + x)$. Deduce that
$$\Gamma(z + 1) \sim \sqrt{2\pi z}\, e^{-z + z \ln z}$$

2. If x is large and positive and n fixed, show that
$$\int_0^{\infty} e^{-x \cosh u} \cosh nu\, du \sim \sqrt{\frac{\pi}{2x}}\, e^{-x}$$

3. If t is large and positive and $n > t$, show that
$$\frac{1}{2}\int_0^{\infty} e^{-t \cosh z + nz}\, dz \sim \sqrt{\frac{\pi}{2}} \frac{e^{-\sqrt{n^2 + t^2}}}{(n^2 + t^2)^{1/4}} \left(\frac{n}{t} + \sqrt{\frac{n^2}{t^2} + 1}\right)^n$$

[Write $-t \cosh z + nz = ng(z)$, where $g(z) = z - (t/n)\cosh z$.]

4. By setting $z = e^{i\varphi}$ and integrating round the unit circle, show that, if n is large and m finite,
$$\int_{-\pi}^{\pi} (\cos \theta + i \sin \theta \cos \varphi)^n e^{-im\varphi}\, d\varphi$$
$$\sim -i\sqrt{\frac{2\pi}{n \sin \theta}} \left\{\exp\left[i\left(n + \frac{1}{2}\right)\theta + \frac{i\pi}{4}\right] + (-1)^{m+1}\exp\left[-i\left(n + \frac{1}{2}\right)\theta - \frac{i\pi}{4}\right]\right\}$$

5. Deduce from (8.5.9) that the equation $hJ_s(z) + zJ_s'(z) = 0$ possesses an infinite number of real roots if h is real.

6. By using a formula analogous to (8.2.6) show that

$$\int_a^\infty r\varphi_s(k,r) H_s^{(1)}(zr)\, dr = \frac{2i H_s^{(1)}(za)}{\pi(z^2 - k^2)}$$

where $I(z) > |I(k)|$ and

$$\varphi_s(k,r) = J_s(kr) H_s^{(1)}(ka) - J_s(ka) H_s^{(1)}(kr)$$

8.6 DIFFRACTION BY A CYLINDER

We saw in Section 7.2 that for the solution of a boundary value problem with the Helmholtz equation in a region extending to infinity it is not sufficient to impose merely a condition of boundedness at infinity. To determine the solution uniquely, it is necessary to state more precisely the asymptotic behavior of the

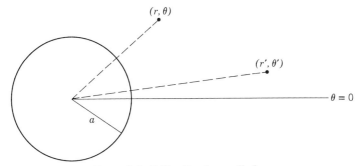

Figure 8.5 Diffraction by a cylinder.

distant field. In (7.2.20) we saw how to prescribe the behavior at infinity by means of a radiation condition. The results of that discussion apply to wave propagation in three space dimensions and a similar argument can be used in two dimensions. In Section 8.4 we constructed the Green's function for the interior of a circle in which case this question does not arise.

Let us now consider the Green's function of the Helmholtz equation for the region exterior to a circle (Figure 8.5). For simplicity, take the condition of vanishing to apply on the internal boundary $r = a$. The equation to be solved is again (8.4.1) but the domain is now $a \leq r \leq \infty$, $0 \leq \theta \leq 2\pi$. If we eliminate the angular dependence by the Fourier series (8.4.2), we find as before the Bessel equation (8.4.4) for which we require the Green's function bounded in the interval $a \leq r < \infty$ and vanishing at $r = a$. We therefore set

$$(8.6.1) \qquad \bar{G}_n(r) = \begin{cases} A[J_n(kr) Y_n(ka) - J_n(ka) Y_n(kr)] & a < r < r' < \infty \\ B J_n(kr) + C Y_n(kr) & a < r' < r < \infty \end{cases}$$

We have written the first of these expressions so as to satisfy the condition of vanishing on $r = a$. In the second expression, which is valid when $r \to \infty$, we have retained both the basic solutions $J_n(kr)$ and $Y_n(kr)$ since both vanish at infinity. Thus we are left with three disposable constants A, B, C, but only two conditions remain to determine them, namely, the continuity of G at r' and the discontinuity of G_r at r'.

The actual wave function ψ_n corresponding to the expression $BJ_n(kr) + CY_n(kr)$ is obtained by multiplying by the factor $e^{-ikct-in\theta}$. We note by virtue of the asymptotic expressions (8.5.9) and (8.5.10) that, if r is large enough, the wave function

$$v \sim \sqrt{\frac{2}{\pi kr}} \left\{ B \cos\left[kr - \left(n + \frac{1}{2}\right)\frac{\pi}{2}\right] + C \sin\left[kr - \left(n + \frac{1}{2}\right)\frac{\pi}{2}\right] \right\} e^{-ikct-in\theta}$$

This function is a linear combination of two standing waves. However it represents outgoing waves if we set $C = iB$, for then we have

$$v \sim \sqrt{\frac{2}{\pi kr}} B \exp\left\{i\left[kr - \left(n + \frac{1}{2}\right)\frac{\pi}{2}\right] - ikct - in\theta\right\}$$

This function now involves only the combination $r - ct$, and so represents a wave traveling out to infinity. We may refer to such a wave as outgoing, or radiated, and we see that the corresponding function v satisfies

$$(8.6.2) \qquad \sqrt{r}\left(\frac{\partial v}{\partial r} - ikv\right) \to 0$$

as $r \to \infty$. This is the *radiation condition* of Helmholtz and Sommerfeld.

If the wave function originates from sources in the finite part of the plane, then we may expect outgoing waves and (8.6.2) is then valid. The corresponding solution of Bessel's equation is then the Hankel function $H_n^{(1)}(kr) = J_n(kr) + iY_n(kr)$ as defined in (8.5.11). We therefore have

$$\bar{G}_n = \begin{cases} A[J_n(kr)Y_n(ka) - J_n(ka)Y_n(kr)] & a < r < r' < \infty \\ BH_n^{(1)}(kr) & a < r' < r < \infty \end{cases}$$

We may evaluate the constants A and B by applying the conditions stated in Section 1.8, and we find that for $r < r'$

$$(8.6.3) \qquad \bar{G} = \frac{i\pi}{2} \frac{H_n^{(1)}(kr')}{H_n^{(1)}(ka)} [J_n(kr)H_n^{(1)}(ka) - J_n(ka)H_n^{(1)}(kr)] e^{in\theta'}$$

with a formula for $r' < r$ in which r and r' are interchanged. In constructing these expressions, we have used the Wronskian identity (8.5.13). Finally, by inserting (8.6.3) into (8.4.3), we find the solution

$$(8.6.4) \quad G(r, \theta; r', \theta') = \frac{i}{4} \sum_{n=-\infty}^{\infty} e^{in(\theta'-\theta)} \frac{H_n^{(1)}(kr)}{H_n^{(1)}(ka)}$$

$$\times [J_n(kr')H_n^{(1)}(ka) - J_n(ka)H_n^{(1)}(kr')]$$

for $a < r' < r < \infty$, and a transposed expression for $a < r < r' < \infty$.

From this expression we may deduce several further results. First we write (8.6.4) for $a < r' < r < \infty$ as

$$(8.6.5) \quad G(r, \theta; r', \theta') = \frac{i}{4} \sum_{n=-\infty}^{\infty} e^{in(\theta-\theta')} H_n^{(1)}(kr) J_n(kr')$$

$$- \frac{i}{4} \sum_{n=-\infty}^{\infty} e^{in(\theta-\theta')} \frac{J_n(ka) H_n^{(1)}(kr) H_n^{(1)}(kr')}{H_n^{(1)}(ka)}$$

In this formula we let $a \to 0$ and thereby deduce an expression for the Green's function G_0 for the entire plane representing outgoing waves at infinity. All the $J_n(z)$ of integral order vanish at the origin except J_0, which reduces to unity there. Furthermore, every function $H_n^{(1)}(z)$ is infinite at the origin, so that in the limit as $a \to 0$ the second series disappears. We find that

$$(8.6.6) \quad G_0(r, \theta; r', \theta') = \frac{i}{4} \sum_{n=-\infty}^{\infty} e^{in(\theta-\theta')} H_n^{(1)}(kr) J_n(kr')$$

for $0 < r' < r < \infty$.

Now let us place the singularity at the origin by setting $r' = 0$ in (8.6.6). Again all the J_n terms vanish except J_0, and we are left with the expression $(i/4)H_0^{(1)}(kr)$. This expression, as we should expect, depends only on the distance separating the field point from the source point. If we restore the source to its former position (r', θ'), then the Green's function will become $(i/4)H_0^{(1)}(kR)$, where R is the separating distance which is now given by (8.4.9). This expression for the Green's function for the whole of the plane should be compared with (8.4.8) for finite regions.

This argument has shown that if $r' < r$ the expression occurring in (8.6.6) is equal to $H_0^{(1)}(kR)/4i$.

ADDITION THEOREM 8.6.1 With $R^2 = r^2 + r'^2 - 2rr' \cos(\theta - \theta')$, we have

$$(8.6.7) \quad H_0^{(1)}(kR) = \sum_{n=-\infty}^{\infty} e^{in(\theta'-\theta)} H_n^{(1)}(kr) J_n(kr') \qquad r > r'$$

For $r < r'$ we must interchange r, r' in the expansion on the right.

Now by (8.1.21), (8.1.28), and (8.5.11) we see that $J_{-n} = (-1)^n J_n$ and $H_{-n} = (-1)^n H_n$, so (8.6.7) may be expressed

$$(8.6.8) \quad H_0^{(1)}(kR) = \sum_{n=0}^{\infty} \varepsilon_n H_n^{(1)}(kr) J_n(kr') \cos n(\theta - \theta') \qquad r > r'$$

where $\varepsilon_0 = 1$ and $\varepsilon_n = 2$ for $n \geq 1$. On recalling the definition (8.5.11) of the Hankel functions and separating the real and imaginary parts in (8.6.8), we find the expansions

$$(8.6.9) \quad J_0(kR) = \sum_{n=0}^{\infty} \varepsilon_n J_n(kr) J_n(kr') \cos n(\theta - \theta')$$

and

$$(8.6.10) \quad Y_0(kR) = \sum_{n=0}^{\infty} \varepsilon_n J_n(kr') Y_n(kr) \cos n(\theta - \theta')$$

The second of these expressions is applicable only when $0 < r' < r$.

Now let us return to (8.6.5) for the Green's function set up by a unit source at the point (r', θ') exterior to the circle. By virtue of the addition formula (8.6.7) we may write the Green's function as

$$(8.6.11) \quad G(r, \theta; r', \theta') = \frac{i}{4} H_0^{(1)}(kR) - \frac{i}{4} \sum_{-\infty}^{\infty} e^{in(\theta-\theta')} \frac{J_n(ka) H_n^{(1)}(kr) H_n^{(1)}(kr')}{H_n^{(1)}(ka)}$$

The leading term on the right represents the incident wave due to the source, while the Fourier expansion represents the *diffracted* wave.

The solution of the related problem of diffraction of a plane wave by a cylinder is more difficult to solve directly. It can be found by modifying (8.6.11). We may obtain the effect of a plane wave by commencing with a source of suitable strength $f(r')$ and allowing r' to tend to infinity. The appropriate intensity factor $f(r')$ can be determined by reference to the asymptotic relation

$$\frac{1}{4i} H_0^{(1)}(kR) \sim \frac{1}{4i} \sqrt{\frac{2}{\pi kR}} e^{i(kR - \pi/4)}$$

which follows from (8.5.12). If r' is large enough, we may approximate the distance R defined by (8.4.9) by the expression $r' - r \cos(\theta - \theta')$ and write

$$\frac{1}{4i} H_0^{(1)}(kR) \sim \frac{1}{4i} \sqrt{\frac{2}{\pi r' k}} e^{i(kr' - \pi/4) - ikr \cos(\theta - \theta')} \qquad r' \to \infty$$

We write this relation in the form

$$(8.6.12) \quad e^{-ikr \cos(\theta - \theta')} \sim 4i \sqrt{\frac{\pi kr'}{2}} e^{-i(kr' - \pi/4)} \frac{H_0^{(1)}(kR)}{4i}$$

The expression on the left is the reduced wave function due to a plane wave inclined at an angle θ' with the direction $\theta = 0$. We see that such a wave can be obtained by taking a source of intensity

$$f(r') = -4i \sqrt{\frac{\pi kr'}{2}} e^{-i(kr' - \pi/4)}$$

and letting $r' \to \infty$.

Now multiply the series (8.6.11) by $f(r')$, proceed to the limit, and recall the asymptotic relation

$$(8.6.13) \quad H_n^{(1)}(kr') \sim \sqrt{\frac{2}{\pi kr'}} \exp\left\{ i \left[kr' - \left(n + \frac{1}{2}\right) \frac{\pi}{2} \right] \right\}$$

We find that

$$(8.6.14) \quad u = e^{-ikr \cos(\theta - \theta')} - \sum_{n=-\infty}^{\infty} e^{in(\theta-\theta') - in\pi/2} \frac{J_n(ka) H_n^{(1)}(kr)}{H_n^{(1)}(ka)}$$

This represents the reduced wave function u when a plane wave is diffracted by a circular cylinder with the "soft" condition $u = 0$ on the surface of the cylinder.

EXERCISES 8.6

1. Show that the Green's function of Helmholtz' equation for the sector region $0 \leq r < \infty$, $0 \leq \theta \leq \alpha \leq 2\pi$, with normal derivative vanishing on both faces, is given by

$$u = \frac{i\pi}{2\alpha}\left[J_0(kr)H_0^{(1)}(kr') + 2\sum_{n=1}^{\infty}\cos\frac{n\pi\theta}{\alpha}\cos\frac{n\pi\theta'}{\alpha}J_{n\pi/\alpha}(kr)H_{n\pi/\alpha}^{(1)}(kr')\right]$$

where $0 < r < r' < \infty$. Deduce that the problem of diffraction of the plane wave $e^{-ikr\cos(\theta-\theta')}$ by the wedge has the solution

$$u = \frac{2\pi}{\alpha}\left[J_0(kr) + 2\sum_{n=1}^{\infty}\cos\frac{n\pi\theta}{\alpha}\cos\frac{n\pi\theta'}{\alpha}J_{n\pi/\alpha}(kr)e^{-in\pi^2/2\alpha}\right]$$

2. Show that the solution of the Sommerfeld problem of diffraction of the plane wave $e^{-ikr\cos(\theta-\theta')}$ by the half plane $\theta = 0$ is given by

$$u = J_0(kr) + 2\sum_{n=1}^{\infty}\cos\frac{n\theta}{2}\cos\frac{n\theta'}{2}J_{n/2}(kr)e^{-in\pi/4}$$

The boundary condition is $\partial u/\partial n = 0$.

3. The rigid cylinder $0 \leq r \leq a$, $0 \leq \theta \leq 2\pi$, $-\infty < z < \infty$, executes small lateral oscillations so that the speed of its center is $V \cos \omega t$ in the direction $\theta = 0$. Show that the wave function of the sound waves generated for $r \geq a$ is

$$v = \frac{V}{k}e^{i\omega t}\frac{H_1^{(1)}(kr)}{H_1^{(1)'}(ka)}\cos\theta$$

where $k = \omega/c$ and c is the speed of sound.

8.7 MODIFIED BESSEL FUNCTIONS

We shall now consider the partial differential equation

(8.7.1) $$\Delta u - k^2 u = 0$$

This equation is sometimes referred to as the damped wave equation since it is satisfied by wave functions which are exponentially damped with respect to the time. For if we set $v = e^{-kt}u(x, y, z)$ in $\Delta v = v_{tt}$, we obtain (8.7.1) at once. This equation also arises when we apply a Laplace transform with respect to the time to either of the equations $\Delta v = v_{tt}$, $\Delta v = v_t$. Furthermore, as we have seen in Section 5.2, the fourth-order vibration equation for plates, namely $\Delta^2 v + v_{tt} = 0$, leads to the fourth order steady-state equation $\Delta^2 u - k^4 u = 0$. In this last equation the operators $\Delta + k^2$ and $\Delta - k^2$ appear as factors, so solutions of both $\Delta u + k^2 u = 0$ and $\Delta u - k^2 u = 0$ are required to describe the vibrations of a plate.

In cylindrical coordinates (8.7.1) is

(8.7.2) $$u_{rr} + \frac{1}{r}u_r + \frac{1}{r^2}u_{\theta\theta} + u_{zz} - k^2 u = 0$$

CYLINDRICAL EIGENFUNCTIONS

Let us first consider two-dimensional solutions independent of the coordinate z. Then

$$(8.7.3) \qquad u_{rr} + \frac{1}{r} u_r + \frac{1}{r^2} u_{\theta\theta} - k^2 u = 0$$

If we extract the θ dependence by suitable Fourier series, we will arrive at the equation

$$(8.7.4) \qquad R_{rr} + \frac{1}{r} R_r - \left(k^2 + \frac{s^2}{r^2}\right) R = 0$$

On comparing with Bessel's equation (8.1.8), we see that its solutions are the Bessel functions of pure imaginary argument $J_{\pm s}(ikr)$. We shall define a new function $I_s(z)$ by the formula

$$(8.7.5) \qquad I_s(z) = e^{-is\pi/2} J_s(ze^{i\pi/2})$$

It then follows from the series (8.1.18) for $J_s(z)$ that

$$(8.7.6) \qquad I_s(z) = \left(\frac{z}{2}\right)^s \sum_{m=0}^{\infty} \frac{1}{m!\,\Gamma(s+m+1)} \left(\frac{z^2}{4}\right)^m$$

If we construct the Wronskian of the two functions $I_s(z)$, $I_{-s}(z)$ by inserting (8.7.5) in (8.1.20), we find that

$$(8.7.7) \qquad I_s(z) I'_{-s}(z) - I_{-s}(z) I'_s(z) = -\frac{2}{\pi z} \sin s\pi$$

This identity shows that $I_s(z)$ and $I_{-s}(z)$ together form the complete solution basis of (8.7.4) provided s is not an integer or zero. These functions are known as the modified Bessel functions of the first kind. When s is an integer (or zero), we find from (8.1.21) and the definition (8.7.5) that

$$(8.7.8) \qquad I_{-n}(z) = I_n(z) \qquad n = 0, 1, 2, 3, \ldots$$

In this case we need another function to form, with $I_n(z)$, a pair of independent solutions.

In view of later applications we introduce the modified Bessel function of the second kind

$$(8.7.9) \qquad K_s(z) = \frac{\pi}{2 \sin s\pi} [I_{-s}(z) - I_s(z)]$$

In terms of this function, the Wronskian relation (8.7.7) becomes

$$(8.7.10) \qquad I_s(z) K'_s(z) - I'_s(z) K_s(z) = -\frac{1}{z}$$

This relation shows that $I_s(z)$ and $K_s(z)$ are independent solutions of Bessel's modified equation for all values of s.

If s is an integer or zero, (8.7.9) becomes indeterminate since, by (8.7.8), the numerator then vanishes. We then define the function $K_n(z)$ of zero or integral order n by taking the limit of (8.7.9) as $s \to n$. Evaluating this limit by differentiation, we obtain

$$(8.7.11) \qquad K_n(z) = \frac{1}{2}(-1)^n \left[\frac{\partial}{\partial s} I_{-s}(z) - \frac{\partial}{\partial s} I_s(z) \right]_{s=n}$$

We also note from the defining equation (8.7.9) that, quite generally,

$$(8.7.12) \qquad K_{-s}(z) = K_s(z)$$

The behavior near the origin of the functions $I_{\pm s}(z)$ resembles that of the Bessel functions $J_{\pm s}(z)$, as we see by comparing their series expansions (8.7.6) and (8.1.18). Therefore $I_s(z)$ vanishes at the origin if Re $(s) > 0$ or whenever s is an integer. The function $I_{-s}(z)$ is singular at the origin like z^{-s} if Re $(s) > 0$ except when s is an integer. The function $I_0(z)$ reduces to unity when $z = 0$. It follows from these remarks and (8.7.9) that the function $K_s(z)$ is singular at the origin for all s.

If s is an integer n, the modified function $K_n(z)$ possesses a pole of order $|n|$ at $z = 0$. The function $K_0(z)$ possesses a logarithmic singularity at $z = 0$. The last two assertions follow from the series expansions which we may obtain as in Section 8.1. We find that

$$(8.7.13) \quad K_0(z) = -I_0(z) \log \frac{z}{2} + \sum_{m=0}^{\infty} \frac{1}{(m!)^2} \left(\frac{z^2}{4}\right)^m \frac{\Gamma'(m+1)}{\Gamma(m+1)}$$

$$(8.7.14) \quad K_n(z) = (-1)^{n+1} I_n(z) \log \frac{z}{2}$$

$$+ \left(\frac{-z}{2}\right)^n \frac{1}{2} \sum_{m=0}^{\infty} \frac{1}{m!(m+n)!} \left(\frac{z^2}{4}\right)^m \left[\frac{\Gamma'(m+1)}{\Gamma(m+1)} + \frac{\Gamma'(n+m+1)}{\Gamma(n+m+1)}\right]$$

$$+ \frac{1}{2} \sum_{m=0}^{n-1} \frac{(-1)^m (n-m-1)!}{m!} \left(\frac{z}{2}\right)^{2m-n}$$

Integral expressions of Schläfli's type may be developed for the function $I_s(z)$. We simply modify the Schläfli integral (8.1.14) for $J_s(z)$ according to the definition (8.7.5) and arrive at the integral representation

$$(8.7.15) \qquad I_s(z) = \frac{1}{2\pi i} \left(\frac{z}{2}\right)^s \int_L w^{-s-1} e^{w+z^2/4w} \, dw$$

where L is the anticlockwise loop round the negative real axis. If $2w = tz$, we find the alternative form

$$(8.7.16) \qquad I_s(z) = \frac{1}{2\pi i} \int_L t^{-s-1} e^{(z/2)(t+1/t)} \, dt$$

valid when Re $(z) > 0$.

By means of this relation we can, by applying the method of steepest descents, obtain the following asymptotic expression valid when $|z|$ is large enough and $-\pi/2 < \arg z < 3\pi/2$:

(8.7.17) $$I_s(z) \sim \frac{1}{\sqrt{2\pi z}}(e^z + e^{-z+i(s+\frac{1}{2})\pi})$$

The corresponding asymptotic expression for the function $K_s(z)$ may be deduced from (8.7.9) which gives

(8.7.18) $$K_s(z) \sim \sqrt{\frac{\pi}{2z}}\, e^{-z} \qquad |\arg z| < \frac{3\pi}{2}.$$

If z is real and positive, we see that $K_s(z)$ tends to zero as $z \to \infty$, whereas $I_s(z) \to \infty$ as $z \to \infty$.

EXERCISES 8.7

1. Show by means of the series (8.1.18) that for $x > 0$
$$J_s(xe^{i\pi}) = e^{is\pi} J_s(x)$$
Deduce that
$$Y_s(xe^{i\pi}) = e^{-is\pi} Y_s(x) + 2i \cos s\pi J_s(x)$$
$$H_s^{(1)}(xe^{i\pi}) = e^{-is\pi}[-J_s(x) + iY_s(x)]$$
Show also that
$$I_s(xe^{i\pi}) = e^{is\pi} I_s(x)$$
$$K_s(xe^{i\pi}) = e^{-is\pi} K_s(x) - i\pi I_s(x)$$

2. Show that
$$Y_s(xe^{i\pi/2}) = -\frac{2}{\pi} e^{-is\pi/2} K_s(x) + ie^{is\pi/2} I_s(x)$$
and deduce that
$$H_s^{(1)}(xe^{i\pi/2}) = \frac{2}{i\pi} e^{-is\pi/2} K_s(x)$$
Hence show that $H_{-s}^{(1)}(x) = e^{is\pi} H_s^{(1)}(x)$.

3. Show that the Green's function for the whole plane $0 \le r < \infty$, $0 \le \theta \le 2\pi$ of the damped wave equation
$$G_{rr} + \frac{1}{r} G_r + \frac{1}{r^2} G_{\theta\theta} - k^2 G = -\frac{1}{r}\delta(r - r')\delta(\theta - \theta')$$
is given by
$$G = \frac{1}{2\pi} \sum_{-\infty}^{\infty} e^{in(\theta' - \theta)} I_n(kr) K_n(kr')$$
for $0 < r < r' < \infty$. Interchange r, r' if $0 < r' < r < \infty$. Show that
$$G = \frac{1}{2\pi} K_0(kR)$$
where
$$R = \sqrt{r^2 + r'^2 - 2rr' \cos(\theta - \theta')}$$
and deduce an addition theorem for $K_0(kR)$.

4. Show that the Green's function of the equation

$$G_{rr} + \frac{1}{r} G_r + \frac{1}{r^2} G_{\theta\theta} + G_{zz} - k^2 G = -\frac{1}{r} \delta(r - r') \delta(\theta - \theta') \delta(z - z')$$

for the infinite slab $0 \le r < \infty$, $0 \le \theta \le 2\pi$, $0 \le z \le a$, vanishing on both planes, is given by

$$G = \frac{1}{a\pi} \sum_{n=1}^{\infty} \sin \frac{n\pi z}{a} \sin \frac{n\pi z'}{a} K_0(sR)$$

where R is defined in Exercise 8.7.3 and $s = \sqrt{k^2 + n^2\pi^2/a^2}$

5. Show that the Green's function of the equation $\Delta G - k^2 G = -\delta(\mathbf{r} - \mathbf{r}')$ vanishing round the whole surface of the cylinder $0 \le r \le a$, $0 \le \theta \le 2\pi$, $0 \le z \le h$ is given by

$$G = \frac{-1}{\pi h} \sum_{m=-\infty}^{\infty} e^{im(\theta-\theta')} \sum_{n=1}^{\infty} \frac{I_m(sr)}{I_m(sa)} [I_m(sr')K_m(sa) - I_m(sa)K_m(sr')] \sin \frac{n\pi z}{h} \sin \frac{n\pi z'}{h}$$

where $s = \sqrt{k^2 + n^2\pi^2/a^2}$ and $0 < r < r' < a$.
Obtain also the equivalent eigenfunction expansion

$$G = \frac{2}{\pi h a^2} \sum_{m=-\infty}^{\infty} e^{im(\theta-\theta')} \sum_{n=1}^{\infty} \sin \frac{n\pi z}{h} \sin \frac{n\pi z'}{h} \sum_{\lambda} \frac{J_m(\lambda r)J_m(\lambda r')}{(\lambda^2 + k^2 + n^2\pi^2/h^2)J_m'(\lambda a)^2}$$

where λ denotes a positive zero of $J_m(\lambda a)$.

6. By taking the path of integration in (8.7.16) to consist of the three segments

(a) $t = e^{u-i\pi}$ $0 < u < \infty$
(b) $t = e^{i\theta}$ $-\pi < \theta < \pi$
(c) $t = e^{u+i\pi}$ $0 < u < \infty$

show that, if $R(z) > 0$,

$$I_s(z) = \frac{1}{\pi} \int_0^{\pi} e^{z \cos \theta} \cos s\theta \, d\theta - \frac{1}{\pi} \sin s\pi \int_0^{\infty} e^{-z \cosh u - su} \, du$$

Deduce that

$$K_s(z) = \int_0^{\infty} e^{-z \cosh u} \cosh su \, du$$

7. By integrating the function $e^{-z \cosh w}$ round the rectangular contour with vertices $w = \pm R$, $w = \pm R + ic$, in the complex w-plane, where $0 < c < \pi/2$, and letting $R \to \infty$, show that if $z > 0$

$$K_0(z) = \int_0^{\infty} e^{-z \cos c \cosh u} \cos (z \sin c \sinh u) \, du$$

Deduce that

$$K_0(\sqrt{x^2 + y^2}) = \int_0^{\infty} e^{-x \cosh u} \cos (y \sinh u) \, du$$

and

$$K_0(y) = \int_0^{\infty} \cos (y \sinh u) \, du$$

8. Show that
$$K_0(\sqrt{x^2+y^2}) = \int_0^\infty e^{-y\sqrt{1+s^2}} \frac{\cos sx}{\sqrt{1+s^2}} ds$$
Where $y > 0$, deduce that
$$\int_0^\infty K_0(\sqrt{x^2+y^2}) \cos sx \, dx = \frac{\pi e^{-y\sqrt{1+s^2}}}{2\sqrt{1+s^2}}$$
Show also that
$$K_0(R\sqrt{k^2+p^2}) = \int_R^\infty e^{-pt} \frac{\cos(k\sqrt{t^2-R^2}) dt}{\sqrt{t^2-R^2}}$$
where $R > 0$, $p > 0$. Deduce the inverse Laplace transform of $K_0(R\sqrt{k^2+p^2})$.

9. By applying an exponential Fourier transform and using the result of Exercise 8.7.3, show that the fundamental singular solution for the whole of space for $\Delta G - k^2 G = -\delta(\mathbf{r} - \mathbf{r}')$ is given by
$$G = \frac{1}{2\pi^2} \int_0^\infty K_0(R\sqrt{k^2+s^2}) \cos s(z-z') ds$$
where $R = \sqrt{(x-x')^2 + (y-y')^2}$. By Exercise 8.7.8 now deduce that
$$G = \frac{e^{-k\rho}}{4\pi\rho}$$
where $\rho = \sqrt{R^2 + (z-z')^2}$.

10. By means of Fourier integrals show that the solution of the Dirichlet problem for $\varphi_{xx} + \varphi_{yy} = k^2\varphi$ on the half-plane $-\infty < x < \infty$, $0 \le y < \infty$, with $\varphi(x, 0) = g(x)$, is given by
$$\varphi(x, y) = \frac{1}{\pi} \int_{-\infty}^\infty g(x') dx' \int_0^\infty e^{-y\sqrt{k^2+s^2}} \cos s(x-x') ds$$
By means of the formula for $K_0(\sqrt{x^2+y^2})$ developed in Exercise 8.7.8, deduce that
$$\varphi(x, y) = -\frac{1}{\pi} \frac{\partial}{\partial y} \int_{-\infty}^\infty g(x') K_0[k\sqrt{y^2+(x-x')^2}] dx'$$

11. Let $u = AJ_s(\lambda r) + BY_s(\lambda r)$ be any Bessel function and $v = CI_s(\mu r) + DK_s(\mu r)$ be any modified Bessel function. Show by analogy with (8.2.6) that
$$(\lambda^2 + \mu^2) \int ruv \, dr = r(uv_r - vu_r)$$
Deduce that
$$\int_0^a rJ_0(\lambda r) I_0(\mu r) dr = \frac{-\lambda a I_0(\mu a) J_0'(\lambda a)}{\lambda^2 + \mu^2}$$
where λ is a zero of $J_0(\lambda a)$.

12. By means of the results given in Exercises 8.7.2 and 8.7.6, deduce *Mehler's integral*
$$H_0^{(1)}(z) = \frac{2}{i\pi} \int_0^\infty e^{iz \cosh u} du \qquad I(z) > 0$$

Show by employing a suitable Fourier integral that the *singular* solution of the wave equation

$$u_{rr} + \frac{1}{r} u_r = u_{tt}$$

for $0 \leq r < \infty$, $-\infty < t < \infty$, such that $\lim_{r \to 0} r u_r = g(t)$ is given by

$$u(r, t) = -\int_{-\infty}^{t-r} \frac{g(t') \, dt'}{\sqrt{(t-t')^2 - r^2}}$$

8.8 THE HANKEL AND WEBER FORMULAS

For infinite regions suitable Hankel transforms may be constructed, but now the inversion formulas involve integrals. For the interval $0 \leq r \leq \infty$, we formulate the following theorem on Hankel transforms of order ν. (Table 5 in the appendix.)

THEOREM 8.8.1 *If $f(r)$ is defined for $0 \leq r < \infty$ and if*

(8.8.1) $$F(k) = \int_0^\infty r J_\nu(kr) f(r) \, dr \qquad \nu, k \geq 0$$

then

(8.8.2) $$f(r) = \int_0^\infty k J_\nu(kr) F(k) \, dk$$

It is possible to obtain these relations from the formal method of eigenfunction expansions outlined in Section 6.7. The argument of that section would have to be modified as the spectrum is not discrete. The eigenfunctions are now the functions $J_\nu(kr)$ for any $k > 0$ and the spectrum is the half line $0 \leq k < \infty$. However, we shall instead derive the above relations from the Fourier integral theorem. This procedure applies only to integral values of ν but is simpler than adapting the analysis of Section 6.7.

We employ the double exponential Fourier integrals taken over the whole x, y-plane:

(8.8.3) $$G(s_1, s_2) = \frac{1}{2\pi} \iint_{-\infty}^{\infty} e^{is_1 x + is_2 y} g(x, y) \, dx \, dy$$

(8.8.4) $$g(x, y) = \frac{1}{2\pi} \iint_{-\infty}^{\infty} e^{-is_1 x - is_2 y} G(s_1, s_2) \, ds_1 \, ds_2$$

In these formulas we change to polar coordinates in both planes by writing $x = r \cos \theta$, $y = r \sin \theta$, $s_1 = k \sin \varphi$, and $s_2 = k \cos \varphi$. Also we set $g(x, y) = e^{-in\theta} f(r)$, where n is an integer or zero. Then we obtain from (8.8.3),

(8.8.5) $$G(s_1, s_2) = \frac{1}{2\pi} \int_0^\infty r f(r) \, dr \int_{-\pi}^{\pi} e^{-in\theta + ikr \sin(\theta + \varphi)} \, d\theta$$

Now by (8.1.9) we see that, since n is an integer or zero,

$$(8.8.6) \qquad e^{in\varphi}J_n(kr) = \frac{1}{2\pi}\int_{-\pi}^{\pi} e^{-in\theta + ikr\sin(\theta+\varphi)}\, d\theta$$

Therefore (8.8.5) can be expressed as

$$(8.8.7) \qquad e^{-in\varphi}G(s_1, s_2) = \int_0^\infty rf(r)J_n(kr)\, dr$$

The integral on the right is merely the Hankel transform $F(k)$ so that $G(s_1, s_2) = e^{in\varphi}F(k)$. Inserting this result together with $g(x, y) = e^{-in\theta}f(r)$ into (8.8.4) gives the equation

$$f(r) = \frac{1}{2\pi}\int_0^\infty kF(k)\, dk \int_{-\pi}^{+\pi} e^{in(\theta+\varphi) - ikr\sin(\theta+\varphi)}\, d\varphi$$

By the complex conjugate of (8.8.6), we find that the last equation is simply

$$f(r) = \int_0^\infty kJ_n(kr)F(k)\, dk$$

This is the required inversion formula to be associated with the integral transform (8.8.1).

In applications it is convenient to use the following result which is established in the same way as the similar relation (8.3.8):

$$(8.8.8) \qquad \int_0^\infty rJ_s(kr)\left(f_{rr} + \frac{1}{r}f_r - \frac{s^2}{r^2}f\right) dr = -k^2 F(k)$$

This law of transformation applies provided the terms

$$(8.8.9) \qquad rf'(r)J_s(kr) - krf(r)J_s'(kr)$$

together vanish in the limits $r \to 0$ and $r \to \infty$. Since $s \geq 0$, this will certainly be true if $f(r), f'(r)$ are bounded at the origin and if $\sqrt{r}f(r)$ and $\sqrt{r}f'(r)$ tend to zero as $r \to \infty$.

Exercise 8.1.9 provides us with the following example of a Hankel transform:

$$(8.8.10) \qquad \int_0^\infty e^{-kz}J_0(kr)\, dk = \frac{1}{\sqrt{r^2 + z^2}} \qquad z, r > 0$$

The integral on the left is the Hankel transform of order zero of the function $k^{-1}e^{-kz}$. If we invert it by Theorem 8.8.1, we find

$$(8.8.11) \qquad \int_0^\infty \frac{rJ_0(kr)\, dr}{\sqrt{r^2 + z^2}} = \frac{1}{k}e^{-kz}$$

We shall illustrate the use of the Hankel transforms by constructing Laplace's integral for the axially symmetric potential function $u(r, z)$ in terms of its values

$f(z)$ on the axis of symmetry. The equation to be solved is

(8.8.12) $$u_{rr} + \frac{1}{r}u_r + u_{zz} = 0$$

and it is supposed that $0 < r < \infty$, $-\infty < z < \infty$. Let us apply the transform (8.8.1) and write

(8.8.13) $$U(k, z) = \int_0^\infty rJ_0(kr)u(r, z)\,dr$$

Then (8.8.12) converts into the equation

(8.8.14) $$U_{zz} - k^2 U = -[ru_r J_0(kr) - kruJ_0'(kr)]_0^\infty$$

The values of u and u_r on the axis are supposed finite, and we shall assume further that $r^{1/2}u$ and $r^{1/2}u_r$ both tend to zero as $r \to \infty$. Then (8.8.14) reduces to $U_{zz} = k^2 U$, so that we have

$$U(k, z) = F(k)e^{-k|z|}$$

Inverting the Hankel transform (8.8.13) and using the value of U just found, we obtain the potential

(8.8.15) $$u(r, z) = \int_0^\infty kF(k)J_0(kr)e^{-k|z|}\,dk$$

The function $F(k)$ can be determined in terms of the values of u on the z-axis. To do this, we set $r = 0$ in the preceding formula; then u reduces to $f(z)$ and we obtain

(8.8.16) $$f(z) = \int_0^\infty kF(k)e^{-kz}\,dk \qquad z > 0$$

This is just a Laplace transform and its inversion is

(8.8.17) $$kF(k) = \frac{1}{2\pi i}\int_L e^{kz}f(z)\,dz$$

Here L is a path $(c - i\infty, c + i\infty)$ in the z-plane, lying to the right of all the singularities of the function $f(z)$. If $f(z)$ is given, then we may calculate $F(k)$ by means of (8.8.17), but it should be noted that to apply this formula the function $f(z)$ must be analytic. Once $F(k)$ has been determined, we may find the potential itself from (8.8.15).

To obtain the Laplace integral referred to, we substitute in (8.8.15) the integral for $J_0(kr)$ given in (8.1.9). This gives

$$u(r, z) = \frac{1}{2\pi}\int_0^{2\pi} d\theta \int_0^\infty kF(k)e^{-k(z - ir\cos\theta)}\,dk \qquad z > 0$$

The k integral occurring here is, by (8.8.16), equal to $f(z - ir \cos \theta)$. Therefore we obtain the explicit solution formula

$$(8.8.18) \qquad u(r, z) = \frac{1}{2\pi} \int_0^{2\pi} f(z - ir \cos \theta) \, d\theta$$

On setting $r = 0$ in this expression, we see that $u(0, z) = f(z)$ as required. Again it should be noted that the function $f(z)$ must be analytic for (8.8.18) to apply.

As an illustration of the use of these formulas, we consider the potential due to electric charge distributed evenly round the circular hoop $r = a$, $z = 0$

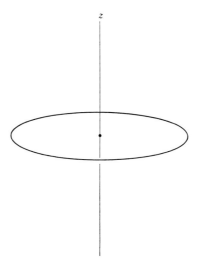

Figure 8.6   A circular hoop.

(Figure 8.6). On the axis the potential is $f(z) = (a^2 + z^2)^{-1/2}$, so by (8.8.16) the corresponding function $F(k)$ is to be determined from the equation

$$\int_0^\infty kF(k)e^{-kz}\,dk = \frac{1}{\sqrt{a^2 + z^2}} \qquad z > 0$$

By comparison with (8.8.10) we see that $kF(k) = J_0(ka)$, so that (8.8.15) yields the solution

$$(8.8.19) \qquad u(r, z) = \int_0^\infty J_0(ka)J_0(kr)e^{-kz}\,dk \qquad z > 0$$

The explicit evaluation of this integral requires elliptic functions. An alternative expression for this potential is developed in Exercise 8.8.2 at the end of this section.

We consider now the truncated interval $0 < a \leq r < \infty$ and form the eigenfunctions of Bessel's equation bounded in this interval and vanishing at $r = a$. These eigenfunctions are readily seen to be the functions

$$(8.8.20) \qquad \varphi_s(k, r) = J_s(kr)H_s^{(1)}(ka) - J_s(ka)H_s^{(1)}(kr)$$

where $k > 0$. The corresponding Hankel transform is expressed by the following theorem:

THEOREM 8.8.2 *If $f(r)$ is defined on $0 < a \leq r < \infty$ and if*

(8.8.21) $$F(k) = \int_a^\infty r\varphi_s(k, r) f(r)\, dr$$

then

(8.8.22) $$f(r) = \frac{1}{2} \int_{ic-\infty}^{ic+\infty} \frac{z H_s^{(1)}(zr) F(z)\, dz}{H_s^{(1)}(za)} \qquad r > a.$$

This result is known as *Weber's integral theorem*. To verify it, we shall assume that $F(k)$ is an even single-valued function of k regular in some strip $|\text{Im}(k)| < \gamma$ and such that $kF(k) \to 0$ as $k \to \infty$ in this strip. The path of integration in (8.8.22) is chosen so that $|c| < \gamma$. If we insert (8.8.22) into (8.8.21) and use the integral given in Exercise 8.5.6, we obtain

(8.8.23) $$\int_a^\infty r\varphi_s(k, r) f(r)\, dr = \frac{i}{\pi} \int_{ic-\infty}^{ic+\infty} \frac{zF(z)\, dz}{z^2 - k^2} \qquad |\text{Im}(k)| < c$$

By virtue of the assumed asymptotic behavior of the function $F(z)$, we may deform the path $(ic - \infty, ic + \infty)$ onto the real axis. The integrand is an odd function of z so the integral along that axis vanishes. Suppose that $\text{Im}(k) > 0$, then by Cauchy's theorem we must take into account the residue at the pole $z = k$. Evaluating this residue, we find that the expression on the right of (8.8.23) reduces to $F(k)$ as required.

In applications it is convenient to use the following formula, which is analogous to (8.8.8),

(8.8.24) $$\int_a^\infty r\varphi_s(k, r) \left(f_{rr} + \frac{1}{r} f_r - \frac{s^2}{r^2} f \right) dr = -k^2 F(k) - \frac{2i}{\pi} f(a),$$

provided that

$$\lim_{r \to \infty} r(\varphi_s f_r - \varphi_s' f) = 0$$

This formula shows that the transform (8.8.21) is effective only when $f(a)$ is known. Similar transforms can be developed to deal with the more general boundary condition in which $hf(a) + f'(a)$ is prescribed. These transforms are useful when applied to regions outside circles and cylinders. For two specific applications of these Weber formulas, we refer the reader to Exercises 8.8.4 and 8.8.5.

EXERCISES 8.8

1. A potential function $\varphi(r, z)$ reduces to z^{-1} on the z-axis. Using (8.8.18), show that the general value is given by

$$\varphi(r, z) = \frac{1}{2\pi} \int_0^{2\pi} \frac{d\theta}{z - ir \cos \theta}$$

Now set $t = e^{i\theta}$ and integrate round the unit circle in the complex t-plane. Deduce that

$$\varphi = \frac{1}{\sqrt{r^2 + z^2}}$$

Show also that

$$\varphi(r, z) = \frac{1}{2\pi i} \int_C \frac{dt}{t\sqrt{r^2 + (t-z)^2}}$$

where C is a contour enclosing the points $t = z \pm ir$ but not enclosing the point $t = 0$. By deforming the contour out to infinity and considering the residue at the origin, show that $\varphi = (r^2 + z^2)^{-1/2}$ as before.

2. By means of the addition formula (8.6.9) show that

$$J_0(ka) J_0(kr) = \frac{1}{2\pi} \int_0^{2\pi} J_0(k\sqrt{a^2 + r^2 - 2ar\cos\theta})\, d\theta$$

Deduce from (8.8.19) that the potential due to the circular ring $r = a$, $z = 0$, is given by

$$u(r, z) = \frac{1}{2\pi} \int_0^{2\pi} \frac{d\theta}{\sqrt{z^2 + a^2 + r^2 - 2ar\cos\theta}}$$

Construct this directly.

3. Use the Hankel transform to construct the Green's function of the equation

$$\varphi_{rr} + \frac{1}{r}\varphi_r + \frac{1}{r^2}\varphi_{\theta\theta} + \varphi_{zz} - k^2\varphi = -\frac{1}{r}\delta(r - r')\,\delta(\theta - \theta')\,\delta(z - z')$$

for the region $0 \le r < \infty$, $0 \le \theta \le 2\pi$, $0 \le z \le a$, with $\partial\varphi/\partial z$ vanishing on both faces $z = 0$, $z = a$. Show that

$$\varphi = \frac{1}{2\pi}\int_0^\infty \frac{\lambda J_0(\lambda R)\cosh\sigma z \cosh\sigma(a - z')\, d\lambda}{\sigma \sinh \sigma a}$$

where $R = \sqrt{r^2 + r'^2 - 2rr'\cos(\theta - \theta')}$ and $\sigma = \sqrt{k^2 + \lambda^2}$. Deduce that

$$\varphi = \frac{1}{2\pi a} K_0(kR) + \frac{1}{\pi a}\sum_{n=1}^\infty \cos\frac{n\pi z}{a} \cos\frac{n\pi z'}{a} K_0(\mu R)$$

where $\mu = \sqrt{k^2 + n^2\pi^2/a^2}$.

4. Use Weber's formula to solve the initial value problem for the diffusion equation

$$\varphi_{rr} + \frac{1}{r}\varphi_r = \varphi_t$$

in the region $a \le r < \infty$ with the boundary condition $\varphi = T_0$ on $r = a$ for $t > 0$ and $\varphi = 0$ on $t = 0$ for $r \ge a$.

Show that

$$\varphi = \frac{-2T_0}{\pi}\int_0^\infty \frac{(1 - e^{-\lambda^2 t})[J_0(\lambda r) Y_0(\lambda a) - J_0(\lambda a) Y_0(\lambda r)]\, d\lambda}{\lambda[J_0(\lambda a)^2 + Y_0(\lambda a)^2]}$$

5. By means of Weber's formula obtain the axially symmetric potential $\varphi(r, z)$ satisfying the equation
$$\varphi_{rr} + \frac{1}{r}\varphi_r + \varphi_{zz} = 0$$
in the exterior region $a \leq r < \infty$, $0 \leq z \leq h$, with the boundary values $\varphi = 0$ on $z = 0$, $\varphi_z = 0$ on $z = h$, and $\varphi = V$ on $r = a$. Show that
$$\varphi = \frac{2V}{\pi}\int_0^\infty \frac{[\cosh \lambda(h-z) - \cosh \lambda h][J_0(\lambda r)Y_0(\lambda a) - J_0(\lambda a)Y_0(\lambda r)]\,d\lambda}{\lambda \cosh \lambda h [J_0(\lambda a)^2 + Y_0(\lambda a)^2]}$$
Deduce that
$$\varphi = \frac{2V}{h}\sum_{n=0}^\infty \frac{1}{\beta}\sin \beta z \frac{K_0(\beta r)}{K_0(\beta a)}$$
where $\beta = (n + \tfrac{1}{2})\pi/h$.

6. If $|\mathrm{Im}\,(\lambda)| < \mathrm{Im}\,(\mu)$, show that
$$\int_0^\infty rJ_s(\lambda r)H_s^{(1)}(\mu r)\,dr = \frac{2i}{\pi(\mu^2 - \lambda^2)}\left(\frac{\lambda}{\mu}\right)^s$$
Assume $s > 0$. Suppose that $F(\lambda)$ is regular in some half-plane $\mathrm{Im}\,(\lambda) > \gamma$ and define
$$f(r) = \frac{1}{2}\int_{ic-\infty}^{ic+\infty} \lambda J_s(\lambda r)F(\lambda)\,d\lambda$$
for $c > \gamma$. Suppose further that $F(\lambda)e^{-i\lambda \alpha} \to 0$ as $\mathrm{Im}\,(\lambda) \to \infty$ for some $\alpha > 0$. Show that
$$F(\lambda) = \int_0^\infty rf(r)H_s^{(1)}(\lambda r)\,dr$$

7. Show that the solution of the equation
$$u_{rr} + \frac{1}{r}u_r + u_{zz} = 0$$
in the region $0 \leq r < \infty$, $-\infty < z < \infty$, such that
$$\lim_{r \to 0} r^2 u = 0 \qquad \lim_{r \to 0} ru_r = g(z)$$
is given by
$$u(r, z) = -\frac{1}{2}\int_{-\infty}^\infty \frac{g(z')\,dz'}{\sqrt{r^2 + (z-z')^2}}$$
[Apply the Hankel transform and use the integral (8.8.10).]

8. By means of the asymptotic expression (8.5.9) show that the zeros of large enough magnitude of $J_s(ka)$ are given approximately by
$$ak_n = \left(s + \frac{1}{2}\right)\frac{\pi}{2} + \left(n + \frac{1}{2}\right)\pi$$
where n is a sufficiently large positive integer. Show also that, for such zeros, $aJ_s'(k_na)^2 \sim 2/\pi k_n$. By letting $a \to \infty$ in the Fourier-Bessel series (Theorem 8.3.1), give a plausible derivation of Hankel's integral transform theorems.

9. The potential function $\varphi(r, z, t)$ for axially symmetric vibrations in water of depth h satisfies for $0 \le r < \infty$, $-h \le z \le 0$, the equation

$$\varphi_{rr} + \frac{1}{r}\varphi_r + \varphi_{zz} = 0$$

with the boundary conditions $\varphi_z = 0$ on $z = -h$ and $\varphi_{tt} + g\varphi_z = 0$ on $z = 0$. If the initial conditions are $\varphi_t = 0$ and $\varphi = f(r)$ for $t = z = 0$, construct the solution

$$\varphi = \int_0^\infty r'f(r')\,dr' \int_0^\infty \lambda J_0(\lambda r)J_0(\lambda r')\,\frac{\cosh \lambda(z+h)\cos \mu t\, d\lambda}{\cosh \lambda h}$$

where $\mu = \sqrt{g\lambda \tanh \lambda h}$. In the case of infinite depth, show that the solution is

$$\varphi = \int_0^\infty r'f(r')\,dr' \int_0^\infty \lambda J_0(\lambda r)J_0(\lambda r')e^{\lambda z} \cos(t\sqrt{g\lambda})\,d\lambda$$

CHAPTER 9

Spherical Eigenfunctions

9.1 LEGENDRE FUNCTIONS

Every region gives rise to a sequence of eigenfunctions which describe its natural modes of vibration. For circles and cylinders we have studied the Bessel functions which are introduced in this way. Now we consider the sphere, the most symmetrical of all three-dimensional regions. With spherical polar coordinates we have one radial variable r with the dimension of length. The eigenfunction factors associated with this coordinate are again Bessel functions. We also have two angle variables which determine latitude and longitude. The longitudinal factors are the familiar trigonometric ones, but the latitude coordinate leads to a new type of special function—the Legendre functions—which we shall study in this chapter.

Like the Bessel functions, which are also particular cases of the hypergeometric function, the Legendre functions satisfy numerous intricate series and integral formulas and recurrence relations. We shall introduce several of these in such a way as to suggest that they are natural consequences of the role which these spherical eigenfunctions must play as solutions of the Laplace and

SPHERICAL EIGENFUNCTIONS

Helmholtz equations. When a wave function can be expressed as a superposition of separated solutions in a different coordinate system, for instance, we will obtain series or integral formulas concerning the appropriate special functions.

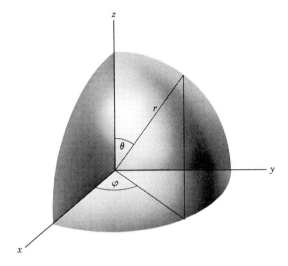

Figure 9.1 Spherical polar coordinates.

We commence by considering the function $u = (z + ix)^n$ which is readily seen to be a solution of the three-dimensional Laplace equation

(9.1.1) $$u_{xx} + u_{yy} + u_{zz} = 0$$

Let us introduce spherical polar coordinates (Figure 9.1)

(9.1.2) $\quad x = r \sin \theta \cos \psi \quad y = r \sin \theta \sin \psi \quad z = r \cos \theta$

where θ is the angle which the radius vector makes with the z-axis and ψ is the azimuthal angle so that $0 \leq \theta \leq \pi$ and $-\pi \leq \psi \leq \pi$; then (9.1.1) becomes

(9.1.3) $$u_{rr} + \frac{2}{r} u_r + \frac{1}{r^2 \sin \theta} \frac{\partial}{\partial \theta} (\sin \theta u_\theta) + \frac{1}{r^2 \sin^2 \theta} u_{\psi\psi} = 0$$

Now $u = r^n (\cos \theta + i \sin \theta \cos \psi)^n = V(\theta, \psi) r^n$, say, so that on substitution into (9.1.3) we obtain the equation for $V(\theta, \psi)$ in the form

(9.1.4) $$\frac{1}{\sin \theta} \frac{\partial}{\partial \theta} (\sin \theta V_\theta) + \frac{1}{\sin^2 \theta} V_{\psi\psi} + n(n+1)V = 0$$

This is known as the equation of *spherical surface harmonics*. Now $V(\theta, \psi)$ is an even function of ψ, and if we expand it as a Fourier series for $-\pi < \psi < \pi$, it is obvious that no sine terms can appear. Therefore we may let

$$V = (\cos \theta + i \sin \theta \cos \psi)^n = \tfrac{1}{2} f_0(\theta) + \sum_{m=1}^{\infty} f_m(\theta) \cos m\psi$$

LEGENDRE FUNCTIONS   343

The functions $f_m(\theta)$, $m \geq 0$ introduced here, can be determined by the customary procedure of multiplying by $\cos m\psi$ and integrating over $-\pi < \psi < \pi$. However, at this point it is convenient to introduce in the accepted notation, the functions $P_n(\cos \theta)$ and $P_n^m(\cos \theta)$. These differ from $f_0(\theta)$ and $f_m(\theta)$ by constant factors such that

$$(9.1.5) \quad (\cos \theta + i \sin \theta \cos \psi)^n = P_n(\cos \theta) + 2 \sum_{m=1}^{\infty} \frac{n!}{(n+m)!} e^{-im\pi/2} P_n^m(\cos \theta) \cos m\psi$$

We shall use this equation to define the functions $P_n(\cos \theta)$ and $P_n^m(\cos \theta)$ for positive integral values of m and n including zero. In fact, on evaluating the Fourier coefficients in (9.1.5), we obtain

$$(9.1.6) \quad P_n^m(\cos \theta) = \frac{e^{im\pi/2}(n+m)!}{2\pi n!} \int_{-\pi}^{\pi} (\cos \theta + i \sin \theta \cos \psi)^n \cos m\psi \, d\psi$$

and

$$(9.1.7) \quad P_n(\cos \theta) = \frac{1}{2\pi} \int_{-\pi}^{\pi} (\cos \theta + i \sin \theta \cos \psi)^n \, d\psi$$

These are known as *Laplace's integrals* (Macmillan, Ref. 42).

If we adopt the convention $P_n(\cos \theta) = P_n^0(\cos \theta)$, we see that (9.1.7) is merely a special case of (9.1.6).

We shall introduce the variable $\mu = \cos \theta$ and will show that the function $P_n(\mu)$ is a polynomial in μ of degree n. These polynomials $P_n(\mu)$ are known as the *Legendre polynomials*. The functions $P_n^m(\mu)$, which for $m \geq 1$ are polynomials in μ and $\sqrt{1-\mu^2}$, are referred to as the *associated Legendre functions*. To establish these results, we shall first write (9.1.6) in the form

$$(9.1.8) \quad P_n^m(\cos \theta) = \frac{e^{im\pi/2}(n+m)!}{2\pi n!} \int_{-\pi}^{\pi} (\cos \theta + i \sin \theta \cos \psi)^n e^{im\psi} \, d\psi$$

Now set $t = \cos \theta + i \sin \theta e^{i\psi}$, which for $-\pi < \psi < \pi$ corresponds in the t-plane, to integrating round a circle C of center $t = \cos \theta$ and radius $|\sin \theta|$. We find after a separate calculation that $\cos \theta + i \sin \theta \cos \psi = (t^2 - 1)/[2(t - \cos \theta)]$. Therefore

$$P_n^m(\cos \theta) = \frac{(-1)^m \sin^m \theta (n+m)!}{2^n n!} \left(\frac{1}{2i\pi}\right) \int_C \frac{(t^2-1)^n \, dt}{(t - \cos \theta)^{n+m+1}}$$

$$= \frac{(-1)^m \sin^m \theta}{2^n n!} \frac{1}{2i\pi} \frac{d^{n+m}}{d\mu^{n+m}} \int_C \frac{(t^2-1)^n \, dt}{t - \mu}$$

where we have written $\mu = \cos \theta$. The pole $t = \mu$ lies inside the contour C, and by Cauchy's theorem (Exercise 4.6.5) we obtain

$$(9.1.9) \quad P_n^m(\mu) = \frac{(-1)^m (1-\mu^2)^{m/2}}{2^n n!} \frac{d^{n+m}}{d\mu^{n+m}} (\mu^2 - 1)^n$$

When $m = 0$, this reduces to $P_n(\mu)$ by definition, so that

$$(9.1.10) \qquad P_n(\mu) = \frac{1}{2^n n!} \frac{d^n}{d\mu^n} (\mu^2 - 1)^n$$

This is known as *Rodrigues' formula*, and from it we deduce at once that $P_n(\mu)$ is a polynomial of degree n in μ. We may also insert (9.1.10) into (9.1.9) and obtain the relation

$$(9.1.11) \qquad P_n^m(\mu) = (-1)^m (1 - \mu^2)^{m/2} \frac{d^m P_n(\mu)}{d\mu^m}$$

From this result we see that the associated Legendre function $P_n^m(\mu)$ vanishes identically if $m > n$. We remind the reader that throughout this section we are confining attention to integral values of m and n. The Fourier series (9.1.5) now reduces to the *finite* sum

$$(9.1.12) \qquad \begin{aligned} & (\cos\theta + i\sin\theta \cos\psi)^n \\ & = P_n(\cos\theta) + 2\sum_{m=1}^{n} \frac{n!}{(n+m)!} e^{-im\pi/2} P_n^m(\cos\theta) \cos m\psi \end{aligned}$$

By the method of construction, each term in this series must satisfy (9.1.4). Since the functions $\cos m\psi$ are linearly independent, we find on substituting into the latter equation that $P_n^m(\cos\theta)$ satisfies

$$(9.1.13) \qquad \frac{1}{\sin\theta} \frac{d}{d\theta}\left(\sin\theta \frac{d\omega}{d\theta}\right) + \left[n(n+1) - \frac{m^2}{\sin^2\theta}\right]\omega = 0$$

On again writing $\mu = \cos\theta$, we obtain *Legendre's associated equation*

$$(9.1.14) \qquad \frac{d}{d\mu}\left[(1 - \mu^2) \frac{d\omega}{d\mu}\right] + \left[n(n+1) - \frac{m^2}{1 - \mu^2}\right]\omega = 0$$

of which $P_n^m(\mu)$ is a solution. When $m = 0$, the polynomial $P_n(\mu)$ satisfies the equation

$$(9.1.15) \qquad \frac{d}{d\mu}\left[(1 - \mu^2) \frac{d\omega}{d\mu}\right] + n(n+1)\omega = 0$$

which is simply known as *Legendre's equation*.

We have so far confined attention to nonnegative values of m. For some purposes it is helpful to express the ψ dependence of a potential function in the exponential form $e^{\pm im\psi}$ instead of the real forms $\cos m\psi$ and $\sin m\psi$; then negative values of m will appear. If we revert to the integral (9.1.8) and substitute $t = \cos\theta + i\sin\theta e^{-i\psi}$ therein, then, by a process similar to that used in developing (9.1.9), we find the formula

$$(9.1.16) \qquad P_n^m(\mu) = \frac{(n+m)!}{(n-m)!} \frac{(1 - \mu^2)^{-m/2}}{2^n n!} \frac{d^{n-m}}{d\mu^{n-m}} (\mu^2 - 1)^n$$

Equation (9.1.9) defines $P_n{}^m(\mu)$ for positive integral values of m such that $0 \leq m \leq n$. As it is undesirable to have different definitions of the function $P_n{}^m(\mu)$ according as m is positive or negative, we shall take (9.1.9) to cover both cases. Thus if $0 \leq m \leq n$, we define

$$P_n{}^{-m}(\mu) = \frac{(-1)^m(1-\mu^2)^{-m/2}}{2^n n!} \frac{d^{n-m}}{d\mu^{n-m}}(\mu^2 - 1)^n \tag{9.1.17}$$

This definition does not introduce a new solution of the associated Legendre equation (9.1.14), since by comparison with (9.1.16) we see that

$$P_n{}^{-m}(\mu) = (-1)^m \frac{(n-m)!}{(n+m)!} P_n{}^m(\mu) \tag{9.1.18}$$

Thus the two functions $P_n{}^m(\mu)$ and $P_n{}^{-m}(\mu)$ differ only by a constant factor. The above formulas will be used when we come to construct the normalized eigenfunctions on the surface of a sphere.

EXERCISES 9.1

1. Show by Rolle's theorem that

$$\frac{d}{d\mu}(\mu^2 - 1)^n$$

possesses at least one real zero in the range $-1 < \mu < +1$. Deduce from (9.1.10) that $P_n(\mu)$ possesses n real zeros in the interval $-1 < \mu < +1$. Deduce also from (9.1.11) that $P_n{}^m(\mu)$ possesses $n - m$ real zeros in the same interval.

2. Show that a homogeneous polynomial $P_n(x, y, z)$ of degree n in three variables contains $\frac{1}{2}(n+1)(n+2)$ independent coefficients. If $\Delta P_n = 0$, how many linear relations are implied? Deduce the number of linearly independent harmonic polynomials of degree n. List these polynomials for $n = 0, 1,$ and 2.

3. Show that every homogeneous derived expression

$$P_n\left(\frac{\partial}{\partial x}, \frac{\partial}{\partial y}, \frac{\partial}{\partial z}\right)\frac{1}{r}$$

is harmonic for $r > 0$. Noting that $\Delta 1/r = 0$ for $r > 0$, show that there are $2n + 1$ independent harmonic functions of degree $-n - 1$. List these homogeneous harmonics for $n = 0, 1,$ and 2.

9.2 EIGENFUNCTIONS OF THE SPHERICAL SURFACE

If we seek solutions of the Laplace equation (9.1.3) of the form $u = R(r)\Theta(\theta)\Psi(\psi)$, we find by the separation of variables method the equations

$$\Psi'' + m^2\Psi = 0$$
$$r^2R'' + 2rR' - s(s+1)R = 0 \tag{9.2.1}$$

and

(9.2.2) $$\frac{1}{\sin\theta}\frac{d}{d\theta}(\sin\theta\Theta') + \left[s(s+1) - \frac{m^2}{\sin^2\theta}\right]\Theta = 0$$

Here m^2 and $s(s+1)$ are separation constants, and primes denote differentiation with respect to the appropriate coordinate.

The first of these equations possesses solutions $e^{\pm im\psi}$, and if the range of variation of ψ is $(-\pi, \pi)$, m must be an integer to obtain a single-valued potential function. Equation (9.2.1) possesses the general solution $R = Ar^s + Br^{-s-1}$. Equation (9.2.2) is the associated Legendre equation and it possesses a solution bounded in the closed interval $0 \leq \theta \leq \pi$ only when $s = n$ (an integer or zero), in which case the bounded solution is the function $P_n{}^m(\cos\theta)$. We postpone until Section 9.6 the justification of this statement. Collecting these results, we may express an harmonic function u as (MacRobert, Ref. 43)

(9.2.3) $$u = \sum_{n=0}^{\infty}\sum_{m=-n}^{n}\left(a_{nm}r^n + \frac{b_{nm}}{r^{n+1}}\right)e^{im\psi}P_n{}^m(\cos\theta)$$

Each term in this expansion is bounded and single-valued in any finite domain which excludes the origin. If the potential function is symmetrical about the z-axis, then ψ does not appear in the above expansion, and we are left with the terms for $m = 0$:

(9.2.4) $$u(r, \theta) = \sum_{n=0}^{\infty}(a_n r^n + b_n r^{-n-1})P_n(\cos\theta)$$

Here only the Legendre polynomials occur. These polynomials are, in fact, the eigenfunctions bounded on the closed interval $0 \leq \theta \leq \pi$ of the equation

$$\frac{1}{\sin\theta}\frac{d}{d\theta}\left(\sin\theta\frac{d\omega}{d\theta}\right) + s(s+1)\omega = 0$$

and the eigenvalues correspond to the values $s = 0, 1, 2, \ldots$. More generally, the eigenfunctions of (9.2.2) bounded in the same interval are the associated Legendre functions $P_n{}^m(\cos\theta)$.

Thus we see that eigenfunctions of the spherical surface equation (9.1.4) bounded and single-valued for $0 \leq \theta \leq \pi$, $-\pi \leq \psi \leq \pi$ arise only when n and m are integers therein, and these eigenfunctions are the functions

(9.2.5) $$P_n{}^m(\cos\theta)e^{im\psi}$$

Since $P_n{}^m(\cos\theta)$ is identically zero for integers m numerically greater than n, we see that there are $2n + 1$ such eigenfunctions for each value of n. That these eigenfunctions form an orthogonal set on the surface of the sphere may be established in the usual manner by referring to the differential equation. Let $\mu = \cos\theta$ and consider the product

$$\int_{-1}^{1}P_n{}^m(\mu)P_{n'}{}^{m'}(\mu)\,d\mu\int_{-\pi}^{\pi}e^{i(m+m')\psi}\,d\psi$$

EIGENFUNCTIONS OF THE SPHERICAL SURFACE

The ψ integral here vanishes unless $m' = -m$. Therefore it is sufficient to show that

(9.2.6) $$\int_{-1}^{+1} P_n^m(\mu) P_{n'}^{-m}(\mu) \, d\mu = 0$$

We write down the differential equations satisfied by P_n^m and $P_{n'}^{-m}$:

$$\frac{d}{d\mu}\left[(1-\mu^2)\frac{dP_n^m(\mu)}{d\mu}\right] + \left[n(n+1) - \frac{m^2}{1-\mu^2}\right]P_n^m(\mu) = 0$$

and

$$\frac{d}{d\mu}\left[(1-\mu^2)\frac{dP_{n'}^{-m}(\mu)}{d\mu}\right] + \left[n'(n'+1) - \frac{m^2}{1-\mu^2}\right]P_{n'}^{-m}(\mu) = 0$$

We multiply the first by $P_{n'}^{-m}(\mu)$, the second by $P_n^m(\mu)$, and subtract and integrate over $-1 \leq \mu \leq 1$. After slight rearrangement we find that

$$(n-n')(n+n'+1)\int_{-1}^{+1} P_n^m(\mu) P_{n'}^{-m}(\mu) \, d\mu$$

$$= \left[(1-\mu^2)\left\{P_n^m(\mu)\frac{dP_{n'}^{-m}(\mu)}{d\mu} - P_{n'}^{-m}(\mu)\frac{dP_n^m(\mu)}{d\mu}\right\}\right]_{-1}^{1}$$

By (9.1.9) and (9.1.18) the functions $P_n^{\pm m}(\mu)$ are finite at $\mu = \pm 1$, so that the terms on the above right vanish. This establishes (9.2.6).

It is necessary also to evaluate the normalizing integral which is

(9.2.7) $$\int_{-1}^{1} P_n^m(\mu)^2 \, d\mu = \frac{1}{n+\frac{1}{2}}\frac{(n+m)!}{(n-m)!}$$

To verify this result, we form the integral on the left by substituting for $[P_n^m(\mu)]^2$ the product of the expressions appearing in (9.1.9) and (9.1.16). Then we obtain

(9.2.8)
$$\int_{-1}^{1} P_n^m(\mu^2) \, d\mu = \frac{(-1)^m(n+m)!}{2^{2n}(n!)^2(n-m)!} \int_{-1}^{1} \left[\frac{d^{n-m}(\mu^2-1)^n}{d\mu^{n-m}}\right]$$
$$\times \left[\frac{d^{n+m}(\mu^2-1)^n}{d\mu^{n+m}}\right] d\mu$$

$$= \frac{(n+m)!}{2^{2n}(n!)^2(n-m)!} \int_{-1}^{+1} \left[\frac{d^n(\mu^2-1)^n}{d\mu^n}\right]$$
$$\times \left[\frac{d^n(\mu^2-1)^n}{d\mu^n}\right] d\mu$$

348   SPHERICAL EIGENFUNCTIONS

after integrating by parts m times and observing that all the integrated terms vanish at $\mu = \pm 1$. Integrating by parts again n times, we find further that

$$\int_{-1}^{1}\left[\frac{d^n(\mu^2-1)^n}{d\mu^n}\right]^2 d\mu = (-1)^n \int_{-1}^{1}(\mu^2-1)^n\left[\frac{d^{2n}(\mu^2-1)^n}{d\mu^{2n}}\right]d\mu$$

$$= (2n)!\,(-1)^n \int_{-1}^{1}(\mu^2-1)^n\,d\mu$$

$$= \frac{2^{2n}(n!)^2}{n+\tfrac{1}{2}}$$

by Exercise 8.1.12. Inserting this result in (9.2.8), we find the normalizing integral (9.2.7) follows at once. If we employ (9.1.18), we may combine (9.2.6) and (9.2.7) in the form

$$(9.2.9) \qquad \int_{-1}^{1} P_n^{\,m}(\mu) P_n^{\,-m}(\mu)\,d\mu = \frac{(-1)^m\,\delta_{nn'}}{n+\tfrac{1}{2}}$$

A special case of this result occurs when $m = 0$ and then we have the orthogonality and normalizing integrals for the Legendre polynomials:

$$(9.2.10) \qquad \int_{-1}^{1} P_n(\mu) P_{n'}(\mu)\,d\mu = \frac{\delta_{nn'}}{n+\tfrac{1}{2}}$$

Now let $f(\theta, \psi)$ be a function defined on the surface of the sphere $0 \leq \theta \leq \pi$, $0 \leq \psi \leq 2\pi$. At the north pole $\theta = 0$ and south pole $\theta = \pi$ the function f has values which must be independent of ψ. We may again appeal to the general expansion theorem of Section 6.4 which suggests the formal expansion

$$(9.2.11) \qquad f(\theta, \psi) = \sum_{n=0}^{\infty} \sum_{m=-n}^{n} c_{nm} P_n^{\,m}(\cos\theta) e^{im\psi}$$

To evaluate the coefficients we multiply by $P_{n'}^{\,-m'}(\cos\theta) e^{-im'\psi} \sin\theta$ and integrate. Then we obtain

$$\int_{-\pi}^{\pi} d\psi \int_{0}^{\pi} f(\theta,\psi) e^{-im'\psi} P_{n'}^{\,-m'}(\cos\theta) \sin\theta\,d\theta$$

$$= \sum_{n=0}^{\infty} \sum_{m=-n}^{n} c_{nm} \int_{-\pi}^{\pi} e^{i(m-m')\psi} d\psi \int_{0}^{\pi} P_n^{\,m}(\cos\theta) P_{n'}^{\,-m'}(\cos\theta) \sin\theta\,d\theta$$

$$= \frac{2\pi(-1)^{m'}}{n'+\tfrac{1}{2}} c_{n'm'}$$

We may state this result as follows (Hobson, Ref. 31, Sommerfeld, Ref. 54):

THEOREM 9.2.1 *Let $f(\theta, \psi)$ be defined on the sphere $0 \leq \theta \leq \pi$, $0 \leq \psi \leq 2\pi$, and let*

$$(9.2.12) \quad c_{nm} = \frac{(-1)^m(n+\tfrac{1}{2})}{2\pi} \int_{-\pi}^{\pi} d\psi \int_0^{\pi} f(\theta, \psi) e^{-im\psi} P_n^{-m}(\cos\theta) \sin\theta \, d\theta$$

Then

$$(9.2.13) \quad f(\theta, \psi) = \sum_{n=0}^{\infty} \sum_{m=-n}^{n} c_{nm} P_n^m(\cos\theta) e^{im\psi}$$

A special case of this theorem occurs when the function to be expanded does not depend on the angle ψ. It is clear then that all the coefficients c_{nm} vanish except those for which $m = 0$. The expansion now involves only the Legendre polynomials, but the result is of sufficient importance to deserve formulation as a separate theorem.

THEOREM 9.2.2 *Let $f(\theta)$ be defined for $0 \leq \theta \leq \pi$ and define*

$$(9.2.14) \quad c_n = (n+\tfrac{1}{2}) \int_0^{\pi} f(\theta) P_n(\cos\theta) \sin\theta \, d\theta$$

Then

$$(9.2.15) \quad f(\theta) = \sum_{n=0}^{\infty} c_n P_n(\cos\theta)$$

A further particular case arises when the function $f(\theta)$ is a polynomial in $\cos\theta$. Let $\mu = \cos\theta$ and suppose that $f(\theta) = g_m(\mu)$, a polynomial of degree m. The $m+1$ polynomials $P_0, P_1, \ldots, P_m$ are linearly independent and it is evidently possible to express $g_m(\mu)$ in terms of them. This means that series (9.2.15) now reduces to the *finite* sum

$$g_m(\mu) = \sum_{n=0}^{m} c_n P_n(\mu)$$

In fact,

$$(9.2.16) \quad \int_{-1}^{+1} g_m(\mu) P_n(\mu) \, d\mu = 0$$

where g_m is any polynomial of degree $m < n$.

As an illustration of the use of these expansion theorems, we shall solve the harmonic Dirichlet problem for the interior of the sphere. Suppose that $u = f(\theta, \psi)$ on the surface $r = a$; then by (9.2.3) we may propose the expansion

$$(9.2.17) \quad u = \sum_{n=0}^{\infty} \sum_{m=-n}^{n} c_{nm} \left(\frac{r}{a}\right)^n P_n^m(\cos\theta) e^{im\psi}$$

The constants c_{nm} are determined by setting $r = a$ and invoking the boundary condition:

$$f(\theta, \psi) = \sum_{n=0}^{\infty} \sum_{m=-n}^{n} c_{nm} P_n^m(\cos\theta) e^{im\psi}$$

350 SPHERICAL EIGENFUNCTIONS

This is exactly the same as (9.2.13), so we see that c_{nm} is given by (9.2.12). Inserting the value there given, we obtain from (9.2.17) the solution

$$u = \frac{1}{2\pi} \int_{-\pi}^{\pi} d\psi' \int_0^{\pi} f(\theta', \psi') \sum_{n=0}^{\infty} \sum_{m=-n}^{n} (-1)^m \left(\frac{r}{a}\right)^n (n + \tfrac{1}{2})$$
$$\times P_n^m(\cos\theta) P_n^{-m}(\cos\theta') e^{im(\psi-\psi')} \sin\theta' \, d\theta'$$

(9.2.18)

This form of the solution can be reconciled with that developed in Section 7.2 by a method outlined in Exercise 9.4.3.

EXERCISES 9.2

1. Show that

$$P_n(\mu) = \frac{1}{2^n} \frac{1}{2\pi i} \int_C \frac{(t^2 - 1)^n \, dt}{(t - \mu)^{n+1}}$$

where C is a simple closed contour enclosing the point $t = \mu$. If $-1 < h < 1$, deduce that

$$\sum_{n=0}^{\infty} h^n P_n(\mu) = \frac{i}{\pi} \int_C \frac{dt}{h(t^2 - 1) + 2(\mu - t)}$$

Hence obtain the generating function

$$\frac{1}{\sqrt{1 - 2\mu h + h^2}} = \sum_{n=0}^{\infty} h^n P_n(\mu)$$

2. By means of the generating function constructed in Exercise 9.2.1 show that

$$P_n(1) = 1 \qquad P_n(-\mu) = (-1)^n P_n(\mu)$$

3. By integration by parts show that if $0 \le m \le n$

$$\int_{-1}^{1} P_n'(\mu) P_m(\mu) \, d\mu = 1 - (-1)^{n+m}$$

By means of the expansion theorem show that

$$P_n'(\mu) = \sum_{m=0}^{n-1} (m + \tfrac{1}{2})[1 - (-1)^{n+m}] P_m(\mu)$$

Deduce that

$$P_{n+1}'(\mu) - P_{n-1}'(\mu) = (2n + 1) P_n(\mu)$$

4. Show that

$$\int_{-1}^{1} \mu P_n(\mu)^2 \, d\mu = 0$$

and that

$$\int_{-1}^{1} \mu P_n(\mu) P_m(\mu) \, d\mu = 0$$

for $m = 0, 1, 2, \ldots, n - 2$.

5. By using the integrals stated in Exercise 9.2.4 and referring to the expansion theorem show that
$$\mu P_n(\mu) = a_{n-1} P_{n-1}(\mu) + a_{n+1} P_{n+1}(\mu)$$
where a_{n-1} and a_{n+1} are suitable constants. By comparing coefficients on either side show that
$$(2n + 1)\mu P_n(\mu) = n P_{n-1}(\mu) + (n + 1) P_{n+1}(\mu)$$

6. By expressing μ^n as a linear combination of Legendre polynomials show that
$$\int_{-1}^{1} \mu^n P_n(\mu) \, d\mu = \frac{2^{n+1}(n!)^2}{(2n + 1)!}$$

7. Show that
$$\int_{-1}^{1} \frac{d\mu}{\sqrt{(1 - \mu)(1 - 2h\mu + h^2)}} = \sqrt{\frac{2}{h}} \log\left(\frac{1 + \sqrt{h}}{1 - \sqrt{h}}\right)$$
for $0 < h < 1$. Deduce that
$$\int_{-1}^{1} \frac{P_n(\mu) \, d\mu}{\sqrt{1 - \mu}} = \frac{2\sqrt{2}}{2n + 1}$$
(Use the generating function and compare coefficients of h.)

8. Show that
$$P_0(\mu) = 1 \qquad P_1(\mu) = \mu \qquad P_2(\mu) = \tfrac{1}{2}(3\mu^2 - 1)$$
$$P_3(\mu) = \tfrac{1}{2}(5\mu^3 - 3\mu)$$
Express $6\mu^2 + 10\mu^3$ as a linear combination of Legendre polynomials. Hence construct the harmonic function for the interior of the sphere $0 \leq r \leq a$ which reduces to $6\cos^2\theta + 10\cos^3\theta$ on $r = a$. Find also the harmonic function for the exterior region $r \geq a$ which reduces to $6\cos^2\theta + 10\cos^3\theta$ on $r = a$ and which vanishes at infinity.

9. Find the harmonic function u in the region $0 < a \leq r \leq b$ between two concentric spheres such that $u(a, \theta) = \cos^3\theta$ and $u(b, \theta) = \cos^2\theta$.

10. Solve the Neumann problem for the interior of the sphere $0 \leq r \leq a$, $0 \leq \theta \leq \pi$, $0 \leq \psi \leq 2\pi$, with the boundary condition $\partial u / \partial n = f(\theta, \psi)$. Show that the problem is correctly formulated if
$$\int_0^{2\pi} \int_0^{\pi} f(\theta, \psi) \sin\theta \, d\theta \, d\psi = 0$$

11. *The angular momentum operator of quantum mechanics.* Express in spherical polar coordinates the operator $\mathbf{L} = \mathbf{x} \times \mathbf{p}$, where $\mathbf{p} = -i\hbar\nabla$, and show that
$$L^2 = L_1^2 + L_2^2 + L_3^2 = -\hbar^2 \left[\frac{1}{\sin\theta} \frac{\partial}{\partial \theta}\left(\sin\theta \frac{\partial}{\partial \theta}\right) + \frac{1}{\sin^2\theta} \frac{\partial^2}{\partial \varphi^2} \right]$$

12. Show that the eigenvalues of L^2 are $l(l + 1)\hbar^2$, where l is an integer, and that the eigenfunctions are surface harmonics of degree l. State the dimension of the eigenspaces.

9.3 EIGENFUNCTIONS FOR THE SOLID SPHERE

In the preceding section we constructed the eigenfunctions on the surface of a sphere and developed the corresponding expansion theorems. We now

352   SPHERICAL EIGENFUNCTIONS

consider the eigenfunctions of the solid sphere $0 \leq r \leq a$, $0 \leq \theta \leq \pi$, $0 \leq \psi \leq 2\pi$. These satisfy the equation of Helmholtz

$$(9.3.1) \quad u_{rr} + \frac{2}{r} u_r + \frac{1}{r^2 \sin\theta} \frac{\partial}{\partial \theta} (\sin\theta u_\theta) + \frac{1}{r^2 \sin^2\theta} u_{\psi\psi} + k^2 u = 0$$

Again we apply the method of separation of variables and seek solutions of the type $u = R(r)\Theta(\theta)\Psi(\psi)$. This process yields the equations

$$\Psi'' + m^2 \Psi = 0$$

$$(9.3.2) \quad R'' + \frac{2}{r} R' + \left[k^2 - \frac{s(s+1)}{r^2}\right] R = 0$$

$$\frac{1}{\sin\theta} \frac{d}{d\theta} (\sin\theta \Theta') + \left[s(s+1) - \frac{m^2}{\sin^2\theta}\right]\Theta = 0$$

The first and last of these equations are exactly those discussed in Section 9.2. To obtain solutions, finite and single-valued, we must as in that section take m and s to be integers. Furthermore (9.3.2) reduces, by means of the substitution $R = r^{-1/2}\omega$, to

$$\omega_{rr} + \frac{1}{r} \omega_r + \left[k^2 - \frac{(n+\tfrac{1}{2})^2}{r^2}\right]\omega = 0$$

This is Bessel's equation of order $n + \tfrac{1}{2}$ and has basic solutions $J_{\pm(n+1/2)}(kr)$. Unlike the Bessel functions of general order, these functions may be expressed as *finite* series involving the trigonometric functions $\cos kr$ and $\sin kr$. Let us first verify this statement for the function $J_{1/2}(kr)$. By (8.1.18) we see that

$$(9.3.3) \quad J_{1/2}(z) = \sqrt{\frac{z}{2}} \sum_{m=0}^{\infty} \frac{1}{m!\,\Gamma(m+\tfrac{3}{2})} \left(\frac{-z^2}{4}\right)^m$$

Now by Exercise 8.1.8,

$$m!\,\Gamma(m+\tfrac{3}{2}) = \sqrt{\pi}\,2^{-2m-1}(2m+1)!$$

so that (9.3.3) becomes, after slight rearrangement,

$$J_{1/2}(z) = \sqrt{\frac{2}{\pi z}} \sum_{m=0}^{\infty} \frac{(-1)^m z^{2m+1}}{(2m+1)!}$$

The series on the right is simply that of $\sin z$, so that

$$(9.3.4) \quad J_{1/2}(z) = \sqrt{\frac{2}{\pi z}} \sin z$$

If we refer to the recurrence relation (Exercise 8.1.1),

$$(9.3.5) \quad zJ_{n+1}(z) = nJ_n(z) - zJ'_n(z)$$

which can be shown to be valid for any value of n, we see by repeated application of this formula that we may express every $J_{n+\frac{1}{2}}(z)$ in terms of the elementary functions. To do this we write (9.3.5) as

$$z^{-(n+1)}J_{n+1}(z) = \frac{-d}{z\,dz}[z^{-n}J_n(z)]$$

By induction we deduce that, if m is a positive integer,

$$z^{-(n+m)}J_{n+m}(z) = (-1)^m\left(\frac{d}{z\,dz}\right)^m[z^{-n}J_n(z)]$$

so that, on setting $n = \frac{1}{2}$ and using (9.3.4), we find

(9.3.6) $$z^{-(m+\frac{1}{2})}J_{m+\frac{1}{2}}(z) = \sqrt{\frac{2}{\pi}}(-1)^m\left(\frac{d}{z\,dz}\right)^m\frac{\sin z}{z}$$

This is the required formula.

We return now to the eigenfunctions of the sphere which must take the form

(9.3.7) $$\frac{1}{\sqrt{r}}J_{n+\frac{1}{2}}(kr)P_n^m(\cos\theta)e^{\pm im\psi} \qquad -n \leq m \leq n$$

For the vibration of air inside a rigid sphere the proper boundary condition is $\partial u/\partial n = 0$ on $r = a$. Here the eigenvalues are the roots of the equation

$$\frac{d}{dr}\left[\frac{1}{\sqrt{r}}J_{n+\frac{1}{2}}(kr)\right]_{r=a} = 0$$

Explicitly,

(9.3.8) $$2kaJ'_{n+\frac{1}{2}}(ka) - J_{n+\frac{1}{2}}(ka) = 0$$

Let $k_{n1}, k_{n2}, \ldots$ denote the positive roots of this equation arranged, as is customary, in ascending order of magnitude. The corresponding eigenfunctions are

(9.3.9) $$u_{nmp} = a_{nmp}\frac{1}{\sqrt{r}}J_{n+\frac{1}{2}}(k_{np}r)P_n^m(\cos\theta)\begin{cases}\cos m\psi \\ \sin m\psi\end{cases} \qquad -n \leq m \leq n$$

It is instructive to study the location of the zeros of the gradient ∇u_{nmp}. At the points (r, θ, ψ) corresponding to these zeros, the air will remain at rest. Such points are called *nodes*. To find them, we must equate to zero the derivative of each of the three factors in (9.3.9). Let us consider first the r factor which yields

(9.3.10) $$\frac{d}{dr}\left[\frac{1}{\sqrt{r}}J_{n+\frac{1}{2}}(k_{np}r)\right] = 0$$

This is essentially the same equation as (9.3.8). Since k_{np} is the pth root of (9.3.8), it is clear that (9.3.10) will vanish on the p spheres of radii $r = ak_{nq}/k_{np}$

for $q = 1, 2, \ldots, p$. For the θ factor in (9.3.9), we recall that for $m \neq 0$ the function $P_n{}^m(\cos \theta)$ of Exercise 9.1.1 vanishes at the poles $\theta = 0, \pi$, and on $n - m$ circles arranged symmetrically about the great circle $\theta = \pi/2$. Thus $(d/d\theta)P_n{}^m(\cos \theta)$ also vanishes on $n - m + 1$ such circles, and possibly at the poles as well. Finally the ψ factors in (9.3.9) possess vanishing derivative on m great circles through the poles (Figure 9.2).

By drawing all these circles on each sphere we may divide the surface of the latter into a series of regions which are known as *tessera*. When $m = 0$, the circles through the poles are absent and the surface of each sphere is divided

Figure 9.2 Nodal lines of a tesseral harmonic, $n = 8, m = 5$.

into *zones* by parallel circles. The corresponding eigenfunctions then involve the Legendre polynomials $P_n(\cos \theta)$ which are sometimes called *zonal harmonics*.

The simplest eigenfunction occurs when $n = m = 0$, in which case the vibrations are entirely radial:

$$u = \frac{1}{\sqrt{r}} J_{1/2}(kr) = \sqrt{\frac{2}{\pi k}} \frac{\sin kr}{r}$$

by (9.3.4). The eigenvalue equation (9.3.8) now reduces to $\tan ka = ka$ of which the smallest zero is approximately $ka = 1.43\pi$.

We shall conclude this section by stating the normalized eigenfunctions for the solid sphere and use them to form the corresponding Green's function of the diffusion equation. To find the normalizing constant a_{nmp} appearing in (9.3.9), we integrate u_{nmp}^2 over the sphere and equate the result to unity. Thus,

$$a_{nmp}^2 \int_0^a r J_{n+1/2}(k_{np}r)^2 \, dr \int_0^\pi P_n{}^m(\cos \theta)^2 \sin \theta \, d\theta \int_0^{2\pi} \begin{Bmatrix} \cos^2 m\psi \\ \sin^2 m\psi \end{Bmatrix} d\psi = 1$$

The first integral occurring here is obtained from (8.2.9) while the second is stated in (9.2.7). The ψ integral equals π if $m \neq 0$. If we take for simplicity the boundary condition to be that of vanishing on $r = a$, we find the normalized eigenfunctions to be

(9.3.11) $$\frac{1}{a}\sqrt{\frac{\varepsilon_m(n+\frac{1}{2})(n-m)!}{\pi(n+m)!}} \frac{J_{n+1/2}(k_{np}r)}{\sqrt{r}\, J'_{n+1/2}(k_{np}a)} P_n^m(\cos\theta) \begin{cases}\cos m\psi \\ \sin m\psi\end{cases}$$

where $\varepsilon_0 = 1$, $\varepsilon_m = 2$ for $m \geq 1$ and $J_{n+1/2}(k_{np}a) = 0$.

If these eigenfunctions are inserted into the bilinear series (7.5.11), we may find the eigenfunction expansion of the Green's function of the diffusion equation for the interior of the sphere. With the Dirichlet boundary condition we find that

$$K(r, r', \theta, \theta', \psi, \psi', t, t') = \frac{1}{2\pi a^2 r} \sum_{m,n,p} \frac{\varepsilon_m(2n+1)(n-m)!}{(n+m)!}$$

$$\times \frac{J_{n+1/2}(k_{np}r)J_{n+1/2}(k_{np}r')P_n^m(\cos\theta)P_n^m(\cos\theta')\cos m(\psi - \psi')}{J'_{n+1/2}(k_{np}a)^2}$$

$$\times e^{-k_{np}^2(t-t')} H(t - t')$$

where k_{np} are the positive zeros of $J_{n+1/2}(ka)$.

EXERCISES 9.3

1. Show that the Green's function of (9.3.2b) for the interval $0 < a \leq r < \infty$, vanishing on $r = a$ and satisfying a radiation condition at infinity, is given by

$$G(r, r') = \frac{i\pi}{2\sqrt{rr'}} \frac{H^{(1)}_{s+1/2}(kr')}{H^{(1)}_{s+1/2}(ka)} [H^{(1)}_{s+1/2}(ka)J_{s+1/2}(kr) - J_{s+1/2}(ka)H^{(1)}_{s+1/2}(kr)]$$

for $0 < a \leq r < r' < \infty$. Interchange r, r' if $0 < a \leq r' < r < \infty$. Deduce that the Green's function of (9.3.2b) for the whole range $0 \leq r < \infty$ and satisfying a radiation condition at infinity is given with $0 \leq r < r' < \infty$ by

$$G(r, r') = \frac{i\pi}{2\sqrt{rr'}} H^{(1)}_{s+1/2}(kr')J_{s+1/2}(kr) = G(r', r)$$

2. Construct the eigenfunction solution of the wave equation $\Delta u = u_{tt}$ in the region $r \geq a$, $0 \leq \theta \leq \pi$, $0 \leq \psi \leq 2\pi$ with the boundary condition $u = f(\theta, \psi)$ on $r = a$ and the initial conditions $u = 0$, $u_t = g(r, \theta, \psi)$ for $t = 0$.

3. Construct the normalized eigenfunctions of the region $0 < a \leq r \leq b$, $0 \leq \theta \leq \pi$, $0 \leq \psi \leq 2\pi$ with the boundary conditions $u = 0$ on $r = a$ and $\partial u/\partial n = 0$ on $r = b$. Hence write down the harmonic Green's function for the same region and same boundary conditions.

4. If k is not an eigenvalue, $J_{n+1/2}(ka) \neq 0$, show that the solution of (9.3.1) in the sphere $0 \leq r \leq a$, $0 \leq \theta \leq \pi$, with the axially symmetric boundary condition $u(a, \theta, \psi) = f(\theta)$, is given by

$$u = \sqrt{\frac{a}{r}} \int_0^\pi f(\theta') \sin\theta'\, d\theta' \sum_{n=0}^\infty (n + \tfrac{1}{2}) \frac{J_{n+1/2}(kr)P_n(\cos\theta)P_n(\cos\theta')}{J_{n+1/2}(ka)}$$

5. Show that the Green's function of the equation $\Delta G - k^2 G = 0$ in the exterior region $0 < a \leq r < \infty$, $0 \leq \theta \leq \pi$ with singularity on $\theta = 0$ and vanishing for $r = a$ is given for $0 < a \leq r' \leq r < \infty$ by

$$G = \frac{1}{2\pi\sqrt{rr'}} \sum_{n=0}^{\infty} (n + \tfrac{1}{2}) P_n(\cos\theta) \frac{K_{n+\frac{1}{2}}(kr)}{K_{n+\frac{1}{2}}(ka)}$$
$$\times [I_{n+\frac{1}{2}}(kr')K_{n+\frac{1}{2}}(ka) - I_{n+\frac{1}{2}}(ka)K_{n+\frac{1}{2}}(kr')]$$

6. Show that the function satisfying the equation $\Delta u = k^2 u$ in the exterior region $0 < a \leq r < \infty$, $0 \leq \theta \leq \pi$ with $u = f(\theta)$ on $r = a$ and u zero at infinity is given by

$$u = \sqrt{\frac{a}{r}} \int_0^\pi f(\theta') \sin\theta'\, d\theta' \sum_{n=0}^{\infty} (n + \tfrac{1}{2}) \frac{K_{n+\frac{1}{2}}(kr) P_n(\cos\theta) P_n(\cos\theta')}{K_{n+\frac{1}{2}}(ka)}$$

7. *Motion in a central field of force.* Construct the Schrödinger equation for an electron in a force field of potential $V(r)$, where r is distance from the nuclear center. Show that the energy levels are eigenvalues E_n of the radial equation

$$U''(r) + \frac{2m}{\hbar^2}\left[E - V(r) - \frac{l(l+1)\hbar^2}{2mr^2}\right] U(r) = 0$$

while the wave functions are

$$r^{-1} U(r) Y_l(\theta, \varphi)$$

8. Show that the Laguerre equation

$$xy'' + (1 - x)y' + \lambda y = 0$$

has polynomial solutions $L_n(x)$ only when λ is a nonnegative integer. Show that $L_n^k(x) = D^k L_n(x)$ is a solution of

$$xy'' + (k + 1 - x)y' + (n - k)y = 0$$

9. *Hydrogen atom.* If $U = rR(r)$ and $V(r) = -e^2/r$, show that the radial equation becomes

$$xR'' + 2R' + \left[n - \frac{x}{4} - \frac{l(l+1)}{x}\right] R = 0$$

Show that

$$R = e^{-x/2} x^l L_{l+n}^{2l+1}(x)$$

is a solution of this radial equation, where $x = (2me^2/n\hbar^2)r$, and deduce that $E_n = -\tfrac{1}{2}me^4/n^2\hbar^2$ are energy levels of the hydrogen atom.

9.4 DIFFRACTION BY A SPHERE: ADDITION THEOREM

Let us construct the Green's function G for the equation $\Delta u + k^2 u = 0$ in the exterior region $r \geq a$, $0 \leq \theta \leq \pi$, $-\pi \leq \psi \leq \pi$, with the condition $u = 0$

DIFFRACTION BY A SPHERE: ADDITION THEOREM

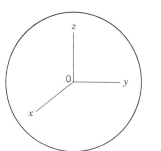

Figure 9.3 Diffraction by a sphere.

on the sphere $r = a$ (Figure 9.3). At infinity we are required to impose a condition of radiation as explained in Section 7.2. We write the formal equation

(9.4.1) $\quad G_{rr} + \dfrac{2}{r} G_r + \dfrac{1}{r^2 \sin\theta} \dfrac{\partial}{\partial\theta}(\sin\theta G_\theta) + \dfrac{1}{r^2 \sin^2\theta} G_{\psi\psi} + k^2 G$

$$= -\frac{\delta(r - r')\delta(\theta - \theta')\delta(\psi - \psi')}{r^2 \sin\theta}$$

for the Green's function $G(r, \theta, \psi; r', \theta', \psi')$. By analogy with (9.2.12) and (9.2.13), we set

(9.4.2) $\quad G(r, \theta, \psi; r', \theta', \psi') = \sum\limits_{n=0}^{\infty} \sum\limits_{m=-n}^{n} G_{nm}(r; r', \theta', \psi') e^{im\psi} P_n{}^m(\cos\theta)$

where

(9.4.3)
$$G_{nm}(r; r', \theta', \psi') = \frac{(-1)^m (2n+1)}{4\pi} \int_{-\pi}^{\pi} e^{-im\psi}\,d\psi$$
$$\times \int_0^{\pi} G(r, \theta, \psi; r', \theta', \psi') P_n{}^{-m}(\cos\theta) \sin\theta\, d\theta$$

Applying to (9.4.1) the integral transform appearing on the right-hand side of the preceding equation, we obtain for G_{nm} the equation

$$\frac{d^2 G_{nm}}{dr^2} + \frac{2}{r}\frac{dG_{nm}}{dr} + \left[k^2 - \frac{n(n+1)}{r^2}\right] G_{nm}$$
$$= -\frac{(-1)^m (2n+1)}{4\pi r^2} e^{-im\psi'} P_n{}^{-m}(\cos\theta')\delta(r - r')$$

This equation defines a Green's function for the radial operator on the left-hand side. The range of interest is $a < r < \infty$ with $G_{nm} = 0$ on $r = a$ and with the radiation requirement at infinity. By Exercise 9.3.1 we see that

$$G_{nm}(r, r', \theta', \psi') = \frac{(-1)^m i(n + \tfrac{1}{2})}{4\sqrt{rr'}} e^{-im\psi'} P_n{}^{-m}(\cos\theta')$$
$$\times \frac{H^{(1)}_{n+1/2}(kr)}{H^{(1)}_{n+1/2}(ka)} [H^{(1)}_{n+1/2}(ka) J_{n+1/2}(kr') - J_{n+1/2}(ka) H^{(1)}_{n+1/2}(kr')]$$

This expression applies for $0 < a \leq r' \leq r < \infty$; otherwise we interchange r, r'. On inserting this result in (9.4.2), we obtain finally

$$G = \frac{i}{8\sqrt{rr'}} \sum_{n=0}^{\infty} \sum_{m=-n}^{+n} (-1)^m (2n+1) e^{im(\psi - \psi')} P_n^m(\cos\theta) P_n^{-m}(\cos\theta')$$

(9.4.4)
$$\times \frac{H_{n+1/2}^{(1)}(kr)}{H_{n+1/2}^{(1)}(ka)} [H_{n+1/2}^{(1)}(ka) J_{n+1/2}(kr') - J_{n+1/2}(ka) H_{n+1/2}^{(1)}(kr')]$$

for $a \leq r' \leq r < \infty$.

From the general expression (9.4.4) we can deduce several important results. As it stands, (9.4.4) solves the problem of diffraction by the sphere of a source located at the point (r', θ', ψ'). Let us set $\theta' = 0$ in this result; this merely corresponds to placing the source on the z-axis. By (9.1.9) and (9.1.18) we note that $P_n^m(1) = 0$ for all integral $m \neq 0$, so that the only surviving terms in (9.4.4) are those for which $m = 0$. Since $P_n(1) = 1$ (Exercise 9.2.2), (9.4.4) now becomes

(9.4.5) $$G_0 = \frac{i}{8\sqrt{rr'}} \sum_{n=0}^{\infty} (2n+1) P_n(\cos\theta) \frac{H_{n+1/2}^{(1)}(kr)}{H_{n+1/2}^{(1)}(ka)}$$

$$\times [H_{n+1/2}^{(1)}(ka) J_{n+1/2}(kr') - J_{n+1/2}(ka) H_{n+1/2}^{(1)}(kr')]$$

for $a \leq r' \leq r < \infty$. This last expression involves only the magnitudes of the radius vectors $\mathbf{r}, \mathbf{r}'$ and the angle θ between them. If we restore the source to its original position (r', θ', ψ'), this angle becomes θ_0, where

(9.4.6) $$\cos\theta_0 = \cos\theta \cos\theta' + \sin\theta \sin\theta' \cos(\psi - \psi')$$

which we see immediately by taking the scalar product of the vectors

$$\mathbf{r} = (r \sin\theta \cos\psi, r \sin\theta \sin\psi, r \cos\theta)$$
$$\mathbf{r}' = (r' \sin\theta' \cos\psi', r' \sin\theta' \sin\psi', r' \cos\theta')$$

Thus on replacing θ by θ_0 in (9.4.5), we find

(9.4.7) $$G = \frac{i}{8\sqrt{rr'}} \sum_{n=0}^{\infty} (2n+1) P_n(\cos\theta_0) \frac{H_{n+1/2}^{(1)}(kr)}{H_{n+1/2}^{(1)}(ka)}$$

$$\times [H_{n+1/2}^{(1)}(ka) J_{n+1/2}(kr') - H_{n+1/2}^{(1)}(kr') J_{n+1/2}(ka)]$$

for $a \leq r' \leq r < \infty$. This expression must be identical with (9.4.4), so that

(9.4.8) $$P_n(\cos\theta_0) = \sum_{m=-n}^{n} (-1)^m e^{im(\psi - \psi')} P_n^m(\cos\theta) P_n^{-m}(\cos\theta')$$

This relation is known as the *addition theorem*. It may be expressed in real form by noting from (9.1.18) that, for $m > 0$,

$$P_n^m(\cos\theta) P_n^{-m}(\cos\theta') = \frac{(-1)^m (n-m)!}{(n+m)!} P_n^m(\cos\theta) P_n^m(\cos\theta')$$

so that

$$P_n(\cos \theta_0) = P_n(\cos \theta)P_n(\cos \theta')$$
(9.4.9)
$$+ 2\sum_{m=1}^{n} \frac{(n-m)!}{(n+m)!} P_n^m(\cos \theta)P_n^m(\cos \theta') \cos m(\psi - \psi')$$

Next we shall deduce the Green's function G_1 for the whole of space by letting the radius a of the sphere tend to zero in (9.4.7). Now as $a \to 0$ each $J_{n+1/2}(ka)$ tends to zero whilst each $H_{n+1/2}^{(1)}(ka) \to \infty$. Thus

(9.4.10) $$G_1 = \frac{i}{8\sqrt{rr'}} \sum_{n=0}^{\infty} (2n+1)P_n(\cos \theta_0)H_{n+1/2}^{(1)}(kr)J_{n+1/2}(kr')$$

for $0 \leq r' \leq r < \infty$. The preceding expansion can be summed in the following way. Let us place the source point at the origin by letting $r' \to 0$ in (9.4.10). Each factor $J_{n+1/2}(kr')$ tends to zero like $(kr')^{n+1/2}$ so that in the limit the only term left is the first. Since $P_0(\cos \theta_0) \equiv 1$ (see Exercise 9.2.8) and

$$J_{1/2}(kr') = \sqrt{\frac{2}{\pi kr'}} \sin kr'$$

while by (8.5.11), (8.1.22), (9.3.4), and Exercise 8.1.1 we have

$$H_{1/2}^{(1)}(kr) = -i\sqrt{\frac{2}{\pi kr}} e^{ikr}$$

we can deduce that G_1 has the limiting value

(9.4.11) $$\frac{e^{ikr}}{4\pi r}$$

This expression depends, of course, only on the distance separating the source and field points. If we restore the source to its original position $\mathbf{r}'$, this distance becomes equal to R, where

(9.4.12) $$R = |\mathbf{r} - \mathbf{r}'| = \sqrt{r^2 + r'^2 - 2rr' \cos \theta_0}$$

The Green's function (9.4.11) now becomes

(9.4.13) $$G_1 = \frac{e^{ikR}}{4\pi R}$$

By combining (9.4.10) and (9.4.13), we obtain for $0 < r' < r$ the identity

(9.4.14) $$\frac{e^{ikR}}{4\pi R} = \frac{i}{8\sqrt{rr'}} \sum_{n=0}^{\infty} (2n+1)P_n(\cos \theta_0)H_{n+1/2}^{(1)}(kr)J_{n+1/2}(kr')$$

This relation expands the fundamental singular solution of the Helmholtz equation as a series of spherical eigenfunctions. By separating the real and imaginary parts of (9.4.14), we obtain further identities (Exercise 9.4.4). If we insert (9.4.14) into (9.4.7), we can express the latter equation as

(9.4.15)
$$G = \frac{e^{ikR}}{4\pi R} + \frac{1}{8i\sqrt{rr'}} \sum_{n=0}^{\infty} (2n+1) P_n(\cos\theta_0) \frac{H^{(1)}_{n+\frac{1}{2}}(kr) H^{(1)}_{n+\frac{1}{2}}(kr') J_{n+\frac{1}{2}}(ka)}{H^{(1)}_{n+\frac{1}{2}}(ka)}$$

In this form the series on the right represents the diffracted part of the field due to the unit source at (r', θ', ψ').

The solution of the problem of diffraction of a *plane* wave by a sphere can be obtained from that of diffraction of a source by a sphere by allowing the position and intensity of the source to tend to infinity in a prescribed manner which we now determine. Suppose that r/r' is small, then we may by (9.4.12) write $R \sim r' - r\cos\theta_0$ and hence deduce that

(9.4.16)
$$e^{-ikr\cos\theta_0} = \lim_{r' \to \infty} \left(4\pi r' e^{-ikr'} \frac{e^{ikR}}{4\pi R} \right)$$

The expression on the left represents a plane wave approaching in the direction θ', ψ' and we can obtain such a wave by taking a source of intensity $4\pi r' e^{-ikr'}$ at the point (r', θ', ψ') and letting $r' \to \infty$. Thus we modify (9.4.15) by multiplying by the stated intensity factor and let $r' \to \infty$. Recalling the asymptotic relation (8.5.12)

(9.4.17)
$$H^{(1)}_{n+\frac{1}{2}}(kr') \sim \sqrt{\frac{2}{\pi kr'}} e^{i[kr' - \overline{n+1}(\pi/2)]}$$

we find the wave function

(9.4.18) $$u = e^{-ikr\cos\theta_0} - \sqrt{\frac{2\pi}{kr}} \sum_{n=0}^{\infty} \frac{(n+\frac{1}{2}) P_n(\cos\theta_0) J_{n+\frac{1}{2}}(ka) H^{(1)}_{n+\frac{1}{2}}(kr) e^{-in\pi/2}}{H^{(1)}_{n+\frac{1}{2}}(ka)}$$

This solves the problem of diffraction of a plane wave by a sphere, the series on the right representing the diffracted part of the wave. We may also obtain a further relation by recalling that the boundary condition imposed on $r = a$ was that of vanishing. Thus on setting $r = a$ in (9.4.18), the function u must vanish and we obtain

$$e^{-ika\cos\theta_0} = \sqrt{\frac{2\pi}{ka}} \sum_{n=0}^{\infty} (n+\tfrac{1}{2}) P_n(\cos\theta_0) J_{n+\frac{1}{2}}(ka) e^{-in\pi/2}$$

Replacing a by r in this formula and omitting the subscript zero on θ, we deduce the identity

(9.4.19)
$$e^{-ikr\cos\theta} = \sqrt{\frac{2\pi}{kr}} \sum_{n=0}^{\infty} (n+\tfrac{1}{2}) P_n(\cos\theta) J_{n+\frac{1}{2}}(kr) e^{-in\pi/2}$$

This identity expands the plane wave function on the left as a series of spherical eigenfunctions.

Finally, by taking the limit as $k \to 0$ in certain of the preceding results, we can deduce corresponding results for the Laplace equation. For example, if we use (9.4.5), we can derive an expression for the harmonic Green's function for the domain external to a sphere with the condition of vanishing on the sphere (see Exercise 9.5.1). We shall carry out this procedure on (9.4.14) only. We recall that, when x is small enough and s positive,

$$J_s(x) \sim \frac{1}{\Gamma(s+1)}\left(\frac{x}{2}\right)^s \qquad H_s^{(1)}(x) \sim \frac{-i}{\pi}\Gamma(s)\left(\frac{2}{x}\right)^s$$

and find by letting $k \to 0$ in (9.4.14) that

$$\frac{1}{4\pi R} = \frac{1}{4\pi r}\sum_{n=0}^{\infty}\left(\frac{r'}{r}\right)^n P_n(\cos\theta_0) \qquad 0 < r' < r$$

Therefore,

(9.4.20) $$\frac{1}{\sqrt{r^2 + r'^2 - 2rr'\cos\theta}} = \begin{cases} \dfrac{1}{r}\sum_{n=0}^{\infty}\left(\dfrac{r'}{r}\right)^n P_n(\cos\theta) & r' < r \\ \dfrac{1}{r'}\sum_{n=0}^{\infty}\left(\dfrac{r}{r'}\right)^n P_n(\cos\theta) & r' > r \end{cases}$$

This relation is often used to define the Legendre polynomials and from it we may derive many of their properties. (see Exercise 9.2.1) The expression on the left of (9.4.20) is known as a *generating* function (Erdelyi, Ref. 18).

EXERCISES 9.4

1. Starting with the relation

$$(1 + h^2 - 2h\cos\theta_0)^{-1/2} = \sum_{n=0}^{\infty} h^n P_n(\cos\theta_0)$$

show by differentiation with respect to h that, if $|h| < 1$,

$$\frac{1 - h^2}{(1 + h^2 - 2h\cos\theta_0)^{3/2}} = \sum_{n=0}^{\infty} (2n + 1)h^n P_n(\cos\theta_0)$$

By using the addition theorem deduce that

$$\int_{-\pi}^{\pi} \frac{(1 - h^2)\,d\psi}{(1 + h^2 - 2h\cos\theta_0)^{3/2}} = 2\pi\sum_{n=0}^{\infty}(2n + 1)h^n P_n(\cos\theta)P_n(\cos\theta')$$

where $\cos\theta_0 = \cos\theta\cos\theta' + \sin\theta\sin\theta'\cos(\psi - \psi')$.

2. Show that the solution of the axially symmetric harmonic Dirichlet problem for the sphere $0 \le r \le a, 0 \le \theta \le \pi, -\pi \le \psi \le \pi$ with the boundary condition $\varphi(a, \theta) = f(\theta)$ is

$$\varphi(r, \theta) = \int_0^{\pi} f(\theta')\sin\theta'\,d\theta'\sum_{n=0}^{\infty}(n + \tfrac{1}{2})\left(\frac{r}{a}\right)^n P_n(\cos\theta)P_n(\cos\theta')$$

3. By means of the addition theorem (9.4.8) and the second result stated in Exercise 9.4.1, show that (9.2.18) can be put in the form

$$u = \frac{a(a^2 - r^2)}{4\pi}\int_{-\pi}^{\pi} d\psi' \int_0^{\pi} \frac{f(\theta', \psi')\sin\theta'\,d\theta'}{(r^2 + a^2 - 2ar\cos\theta_0)^{3/2}}$$

4. Deduce from (9.4.14) the expansions

$$\frac{\cos kR}{\pi R} = \frac{-1}{\sqrt{rr'}} \sum_{n=0}^{\infty} (n + \tfrac{1}{2}) P_n(\cos \theta_0) J_{n+1/2}(kr') Y_{n+1/2}(kr)$$

for $0 < r' < r$,

$$\frac{\sin kR}{\pi R} = \frac{1}{\sqrt{rr'}} \sum_{n=0}^{\infty} (n + \tfrac{1}{2}) P_n(\cos \theta_0) J_{n+1/2}(kr) J_{n+1/2}(kr')$$

where R is defined by (9.4.12).

5. Show by applying Theorem 9.2.1 that the harmonic Green's function in the sphere $0 \le r \le a$ vanishing on the surface is given by

$$G = \frac{-1}{4\pi a} \sum_{n=0}^{\infty} \left(\frac{r}{a}\right)^n \left[\left(\frac{r'}{a}\right)^n - \left(\frac{a}{r'}\right)^{n+1}\right] P_n(\cos \theta_0)$$

for $0 \le r \le r' \le a$. By letting $a \to \infty$, verify the generating series (9.4.20).

6. Apply Theorem 9.2.1 to show that the fundamental singular solution of the equation $\Delta G - k^2 G = -\delta(\mathbf{r} - \mathbf{r}')$ for the whole of space is given by

$$G = \frac{1}{2\pi\sqrt{rr'}} \sum_{n=0}^{\infty} (n + \tfrac{1}{2}) I_{n+1/2}(kr') K_{n+1/2}(kr) P_n(\cos \theta_0)$$

for $0 < r' < r < \infty$. By letting $r' \to 0$, deduce the identity

$$\frac{e^{-kR}}{R} = \frac{1}{\sqrt{rr'}} \sum_{n=0}^{\infty} (2n+1) I_{n+1/2}(kr') K_{n+1/2}(kr) P_n(\cos \theta_0)$$

where R is defined by (9.4.12) and $0 < r' < r < \infty$. By letting $r' \to \infty$, show also that

$$e^{kr \cos \theta_0} = \sqrt{\frac{2\pi}{kr}} \sum_{n=0}^{\infty} (n + \tfrac{1}{2}) I_{n+1/2}(kr) P_n(\cos \theta_0)$$

9.5 INTERIOR AND EXTERIOR EXPANSIONS

In Section 9.1 we saw that there are exactly $2n + 1$ distinct functions $P_n^m(\cos \theta) e^{\pm i m \psi}$ for each positive integer n. The corresponding harmonic functions are given by $r^n P_n^m(\cos \theta) e^{\pm i m \psi}$ and $r^{-n-1} P_n^m(\cos \theta) e^{\pm i m \psi}$. Let us for the moment consider only the first of these two sets of solutions. If we revert to Cartesian coordinates x, y, z by means of the relations (9.1.2), it is clear that we will obtain $2n + 1$ harmonic polynomials homogeneous in x, y, z of degree n. Furthermore there are no more than $2n + 1$ such independent harmonic polynomials. For such a polynomial necessarily corresponds to a surface harmonic of degree n bounded on the unit sphere, and we know by Exercise 9.1.2 that there are only $2n + 1$ such surface harmonics. Thus if $Y_n(\theta, \psi)$ denotes *any* surface harmonic of degree n, it must be expressible as a linear combination of the $2n + 1$ fundamental surface harmonics:

(9.5.1) $$Y_n(\theta, \psi) = \sum_{m=-n}^{+n} a_m P_n^m(\cos \theta) e^{i m \psi}$$

INTERIOR AND EXTERIOR EXPANSIONS 363

Before proceeding, we note that we have already encountered a particular example of this type of expansion in the addition theorem [equation (9.4.8)]:

$$(9.5.2) \qquad P_n(\cos \theta_0) = \sum_{m=-n}^{n} (-1)^m e^{im(\psi - \psi')} P_n{}^m(\cos \theta) P_n{}^{-m}(\cos \theta')$$

where $\cos \theta_0$ is defined by (9.4.6).

Let us denote the homogeneous solid harmonic of degree n associated with (9.5.1) by $V_n(x, y, z)$ so that

$$V_n(x, y, z) = r^n Y_n(\theta, \psi)$$

We know already that a second solid harmonic is given by $r^{-n-1} Y_n(\theta, \psi)$, and this is just $r^{-2n-1} V_n(x, y, z)$. Thus we have the following theorem of Kelvin:

THEOREM 9.5.1 *If $V_n(x, y, z)$ is a solid spherical harmonic of degree n, then $r^{-2n-1} V_n(x, y, z)$ is a solid spherical harmonic of degree $-(n+1)$.*

This result is actually a particular instance of a more general proposition.

THEOREM 9.5.2 *If*

$$U(x, y, z) = u(r, \theta, \psi)$$

is harmonic, so is

$$(9.5.3) \qquad V(x, y, z) \equiv \frac{1}{r} U\left(\frac{xa^2}{r^2}, \frac{ya^2}{r^2}, \frac{za^2}{r^2}\right) = \frac{1}{r} u\left(\frac{a^2}{r}, \theta, \psi\right)$$

where a is any constant.

In the proof we shall confine attention to potentials which can be expanded in the form (9.2.3), which we rewrite as

$$(9.5.4) \qquad u(r, \theta, \psi) = \sum_{n=0}^{\infty} \sum_{m=-n}^{n} (a_{nm} r^n + b_{nm} r^{-n-1}) P_n{}^m(\cos \theta) e^{im\psi}$$

From this expansion we see that

$$(9.5.5) \quad v(r, \theta, \psi) = \frac{1}{r} u\left(\frac{a^2}{r}, \theta, \psi\right) = \sum_{n=0}^{\infty} \sum_{m=-n}^{n} (a'_{nm} r^{-n-1} + b'_{nm} r^n) P_n{}^m(\cos \theta) e^{im\psi}$$

where a'_{nm} and b'_{nm} are further constants.

This is of the general form (9.5.4) and is therefore harmonic. The transformation from (r, θ, ψ) to $(a^2/r, \theta, \psi)$ leaves the angles θ and ψ unaltered but multiplies the vector **r** by the factor a^2/r^2. Thus the point with Cartesian coordinates x, y, and z becomes $\left(\frac{xa^2}{r^2}, \frac{ya^2}{r^2}, \frac{za^2}{r^2}\right)$ and the formula is proved.

This theorem is the basis of the method of inversion which by the above transformation maps one region D of space into another region D'. By varying the origin of the coordinate system and adjusting the value of the parameter a, a variety of regions can be obtained from a given initial region by successive inversions. By means of the transformation (9.5.3) harmonic functions defined in the original domain D yield new harmonic functions defined in D'. Also, if the harmonic function u vanishes on any surface S of D, then by (9.5.3) the corresponding function V will vanish on the surface S' into which S is mapped. Likewise, singular points are mapped into singular points.

As an illustration of the method of inversion, we shall derive the solution of the exterior harmonic Dirichlet problem for the sphere from that of the corresponding interior problem. The solution of the latter problem was found in (9.2.18) and Exercise 9.4.3. If we invert with respect to the center of the sphere, then the surface of the sphere is left unchanged, the interior being mapped into the exterior. By (9.5.5) the values on $r = a$ are related by $u(a, \theta, \psi) = av(a, \theta, \psi)$, and on applying the rule (9.5.5) to the solution given in the exercise we find for $r \geq a$ the potential

$$(9.5.6) \quad v(r, \theta, \psi) = \frac{a(r^2 - a^2)}{4\pi} \int_{-\pi}^{\pi} d\psi'$$
$$\times \int_0^\pi \frac{v(a, \theta', \psi') \sin \theta' \, d\theta'}{(a^2 + r^2 - 2ar \cos \theta_0)^{3/2}}$$

We shall now consider an harmonic boundary value problem or *transmission* problem for a region consisting of the interior and the exterior of a sphere. Certain conditions of continuity are imposed on the spherical surface and different expressions for the potential are required on the two sides of this surface.

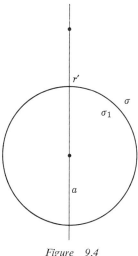

Figure 9.4
Electrode and sphere.

An electrode discharging a steady unit current is placed outside a conducting sphere of conductivity σ_1 (Figure 9.4). The sphere and the electrode are situated in an infinite medium of conductivity σ. Let a be the radius of the sphere and suppose that the electrode is distant $r' > a$ from the center. If we represent the electrode by means of a source, then the potential $u(r, \theta)$ outside the sphere satisfies the source function equation,

$$(9.5.7) \quad u_{rr} + \frac{2}{r} u_r + \frac{1}{r^2 \sin \theta} \frac{\partial}{\partial \theta} (\sin \theta u_\theta) = - \frac{\delta(r - r')\delta(\theta)}{2\pi\sigma(r')^2 \sin \theta}$$

for $0 < a \leq r < \infty$. We shall denote the potential inside the sphere by $u_1(r, \theta)$. It satisfies the equation

$$(9.5.8) \quad u_{1rr} + \frac{2}{r} u_{1r} + \frac{1}{r^2 \sin \theta} \frac{\partial}{\partial \theta} (\sin \theta u_{1\theta}) = 0$$

for $0 \leq r \leq a$.

The boundary conditions at the surface of separation are found by noting that the potential u and the normal component of current $J_n = \sigma E_n = \sigma \nabla u \cdot \mathbf{n}$

INTERIOR AND EXTERIOR EXPANSIONS

must be continuous. Therefore

(9.5.9) $$u(a, \theta) = u_1(a, \theta)$$

(9.5.10) $$\sigma u_r(a, \theta) = \sigma_1 u_{1r}(a, \theta)$$

We shall reduce (9.5.7) and (9.5.8) to ordinary differential equations by applying the finite transforms

(9.5.11) $$C_n(r) = \int_0^\pi u(r, \theta) P_n(\cos \theta) \sin \theta \, d\theta$$

(9.5.12) $$C_{1n}(r) = \int_0^\pi u_1(r, \theta) P_n(\cos \theta) \sin \theta \, d\theta$$

We then obtain the equations

(9.5.13) $$\frac{d^2 C_n}{dr^2} + \frac{2}{r} \frac{dC_n}{dr} - \frac{n(n+1)C_n}{r^2} = -\frac{\delta(r - r')}{2\pi \sigma (r')^2} \qquad r \geq a$$

(9.5.14) $$\frac{d^2 C_{1n}}{dr^2} + \frac{2}{r} \frac{dC_{1n}}{dr} - \frac{n(n+1)C_{1n}}{r^2} = 0 \qquad 0 \leq r \leq a$$

Basic solutions of these equations are r^n and r^{-n-1}. From the conditions of boundedness in the respective intervals, we are led to the forms

(9.5.15) $$C_n(r) = \begin{cases} Br^n + Cr^{-n-1} & 0 < a \leq r \leq r' < \infty \\ Dr^{-n-1} & 0 < a \leq r' \leq r < \infty \end{cases}$$

$$C_{1n}(r) = Ar^n \qquad 0 \leq r \leq a$$

Here we have four disposable constants A, B, C, and D. To determine them, four conditions are required. From (9.5.9) and (9.5.10) we obtain the two conditions $C_n(a) = C_{1n}(a)$, $\sigma C_n'(a) = \sigma_1 C_{1n}'(a)$. Two further equations arise from the requirements that the Green's function $C_n(r)$ defined by (9.5.15) should be continuous at $r = r'$ and that $C_n'(r)$ should possess a negative jump of $[2\pi\sigma(r')^2]^{-1}$ at r'. On applying these conditions, we find the solution

(9.5.16) $$C_n(r) = \begin{cases} \dfrac{r^n}{2\pi\sigma(2n+1)(r')^{n+1}} + \dfrac{n(\sigma - \sigma_1)a^{2n+1}}{2\pi\sigma(2n+1)[n\sigma_1 + (n+1)\sigma](rr')^{n+1}} \\ \qquad\qquad\qquad\qquad\qquad\qquad\qquad\qquad a < r < r' < \infty \\ \dfrac{(r')^n}{2\pi\sigma(2n+1)r^{n+1}} + \dfrac{n(\sigma - \sigma_1)a^{2n+1}}{2\pi\sigma(2n+1)[n\sigma_1 + (n+1)\sigma](rr')^{n+1}} \\ \qquad\qquad\qquad\qquad\qquad\qquad\qquad\qquad a < r' < r < \infty \end{cases}$$

(9.5.17) $$C_{1n}(r) = \frac{r^n}{2\pi[n\sigma_1 + (n+1)\sigma](r')^{n+1}} \qquad 0 \leq r \leq a$$

By Theorem 9.2.2 the inversion formula corresponding to (9.5.11) is the Legendre expansion

(9.5.18) $$u(r, \theta) = \sum_{n=0}^{\infty} (n + \tfrac{1}{2}) C_n(r) P_n(\cos \theta) \qquad 0 \le \theta \le \pi$$

If we insert (9.5.16) into this series, we find, by virtue of the generating series (9.4.20), the combined formula

(9.5.19) $$u(r, \theta) = \frac{1}{4\pi\sigma\sqrt{r^2 + r'^2 - 2rr'\cos\theta}} + \frac{\sigma - \sigma_1}{4\pi\sigma} \sum_{n=0}^{\infty} \frac{na^{2n+1} P_n(\cos\theta)}{[n\sigma_1 + (n+1)\sigma](rr')^{n+1}}$$

which is valid for $0 < a \le r < \infty$. Similarly, the potential inside the sphere is obtained by substituting (9.5.17) in (9.5.18), giving

(9.5.20) $$u_1(r, \theta) = \frac{1}{4\pi r'} \sum_{n=0}^{\infty} \frac{(2n+1)}{n\sigma_1 + (n+1)\sigma} \left(\frac{r}{r'}\right)^n P_n(\cos\theta)$$

for $0 \le r \le a$.

EXERCISES 9.5

1. Show that the harmonic Green's function for the exterior region $0 < a \le r < \infty$ vanishing on $r = a$ and with singularity at the point $r = r'$, $\theta = 0$ is given for $a \le r \le r'$ by

$$G(r, r', \theta) = \frac{1}{4\pi a} \sum_{n=0}^{\infty} \left(\frac{a}{r'}\right)^{n+1} \left[\left(\frac{r}{a}\right)^n - \left(\frac{a}{r}\right)^{n+1}\right] P_n(\cos\theta)$$

Express this result in closed form.

2. Let $Y_n(\theta, \psi)$ denote any surface spherical harmonic of degree n. Show by means of the addition theorem (9.4.8) that

$$\int_{-\pi}^{\pi} d\psi \int_0^{\pi} Y_n(\theta, \psi) P_n(\cos\theta_0) \sin\theta \, d\theta = \frac{4\pi}{2n+1} Y_n(\theta', \psi')$$

where

$$\cos\theta_0 = \cos\theta\cos\theta' + \sin\theta\sin\theta'\cos(\psi - \psi')$$

3. Show that $P_2(\cos\theta) = \tfrac{1}{2}(3\cos^2\theta - 1)$, $P_2^2(\cos\theta) = 3\sin^2\theta$, and hence express $\sin^2\theta\cos^2\psi$ as a linear combination of Legendre functions. Deduce that the harmonic function inside the sphere $0 \le r \le a$ which reduces to $6x^2$ on the sphere is given by

$$u = 2a^2 - 2r^2 P_2(\cos\theta) + r^2 P_2^2(\cos\theta)\cos 2\psi$$

Show also that $u = 2(a^2 + 2x^2 - y^2 - z^2)$.

4. By means of inversion, deduce from Exercise 9.5.1 the harmonic Green's function for the sphere $0 \le r \le a$ vanishing on $r = a$. Reconcile your result with that given in Exercise 9.4.5 and with (7.2.13).

5. Find the series analogous to (9.5.19) and (9.5.20) when the electrode lies within the sphere $r' < a$.

6. From (9.5.19), (9.5.20), and Exercise 9.5.5 specify the Green's function for the three-dimensional space which:
   (a) is harmonic for $r \neq a$;
   (b) satisfies the surface conditions (9.5.9) and (9.5.10).

9.6 FUNCTIONS OF NONINTEGRAL ORDER

In this chapter we have confined attention to potentials and wave functions which are one-valued on the whole range $0 \leq \psi \leq 2\pi$ of angles. Consequently, the separation constant m was restricted to be an integer (or zero). Furthermore we stated in Section 9.2 that the associated Legendre equation (9.2.2) possessed solutions bounded in the closed interval $0 \leq \theta \leq \pi$ only when the order s was also restricted to be an integer. One such solution was the function $P_n{}^m(\cos \theta)$ which by (9.1.11) is a polynomial in $\sin \theta$, $\cos \theta$. Now the complete solution of (9.1.13) involves a second solution independent of $P_n{}^m(\cos \theta)$. This solution is denoted by $Q_n{}^m(\cos \theta)$. We shall not give a detailed discussion of this function but shall merely demonstrate that it is singular on both the axes $\theta = 0, \pi$. When the prescribed solution is to be finite at either or both of these values, the function $Q_n{}^m(\cos \theta)$ must be excluded.

To verify that $P_n{}^m(\cos \theta)$ is the only solution of (9.2.2) bounded for $0 \leq \theta \leq \pi$, let $w_n{}^m(\cos \theta)$ be any other solution, independent of $P_n{}^m$. Then we may write

$$\frac{1}{\sin \theta} \frac{d}{d\theta}\left(\sin \theta \frac{dP_n{}^m}{d\theta}\right) + \left(n(n+1) - \frac{m^2}{\sin^2 \theta}\right) P_n{}^m = 0$$

$$\frac{1}{\sin \theta} \frac{d}{d\theta}\left(\sin \theta \frac{dw_n{}^m}{d\theta}\right) + \left(n(n+1) - \frac{m^2}{\sin^2 \theta}\right) w_n{}^m = 0$$

Multiply the first of these equations by $\sin \theta w_n{}^m$, the second by $\sin \theta P_n{}^m$, and subtract. Then we find after slight rearrangement that

$$\frac{d}{d\theta}\left[\sin \theta \left(w_n{}^m \frac{dP_n{}^m}{d\theta} - P_n{}^m \frac{dw_n{}^m}{d\theta}\right)\right] = 0$$

On integration we obtain the Wronskian

(9.6.1) $$w_n{}^m \frac{dP_n{}^m}{d\theta} - P_n{}^m \frac{dw_n{}^m}{d\theta} = \frac{A}{\sin \theta}$$

Here A is some constant whose value cannot be zero since the functions $w_n{}^m$, $P_n{}^m$ are independent. Since the terms $P_n{}^m$ and $dP_n{}^m/d\theta$ are bounded for $0 \leq \theta \leq \pi$, it is clear from (9.6.1) that $w_n{}^m$ or its derivative must be singular at $\theta = 0$ and $\theta = \pi$. A closer examination of the nature of the singularity reveals that $w_n{}^m$ is itself necessarily singular at $\theta = 0, \pi$. For if not, $dw_n{}^0(\theta)/d\theta \sim C \operatorname{cosec} \theta$ for θ close to 0 or π therefore $w_n{}^0(\theta) \sim C \log (\tan \tfrac{1}{2}\theta)$. This is singular. The singularity is of higher order if m is a nonzero integer for then $P_n{}^m(\pm 1) = 0$.

368 SPHERICAL EIGENFUNCTIONS

We shall now construct solutions of the Legendre equation

(9.6.2) $$\frac{d}{d\mu}\left[(1-\mu^2)\frac{dw}{d\mu}\right] + s(s+1)w = 0$$

for general values of s. (For simplicity we consider only the case $m = 0$.) To do this, we resort to the generating function (9.4.20) which we rewrite in the form

$$\frac{1}{\sqrt{1 - 2\mu t + t^2}} = \sum_{n=0}^{\infty} t^n P_n(\mu) \qquad |t| < 1$$

Multiply this equation by t^{-s-1} and integrate around the circle $C: |t| = R$ where $0 < R < 1$. Then

$$\int_C \frac{dt}{t^{s+1}\sqrt{1 - 2\mu t + t^2}} = \sum_{n=0}^{\infty} P_n(\mu) \int_C t^{n-s-1}\, dt$$

If s is an integer, each of the integrals on the right vanishes except that one for which $n = s$. Thus if s is an integer, we obtain the formula

(9.6.3) $$P_s(\mu) = \frac{1}{2\pi i} \int_C \frac{dt}{t^{s+1}\sqrt{1 - 2\mu t + t^2}}$$

We shall adopt this integral as our definition of the function $P_s(\mu)$ for all values of s such that Re $(s) > -1$. It is however necessary to modify the path of integration in (9.6.3) since the integrand now possesses a branch point at the origin as well as at $t = e^{\pm i\theta}$, $\mu = \cos\theta$. Branch cuts are inserted along the negative real axis from $-\infty$ to the origin and along the arc $t = e^{i\varphi}$ for $-\theta < \varphi < \theta$. The path C is chosen to be a loop around the negative real axis crossing the positive real axis to the left of the second cut (Figure 9.5). The function t^{-s-1} is given its principal value, and that branch of $\sqrt{1 - 2\mu t + t^2}$ is chosen which reduces to unity at the origin.

We shall now show that $P_s(\mu)$ becomes infinite as $\mu \to -1$ unless s is an integer. To do this, we first put (9.6.3) into a more convenient form. If Re $(s) > -1$, we may deform the path C onto both sides of the cut joining the points $e^{\pm i\theta}$ together with small circles around these points. The contributions from these circles vanish as their radii tend to zero. On the arcs we set $t = e^{i\varphi}$, $-\theta < \varphi < \theta$, and find

(9.6.4) $$P_s(\cos\theta) = \frac{2}{\pi}\int_0^\theta \frac{\cos(s + \tfrac{1}{2})\varphi\, d\varphi}{\sqrt{2\cos\varphi - 2\cos\theta}}$$

As $\mu \to -1$, $\theta \to \pi$, and in the neighborhood of this point the integrand behaves like $(\sin s\pi)/(\pi - \varphi)$. Thus $P_s(\mu)$ possesses a logarithmic singularity at $\mu = -1$ unless s is an integer. There is no singularity at $\mu = +1$, however, since when

θ is small, we may approximate (9.6.4) by

$$(9.6.5) \qquad P_s(\cos\theta) \approx \frac{2}{\pi}\int_0^\theta \frac{d\varphi}{\sqrt{\theta^2 - \varphi^2}} = 1$$

The function $P_s(-\mu)$ evidently satisfies Legendre's equation. Also it is singular at $\mu = +1$ whereas $P_s(\mu)$ is not. Therefore $P_s(\mu)$ and $P_s(-\mu)$ are linearly independent, except when s is an integer, in which case $P_s(-\mu) = (-1)^s P_s(\mu)$.

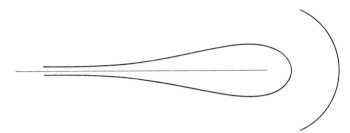

Figure 9.5 Contour of integration for $P_s(\mu)$.

To form the complete solution of Legendre's equation (9.6.2) when s is not an integer, we need only take the combination $AP_s(\mu) + BP_s(-\mu)$.

For $\mu = \cos\theta$, we see that when s is not an integer there is no solution of (9.6.2) bounded in the closed interval $0 \leq \theta \leq \pi$. The solution bounded in the subinterval $0 \leq \theta \leq \alpha < \pi$ is $P_s(\cos\theta)$ and that bounded for $0 < \alpha \leq \theta \leq \pi$ is $P_s(-\cos\theta)$.

EXERCISES 9.6

1. Verify that (9.6.3) is a solution of Legendre's equation (9.6.2) for general values of s.
2. Show by means of (9.6.3) that, if Re $(s) > -1$,

$$P_s(0) = \frac{1}{2\pi i}\int_C \frac{dt}{t^{s+1}\sqrt{1+t^2}}$$

By integrating along both sides of a straight cut joining the points $\pm i$ show that, if Re $(s) < 0$

$$P_s(0) = \frac{-2}{\pi}\sin\frac{\pi s}{2}\int_0^1 \frac{dy}{y^{s+1}\sqrt{1-y^2}}$$

Deduce that

$$P_s(0) = \frac{\sqrt{\pi}}{\Gamma(1+s/2)\Gamma(1/2 - s/2)}$$

3. By means of (9.6.3) show that

$$(\mu^2 - 1)P_s'(\mu) = s[\mu P_s(\mu) - P_{s-1}(\mu)]$$

Deduce that
$$P'_s(0) = \frac{-2\sqrt{\pi}}{\Gamma(-s/2)\Gamma(1/2 + s/2)}$$

Show that $P'_s(0) = 0$ when $s = 2n$ or $s = -(2n + 1)$ where $n = 0, 1, 2, 3 \ldots$.

4. Replace t by t^{-1} in (9.6.3) and deduce that
$$P_s(\mu) = \frac{1}{2\pi i} \int_C \frac{t^s \, dt}{\sqrt{1 + t^2 - 2\mu t}}$$

where C is a circuit enclosing the points $t = e^{\pm i\theta}$ (where $\mu = \cos \theta$) and not crossing the negative real axis. Deduce that
$$P_s(\mu) = P_{-s-1}(\mu).$$

5. Show that
$$P_s(\mu) \frac{d}{d\mu} P_s(-\mu) - P_s(-\mu) \frac{d}{d\mu} P_s(\mu) = \frac{C}{1 - \mu^2}$$

By utilizing the values of $P_s(0)$, $P'_s(0)$ stated in Exercises 9.6.2 and 9.6.3 show that $C = (-2/\pi) \sin s\pi$. Deduce that $P_s(\mu)$ and $P_s(-\mu)$ are independent solutions of Legendre's equation except when s is an integer or zero.

6. If n is a positive integer, show that
$$P_{2n+1}(0) = 0 \qquad P'_{2n+1}(0) = \frac{(-1)^n(2n+1)!}{2^{2n}(n!)^2} \qquad P_{2n}(0) = \frac{(-1)^n(2n)!}{2^{2n}(n!)^2}$$

7. Show that when s is not an integer, the Green's function bounded in the interval $0 \leq \theta \leq \pi$ of the equation
$$\frac{1}{\sin \theta} \frac{d}{d\theta}(\sin \theta G_\theta) + s(s+1)G = \frac{\delta(\theta - \theta')}{\sin \theta}$$
is
$$G(\theta, \theta') = \begin{cases} \dfrac{\pi P_s(\cos \theta) P_s(-\cos \theta')}{2 \sin s\pi} & 0 < \theta < \theta' < \pi \\[2mm] \dfrac{\pi P_s(-\cos \theta) P_s(\cos \theta')}{2 \sin s\pi} & 0 < \theta' < \theta < \pi \end{cases}$$

8. Show that the eigenfunction expansion of the Green's function defined in Exercise 9.6.7 is
$$G(\theta, \theta') = \sum_{n=0}^{\infty} \frac{(n + \tfrac{1}{2})P_n(\cos \theta)P_n(\cos \theta')}{(s - n)(s + n + 1)}$$

9. The harmonic Green's function for the whole of space with singularity on the axis $\theta = 0$ satisfies the equation
$$\sin \theta \frac{\partial}{\partial r}\left(r^2 \frac{\partial G}{\partial r}\right) + \frac{\partial}{\partial \theta}\left(\sin \theta \frac{\partial G}{\partial \theta}\right) = -\frac{\delta(r - r')\delta(\theta)}{2\pi}$$

Show that the Mellin transform

$$\bar{G}(s, \theta) = \int_0^\infty r^s G(r, \theta)\, dr$$

satisfies the equation

$$\frac{d}{d\theta}\left(\sin\theta \frac{d\bar{G}}{d\theta}\right) + s(s+1)\bar{G}\sin\theta = -\frac{(r')^s \delta(\theta)}{2\pi}$$

Deduce that

$$G = \frac{1}{4\pi\sqrt{rr'}} \int_0^\infty \frac{\cos(\lambda \log r'/r) P_{-1/2+i\lambda}(-\cos\theta)\, d\lambda}{\cosh \lambda\pi}$$

10. (a) Show that the substitutions $v = ur^{1/2}$, $\sigma = \ln\left(\dfrac{a}{r}\right)$ convert the axially symmetric potential equation

$$r^2 u_{rr} + 2r u_r + \frac{1}{\sin\theta}\frac{\partial}{\partial\theta}(\sin\theta\, u_\theta) = 0$$

into

$$\frac{1}{\sin\theta}\frac{\partial}{\partial\theta}(\sin\theta\, v_\theta) + v_{\sigma\sigma} - \frac{v}{4} = 0$$

(b) Hence show that the potential $u(r, \theta)$ in the region $0 \le r \le a$, $0 \le \theta \le \alpha < \pi$, with the conditions $u = 0$ on $r = a$ and $u = g(r)$ on $\theta = \alpha$, is given by

$$u = \frac{2}{\pi}\int_0^a \frac{g(r')\, dr'}{\sqrt{rr'}} \int_0^\infty \sin\left(s\log\frac{r}{a}\right)\sin\left(s\log\frac{r'}{a}\right) \frac{P_{-1/2+is}(\cos\theta)}{P_{-1/2+is}(\cos\alpha)}\, ds$$

[Apply a suitable Fourier integral in the variable s.]

(c) For the region $0 < a \le r \le b$, $0 \le \theta \le \alpha < \pi$, with $u = 0$ on $r = a$, $u = 0$ on $r = b$ and $u = g(r)$ on $\theta = \alpha$, show that

$$u = \frac{2}{h}\sum_{n=1}^\infty \sin\left(\lambda_n \log\frac{r}{a}\right) \frac{P_{-1/2+i\lambda_n}(\cos\theta)}{P_{-1/2+i\lambda_n}(\cos\alpha)} \int_a^b \frac{g(r')}{\sqrt{rr'}} \sin\left(\lambda_n \log\frac{r'}{a}\right) dr'$$

where $h = \log(b/a)$ and $\lambda_n = n\pi/h$.

CHAPTER *10*

Wave Propagation in Space

10.1 CHARACTERISTIC SURFACES

In this concluding chapter we shall discuss the three-dimensional wave equation and some of its numerous applications. When studying the one-dimensional equation of the string in Chapter 2, we saw that there were two possible viewpoints—those of traveling waves and of Fourier series expansions. Here we shall find that a judicious combination of these methods will lead to the most effective understanding of the three-dimensional wave equation

(10.1.1) $$\Box u = u_{tt} - c^2 \Delta u = 0$$

In one space and one time dimension the locus of a traveling wave is a characteristic *curve*. The higher-dimensional characteristics are *surfaces*. With appropriate modification for the higher dimension the characteristic surfaces share the same significant properties discussed in Chapter 2.

The most general *plane wave* of the form

(10.1.2) $$f(\mathbf{x} \cdot \mathbf{s} - ct) = f(x_1 s_1 + x_2 s_2 + x_3 s_3 - ct)$$

where f is an arbitrary function of one variable, will satisfy the wave equation if the vector **s** has unit length: $\mathbf{s}^2 = s_1^2 + s_2^2 + s_3^2 = 1$. Such a solution has the same value at every point of the plane

(10.1.3) $$\mathbf{x} \cdot \mathbf{s} - ct = \text{const.}$$

and, in fact, any singularity of a plane-wave solution will lie on such a plane. The planes (10.1.3) are the *characteristic planes* of the wave equation.

However there are other characteristic surfaces in addition to the characteristic planes. For a general definition of a characteristic surface, it is best to proceed by considering the initial value problem with data given on a surface S. For a second-order equation, the values of u and of its first derivative transverse to the surface are assigned as data on S. For instance, the initial problem on the surface $t = \text{constant}$ involves data for u and u_t. When these two quantities are known on the surface, the differential equation enables us to calculate the value of u_{tt} on the surface:

$$u_{tt} = c^2 \Delta u$$

By differentiating with respect to t, we find, in succession,

$$u_{ttt} = c^2 \Delta u_t$$
$$u_{tttt} = c^2 \Delta u_{tt} = c^4 \Delta^2 u$$
$$\cdots$$

Thus the derivatives of u with respect to t of every order can be calculated on the surface when the data for the initial value problem are given as functions of x_1, x_2, x_3 at $t = \text{constant}$. If the solution is an analytic function, its values for later times can now be found from its Taylor series expansion as a function of t about the initial time. In this case, therefore, the process of computing higher derivatives enables us to construct the solution for later times.

If the datum surface is $\varphi = \text{const.}$, we have to find out when this process will succeed. Let u be regarded as a function of φ and of three other quantities x_1', x_2', x_3' which play the role of coordinates. Then

$$\frac{\partial u}{\partial x_i} = \frac{\partial u}{\partial \varphi} \frac{\partial \varphi}{\partial x_i} + \sum_k \frac{\partial u}{\partial x_k'} \frac{\partial x_k'}{\partial x_i}$$

$$\frac{\partial^2 u}{\partial x_i^2} = \frac{\partial^2 u}{\partial \varphi^2} \cdot \left(\frac{\partial \varphi}{\partial x_i}\right)^2 + \frac{\partial u}{\partial \varphi} \frac{\partial^2 \varphi}{\partial x_i^2} + \cdots$$

while similar formulas hold for u_t and u_{tt}. The terms omitted do not involve $u_{\varphi\varphi}$, u_φ or derivatives of the function φ. Thus

$$u_{tt} - c^2 \Delta u = \frac{\partial^2 u}{\partial \varphi^2}\left[\left(\frac{\partial \varphi}{\partial t}\right)^2 - c^2 \sum_i \left(\frac{\partial \varphi}{\partial x_i}\right)^2\right]$$

(10.1.4)
$$+ \frac{\partial u}{\partial \varphi}\left[\frac{\partial^2 \varphi}{\partial t^2} - c^2 \sum_i \frac{\partial^2 \varphi}{\partial x_i^2}\right] + \cdots = 0$$

In the construction of the solution, therefore, we can determine the second derivative $\partial^2 u/\partial \varphi^2 \equiv u_{\varphi\varphi}$ only if φ satisfies

(10.1.5) $$\left(\frac{\partial \varphi}{\partial t}\right)^2 - c^2(\nabla \varphi)^2 \neq 0$$

The surfaces for which this condition fails are called characteristic surfaces.

DEFINITION 10.1.1 *A surface $\varphi(x_1, x_2, x_3, t) = $ constant is characteristic for (10.1.1) if*

(10.1.6) $$\varphi_t^2 - c^2(\nabla \varphi)^2 \equiv \varphi_t^2 - c^2(\varphi_{x_1}^2 + \varphi_{x_2}^2 + \varphi_{x_3}^2) = 0$$

A complete study of characteristic surfaces would lead us far into the mathematical theory of nonlinear partial differential equations of the first order. Here we shall only remark that a surface which at each point is tangent to a characteristic surface is itself characteristic. Therefore, for example, the *envelope* of a family of such surfaces is characteristic. Now the characteristic planes

Figure 10.1
The characteristic cone.

(10.1.7) $$\mathbf{x} \cdot \mathbf{s} - ct = 0$$

pass through the origin and have as envelope the circular cone (Figure 10.1)

(10.1.8) $$|\mathbf{x}|^2 - c^2t^2 = x_1^2 + x_2^2 + x_3^2 - c^2t^2 = 0$$

This cone is known as the *characteristic cone* with vertex at the origin. It has a *forward* sheet extending into the future $t > 0$, and a *retrograde* sheet extending into the past $t < 0$. (Courant-Hilbert, Ref. 13, vol. 2.)

THEOREM 10.1.1 *The locus of a singularity of a wave function satisfying (10.1.1) is a characteristic surface.*

Proof. Suppose $u = f(\varphi, x_1, x_2, x_3)$, where f is singular for $\varphi = 0$, say. The calculation (10.1.4) shows that the highest singularity of $u_{tt} - c^2 \Delta u$ is the term $f''(\varphi)$, the coefficient of which must therefore vanish. That coefficient is the four-dimensional squared gradient (10.1.6), so $\varphi = 0$ is a characteristic surface, as required.

One consequence of this theorem is that the Green's function for a wave equation cannot have a point singularity as does the elliptic (or harmonic) Green's function of Chapter 7. Instead, as we have seen in Chapter 2 for the one-dimensional case, the singular locus of the Green's function is the characteristic cone with vertex at the source point. This will be established later by explicit computation.

The characteristic cone also determines the domain of dependence and the region of influence in this higher-dimensional case. We shall establish the domain of dependence result by proving the following theorem of uniqueness (Figure 10.2).

THEOREM 10.1.2 *Let the retrograde characteristic cone with vertex* $P_0 = (x_{10}, x_{20}, x_{30}, t_0)$ *in space-time intersect the initial surface* $t = 0$ *in a sphere S. If the initial data within and on S are given, there is at most one solution of*

$$u_{tt} = c^2 \Delta u$$

which is continuously differentiable within the cone and at the vertex P_0.

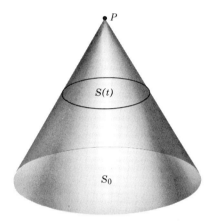

Figure 10.2 Retrograde cone with vertex *P*.

Proof. If there is more than one solution, we shall form the difference of two. By linearity, this difference must have zero data on the initial surface within and on the sphere *S*. Let *u* denote the difference, and consider the energy integral

(10.1.9) $$\mathscr{E}(t) = \int_{S(t)} [u_t^2 + c^2(\nabla u)^2] \, dV$$

A similar integral was employed in Chapter 6, Section 6.6, but here we shall take the domain of integration to be the sphere $S(t)$, which is the three-dimensional solid intersection of the retrograde cone with the plane $t =$ constant. The radius of $S(t)$ is $r = c(t_0 - t)$.

We form the time derivative of this integral, observing that the decrease in radius introduces a surface integral term. Thus

(10.1.10)
$$\frac{d}{dt} \mathscr{E}(t) = 2 \int_{S(t)} (u_t u_{tt} + c^2 \nabla u \cdot \nabla u_t) \, dV - c \int_{r=c(t_0-t)} [u_t^2 + c^2(\nabla u)^2] \, dS$$

An integration by parts contributes to the surface integral terms, and we find

(10.1.11)
$$\frac{d}{dt}\mathcal{E}(t) = 2\int_{S(t)} u_t(u_{tt} - c^2 \Delta u)\, dV - c\int_{r=c(t_0-t)} [u_t^2 + c^2(\nabla u)^2 - 2cu_t u_n]\, dS$$

The volume term vanishes by (10.1.1). Since the magnitude of u_n is at most the magnitude of the gradient ∇u, we have

(10.1.12) $\quad u_t^2 - 2cu_t u_n + c^2(\nabla u)^2 \geq u_t^2 - 2cu_t u_n + c^2 u_n^2 = (u_t - cu_n)^2 \geq 0$

Therefore the surface integral contribution in (10.1.11) is not positive, so that

(10.1.13)
$$\frac{d}{dt}\mathcal{E}(t) \leq 0$$

Therefore, if $\mathcal{E}(0) = 0$, we can conclude that the nonnegative quadratic energy functional $\mathcal{E}(t)$ is identically zero for $0 \leq t \leq t_0$. As the first derivatives of u are assumed continuous, this implies that $u =$ constant within the conical region and at the vertex. From the initial value zero for u, we conclude that u vanishes identically. This completes the proof.

Therefore, if a wave function u has assigned data on the sphere S, its value at P is uniquely determined. That is, the domain of dependence of $u(P)$ on the initial data is contained in the sphere. Likewise, the domain of dependence of $u(P)$ upon a forcing term in the differential equation is contained in the cone with vertex at P.

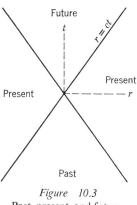

Figure 10.3
Past, present, and future.

COROLLARY 10.1.1 *The domain of dependence is bounded by the retrograde characteristic cone.*

Similarly we can show that the region of influence is bounded by the forward cone with vertex at P.

The physical interpretation of these results is, of course, that the upper limit of velocity of wave propagation is c. Thus the wave equation plays a central role in relativity and electromagnetic theory, in which the velocity c of light is the highest physically attainable speed. Some of the quite graphic terminology of this theory has been widely adopted in the study of hyperbolic differential equations (Figure 10.3). Thus the characteristic cone is called the light cone or wave cone. It divides the four-dimensional space-time into three regions: the *past*

$$ct < -\sqrt{x_1^2 + x_2^2 + x_3^2}$$

the *present*

$$-\sqrt{x_1^2 + x_2^2 + x_3^2} < ct < \sqrt{x_1^2 + x_2^2 + x_3^2}$$

and the *future*

$$\sqrt{x_1^2 + x_2^2 + x_3^2} < ct$$

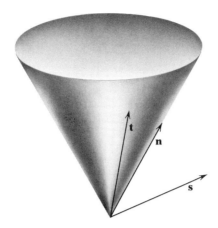

Figure 10.4 Timelike, null, and spacelike vectors.

A four-vector in space-time is called *timelike* if it is directed into the future (or past), and *spacelike* if it is directed towards the present. A vector parallel to a generator of the characteristic cone is called a *null* vector (Figure 10.4). If we adopt the indefinite or Lorentzian definition of length,

$$v^2 = v_0^2 - c^2 v_1^2 - c^2 v_2^2 - c^2 v_3^2$$

for a vector with time and space components (v_0, v_1, v_2, v_3), we see that a null vector has length zero. In particular, the four-gradient of a function φ is a null vector if and only if the surface $\varphi = \text{constant}$ is characteristic:

$$(\text{grad}_4 \varphi)^2 = \varphi_t^2 - c^2 (\nabla \varphi)^2 = 0$$

If $(\text{grad}_4 \varphi)^2 > 0$, the surface is *spacelike* and if $(\text{grad}_4 \varphi)^2 < 0$, it is a *timelike* surface.

EXERCISES 10.1

1. Find the characteristic planes and surfaces for the equations
   (a) $u_t = \Delta u$
   (b) $u_{tt} = \Delta u + k^2 u$
   (c) $u_{tt} = a u_{xx} + 2h u_{xy} + b u_{yy}$ $\quad ab > h^2$

2. Show that $1/R$, where $R^2 = c^2(t - t_0)^2 - (x - x_0)^2 - (y - y_0)^2$, is a solution of $u_{tt} = c^2(u_{xx} + u_{yy})$.

3. Show that $1/R^2$, where $R^2 = c^2(t - t_0)^2 - (x - x_0)^2 - (y - y_0)^2 - (z - z_0)^2$, is a solution of $u_{tt} = c^2(u_{xx} + u_{yy} + u_{zz})$.

4. Establish a uniqueness and domain of dependence theorem for the equation $u_{tt} = c^2 \Delta u - k^2 u$, k real. *Hint:* Modify $\mathscr{E}(t)$.

5. Show that $u_{tt} = c^2 \Delta u - k^2 u$ has plane-wave solutions $u = f(\mathbf{x} \cdot \mathbf{s} - ct)$ provided that $|\mathbf{s}| \neq 1$, and determine the form of the function f. Explain the paradox that the phase velocity is greater than c if $|\mathbf{s}| < 1$.

378   WAVE PROPAGATION IN SPACE

6. Sound waves having a velocity potential φ travel through a gas of density ρ. Show that the energy flow through a unit area of surface is

$$\rho \frac{\partial \varphi}{\partial t} \frac{\partial \varphi}{\partial n}$$

where n is the normal to the surface.

7. Show that solutions of (10.1.1) depending only on t and the distance r from the origin satisfy $c^2(u_{rr} + (2/r)u_r) = u_{tt}$. Deduce that, if $v = ru$, then v satisfies a one-dimensional wave equation, and hence that

$$u = \frac{1}{r}[f(r - ct) + g(r + ct)]$$

is the general solution for radial waves.

8. Show that the flow $\int (\partial \varphi / \partial n)\, dS$ through a small sphere with the center at the origin, of wave motion with velocity potential

$$\varphi = \frac{-1}{4\pi r} f\left(t - \frac{r}{c}\right)$$

is the source strength $f(t)$. Find the velocity potential due to a small explosion at the origin and describe its locus of singularities.

9. Show that radial oscillations of air in a rigid sphere of radius a have the frequencies ck_j, where k_j is a root of $\tan(ka) = ka$ and c is the sound velocity.

10. Show that the envelope of the planes $\mathbf{x} \cdot \mathbf{s} - ct = 0$, as $\mathbf{s}$ varies with $|\mathbf{s}| = 1$, is the cone $\mathbf{x}^2 - c^2 t^2 = 0$.

10.2 SOURCE FUNCTION FOR THE WAVE EQUATION

To express the solutions of the wave equation it is again convenient to use a source function or Green's function. The corresponding one-dimensional function, or rather distribution, was introduced in Chapter 2, Section 2.4. Likewise, the elliptic and parabolic equations in three space dimensions were treated in Chapter 7, where we made use of the eigenfunctions $u_n(P)$ of a general region R. We shall now extend this method to the hyperbolic case.

The reader will recall that the Green's functions for the elliptic and parabolic equations have a singularity at one point only—the source point—and that elsewhere they are smooth and well-behaved functions.

In contrast, the hyperbolic source functions are distributions, with singular support the entire light cone. Also, by the preceding uniqueness theorem, the hyperbolic Green's functions must vanish outside the light cone. Therefore, at a later time, their spatial support is bounded by a sphere.

The problem at hand, then, is to find the eigenfunction expansion of the solution of

(10.2.1) $$u_{tt} - c^2 \Delta u = \delta(P, Q)\delta(t)$$

which is identically zero at earlier times $t < 0$. We denote this source solution by $K(P, Q, t)$, where P and Q are points of a given region R. A boundary condition, one of the types studied in Chapter 6, will be imposed on $K(P, Q, t)$ for P on the surface S. The appropriate eigenfunctions we write as $u_n(P)$ and we look for the Fourier coefficients of K.

Setting

(10.2.2) $$K(P, Q, t) = \sum_{n=1}^{\infty} c_n(Q, t) u_n(P)$$

where $c_n(Q, t) = \int_R K(P, Q, t) u_n(P) \, dV_P$, we multiply (10.2.1) by $u_n(P)$ and integrate over R. By a calculation that is now familiar, we find

(10.2.3) $$\frac{d^2 c_n(Q, t)}{dt^2} + c^2 \lambda_n c_n(Q, t) = u_n(Q) \, \delta(t)$$

Therefore we can take

(10.2.4) $$c_n(Q, t) = u_n(Q) c_n(t)$$

where

(10.2.5) $$\frac{d^2 c_n(t)}{dt^2} + c^2 \lambda_n c_n(t) = \delta(t)$$

and $c_n(t) \equiv 0$ for $t < 0$. The solution of the time equation (10.2.5) was described in Chapter 1, Section 1.7, and it is

(10.2.6) $$c_n(t) = \frac{\sin c k_n t}{c k_n} H(t)$$

where $k_n^2 = \lambda_n$.

Assembling these formulas, we find for the source function the formal series

(10.2.7) $$K(P, Q, t) = \sum_{n=1}^{\infty} \frac{\sin c k_n t}{c k_n} u_n(P) u_n(Q) H(t)$$

Once again we emphasize that this is to be interpreted as a series of distributions and that its convergence is in the weak or distribution sense.

THEOREM 10.2.1 *The Green's function $K(P, Q, t)$ for the wave equation on R formally satisfies the differential equation (10.2.1) and the boundary condition for P on S. It also satisfies the initial conditions*

(10.2.8) $$K(P, Q, 0+) = 0$$

(10.2.9) $$\frac{\partial}{\partial t} K(P, Q, 0+) = \delta(P, Q)$$

and is symmetric in P and Q.

For the proof, we shall limit ourselves to the formal relations (10.2.8) and (10.2.9). A detailed proof of the latter is suggested as an exercise. For $t = 0$, the Fourier coefficients of $K(P, Q, t)$ are all zero, and so the kernel itself is zero.

380   WAVE PROPAGATION IN SPACE

For $t = 0$, the time derivative $K_t(P, Q, t)$ has the time factor equal to unity in (10.2.7) as t tends to zero from positive values. Therefore the series becomes the bilinear expansion of the Dirac distribution, which establishes (10.2.9).

When the source function is known, we can use it, as in Chapter 7, to represent solutions of general problems in wave propagation by potentials or integrals.

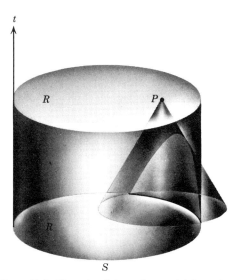

Figure 10.5 Domain of dependence with boundary.

For the product domain $R \times I_+$, where I_+ is the time interval $0 < \tau < t$, we construct a suitable Green's formula by integrating the product

$$[u_{tt}(Q, \tau) - c^2 \Delta u(Q, \tau)] K(P, Q, t - \tau)$$

over the domain and then integrating by parts. Let T_0 be the initial surface $\tau = 0$ and T the "final" surface $\tau = t$. Then, with $u = u(Q, T)$

$$\int_R \int_0^{t-\varepsilon} (u_{tt} - c^2 \Delta u) K(P, Q, t - \tau) \, dV_Q \, d\tau$$

(10.2.10) $\quad = \int_R \int_0^{t-\varepsilon} u K_{tt}(P, Q, t - \tau) \, dV_Q \, d\tau + \int_R [u_t K + K_t u]_{\tau=0}^{\tau=t-\varepsilon} \, dV_Q$

$$- \int_R \int_0^{t-\varepsilon} u c^2 \, \Delta K(P, Q, t - \tau) \, dV_Q \, d\tau - c^2 \int_S \int_0^{t-\varepsilon} (u_n K - u K_n) \, dS \, d\tau$$

Suppose that the problem is (Figure 10.5):

(10.2.11) $\quad\quad\quad\quad u_{tt} - c^2 \, \Delta u = f(P, t)$

that the boundary values of u are

(10.2.12) $\quad\quad\quad\quad u(P, t) = g(P, t) \quad\quad\quad\quad P$ on S

and that
(10.2.13)
$$u(P, 0) = \varphi(P)$$
$$u_t(P, 0) = \psi(P)$$

initially on T_0. Then the integral on the left in (10.2.10) becomes a volume potential with volume density $f(Q, \tau)$. We combine the two fourfold integrals on the right and obtain zero since $\delta(t - \tau)$ vanishes for $0 \leq \tau < t$. In the second term on the right there are four contributions. For $\tau = 0$ we substitute the two initial data (10.2.13), while for $\tau = t - \varepsilon$ we apply (10.2.8) and (10.2.9), obtaining zero in the first case and $u(P, t)$ in the second as $\varepsilon \to 0$. In the first of the surface integral terms K vanishes and in the second we substitute (10.2.12). The final result is obtained by making ε tend to zero.

THEOREM 10.2.2 *The solution of the initial and boundary value problem* (10.2.11) *through* (10.2.13) *has the representation*

(10.2.14)
$$u(P, t) = \int_R \int_0^t K(P, Q, t - \tau) f(Q, \tau) \, dV_Q \, d\tau$$
$$- c^2 \int_S \int_0^t g(Q, \tau) \frac{\partial}{\partial n_Q} K(P, Q, t - \tau) \, dS_Q \, d\tau$$
$$+ \int_R [K(P, Q, t)\psi(Q) + K_t(P, Q, t)\varphi(Q)] \, dV_Q$$

There are many special cases and particular applications of this theorem, some of which are included as exercises. We mention here the point source of radiation or sound of variable strength $f(t)$. The source point Q_0 corresponds to the forcing function

(10.2.15)
$$f(P, t) = f(t)\delta(P, Q_0)$$

in (10.2.11) and, by (10.2.14), the solution is

(10.2.16)
$$u(P, t) = \int_0^t K(P, Q_0, t - \tau) f(\tau) \, d\tau$$

Since $K(P, Q, t)$ vanishes for Q outside the retrograde cone with vertex (P, t), the integrals in (10.2.14) are actually extended over only a part of the region R on the boundary surface S, namely, the part common to the cone with vertex P (see Figure 10.5).

We shall now evaluate K explicitly in the important case where R is the entire three-dimensional space. It will appear that the support of K is the surface of the cone only, and that K vanishes in the interior of the cone as well as the exterior.

Without loss of generality we can choose the source point Q at the origin and select the normalized eigenfunctions [see (3.6.18)]

$$\frac{e^{is \cdot x}}{(2\pi)^{3/2}}$$

where P has coordinate vector $\mathbf{x}$. The normalized functions of Q contribute the further factor $[(2\pi)^{-3/2} e^{-i\mathbf{s}\cdot\mathbf{y}}]_{\mathbf{y}=0} = (2\pi)^{-3/2}$ since $\mathbf{y} = 0$ is the position vector of Q. The eigenvalues are $s = |\mathbf{s}| = (s_1^2 + s_2^2 + s_3^2)^{1/2}$ and the formal expression for K is

(10.2.17) $$K(\mathbf{x}, t) = \frac{1}{(2\pi)^3} \iiint_{-\infty}^{\infty} e^{i\mathbf{s}\cdot\mathbf{x}} \frac{\sin sct}{sc} \, ds_1 \, ds_2 \, ds_3$$

The convergence of this integral, like that of the bilinear series (10.2.7), is to be understood in the distribution sense of Section 7.7. Such integrals have appeared as Fourier transforms in Chapter 3, Section 3.6, and we recall the representation (3.6.18) of the Dirac distribution.

To evaluate the integral, we choose suitably oriented spherical polar coordinates s, θ, φ in the s-space, and write for $t > 0$,

(10.2.18)
$$K(\mathbf{x}, t) = \frac{1}{(2\pi)^3} \int_0^\infty \int_0^\pi \int_0^{2\pi} e^{is|\mathbf{x}|\cos\theta} \frac{\sin sct}{sc} s^2 \sin\theta \, ds \, d\theta \, d\varphi$$

$$= \frac{1}{(2\pi)^2} \int_0^\infty \frac{\sin sct}{c} s \, ds \int_0^\pi e^{is|\mathbf{x}|\cos\theta} \sin\theta \, d\theta$$

The inner integral is elementary, and yields $2 \sin s|\mathbf{x}|/s|\mathbf{x}|$. Substitution into the expression for $K(\mathbf{x}, t)$ now gives

(10.2.19)
$$K(\mathbf{x}, t) = \frac{1}{2\pi^2 c |\mathbf{x}|} \int_0^\infty \sin sct \sin s|\mathbf{x}| \, ds$$

$$= \frac{1}{4\pi^2 c |\mathbf{x}|} \int_0^\infty [\cos s(ct - |\mathbf{x}|) - \cos s(ct + |\mathbf{x}|)] \, ds$$

$$= \frac{1}{8\pi^2 c |\mathbf{x}|} \int_{-\infty}^\infty (e^{is(ct-|\mathbf{x}|)} - e^{is(ct+|\mathbf{x}|)}) \, ds$$

$$= \frac{1}{4\pi c |\mathbf{x}|} [\delta(ct - |\mathbf{x}|) - \delta(ct + |\mathbf{x}|)]$$

The second Dirac function is zero for $t > 0$ since $|\mathbf{x}| > 0$, and so we conclude that

(10.2.20) $$K(\mathbf{x}, t) = \frac{H(t)}{4\pi c |\mathbf{x}|} \delta(ct - |\mathbf{x}|)$$

THEOREM 10.2.3 *The source function (10.2.20) of the wave equation in three-dimensional space has as support the surface of the half-cone with vertex at the source point.*

More than a century before distributions were invented, this result was found by Poisson. We have outlined his method in a sequence of exercises (Exercise 10.2.9) which involve the average values of a solution over the surface of a sphere. Let

(10.2.21) $$M_{ct}u(P) = \frac{1}{4\pi} \int_{r=ct} u(Q) \, d\Omega_Q$$

denote the average, or mean, value of $u(Q)$ on the surface of the sphere with center P and radius $r = ct$. Then the term containing $\psi(P) = u_t(P, 0)$ in (10.2.14) becomes

(10.2.22)
$$\frac{1}{4\pi c} \int_{-\infty}^{\infty} \frac{1}{r} \delta(ct - r)\psi(Q) \, dV_Q = \frac{1}{4\pi c} \int_\Omega \int_0^\infty \frac{1}{r} \delta(ct - r)\psi(Q) r^2 \, dr \, d\Omega$$
$$= \frac{1}{4\pi c} \int_\Omega [r\psi(Q)]_{r=ct} \, d\Omega$$
$$= \frac{t}{4\pi} \int_\Omega [\psi(Q)]_{r=ct} \, d\Omega = tM_{ct}\psi(P)$$

Therefore by Stokes' rule (see Theorem 2.5.1), which is also evident from the last term in (10.2.14), the solution $u(P)$ of the initial problem in three-dimensional space with data (10.2.13) is

(10.2.23) $$u(P, t) = tM_{ct}\psi(P) + \frac{\partial}{\partial t}[tM_{ct}\varphi(P)]$$

EXERCISES 10.2

1. Show that the Fourier transform of $H(x + ct) - H(x - ct)$ is
$$\sqrt{\frac{2}{\pi}} \frac{\sin cts}{s}$$

2. Show that the bilinear expression for the vibrating string equation $u_{tt} = c^2 u_{xx}$ on $-\infty < x < \infty$ is
$$K(x, t) = \frac{1}{2\pi} \int_{-\infty}^{\infty} e^{ixs} \frac{\sin cst}{cs} H(t) \, ds$$
Evaluate the integral and compare with (2.5.5).

3. If $|a| < l$, $|b| < l$, show that, for $|x| < l$,
$$H(x - a) - H(x - b) - H(-a - x) + H(-b - x)$$
$$= \frac{2}{\pi} \sum_{n=1}^{\infty} \frac{1}{n}\left(\cos\frac{n\pi a}{l} - \cos\frac{n\pi b}{l}\right) \sin\frac{n\pi x}{l}$$

4. Show that the bilinear series for the string equation on $-l \leq x \leq l$, with Dirichlet end conditions, is
$$K(x, x_1, t) = \frac{2}{\pi c} \sum_{n=1}^{\infty} \frac{1}{n} \sin\frac{n\pi ct}{l} \sin\frac{n\pi x}{l} \sin\frac{n\pi x_1}{l}$$

Transform the series into the expression
$$\frac{1}{2c} \sum_{m=-\infty}^{\infty} [H(ct + x - x_1 - 2ml) - H(ct - x - x_1 - 2ml)$$
$$- H(-ct + x - x_1 - 2ml) + H(-ct - x - x_1 - 2ml)]$$
and interpret.

5. By means of Weber's integral

$$\int_0^\infty J_0(xs) \sin cst\, ds = \frac{H(c^2t^2 - x^2)}{\sqrt{c^2t^2 - x^2}} \qquad ct > 0$$

show that the bilinear series for the two-dimensional wave equation $u_{tt} = c^2(u_{xx} + u_{yy})$, $-\infty < x, y < \infty$, gives

$$K(x, y, t) = \frac{1}{2\pi c} \frac{H(c^2t^2 - x^2 - y^2)}{\sqrt{c^2t^2 - x^2 - y^2}} H(t)$$

6. State analogues of Theorem 10.2.2 if the boundary conditions on S are of the second or third kind.

7. Discuss the series which would appear in Exercise 10.2.4 if the Neumann end conditions $u_x(-l, t) = u_x(l, t) = 0$ are assigned.

8. *Mean values.* If $v(r, t)$ denotes the mean value of $u(P, t)$, where $u_{tt} = c^2 \Delta u$, over the surface of a sphere of radius r and center P, show that

$$v_{tt} = c^2 \left(v_{rr} + \frac{2}{r} v_r \right)$$

Deduce that

$$rv = f(ct - r) + g(ct + r)$$

9. *Poisson's solution.* (a) If the solution u of Exercise 10.2.8 is bounded, show that $f(ct) + g(ct) = 0$.
  (b) Show that $v(0, t) = 2g'(ct)$.
  (c) Show that $rv_t(r, t) = cf'(ct - r) + cg'(ct + r)$ and $[rv(r, t)]_r = -f'(ct - r) + g'(ct + r)$.
  (d) Show that $2g'(r) = [rv_0]_r + (r/c)v_{0t}$, where the subscript zero indicates the insertion of initial data for v and v_t.
  (e) Deduce the Poisson solution (10.2.23).

10. Deduce the rules of Stokes and Duhamel (Theorems 2.5.1 and 2.5.2) from the representation formula of Theorem 10.2.2.

11. Find the Green's function for wave propagation in the half-space $z > 0$:
  (a) When the wave is zero on $z = 0$.
  (b) When the derivative $\partial u/\partial z$ is zero on $z = 0$.

12. Show that the solution of the homogeneous wave equation in the half-space $z > 0$, with $u(x, y, 0, t) = \delta(x)\delta(y)\delta(t)$ is

$$-H(t) \frac{cz}{4\pi r^2} \left[\frac{\delta(ct - r)}{r} + \delta'(ct - r) \right]$$

where $r^2 = x^2 + y^2 + z^2$.

13. For an arbitrary function f of several variables, show that the integrals

$$\int_0^{2\pi} \int_0^\pi f(x \sin\theta \cos\varphi + y \sin\theta \sin\varphi + z \cos\theta - ct; \theta, \varphi)\, d\theta\, d\varphi$$

and

$$\int_0^{2\pi} \int_0^\pi f(x \cos\theta + y \sin\theta + iz, x \sin\varphi + y \cos\varphi - ct, \theta, \varphi)\, d\theta\, d\varphi$$

are solutions of the wave equation.

10.3 APPLICATIONS. HUYGENS' PREMISE

The source solution for the wave equation has numerous applications and extensions. Some of the more immediate instances will be given here, together with a discussion of Huygens' premise, and of the difference between two- and three-dimensional wave propagation.

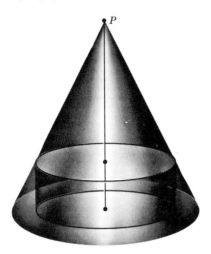

Figure 10.6 Integration domain for retarded potential.

The volume potential term of the main representation theorem (Theorem 10.2.2) appears repeatedly in electrodynamics. For the three-dimensional wave equation with a space-time source density $f(\tau, Q)$, this term is

$$(10.3.1) \quad \int_0^t \int_{-\infty}^{\infty} K(P, Q, t - \tau) f(\tau, Q) \, d\tau \, dV_Q$$

$$= \int_0^t \int_0^{\infty} \int_{\Omega} \frac{\delta(c(t - \tau) - r)}{4\pi c r} f(\tau, Q) \, d\tau \, r^2 dr \, d\Omega$$

where $r = r(P, Q)$ is measured from P. Let us carry out the time integration by using the substitution property of the delta function, and by observing that an additional factor c will thereafter appear in the denominator. We obtain the *retarded potential*

$$(10.3.2) \quad \frac{1}{4\pi c^2} \int_0^{ct} \int_{\Omega} f\left(t - \frac{r}{c}, Q\right) r \, dr \, d\Omega = \frac{1}{4\pi c^2} \int_{r \leq ct} f\left(t - \frac{r}{c}, Q\right) \frac{dV}{r}$$

This potential, which was found by Lorentz, expresses the solution at (P, t) as a sum of contributions from all points on the retrograde light cone with vertex at (P, t). Each contribution was "emitted" at a time r/c earlier, and is attenuated with distance by the factor $1/r$ (Figure 10.6).

When combined with Poisson's solution (10.2.23), the retarded potential gives the solution of the general initial value problem in three-dimensional space. All of the terms in this solution exhibit the same property of "clean-cut" wave propagation. The domain of dependence is the surface of the light cone only. Thus if a source is terminated at time t_1, the signal from that source, when received at a distance r, will terminate at the time $t_1 + r/c$. In fact, if the signal is emitted precisely at instant $t_1[f(Q, t) = \delta(Q, Q_0)\delta(t - t_1)]$, then at time t the signal lies on a spherical shell (or surface) of radius $r = c(t - t_1)$. (Figure 10.7.)

Wave propagation of this nature is said to satisfy *Huygens' premise*. As the traveling wave front leaves no residual wave behind, the propagation is also

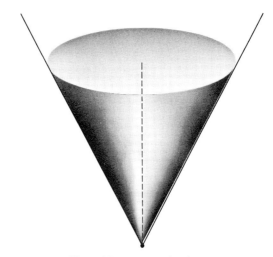

Figure 10.7 Locus of a signal.

described as clean-cut. Since the clarity of signals by light and sound depends upon this clean-cut property, it is one of the most significant basic facts of the universe of sight and sound in which we live. (Baker and Copson, Ref. 2.)

The reader may be tempted to suppose that clean-cut wave propagation is an inevitable property of the solutions of wave equations. We will now show that it does *not* hold for the two-dimensional wave equation

(10.3.3) $$u_{tt} = c^2(u_{xx} + u_{yy})$$

The source solution for this equation could be found from its bilinear expansion (Exercise 10.2.5). However, we will calculate it here by the *method of descent* of dimension, starting from the three-dimensional solution (10.2.20). For the three-dimensional wave equation, let us postulate a *line source* parallel to the z-axis:

(10.3.4) $$f(Q, \tau) = \delta(x)\delta(y)\delta(t)$$

The solution arising from this source must be independent of z, because it is unique and is also unchanged by translation parallel to the z-axis. Therefore the z derivative in the wave equation vanishes and the solution satisfies the two-dimensional equation (10.3.3), but with the right-hand term (10.3.4). Therefore, by definition, it is the source solution for two dimensions.

To calculate this solution, we must integrate (10.2.20) over the entire z-axis. Thus, with subscripts to specify dimension,

$$K_2(x, y, t) = \int_{-\infty}^{\infty} K_3(x, y, z, t)\, dz$$
(10.3.5)
$$= \int_{-\infty}^{\infty} \frac{H(t)\, \delta(ct - r)}{4\pi cr}\, dz$$

According to Figure 10.8, there will be no contribution from the line source until $ct = \sqrt{x^2 + y^2} = \rho$, say. Thus we should adjoin the Heaviside factor $H(c^2 t^2 - \rho^2)$. At the instant $t = \rho/c$ the expanding cylindrical wave front meets the point $(x, y, 0)$. Subsequently, signals are received there from increasingly distant points of the infinite line source.

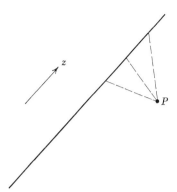

Figure 10.8
Descent to two dimensions.

Since
$$r^2 = x^2 + y^2 + z^2 = \rho^2 + z^2$$
we have
$$\frac{dz}{r} = \frac{dr}{z} = \frac{dr}{\pm\sqrt{r^2 - \rho^2}}$$

The positive and negative halves of the z-axis each contribute at one point, namely, $z = \pm\sqrt{c^2 t^2 - \rho^2}$. The two contributions are equal, and the solution (10.3.5) becomes

$$K_2(x, y, t) = \frac{H(c^2 t^2 - \rho^2)}{4\pi c} 2 \int_0^\infty \delta(ct - r) \frac{dr}{\sqrt{r^2 - \rho^2}} H(t)$$
(10.3.6)
$$= \frac{H(c^2 t^2 - \rho^2)}{2\pi c} \frac{H(t)}{\sqrt{c^2 t^2 - \rho^2}}$$

Since this source solution is not zero within the light cone, the propagation of waves in two dimensions is accompanied by residual, or lingering, waves. We summarize this as follows:

THEOREM 10.3.1 *The support of the source solution for two-dimensional wave propagation is the interior of the retrograde light cone.*

Conversely, the region of influence of a signal is the interior of the forward light cone based at the point of emission. It is interesting to speculate on the practical difficulties of radio or voice communication in a two-dimensional world.

388   WAVE PROPAGATION IN SPACE

Returning now to three dimensions of space, we shall calculate the source solution for the damped wave equation

(10.3.7) $$\Box u = u_{tt} - c^2 \Delta u = lu$$

The one-dimensional equation of this form was treated in Chapter 2, Section 2.5. We again assume that l is a constant.

The bilinear expansion for the source function $K_l(P, Q, t)$ of a region R differs from (10.2.7) in only one respect. The coefficients $c_n(t)$ now satisfy an equation

$$\frac{d^2}{dt^2} c_n(t) + (c^2 \lambda_n - l) c_n(t) = \delta(t)$$

with $c_n(t) \equiv 0$ for $t < 0$ and therefore,

(10.3.8) $$c_n(t) = \frac{\sin c k'_n t}{c k'_n}$$

where

(10.3.9) $$k'^2_n = \lambda_n - \frac{l}{c^2}$$

The results of Theorems 10.2.1 and 10.2.2 are still valid, as the reader can easily verify. (Courant-Hilbert, Ref. 13, vol 2, Duff, Ref. 15, Chapter 10.)

We now construct the source solution for the full three-dimensional space. By analogy with (10.2.17), it is the integral

(10.3.10) $$K_l^{(3)}(\mathbf{x}, t) = \frac{H(t)}{(2\pi)^3} \iiint_{-\infty}^{\infty} e^{i\mathbf{s}\cdot\mathbf{x}} \frac{\sin(\sqrt{c^2 s^2 - l}\, t)}{\sqrt{c^2 s^2 - l}} \, ds_1 \, ds_2 \, ds_3$$

As the straightforward evaluation of this integral is difficult, we shall instead compare it with the one-dimensional Green's function for the damped-wave equation, which we calculated in Chapter 2, Section 2.5. The bilinear expression for this one-dimension solution is

(10.3.11)
$$K_l^{(1)}(x, t) = \frac{H(t)}{2\pi} \int_{-\infty}^{\infty} e^{isx} \frac{\sin(\sqrt{c^2 s^2 - l}\, t)}{\sqrt{c^2 s^2 - l}} \, ds$$

$$= \frac{H(t)}{\pi} \int_0^{\infty} \cos sx \, \frac{\sin(\sqrt{c^2 s^2 - l}\, t)}{\sqrt{c^2 s^2 - l}} \, ds$$

Now we express the three-dimensional integral as

(10.3.12)
$$K_l^{(3)}(\mathbf{x}, t) = \frac{H(t)}{(2\pi)^3} \int_0^{\infty} \int_0^{2\pi} \int_0^{\pi} e^{is|\mathbf{x}|\cos\theta} \frac{\sin(\sqrt{c^2 s^2 - l}\, t)}{\sqrt{c^2 s^2 - l}} s^2 \sin\theta \, ds \, d\theta \, d\varphi$$

$$= \frac{H(t)}{(2\pi)^2} \int_0^{\infty} \left(\int_0^{\pi} e^{is|\mathbf{x}|\cos\theta} \sin\theta \, d\theta \right) \frac{\sin(\sqrt{c^2 s^2 - l}\, t)}{\sqrt{c^2 s^2 - l}} s^2 \, ds$$

$$= \frac{2H(t)}{(2\pi)^2 |\mathbf{x}|} \int_0^{\infty} \sin|\mathbf{x}| s \, \frac{\sin(\sqrt{c^2 s^2 - l}\, t)}{\sqrt{c^2 s^2 - l}} s \, ds$$

$$= \frac{-H(t)}{2\pi |\mathbf{x}|} \frac{d}{d|\mathbf{x}|} K_l^{(1)}(|\mathbf{x}|, t)$$

APPLICATIONS. HUYGENS' PREMISE   389

By (2.5.5), (2.5.35), and Exercise 2.5.11 this becomes

$$K_l^{(3)}(\mathbf{x}, t) = \frac{-H(t)}{4\pi c\, |\mathbf{x}|}\frac{d}{d\,|\mathbf{x}|}\left\{I_0\left(\sqrt{l\left(t^2 - \frac{|\mathbf{x}|^2}{c^2}\right)}\right)H(c^2 t^2 - |\mathbf{x}|^2)\right\}$$

$$= \frac{H(t)}{4\pi c\, |\mathbf{x}|}\,\delta(ct - |\mathbf{x}|)I_0\left(\sqrt{l\left(t^2 - \frac{|\mathbf{x}|^2}{c^2}\right)}\right)$$

(10.3.13)

$$+ \frac{H(t)H(c^2 t^2 - |\mathbf{x}|^2)}{4\pi c^2}\,\frac{\sqrt{l}}{\sqrt{c^2 t^2 - |\mathbf{x}|^2}}\,I_0'\left(\sqrt{l\left(t^2 - \frac{|\mathbf{x}|^2}{c^2}\right)}\right)$$

$$= \frac{H(t)\delta(ct - |\mathbf{x}|)}{4\pi c\,|\mathbf{x}|} + \frac{H(t)H(c^2 t^2 - |\mathbf{x}|^2)}{4\pi c^2\sqrt{c^2 t^2 - |\mathbf{x}|^2}}\,\sqrt{l}\,I_1\left(\sqrt{l\left(t^2 - \frac{|\mathbf{x}|^2}{c^2}\right)}\right)$$

In this last step we use the property $I_0(0) = 1$ to drop the Bessel function in the leading term, and the relation $I_0'(z) = I_1(z)$.

For $l = 0$ this Green's function reduces, as it must, to the solution (10.2.20) for the wave equation. However, for $l \neq 0$, the second term has nonzero values throughout the interior of the wave cone.

THEOREM 10.3.2   *The support of the source solution for three-dimensional damped wave propagation is the surface and interior of the retrograde light cone.*

EXERCISES 10.3

1. Construct the Lorentz retarded potential for the problem with initial time $t_0 = -\infty$.

2. *Olbers' paradox*. Show that in a static universe in three-dimensional Euclidean space, with a uniform density of stars, the intensity of light must be infinite.

3. Show that the retarded potential (10.3.2) can be expressed as

$$\int_0^t M_{c(t-\tau)}(f(\tau, P))\, d\tau$$

4. Construct the full solution formula for the wave equation $u_{tt} = c^2(u_{xx} + u_{yy}) + f(x, y, t)$, $t > 0$, with $u(x, y, 0) = \varphi(x, y)$ and $u_t(x, y, 0) = \psi(x, y)$.

5. A pebble falls into a pond, creating surface waves which satisfy the two-dimensional wave equation. When does the surface return to equilibrium? How can you explain the appearance of several concentric crests of the wave?

6. Do solutions of the equation of the vibrating string satisfy Huygens' premise? Do the first derivatives of the solutions satisfy it?

7. Verify that Theorem 10.2.2 holds for the damped wave equation.

8. Find a source solution for the equation of damped waves in two dimensions by descent, using a line source of density $e^{z\sqrt{l}}\delta(x)\delta(y)\delta(t)$.

9. Find the source function for the equation $u_{tt} + \alpha u_t = c^2\,\Delta u$ in three space dimensions.

390   WAVE PROPAGATION IN SPACE

10. By taking Fourier transforms with respect to the time, or otherwise, show that the two-dimensional wave equation with axial symmetry: $u_{rr} + (1/r)u_r = u_{tt}$ has the solution

$$u = \int_0^{2\pi} f(t - r\cos\theta)\, d\theta$$

11. Show that

$$u_{tt} - u_{rr} - \frac{1}{r}u_r = \frac{\delta(r)}{r}H(t)$$

has the solution $u = \cosh^{-1}\left(\dfrac{t}{r}\right) H(t - r)$ and interpret the result.

12. Show that the solution of the problem of the moving source:

$$u_{tt} - c^2(u_{xx} + u_{yy}) = \delta(x - Ut)\delta(y)H(t)$$

is

$$u = \frac{1}{2\pi c}\int_0^t \frac{H(c^2(t-\tau)^2 - \rho^2)}{\sqrt{c^2(t-\tau)^2 - \rho^2}}\, d\tau$$

where $\rho^2 = (x - U\tau)^2 + y^2$, and U is a constant. Describe the singularities of this solution for:
  (a) subsonic motion $U < c$;
  (b) sonic speed $U = c$;
  (c) supersonic speed $U > c$.

13. The problem of supersonic flow of Mach number $M > 1$ past a wing lying in the plane $y = 0$ gives rise to this problem: Find the velocity perturbation potential φ, satisfying

$$\varphi_{xx} + \varphi_{yy} - B^2\varphi_{zz} = 0 \qquad B = \sqrt{M^2 - 1} > 0 \qquad y > 0$$

with $\partial\varphi/\partial y = g(x, z)$ for $y = 0$, where $g(x, z)$ is given on the wing and is zero elsewhere on the plane $y = 0$. Show that the solution is

$$\varphi(x, y, z) = -\frac{1}{\pi}\iint_{-\infty}^{\infty} g(x_1, z_1)\, dx_1\, dz_1\, \frac{H[(z - z_1)^2 - B^2(x - x_1)^2 - B^2 y^2]}{\sqrt{(z - z_1)^2 - B^2(x - x_1)^2 - B^2 y^2}}$$

10.4 ELECTROMAGNETIC AND ELASTIC WAVES

To illustrate the use of the Green's function (10.2.20) and the widespread occurrence of the scalar wave equation (10.1.1) in physical applications, we shall construct solutions of two important systems of wave equations. For simplicity, we consider the initial value problem in an unbounded three-dimensional space.

With a suitable choice of units, Maxwell's equations (see Chapter 5, Section 5.6) can be written

(10.4.1)
$$\operatorname{div} \mathbf{E} = 0 \qquad \operatorname{div} \mathbf{H} = 0$$
$$\frac{\partial \mathbf{E}}{\partial t} = \operatorname{curl} \mathbf{H} \qquad \frac{\partial \mathbf{H}}{\partial t} = -\operatorname{curl} \mathbf{E}$$

These homogeneous equations govern the electric and magnetic fields in empty space free of charge or current. Let initial values

(10.4.2) $\quad\quad\quad\quad \mathbf{E}(P, 0) = \mathbf{E}_0(P) \quad\quad \mathbf{H}(P, 0) = \mathbf{H}_0(P)$

be assigned throughout space, and let us calculate the electromagnetic field at a subsequent time $t > 0$.

In Chapter 5 it was shown that $\mathbf{E}$ and $\mathbf{H}$ satisfy wave equations which involve their Cartesian components separately. For the present case we have

$$\mathbf{E}_{tt} = \operatorname{curl} \mathbf{H}_t = -\operatorname{curl} \operatorname{curl} \mathbf{E}$$

By the first of (10.4.1) and the identity (5.3.11) for the vector Laplacian, we find

(10.4.3) $\quad\quad\quad\quad\quad\quad \mathbf{E}_{tt} = \Delta \mathbf{E}$

For this second-order wave equation we must specify the initial time derivative of the electric-field vector. From the Maxwell equations (10.4.1) we see that

(10.4.4) $\quad\quad\quad \mathbf{E}_t(P, 0) = \operatorname{curl} \mathbf{H}(P, 0) = \operatorname{curl} \mathbf{H}_0(P)$

Therefore, by Poisson's solution (10.2.23), we have the solution formula

(10.4.5) $\quad\quad\quad \mathbf{E}(P, t) = \dfrac{\partial}{\partial t}[tM_t(\mathbf{E}_0(P))] + tM_t(\operatorname{curl} \mathbf{H}_0(P))$

A similar calculation leads to

(10.4.6) $\quad\quad\quad \mathbf{H}(P, t) = \dfrac{\partial}{\partial t}[tM_t(\mathbf{H}_0(P))] - tM_t(\operatorname{curl} \mathbf{E}_0(P))$

To complete our solution we must show that the subsidiary conditions div $\mathbf{E} = 0$ and div $\mathbf{H} = 0$ remain valid at later times provided they are true initially. We therefore require

(10.4.7) $\quad\quad\quad\quad \operatorname{div} \mathbf{E}(P, 0) = \operatorname{div} \mathbf{E}_0(P) = 0$
$\quad\quad\quad\quad\quad\quad \operatorname{div} \mathbf{H}(P, 0) = \operatorname{div} \mathbf{H}_0(P) = 0$

From the Maxwell equation $\partial \mathbf{E}/\partial t = \operatorname{curl} \mathbf{H}$, we obtain

(10.4.8) $\quad\quad\quad\quad \dfrac{\partial \theta}{\partial t} = \dfrac{\partial \operatorname{div} \mathbf{E}}{\partial t} = \operatorname{div} \operatorname{curl} \mathbf{H} \equiv 0$

by taking the divergence. Therefore $\theta_E(t) = \theta_E(0) = 0$ and the electric-field vector is always solenoidal. Similarly, we can show that $\theta_H \equiv \operatorname{div} \mathbf{H}$ is identically zero. Verification of the remaining two Maxwell equations is straightforward, and has been set as an exercise.

Inspection of the solution formulas shows that Huygens' premise is satisfied by electromagnetic wave propagation.

Next we shall study the initial value problem for the elastic-wave system (5.4.17). It is sufficient to consider the homogeneous differential equations with vanishing initial values and arbitrary initial velocities. For the three-dimensional vector $u_r(P, t)$, we therefore have

(10.4.9) $$\rho u_{r,tt} = (\lambda + \mu)u_{s,sr} + \mu \Delta u_r$$

together with initial data

(10.4.10) $$u_r(P, 0) = 0 \quad u_{r,t}(P, 0) = g_r(P)$$

As we saw in Chapter 5, Section 5.4, the elastic wave system leads to scalar and vector wave equations with differing propagation velocities. The divergence $\theta = u_{r,r}$, according to (5.4.18), satisfies the scalar wave equation

(10.4.11) $$\theta_{tt} = c_1^2 \Delta \theta \quad c_1^2 = \frac{\lambda + 2\mu}{\rho}$$

with pressure-wave velocity c_1. Since the initial value for θ is zero while the initial time derivative is

(10.4.12) $$\theta_t(P, 0) = \operatorname{div} u_{r,t}(P, 0) = u_{r,rt}(P, 0) = g_{r,r}(P)$$

we can with the aid of (10.2.23) write down the solution for θ in the mean value form

(10.4.13) $$\theta(P, t) = tM_{c_1 t}(g_{r,r}(P)) = \frac{t}{4\pi} \iint_\Omega g_{r,r}(x_s + c_1 t l_s)\, d\Omega$$

where $P = (x_1, x_2, x_3)$ and (l_1, l_2, l_3) is a unit vector ranging over all directions. We see that

(10.4.14) $$\theta(P, t) = v_{m,m}(P, t)$$

where

(10.4.15) $$v_m(P, t) = \frac{t}{4\pi} \iint_\Omega g_m(x_s + c_1 t l_s)\, d\Omega$$

Since θ is known, we can return to the elastic wave system itself and write

(10.4.16) $$\begin{aligned} u_{r,tt} - c_2^2 \Delta u_r &= (c_1^2 - c_2^2)\theta_r \\ &= (c_1^2 - c_2^2)v_{m,mr} \end{aligned}$$

where $c_2^2 = \mu/\rho$ as in (5.4.21), and $(\lambda + \mu)/\rho = c_1^2 - c_2^2$. Rather than introduce a conventional vector potential, we shall work directly with (10.4.16), regarding the known right-hand side as a nonhomogeneous term. Solution of such a wave equation for each component u_r is possible by the use of the retarded potential (10.3.2).

Before launching into the calculation that will be required, we shall comment on the interpretation of this method and the nature of the solution to be found. The wave equation (10.4.11) describes the intensity of compression of the solid medium, and the waves of compression are transmitted with pressure-wave velocity $c_1 > c_2$. These waves spread from a source along the characteristic cone of propagation speed c_1, which is therefore the support of θ and v_r. From (10.4.16) we see that the compression θ acts as a scalar potential generating waves u_r which travel with the shear-wave velocity c_2. Since $c_2 < c_1$, these

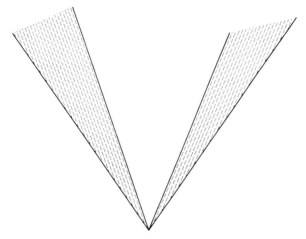

Figure 10.9 Elastic waves.

shear waves lag behind the pressure-wave front in space (Figure 10.9) and their support is a region contained within that spherical front. As we shall see, this region lies between the pressure-wave front $r = c_1 t$ and the shear-wave front $r = c_2 t$, where r is distance from the source.

To satisfy (10.4.16), we shall use the particular integral $w_{m,mr}$, where w_m satisfies the equation

$$w_{m,tt} - c_2^2 \Delta w_m = (c_1^2 - c_2^2) v_m$$

together with vanishing initial conditions. The vector w_m is actually a vector potential especially adapted to the elastic wave system.

Now, by (10.3.2),

(10.4.17) $$w_m(P, t) = \frac{c_1^2 - c_2^2}{4\pi c_2^2} \iiint_{r \leq c_2 t} v_m\left(Q, t - \frac{r}{c_2}\right) \frac{dV_Q}{r}$$

where $r = r(P, Q)$ is the distance from P to Q. Before substituting (10.4.15) for v_m, we shall express it as a surface integral:

(10.4.18) $$v_m(Q, t) = \frac{1}{4\pi c_1^2 t} \iint g_m(y_s + c_1 t l_s) \, dS$$

394 WAVE PROPAGATION IN SPACE

Inserting a delta function of radial distance, we can then write (10.4.18) as the volume integral

(10.4.19) $$v_m(Q, t) = \frac{1}{4\pi c_1^2 t} \iiint g_m(y_s + z_s) \delta(\rho - c_1 t) \, dV_Z$$

where $dV_z = \rho^2 \, d\rho \, d\Omega$, and the formal integration extends over the entire space. Here ρ denotes the radial distance from Q to Z.

We shall also extend the formal integration in (10.4.17) over the complete space by introducing a Heaviside factor. Thus, combining (10.4.17) and (10.4.19), we have

(10.4.20)
$$w_m(P, t) = \frac{c_1^2 - c_2^2}{4\pi c_2^2} \iiint v_m\!\left(Q, t - \frac{r}{c_2}\right) H\!\left(t - \frac{r}{c_2}\right) \frac{dV_Q}{r}$$
$$= \frac{c_1^2 - c_2^2}{4\pi c_2^2} \iiint \frac{H(t - r/c_2)}{4\pi c_1^2(t - r/c_2)} \frac{dV_Q}{r} \iiint g_m(y_s + z_s)$$
$$\times \delta\!\left(\rho - c_1\!\left(t - \frac{r}{c_2}\right)\right) dV_Z$$

Now let Z be the point with coordinates $y_s + z_s$, so that $\rho = |z_s|$ is the distance from Q to Z. Then we can interchange the order of the volume integrations and so find

(10.4.21)
$$w_m(P, t) = \frac{c_1^2 - c_2^2}{16\pi^2 c_1^2 c_2^2} \iiint g_m(Z) \, dV_Z$$
$$\times \iiint \frac{H(t - r/c_2)}{t - r/c_2} \delta\!\left(\rho - c_1\!\left(t - \frac{r}{c_2}\right)\right) \frac{dV_Q}{r}$$

We shall evaluate the inner integral, which expresses the generation of shear waves from the more rapid pressure waves. Let

(10.4.22) $$F(P, Z, t) = \iiint \frac{H(t - r/c_2)}{t - r/c_2} \delta\!\left(\rho - c_1\!\left(t - \frac{r}{c_2}\right)\right) \frac{dV_Q}{r}$$

and choose spherical polar coordinates r, θ, φ with origin at P and north polar axis PZ. Denoting the distance PZ by R, we have

(10.4.23) $$\rho^2 = R^2 + r^2 - 2Rr \cos \theta = R^2 + r^2 - 2Rr\mu$$

where $\mu = \cos \theta$, and

$$dV = r^2 \sin \theta \, dr \, d\theta \, d\varphi = -r^2 \, dr \, d\mu \, d\varphi$$

Thus $\rho \, d\rho = -Rr \, d\mu$, when r and φ are fixed, and we have

(10.4.24)
$$F(P, Z, t) = \int_0^\infty \int_0^{2\pi} \int_{-1}^1 \frac{r^2 \, dr \, d\mu \, d\varphi}{r} \frac{H(t - r/c_2)}{t - r/c_2} \delta\!\left(\rho - c_1\!\left(t - \frac{r}{c_2}\right)\right)$$
$$= 2\pi \int_0^\infty \int \frac{r \, dr}{r} \frac{\rho \, d\rho}{R} \frac{H(t - r/c_2)}{t - r/c_2} \delta\!\left(\rho - c_1\!\left(t - \frac{r}{c_2}\right)\right)$$
$$= \frac{2\pi}{R} \int_0^{c_2 t} \frac{dr}{t - r/c_2} \int \rho \, d\rho \, \delta\!\left(\rho - c_1\!\left(t - \frac{r}{c_2}\right)\right)$$

The limits of integration for ρ are found from (10.4.23) to be $|R - r| \leq \rho \leq R + r$. In the integration over ρ, we have a contribution only when $\rho = c_1(t - r/c_2)$, and therefore the quotient of the two linear factors in the integrand is just c_1. Therefore,

(10.4.25)
$$F(P, Z, t) = \frac{2\pi c_1}{R} \int_0^{c_2 t} dr \int_{|R-r|}^{R+r} \delta\left(\rho - c_1\left(t - \frac{r}{c_2}\right)\right) d\rho$$

$$= \frac{2\pi c_1}{R} \int_0^{c_2 t} \left[H\left(R + r - c_1\left(t - \frac{r}{c_2}\right)\right) - H\left(|R - r| - c_1\left(t - \frac{r}{c_2}\right)\right)\right] dr$$

Evaluation of this integral is elementary, and the result is

(10.4.26)
$$F(P, Z, t) = \begin{cases} \dfrac{4\pi c_1 c_2}{c_1 + c_2} & 0 < R < c_2 t \\ \dfrac{4\pi c_1 c_2^2}{c_1^2 - c_2^2}\left(\dfrac{c_1 t}{R} - 1\right) & c_2 t < R < c_1 t \\ 0 & c_1 t < R \end{cases}$$

This can also be written as

(10.4.27) $\quad F(P, Z, t) = F(R, t) = \dfrac{4\pi c_1^2 c_2^2}{c_1^2 - c_2^2}\left\{\dfrac{1}{R}\left(t - \dfrac{R}{c_1}\right)_+ - \dfrac{1}{R}\left(t - \dfrac{R}{c_2}\right)_+\right\}$

with the $+$ notation of the power distribution of Chapter 2, Section 2.4. From (10.4.21) we have

(10.4.28) $\quad w_m(P, t) = \dfrac{1}{4\pi c_1 c_2} \iiint \left\{c_2\left(\dfrac{c_1 t}{R} - 1\right)_+ - c_1\left(\dfrac{c_2 t}{R} - 1\right)_+\right\} g_m(Z) \, dV_Z$

where $R = R(P, Z)$

We leave to the reader the details of the differentiations necessary to satisfy (10.4.16), and we quote the final solution. To the volume potentials arising from (10.4.16) must be added the solution for the initial data (10.4.10). The source solution incorporating both contributions, which is the Green's matrix of the elastic wave system, is

(10.4.29)
$$K_{sm}(P, Z, t) = \frac{t}{4\pi R^2}[l_s l_m \delta(c_1 t - R) + (\delta_{sm} - l_s l_m) \delta(c_2 t - R)]$$
$$+ \frac{t}{4\pi R^3}(3 l_s l_m - \delta_{sm})[H(c_1 t - R) - H(c_2 t - R)]$$

where l_s is the unit vector parallel to PZ.

The waves emanating from a point source contain two surface contributions or "sharp" waves. Between the arrival of the faster pressure wave and the

These space-time curves are known in the theory of partial differential equations as *bicharacteristic curves* of the wave equation. The characteristic surfaces, which satisfy the first order equation (10.1.6), can in fact be constructed by families of the bicharacteristic curves. (Friedlander, Ref. 21.)

By means of the eikonal equation we can find differential equations that completely determine the bicharacteristic curves. From (10.5.3) we have

$$\frac{d}{ds}\left(\frac{\partial \psi}{\partial x_k}\right) = \frac{\partial^2 \psi}{\partial x_k \, \partial x_r} \frac{dx_r}{ds} = \frac{\partial^2 \psi}{\partial x_k \, \partial x_r} \frac{\partial \psi}{\partial x_r}$$

$$= \frac{1}{2}\frac{\partial}{\partial x_k}\left(\frac{\partial \psi}{\partial x_r}\frac{\partial \psi}{\partial x_r}\right) = \frac{1}{2}\frac{\partial}{\partial x_k}\frac{1}{c^2}$$

according to (10.5.2). We shall assume that the wave velocity c is constant in space and time and therefore

(10.5.5) $$\frac{d}{ds}\left(\frac{\partial \psi}{\partial x_k}\right) = 0$$

THEOREM 10.5.1 *The rays are straight lines.*
Proof. Along a ray the direction ratios (10.5.3) are constant, by (10.5.5).

The eikonal equation (10.5.2) has a family of linear solutions which can easily be found by an additive separation of variables, namely,

(10.5.6) $$\psi = \frac{1}{c}(l_1 x_1 + l_2 x_2 + l_3 x_3)$$

where

(10.5.7) $$l^2 = l_1^2 + l_2^2 + l_3^2 = 1$$

The corresponding characteristic surfaces

$$t = \psi = \frac{1}{c}(l_1 x_1 + l_2 x_2 + l_3 x_3) + \text{const.}$$

represent plane wave fronts traveling with velocity c in the direction specified by the vector (l_1, l_2, l_3). In the theory of first order partial differential equations it is shown that every characteristic surface is an envelope of the plane wave fronts. Correspondingly, every solution of the wave equation can be represented by a superposition of plane waves traveling in various directions. Actually, the resolving of a wave function into its Fourier components shows that the plane waves can even be taken as pure imaginary exponentials.

Consider a wave solution which is singular on a certain characteristic surface and which represents a wave with a sharp wave front advancing along a two-parameter family of rays. We will show that on each of these rays the magnitude of the singularity is governed by an ordinary differential equation.

For simplicity we assume that our wave is zero on one side of the characteristic surface. The general case can be reduced to this by subtraction of a nonsingular solution. We therefore write

(10.5.8) $$u = vH(\varphi)$$

where the Heaviside function is zero on the side $\varphi < 0$ of the characteristic surface $\varphi = 0$. By differentiation, we find

$$u_t = v_t H(\varphi) + v\delta(\varphi)\varphi_t$$
$$u_{tt} = v_{tt}H(\varphi) + 2v_t\varphi_t\delta(\varphi) + v\varphi_{tt}\delta(\varphi) + v\varphi_t^2\delta'(\varphi),$$

together with similar equations for the space derivations.

Combining these second derivatives in the manner suggested by the wave equation, we find

(10.5.9) $$0 = \Box u = \Box vH(\varphi) + [2(v_t\varphi_t - c^2\nabla v \cdot \nabla \varphi) + v\Box(\varphi)]\delta(\varphi)$$

Here the term in $\delta'(\varphi)$ has coefficient zero since $\varphi = $ constant is a characteristic surface. Since v is a smooth function, by hypothesis, the relation (10.5.9) can only hold if, for $\varphi = 0$, the coefficient of $\delta(\varphi)$ vanishes. Thus we deduce that on the surface

(10.5.10) $$2(v_t\varphi_t - c^2\nabla v \cdot \nabla \varphi) + v\Box(\varphi) = 0$$

The four-dimensional derivative of v appearing here is a differentiation along the bicharacteristic curve, for $\varphi = t - \psi(x)$ and

$$\frac{dv}{ds} = \frac{\partial v}{\partial t}\frac{dt}{ds} + \frac{\partial v}{\partial x_k}\frac{dx_k}{ds}$$

(10.5.11) $$= v_t \frac{1}{c^2} + v_k\psi_k$$

$$= \frac{1}{c^2}(v_t\varphi_t - c^2 v_k\varphi_k)$$

Here summation over the values 1, 2, 3 is understood for the index k.

Therefore v satisfies the *transport equation*

(10.5.12) $$2c^2\frac{dv}{ds} + v\Box(\varphi) = 0$$

along the ray. This homogeneous first order ordinary differential equation shows that if v does not vanish at one point of a ray, it cannot vanish anywhere on the ray.

Let us study the variation of v along the ray. It is constant for a plane wave and varies as $1/r$ for a spherical wave, as examples already known to us indicate. In general, the variation of v will depend on the curvature of the wave front. Let $v(x_k, t) = v(x_k, \psi(x))$ be constructed as a compounded function of x_k. Then, since $\varphi = t - \psi(x)$, we have

$$v_t \varphi_t - c^2 \nabla v \cdot \nabla \varphi = v_t + c^2 v_k \psi_k$$
$$= c^2 \left(\frac{1}{c^2} v_t + v_k \psi_k \right) = c^2 (\psi_k \psi_k v_t + v_k \psi_k)$$
$$= c^2 \psi_k (v_k + \psi_k v_t) = c^2 \psi_k \frac{dv}{dx_k}$$

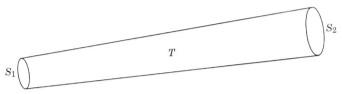

Figure 10.12 A tube of rays.

where the straight d denotes a total derivative. Inserting this in (10.5.10), we can write $\Box \varphi = -c^2 \Delta \varphi = c^2 \Delta \psi$ and

$$2c^2 \psi_k \frac{dv}{dx_k} + c^2 v \frac{d}{dx_k} \psi_k = 0$$

This relation has the integrating factor v and is equivalent to

(10.5.13) $$\frac{d}{dx_k} (\psi_k v^2) = \text{div} \, (v^2 \, \text{grad} \, \psi) = 0$$

We integrate this equation over a region R defined by a tube T of rays (Figure 10.12), with an initial surface S_1 and a terminal surface S_2. By the divergence theorem

$$\int_{S_2} v^2 \frac{\partial \psi}{\partial n} dS_2 + \int_{S_1} v^2 \frac{\partial \psi}{\partial n} dS_1 + \int_T v^2 \frac{\partial \psi}{\partial n} dS_T = \int_R \text{div} \, (v^2 \, \text{grad} \, \psi) \, dV = 0$$

However, on the tube $\partial \psi / \partial n = 0$ because the rays are orthogonal to the level surfaces $\psi = $ constant, which are the wave fronts. On S_1 we have

$$\frac{\partial \psi}{\partial n} = \frac{\partial t}{\partial n} = \frac{1}{c}$$

since the front advances with velocity c, and on S_2 we find

$$\frac{\partial \psi}{\partial n} = \frac{\partial t}{\partial n} = -\frac{1}{c}$$

due to the retrograde orientation of the normal.

Thus

$$\int_{S_2} v^2 \, dS_2 = \int_{S_1} v^2 \, dS_1$$

and we have proved the following result.

THEOREM 10.5.2 *A discontinuity or singularity of a wave solution is propagated along rays, the squared magnitude of the intensity v varying inversely as the area of the ray tube.*

The tracing of rays plays a prominent role in optics, and the topic we have introduced here is called *geometrical optics*.

As singularities of a wave motion cannot originate or terminate along any single ray, we can enumerate the various possible singularities in any given problem. A singularity may originate from the source density (as with the Green's function), from a singularity of the initial data, or from a singularity such as an edge or corner of the spatial domain. Finally, a solution may also become singular by the focusing of a family of rays.

EXERCISES 10.5

1. Show that the equations of a ray in a medium with variable velocity $c = c(x)$ can be written as a Hamiltonian system with $2H = (\nabla \psi)^2 - 1/c^2(x)$.

2. If a solution u together with its derivatives up to an order $m-1$ is continuous across S, but the mth derivative is discontinuous, show that its discontinuity satisfies the transport equation.

3. Sound waves in water travel at speed $c(z)$ depending on depth z. Show that the wave fronts are envelopes of the surfaces

$$\psi = \alpha x + \beta y + \int \sqrt{\frac{1}{c(z)^2} - \alpha^2 - \beta^2}\, dz = t + \text{const.}$$

4. Find a two-parameter family of wave fronts for elastic pressure waves when $c_1 = c_1(r)$, r being distance from the center of the earth. The eikonal equation is

$$\psi_r^2 + \frac{1}{r^2}\psi_\theta^2 + \frac{1}{r^2 \sin^2 \theta}\psi_\varphi^2 = \frac{1}{c_1^2(r)}$$

5. *Snell's law (acoustic case).* A plane wave of sound traveling at velocity c_1 in the medium $x > 0$, is incident at angle α on the interface $x = 0$ with a second medium of propagation velocity c_2. Show that the angle of refraction α_2 satisfies $c_2 \sin \alpha_1 = c_1 \sin \alpha_2$, the boundary conditions being u_x and ρu_t continuous.

10.6 REFLECTION AND DIFFRACTION

Up to this point we have been mainly concerned with wave propagation in an unlimited space. However, a majority of the problems met in actual practice involve the interaction of waves with boundary surfaces. We have treated numerous examples of *reflection* problems for the string and diffusion equation in our earlier chapters, and have given examples of the image methods for equations of all three types. For boundary conditions of the first or second kinds in a half-space or other suitable domain, the construction of Green's functions by placement of image sources will be successful. The reflection of a spherical wave front by a plane surface leads to a spherical reflected wave front carrying a singularity of the same order.

402   WAVE PROPAGATION IN SPACE

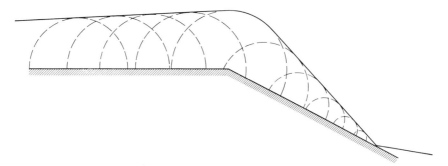

Figure 10.13 Secondary sources on reflecting wall with corner.

The transmission of waves through a boundary surface into a second medium having a different velocity of propagation results in the *refraction* or bending of the rays at the surface. In this case, a spherical wave front will acquire a more complicated structure and additional wave fronts having rectilinear generating lines may appear. As the study of these phenomena is beyond our scope, we shall restrict attention to a third type of problem, namely, *diffraction*.

According to Huygens' theory of light, every point of a wave front can act as a secondary source, emitting spherical waves of which the envelope at a later instant forms the continuing wave front. It follows that points of a reflecting surface also act as secondary sources, the reflected wave enveloping their secondary wave fronts (Figure 10.13). At an edge or corner of the surface, the secondary source may contribute a finite portion to the envelope on account of the change in direction of the tangent planes there. Such a secondary wave from a singular point of a boundary surface is said to be diffracted.

We shall study the diffraction of a wave in two space dimensions by a wedge and, in particular, by a half-line or wedge of angle zero. That is, we construct the Green's function of the two-dimensional wave equation in a sector $0 \leq \theta \leq \alpha < 2\pi$ of infinite radius, the angle of the wedge being $2\pi - \alpha < \pi$ as we shall take $\pi \leq \alpha \leq 2\pi$ (Figure 10.14).

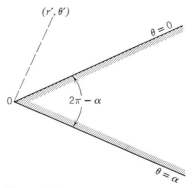

Figure 10.14 Wedge and source point.

For a source at (r', θ'), the wave equation is

(10.6.1) $$\frac{\partial^2 K}{\partial t^2} - c^2 \Delta K = \frac{1}{r} \delta(r - r') \delta(\theta - \theta') \delta(t)$$

where $K = K(r, \theta, r', \theta', t)$, and the origin is taken at the vertex of the wedge. We adopt the second boundary condition, which is appropriate for the reflection of sound waves, so that

(10.6.2) $$\frac{\partial K}{\partial \theta} = 0 \quad \text{for} \quad \theta = 0, \theta = \alpha$$

To construct the Green's function by the bilinear formula (7.1.9), we require the normalized eigenfunctions of the sector region. At any instant t our solution will vanish at distances greater than ct from the source, so we could select any circle of radius $b > r' + ct$ and use its eigenfunctions. However, it is simplest to take $b = +\infty$ and derive the normalized radial functions from the Hankel transform. Insertion of a test function $\varphi(k)$ $(0 \le k < \infty)$ in (8.10.1) and (8.10.2) shows us that

$$\int_0^\infty \sqrt{k}\, J_\nu(kr) \sqrt{k'}\, J_\nu(k'r) r\, dr = \delta(k - k')$$

so the radial factor in the normalized eigenfunctions is $\sqrt{k}\, J_\nu(kr)$. The values of ν are determined from the boundary conditions (10.6.2) as $\nu = n\pi/\alpha$, and thus the normalized eigenfunctions are

(10.6.3) $$\sqrt{\frac{2\varepsilon_n}{\alpha}}\, k\, J_{n\pi/\alpha}(kr) \cos \frac{n\pi\theta}{\alpha}$$

Here ε_n denotes 1 for $n > 0$ and $\tfrac{1}{2}$ for $n = 0$.

From the bilinear series (10.2.7), which now contains an integration over k, we see that

(10.6.4) $$K(r, \theta, r', \theta', t) = \frac{2}{\alpha} \sum_{n=0}^\infty \varepsilon_n \int_0^\infty \sqrt{k}\, J_{n\pi/\alpha}(kr) \sqrt{k}\, J_{n\pi/\alpha}(kr')$$
$$\times \cos \frac{n\pi\theta}{\alpha} \cos \frac{n\pi\theta'}{\alpha} \frac{\sin kct}{kc} dk$$
$$= \frac{2}{\alpha c} \sum_{n=0}^\infty \varepsilon_n \cos \frac{n\pi\theta}{\alpha} \cos \frac{n\pi\theta'}{\alpha}$$
$$\times \int_0^\infty J_{n\pi/\alpha}(kr) J_{n\pi/\alpha}(kr') \sin kct\, dk$$

For brevity we quote the value of the radial integral (Table 3; Erdelyi, Ref. 19, vol. I, p. 102). It is given by three separate analytical formulas according to which of the three intervals $(0, |r - r'|)$, $(|r - r'|, r + r')$, $(r + r', \infty)$ contains ct. Thus

(10.6.5)
$$\int_0^\infty J_\nu(kr)J_\nu(kr') \sin kct \, dk = \begin{cases} 0 & 0 < ct < |r - r'| \\ \dfrac{1}{2\sqrt{rr'}} P_{\nu-\frac{1}{2}}(A) & |r - r'| < ct < r + r' \\ \dfrac{\cos \nu\pi}{\pi\sqrt{rr'}} Q_{\nu-\frac{1}{2}}(-A) & r + r' < ct \end{cases}$$

where $P_{\nu-\frac{1}{2}}$ and $Q_{\nu-\frac{1}{2}}$ are Legendre functions to be given below and

(10.6.6)
$$A = \frac{r^2 + r'^2 - c^2t^2}{2rr'}$$

In the first of these three ranges, no signal from the source can possibly be present. For the intermediate range we use formula (9.6.4), observing that

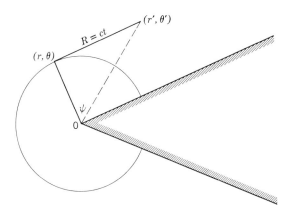

Figure 10.15 Source, field point, and diffracted wave.

$-1 < A < 1$ so that $A = \cos \psi$ for some angle ψ between zero and π:

(10.6.7)
$$P_{(n\pi/\alpha)-\frac{1}{2}}(\cos \psi) = \frac{2}{\pi} \int_0^\psi \frac{\cos n\pi\varphi/\alpha \, d\varphi}{\sqrt{2\cos\varphi - 2\cos\psi}}$$

The physical interpretation of ψ is as follows (Figure 10.15). For $|r - r'| < ct < r - r'$ the wave front of radius ct centered at the source meets the circle of radius r centered at the vertex at a point with polar coordinates $(r, \theta' + \psi)$. As ct increases from $|r - r'|$ to $r + r'$, ψ runs from zero to π.

Now for $|r - r'| < ct < r + r'$,

(10.6.8) $\quad K(r, \theta, r', \theta', t) = \dfrac{1}{\alpha c \sqrt{rr'}} \sum\limits_{n=0}^{\infty} \varepsilon_n \cos \dfrac{n\pi\theta}{\alpha} \cos \dfrac{n\pi\theta'}{\alpha} P_{(n\pi/\alpha)-\frac{1}{2}}(\cos \psi)$

$\qquad = \dfrac{2}{\alpha c \pi \sqrt{rr'}} \displaystyle\int_0^{\psi} \dfrac{d\varphi}{\sqrt{2\cos\varphi - 2\cos\psi}}$

$\qquad \times \sum\limits_{n=0}^{\infty} \varepsilon_n \cos \dfrac{n\pi\theta}{\alpha} \cos \dfrac{n\pi\theta'}{\alpha} \cos \dfrac{n\pi\varphi}{\alpha}$

The nature of the summation over n is best revealed if we recall the bilinear cosine expansion of the Dirac function for the interval $0 \leq \theta \leq \alpha$;

(10.6.9) $\quad \delta(\theta - \theta') = \delta_{(\alpha)}(\theta - \theta') = \dfrac{2}{\alpha} \sum\limits_{n=0}^{\infty} \varepsilon_n \cos \dfrac{n\pi\theta}{\alpha} \cos \dfrac{n\pi\theta'}{\alpha}$

This distribution is even in θ and θ' and is periodic of period 2α. If the angles θ, θ' are unrestricted, there will be delta function singularities for $\theta = \theta' + 2m\alpha$ and $\theta = -\theta' + 2m\alpha$, where m is any integer.

Since the terms of the sum in (10.6.8) contain three factors, we first apply the trigonometric transformation rules and so obtain

(10.6.10) $\quad \dfrac{2}{\alpha} \sum\limits_{n=0}^{\infty} \varepsilon_n \cos \dfrac{n\pi\theta}{\alpha} \cos \dfrac{n\pi\theta'}{\alpha} \cos \dfrac{n\pi\varphi}{\alpha}$

$\qquad = \dfrac{1}{\alpha} \sum\limits_{n=0}^{\infty} \varepsilon_n \left(\cos \dfrac{n\pi}{\alpha}(\theta - \theta') + \cos \dfrac{n\pi}{\alpha}(\theta + \theta') \right) \cos \dfrac{n\pi\varphi}{\alpha}$

$\qquad = \tfrac{1}{2}[\delta_{(\alpha)}(\theta - \theta' - \varphi) + \delta_{(\alpha)}(\theta + \theta' - \varphi)]$

This expression yields a contribution when φ is equal to any of the values $\theta - \theta', \theta' - \theta, \theta + \theta', -\theta - \theta'$, which fall in the range $0 \leq \varphi \leq \psi$ modulo 2α. Since $0 < \theta < \alpha$, $0 < \theta' < \alpha$, and $0 \leq \psi \leq \pi$, this will happen only for φ equal to $\theta - \theta', \theta' - \theta, \theta' + \theta$, and $2\alpha - \theta - \theta'$, provided these angles lie in the range $(0, \psi)$. For our purposes, therefore, quantity (10.6.10) can be written as

(10.6.11)
$\tfrac{1}{2}[\delta(\theta - \theta' - \varphi) + \delta(\theta' - \theta - \varphi) + \delta(\theta + \theta' - \varphi) + \delta(2\alpha - \theta - \theta' - \varphi)]$

Expression (10.6.8) for the Green's kernel now separates into four integrals of which the first is

(10.6.12) $\quad K_1 = \dfrac{1}{2\pi c \sqrt{rr'}} \displaystyle\int_0^{\psi} \dfrac{\delta(\theta - \theta' - \varphi) \, d\varphi}{\sqrt{2\cos\varphi - 2\cos\psi}}$

$\qquad = \dfrac{1}{2\pi c \sqrt{rr'}} \dfrac{1}{\sqrt{2\cos(\theta - \theta') - 2\cos\psi}} \displaystyle\int_0^{\psi} \delta(\theta - \theta' - \varphi) \, d\varphi$

$\qquad = \dfrac{H(\theta - \theta') - H(\theta - \theta' - \psi)}{2\pi c \sqrt{2rr' \cos(\theta - \theta') - 2rr' \cos\psi}}$

since the Heaviside function is the indefinite integral of the Dirac function. Since $\cos \psi = A$, we see from (10.6.6) that

(10.6.13)
$$K_1 = \frac{H(\theta - \theta') - H(\theta - \theta' - \psi)}{2\pi c \sqrt{c^2 t^2 - [r^2 + r'^2 - 2rr' \cos(\theta - \theta')]}}$$
$$= \frac{H(\theta - \theta') - H(\theta - \theta' - \psi)}{2\pi c \sqrt{c^2 t^2 - R^2}}$$

where R is the distance from the source point (r', θ') to the field point (r, θ). The second term in (10.6.11) yields a similar expression with θ and θ' interchanged.

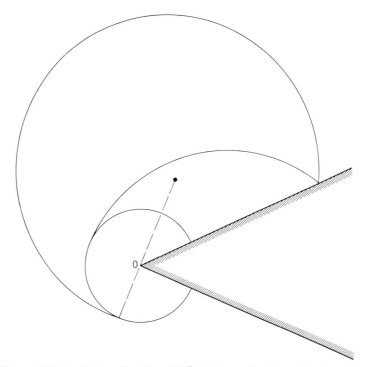

Figure 10.16 Incident, reflected, and diffracted wave fronts, and shadow zone.

Since $H(x) + H(-x) = 1$ for all $x \neq 0$, and $H(x - \psi) + H(-x - \psi) = H(|x| - \psi)$ when $\psi > 0$, we can combine these two contributions. From Figure 10.15 it will be seen that ψ is the angle at a vertex of the triangle of sides r', r, and $R = ct$. Therefore, since also $0 \leq \psi \leq \pi$, we have

$$H(\theta - \theta') - H(\theta - \theta' - \psi) + H(\theta' - \theta) - H(\theta' - \theta - \psi)$$
$$= 1 - H(|\theta - \theta'| - \psi) = H(ct - R)H(\pi - |\theta - \theta'|)$$

The factor $H(\pi - |\theta - \theta'|)$, which is present because the range of ψ is restricted to $(0, \pi)$, is responsible for the shadow zone of the wedge (Figure 10.16).

The third term of (10.6.11) has a contribution for $\varphi = \theta + \theta'$ and the corresponding distance function is R_1, where

$$R_1^2 = r^2 + r'^2 - 2rr' \cos(\theta + \theta')$$

This gives distance from the image point in the line $\theta = 0$, and this term yields the wave reflected from that side of the wedge. The remaining term in (10.6.11) gives rise to a similar wave depending on $R_2^2 = r^2 + r'^2 - 2rr' \cos(2\alpha - \theta - \theta')$ and reflected from the side $\theta = \alpha$. Collecting these contributions, we find that in the intermediate range $|r - r'| < ct < r + r'$,

(10.6.14)
$$K(r, \theta, r', \theta', t) = \frac{H(ct - R)H(\pi - |\theta - \theta'|)}{2\pi c \sqrt{c^2 t^2 - R^2}}$$
$$+ \frac{H(\theta + \theta') - H(\theta + \theta' - \psi)}{2\pi c \sqrt{c^2 t^2 - R_1^2}}$$
$$+ \frac{H(2\alpha - \theta - \theta') - H(2\alpha - \theta - \theta' - \psi)}{2\pi c \sqrt{c^2 t^2 - R_2^2}}$$

The wave front diffracted at the vertex lies at time t on a circle of radius $r = ct - r'$. As we cross this front, we enter the zone of the residual wave and must use the third of the analytic expressions given in (10.6.5). For $ct > r + r'$ the value of A in (10.6.6) is less than -1, so there is a positive real number Ψ with $-A = \cosh \Psi$.

For the Legendre function of the second kind, which is singular at $-A = 1$ when the order $(n\pi/\alpha) - \frac{1}{2}$ is an integer, we have a formula similar to (10.6.7), namely,

(10.6.15)
$$Q_{(n\pi/\alpha) - \frac{1}{2}}(\cosh \Psi) = \int_{\Psi}^{\infty} \frac{e^{-n\pi \xi/\alpha} \, d\xi}{\sqrt{2 \cosh \xi - 2 \cosh \Psi}}$$

(See Erdelyi, Ref. 18, vol. 1, p. 155.)

The series for K is now

(10.6.16)
$$K(r, \theta, r', \theta', t)$$
$$= \frac{2}{\pi \alpha c \sqrt{rr'}} \sum_{n=0}^{\infty} \varepsilon_n \cos \frac{n\pi\theta}{\alpha} \cos \frac{n\pi\theta'}{\alpha} \cos \frac{n\pi^2}{\alpha} Q_{(n\pi/\alpha) - \frac{1}{2}}(\cosh \Psi)$$
$$= \frac{2}{\pi \alpha c \sqrt{rr'}} \int_{\Psi}^{\infty} \frac{d\xi}{\sqrt{2 \cosh \xi - 2 \cosh \Psi}}$$
$$\times \sum_{n=0}^{\infty} \varepsilon_n \cos \frac{n\pi\theta}{\alpha} \cos \frac{n\pi\theta'}{\alpha} \cos \frac{n\pi^2}{\alpha} e^{-n\pi\xi/\alpha}$$

The summation over n now involves an exponentially diminishing factor and its character resembles the sum in the elliptic problem for Poisson's integral in Chapter 4 rather than the hyperbolic solution of Chapter 2. The sum actually divides into four terms of the Poisson type as in (4.3.17).

The Green's kernel can now be expressed as

(10.6.17) $\quad K(r, \theta, r', \theta', t) = K_2(\theta - \theta' - \pi) + K_2(\theta - \theta' + \pi)$
$$+ K_2(\theta + \theta' - \pi) + K_2(\theta + \theta' + \pi)$$

where

(10.6.18)
$$K_2(\varphi) = \frac{1}{4\pi\alpha c\sqrt{rr'}} \int_\Psi^\infty \frac{\sinh\frac{\pi\xi}{\alpha}}{\left(\cosh\frac{\pi\xi}{\alpha} - \cos\frac{\pi\varphi}{\alpha}\right)} \frac{d\xi}{\sqrt{2\cosh\xi - 2\cosh\Psi}}$$

$$= \frac{1}{4\pi\alpha c} \int_\Psi^\infty \frac{\sinh\frac{\pi\xi}{\alpha}}{\left(\cosh\frac{\pi\xi}{\alpha} - \cos\frac{\pi\varphi}{\alpha}\right)} \frac{d\xi}{\sqrt{r^2 + r'^2 + 2rr'\cosh\xi - c^2t^2}}$$

There are no further singularities within the residual range $ct > r + r'$.

For general sector angles α these integrals are too difficult for us to evaluate explicitly. A detailed discussion may be found in Friedlander, Ref. 21, Chapter 5. However in the interesting particular case $\alpha = 2\pi$, when the wedge is a half-line, the integrals are elementary and we shall calculate them. The first two terms of (10.6.17) then combine to give

(10.6.19)
$$\frac{1}{8\pi c\sqrt{rr'}} \int_\Psi^\infty \left[\frac{\sinh\tfrac{1}{2}\xi}{\cosh\tfrac{1}{2}\xi - \cos\tfrac{1}{2}(\theta - \theta' - \pi)} + \frac{\sinh\tfrac{1}{2}\xi}{\cosh\tfrac{1}{2}\xi - \cos\tfrac{1}{2}(\theta - \theta' + \pi)}\right]$$
$$\times \frac{d\xi}{\sqrt{2\cosh\xi - 2\cosh\Psi}}$$

$$= \frac{1}{8\pi c\sqrt{rr'}} \int_\Psi^\infty \frac{2\sinh\xi}{\cosh\xi + \cos(\theta - \theta')} \frac{d\xi}{\sqrt{2\cosh\xi - 2\cosh\Psi}}$$

Setting $\tau^2 = \cosh\xi - \cosh\Psi$, we obtain an integral of the inverse tangent type which is readily computed. The remaining two terms of (10.6.17) yield a similar expression containing the angle $\theta + \theta'$ which refers to distance from the image source. Finally, therefore, the Green's function for diffraction by a half-line is

(10.6.20) $\quad K(r, \theta, r', \theta', t) = \dfrac{1}{4\pi c\sqrt{c^2t^2 - R^2}} + \dfrac{1}{4\pi c\sqrt{c^2t^2 - R_1^2}}$

for $ct > r + r'$. Comparing this with the Green's function (10.3.6) for the full space, we see that the source and its image each contribute with half strength.

As we have used separate formulas on either side of the diffracted wave front $ct = r + r'$, the actual singularity on that front must be computed by comparing the two expressions. In the half-line case $\alpha = 2\pi$, we have a finite discontinuity of the wave amplitude across the diffracted front, and this is also the behavior for other wedge angles. Thus the diffraction at the endpoint reduces the singularity from the order $\tfrac{1}{2}$ of the incident wave to a finite discontinuity of order zero.

EXERCISES 10.6

1. What part of the locus $|r - r'| = ct$ belongs to a wave front? How is it possible that the remaining portion does not?

2. Using the asymptotic formula (8.6.9) for $J_\nu(ka)$, show that the normalized sector eigenfunctions of (8.2.14) yield (10.6.3) in the limit $a \to \infty$.

3. Employing Heaviside functions where necessary, write the entire solution for diffraction by a half-line in one formula valid for all places and times.

4. Describe the changes necessary in the wedge diffraction solution if the boundary condition is $u = 0$.

5. Show that $\delta_{(\alpha)}(\theta) = \sum_{m=-\infty}^{\infty} [\delta(\theta + 2m\alpha) + \delta(-\theta + 2m\alpha)]$

6. For the half-space problem $\alpha = \pi$, calculate the explicit solution at all times and check the result.

7. What further wave fronts appear if $\alpha < \pi$? Show how to obtain them from (10.6.10).

8. Find the explicit solution for the quadrant $\alpha = \pi/2$ and sketch the wave fronts.

9. Write down normalized eigenfunctions, and Green's functions for the wave equation, for diffraction with boundary condition $\partial u/\partial n = 0$ in three space dimensions:
   (a) For an infinite straight wedge of angular width $\beta < \pi$.
   (b) For an infinite circular cone of angle $\beta < \pi$.

TABLE 1 Fourier Transforms

$f(x) = \dfrac{1}{\sqrt{2\pi}} \int_{-\infty}^{\infty} F(s)e^{isx}\, ds$	$F(s) = \dfrac{1}{\sqrt{2\pi}} \int_{-\infty}^{\infty} f(x)e^{-isx}\, dx$
$F(x)$	$f(-s)$
$\overline{f(x)}$	$\overline{F(-s)}$
$f(x) = f(-x)$	$F_c(s) = \sqrt{\dfrac{2}{\pi}} \int_0^{\infty} \cos xs\, f(x)\, dx$
$f(x) = -f(-x)$	$-iF_s(s) = -i\sqrt{\dfrac{2}{\pi}} \int_0^{\infty} \sin xs\, f(x)\, dx$
$f(a^{-1}x + b)e^{icx}$ $\quad a, c$ real	$|a|\, e^{iab(s-c)} F(a(s-c))$
$x^n f(x)$	$i^n \dfrac{d^n F(s)}{ds^n}$
$\dfrac{d^n}{dx^n} f(x)$	$i^n s^n F(s)$
$\delta(x)$	$\dfrac{1}{\sqrt{2\pi}}$
$\dfrac{d^n}{dx^n} \delta(x-a)$	$\dfrac{1}{\sqrt{2\pi}} i^n s^n e^{-ias}$
$f*g(x) = \displaystyle\int_{-\infty}^{\infty} f(x-y)g(y)\, dy$	$\sqrt{2\pi} F(s)G(s)$
$\cos bx$	$\sqrt{\dfrac{\pi}{2}}\,[\delta(s+b) + \delta(s-b)]$
$\sin bx$	$i\sqrt{\dfrac{\pi}{2}}\,[\delta(s+b) - \delta(s-b)]$
x^{-1}	$-i\sqrt{\dfrac{\pi}{2}}\, \text{sgn}\, s$
x^{-2}	$-\sqrt{\dfrac{\pi}{2}}\, |s|$
$H(x)$	$\dfrac{-i}{s\sqrt{2\pi}} + \sqrt{\dfrac{\pi}{2}}\, \delta(s)$
x_+^n	$\dfrac{(-i)^{n+1} n!}{\sqrt{2\pi}}\, s^{-n-1} + \sqrt{\dfrac{\pi}{2}}\, \delta^{(n)}(s)$
x_+^λ	$\dfrac{-i}{\sqrt{2\pi}} (e^{-i\lambda\pi/2} s_+^{-\lambda-1} - e^{i\lambda\pi/2} s_-^{-\lambda-1})\Gamma(\lambda+1)$
$[1-x^2]_+^\lambda$	$\dfrac{1}{\sqrt{2}}\, \Gamma(\lambda+1) \left(\dfrac{s}{2}\right)^{-\lambda-1} J_{\lambda+1/2}(s)$
$e^{-(1/2)x^2} H_n(x) = (-1)^n e^{(1/2)x^2} \dfrac{d^n}{dx^n} e^{-x^2}$	$i^n e^{-(1/2)s^2} H_n(s)$

TABLE 2 Fourier Cosine Transforms

$f(x) = \sqrt{\dfrac{2}{\pi}} \displaystyle\int_0^\infty F_c(s)\cos sx\, ds$	$F_c(s) = \sqrt{\dfrac{2}{\pi}} \displaystyle\int_0^\infty f(x)\cos sx\, dx$
$x^{\nu-1}e^{-ax}$	$\sqrt{\dfrac{2}{\pi}}\,\Gamma(\nu)(a^2+s^2)^{-\nu/2}\cos\left(\nu\tan^{-1}\dfrac{s}{a}\right)$
$J_0(ax)$	$\sqrt{\dfrac{2}{\pi}}\,\dfrac{H(a^2-s^2)}{\sqrt{a^2-s^2}}$
$Y_0(ax)$	$-\sqrt{\dfrac{2}{\pi}}\,\dfrac{H(s^2-a^2)}{\sqrt{s^2-a^2}}$
$K_0(ax)$	$\sqrt{\dfrac{\pi}{2}}\,\dfrac{1}{\sqrt{a^2+s^2}}$
$J_0(\sqrt{x^2+a^2})$	$\sqrt{\dfrac{2}{\pi}}\,\dfrac{\cos(a\sqrt{1-s^2})}{\sqrt{1-s^2}}H(1-s^2)$
$J_0(\sqrt{a^2-x^2})H(a-x)$	$\sqrt{\dfrac{2}{\pi}}\,\dfrac{\sin(a\sqrt{1+s^2})}{\sqrt{1+s^2}}$
$I_0(\sqrt{a^2-x^2})H(a-x)$	$\sqrt{\dfrac{2}{\pi}}\,\dfrac{\sin(a\sqrt{s^2-1})}{\sqrt{s^2-1}}$
$Y_0(b\sqrt{x^2+a^2})$	$\begin{cases}\sqrt{\dfrac{2}{\pi}}\,\dfrac{\sin(a\sqrt{b^2-s^2})}{\sqrt{b^2-s^2}} & b>s \\ -\sqrt{\dfrac{2}{\pi}}\,\dfrac{e^{-a\sqrt{s^2-b^2}}}{\sqrt{s^2-b^2}} & 0<b<s\end{cases}$
$K_0(b\sqrt{x^2+a^2})$	$\sqrt{\dfrac{\pi}{2}}\,\dfrac{e^{-a\sqrt{b^2+s^2}}}{\sqrt{b^2+s^2}}$
$K_0(b\sqrt{x^2-a^2})H(x-a)$ $\;-\dfrac{\pi}{2}Y_0(b\sqrt{a^2-x^2})H(a-x)$	$\sqrt{\dfrac{\pi}{2}}\,\dfrac{\cos(a\sqrt{b^2+s^2})}{\sqrt{b^2+s^2}}\quad a>0,\,b>0$
$J_0(\sqrt{x^2-a^2})$	$\sqrt{\dfrac{2}{\pi}}\,\dfrac{\cosh(a\sqrt{1-s^2})H(1-s^2)}{\sqrt{1-s^2}}$
$e^{-a^2x^2}$	$\dfrac{1}{\sqrt{2}a}e^{-s^2/4a^2}\quad a>0$
$\dfrac{1}{b^2+x^2}$	$\dfrac{1}{b}\sqrt{\dfrac{\pi}{2}}e^{-bs}\quad b>0$

TABLE 2 (continued)

$\dfrac{\sinh ax}{\sinh bx}$	$\dfrac{1}{b}\sqrt{\dfrac{\pi}{2}}\sin\dfrac{\pi a}{b}\left(\cosh\dfrac{\pi s}{b}+\cos\dfrac{\pi a}{b}\right)^{-1}$ $0<a<b$
$H(a-x)$	$\sqrt{\dfrac{2}{\pi}}\dfrac{\sin as}{s}$
$\dfrac{\cosh ax}{\cosh bx}$	$\dfrac{\sqrt{2\pi}}{b}\cosh\dfrac{\pi s}{2b}\cos\dfrac{\pi a}{2b}\left(\cosh\dfrac{\pi s}{b}+\cos\dfrac{\pi a}{b}\right)^{-1}$ $0<a<b$
$\dfrac{1}{\sqrt{2\pi}}\log\left(1+\dfrac{y^2}{x^2}\right)$	$\dfrac{1}{s}(1-e^{-sy})$ $y>0$
$\dfrac{1}{2\sqrt{2\pi}}\log\left[\dfrac{x^2+(y+y')^2}{x^2+(y-y')^2}\right]$	$\dfrac{1}{s}e^{-sy}\sinh sy'$ $0<y'<y$

TABLE 3 Fourier Sine Transforms

$f(x) = \sqrt{\dfrac{2}{\pi}} \displaystyle\int_0^\infty F_s(s) \sin xs\, ds$	$F_s(s) = \sqrt{\dfrac{2}{\pi}} \displaystyle\int_0^\infty f(x) \sin xs\, ds$
$x^{\nu-1} e^{-ax}$	$\sqrt{\dfrac{2}{\pi}} \Gamma(\nu)(a^2+s^2)^{-\nu/2} \sin\left(\nu \tan^{-1}\dfrac{s}{a}\right)$ $\nu > 0$
$J_0(ax)$	$\sqrt{\dfrac{2}{\pi}} \dfrac{H(s^2-a^2)}{\sqrt{s^2-a^2}}$ $a > 0$
$K_0(ax)$	$\sqrt{\dfrac{2}{\pi}} \dfrac{\log\left(\dfrac{s}{a}+\sqrt{1+\dfrac{s^2}{a^2}}\right)}{\sqrt{s^2+a^2}}$ $a > 0$
$J_0(b\sqrt{x^2-a^2})H(x-a)$	$\sqrt{\dfrac{2}{\pi}} \dfrac{\cos(a\sqrt{s^2-b^2})}{\sqrt{s^2-b^2}} H(s-b)$ $b > 0$
$\dfrac{\sinh ax}{\cosh bx}$	$\dfrac{\sqrt{2\pi}}{b} \sinh\dfrac{\pi s}{2b} \sin\dfrac{\pi a}{2b} \left(\cosh\dfrac{\pi s}{b} + \cos\dfrac{\pi a}{b}\right)^{-1}$ $0 < a < b$
$\dfrac{\cosh ax}{\sinh bx}$	$\dfrac{1}{b}\sqrt{\dfrac{\pi}{2}} \sinh\dfrac{\pi s}{b} \left(\cosh\dfrac{\pi s}{b} + \cos\dfrac{\pi a}{b}\right)^{-1}$ $0 < a < b$
$\dfrac{\sqrt{2\pi}}{a} \displaystyle\sum_{n=0}^{\infty} e^{-x\sqrt{k^2+(n+1/2)^2\pi^2/a^2}}$ $\times \sin\left(n+\dfrac{1}{2}\right)\dfrac{\pi y}{a} \sin\left(n+\dfrac{1}{2}\right)\dfrac{\pi y'}{a}$	$\dfrac{s \sinh(y\sqrt{k^2+s^2}) \cosh[(a-y')\sqrt{k^2+s^2}]}{\sqrt{k^2+s^2} \cosh(a\sqrt{k^2+s^2})}$ $0 < y < y' < a$
$\dfrac{x}{b^2+x^2}$	$\sqrt{\dfrac{\pi}{2}} e^{-bs}$ $b > 0$
$\dfrac{x}{b^2-x^2}$	$-\sqrt{\dfrac{\pi}{2}} \cos bs$ $b > 0$
$\dfrac{1}{2\sqrt{2\pi}} \log\left[\dfrac{y^2+(x+x')^2}{y^2+(x-x')^2}\right]$	$\dfrac{1}{s} e^{-sy} \sin sx'$ $y > 0$
$\dfrac{1}{x} e^{-ax}$	$\sqrt{\dfrac{2}{\pi}} \tan^{-1}\dfrac{s}{a}$ $a > 0$
$H(a-x)$	$\sqrt{\dfrac{2}{\pi}}\left(\dfrac{1-\cos as}{s}\right)$
$J_\nu(xr)J_\nu(xr')$ $r > 0, r' > 0$	$\begin{cases} 0 & 0 < s < \|r-r'\| \\ \dfrac{1}{\sqrt{2\pi rr'}} P_{\nu-1/2}\left(\dfrac{r^2+r'^2-s^2}{2rr'}\right) & \|r-r'\| < s < r+r' \\ \sqrt{\dfrac{2}{\pi^3 rr'}} Q_{\nu-1/2}\left(\dfrac{s^2-r^2-r'^2}{2rr'}\right) \cos\nu\pi & r+r' < s \end{cases}$

TABLE 4 Laplace Transforms

$f(t) = \dfrac{1}{2\pi i}\displaystyle\int_{c-i\infty}^{c+i\infty} e^{pt}L_1(p)\,dp$	$L_1(p) = \displaystyle\int_0^\infty e^{-pt}f(t)\,dt$	
$e^{-at}f(t)$	$L_1(p+a)$	
$t^n f(t)$	$(-1)^n \dfrac{d^n}{dp^n} L_1(p)$	
$\dfrac{d^n}{dt^n} f(t)$	$p^n L_1(p) - p^{n-1}f(0) - p^{n-2}f'(0) - \cdots - f^{(n-1)}(0)$	
$\displaystyle\int_0^t f_1(\tau) f_2(t-\tau)\,d\tau$	$L_1(p) L_2(p)$	
$f_1(t) f_2(t)$	$\dfrac{1}{2\pi i}\displaystyle\int_{c-i\infty}^{c+i\infty} L_1(z) L_2(p-z)\,dz$	
$t^{\nu-1} e^{-at}$	$\Gamma(\nu)(p+a)^{-\nu}$	$\operatorname{Re} p > -\operatorname{Re} a,\ \operatorname{Re}\nu > 0$
$\cos at$	$\dfrac{p}{p^2 + a^2}$	$\operatorname{Re} p > 0$
$\sin at$	$\dfrac{a}{p^2 + a^2}$	$\operatorname{Re} p > 0$
$\dfrac{H(t-a)}{\sqrt{t^2 - a^2}}$	$K_0(ap)$	$a > 0$
$\delta(t-a)$	e^{-ap}	$a > 0$
$H(t-a)$	$\dfrac{1}{p} e^{-ap}$	$a > 0$
$J_0(at)$	$\dfrac{1}{\sqrt{a^2 + p^2}}$	$\operatorname{Re} p > 0$
$I_0(at)$	$\dfrac{1}{\sqrt{p^2 - a^2}}$	$\operatorname{Re} p > 0$
$Y_0(at)$	$-\dfrac{2}{\pi} \dfrac{\sinh^{-1}(p/a)}{\sqrt{p^2 + a^2}}$	$\operatorname{Re} p > 0,\ a > 0$
$K_0(at)$	$\begin{cases}\dfrac{\cosh^{-1}(p/a)}{\sqrt{p^2 - a^2}} & p > a \\ \dfrac{\cos^{-1}(p/a)}{\sqrt{a^2 - p^2}} & 0 < p < a\end{cases}$	
$J_0(a\sqrt{t^2 - b^2}) H(t-b)$	$\dfrac{e^{-b\sqrt{p^2 + a^2}}}{\sqrt{p^2 + a^2}}$	$b > 0$
$J_0(2b\sqrt{at})$	$\dfrac{1}{p} e^{-b^2 a/p}$	$\operatorname{Re} p > 0$
$\dfrac{t}{a} + \dfrac{2}{\pi}\displaystyle\sum_{n=1}^\infty \dfrac{(-1)^n}{n} \sin\dfrac{n\pi t}{a}$	$\dfrac{1}{p\sinh ap}$	

TABLE 5 Hankel Transforms Order

$f(r) = \int_0^\infty \lambda J_\nu(\lambda r) F(\lambda)\, d\lambda$		$F(\lambda) = \int_0^\infty r J_\nu(\lambda r) f(r)\, dr$		ν
$f(ar)$	$a > 0$	$\dfrac{1}{a^2} F\!\left(\dfrac{\lambda}{a}\right)$		ν
$r^{\mu-1}$	$\mathrm{Re}\,(\nu+1) < \mathrm{Re}\,(\mu) < \tfrac{1}{2}$	$\dfrac{2^\mu \Gamma(\tfrac{1}{2} + \tfrac{1}{2}\mu + \tfrac{1}{2}\nu)}{\lambda^{\mu+1}\Gamma(\tfrac{1}{2} - \tfrac{1}{2}\mu + \tfrac{1}{2}\nu)}$		
$r^\nu e^{-ar}$	$\mathrm{Re}\,a > 0$	$\dfrac{1}{\sqrt{\pi}}\dfrac{2^{\nu+1}\Gamma(\nu+\tfrac{3}{2})a\lambda^\nu}{(a^2+\lambda^2)^{\nu+3/2}}$		$\nu > -1$
$r^\nu e^{-pr^2}$	$\mathrm{Re}\,p > 0$	$\dfrac{\lambda^\nu}{(2p)^{\nu+1}} e^{-\lambda^2/4p}$		$\nu > -1$
$\begin{cases}\tfrac{\pi}{2} J_\nu(kr)\, Y_\nu(kr') & 0 < r < r' \\ \tfrac{\pi}{2} J_\nu(kr')\, Y_\nu(kr) & 0 < r' < r\end{cases}$		$\dfrac{J_\nu(\lambda r')}{k^2 - \lambda^2}$		$\nu > 0$
$\begin{cases} I_\nu(kr) K_\nu(kr') & 0 < r < r' \\ I_\nu(kr') K_\nu(kr) & 0 < r' < r \end{cases}$		$\dfrac{J_\nu(\lambda r')}{k^2 + \lambda^2}$		$\nu > 0$
$\begin{cases} \dfrac{1}{2\nu}\left(\dfrac{r}{r'}\right)^\nu & 0 < r < r' \\ \dfrac{1}{2\nu}\left(\dfrac{r'}{r}\right)^\nu & 0 < r' < r \end{cases}$		$\dfrac{1}{\lambda^2} J_\nu(\lambda r')$		$\nu > 0$
$\tfrac{\pi}{2} Y_0(kr)$		$\dfrac{1}{k^2 - \lambda^2}$		$\nu = 0$
$K_0(kr)$		$\dfrac{1}{k^2 + \lambda^2}$		$\nu = 0$
$\dfrac{2\pi}{a^2}\sum_{n=1}^{\infty} n \sin\dfrac{n\pi z}{a} K_0\!\left(\dfrac{n\pi r}{a}\right)$		$\dfrac{\sinh \lambda(a-z)}{\sinh \lambda a}$	$0 < z < a$	$\nu = 0$
$\dfrac{2}{a}\sum_{n=0}^{\infty} (-1)^n \sin(n+\tfrac{1}{2})\dfrac{\pi z}{a} K_0\!\left((n+\tfrac{1}{2})\dfrac{\pi r}{a}\right)$		$\dfrac{\sinh \lambda z}{\lambda \cosh \lambda a}$	$0 < z < a$	$\nu = 0$
$\dfrac{1}{r} e^{-br}$		$\dfrac{1}{\sqrt{b^2 + \lambda^2}}$		$\nu = 0$
$\dfrac{1}{r}\cos br$		$\dfrac{H(\lambda^2 - b^2)}{\sqrt{\lambda^2 - b^2}}$		$\nu = 0$
$\dfrac{1}{r}\sin br$		$\dfrac{H(b^2 - \lambda^2)}{\sqrt{b^2 - \lambda^2}}$		$\nu = 0$
$\dfrac{1}{r^2}(1 - \cos br)$		$\cosh^{-1}\dfrac{b}{\lambda} H(b-\lambda)$		$\nu = 0$
$\dfrac{\cos br}{r(r^2 + a^2)}$		$\cosh ab\, K_0(a\lambda)$	$\lambda > b$	$\nu = 0$

Bibliography

1. Ahlfors, L., *Complex Analysis*, McGraw-Hill, New York, 1953.
2. Baker, B. B., and E. T. Copson, *The Mathematical Theory of Huyghens' Principle*, Oxford University Press, London, 1939.
3. Bateman, H., *Partial Differential Equations*, Macmillan, New York, 1932 (reprinted Dover, New York, 1944).
4. Bergman, S., and M. Schiffer, *Kernel Functions and Elliptic Differential Equations in Mathematical Physics*, Academic, New York, 1953.
5. Birkhoff, G., and S. MacLane, *A Survey of Modern Algebra*, Macmillan, New York, 1941.
6. Birkhoff, G., and G. C. Rota, *Ordinary Differential Equations*, Ginn, Boston, 1960.
7. Carslaw, H., and J. C. Jaeger, *Conduction of Heat in Solids*, 2nd ed., Oxford University Press, London, 1959.
8. Chapman, S., and T. G. Cowling, *The Mathematical Theory of Non-Uniform Gases*, Cambridge University Press, London (reprinted Dover, 1961).
9. Coddington, E. A., *An Introduction to Ordinary Differential Equations*, Prentice-Hall, Englewood Cliffs, N.J., 1961.
10. Coddington, E. A., and N. Levinson, *Ordinary Differential Equations*, McGraw-Hill, New York, 1955.
11. Copson, E. T., *Functions of a Complex Variable*, Oxford University Press, London, 1935.
12. Courant, R., and K. Friedrichs, *Supersonic Flow and Shock Waves*, Interscience, New York, 1948.
13. Courant, R., and D. Hilbert, *Methods of Mathematical Physics*, 2 vols., 2nd ed., Interscience, New York, 1953, 1962.
14. Davis, H., *Fourier Series and Orthogonal Functions*, Allyn and Bacon, Boston, 1964.
15. Duff, G. F. D., *Partial Differential Equations*, University of Toronto Press, Toronto, 1956.
16. Dunford, N., and J. T. Schwartz, *Linear Operators*, Interscience, New York, vol. 1, 1958; vol. 2, 1963.
17. Erdelyi, A., *Asymptotic Expansions*, Dover, New York, 1956.
18. Erdelyi, A., W. Magnus, F. Oberhettinger, F. G. Tricomi, *Higher Transcendental Functions*, 3 vols., McGraw-Hill, New York, 1953.
19. Erdelyi, A., W. Magnus, F. Oberhettinger, F. G. Tricomi, *Tables of Integral Transforms*, 2 vols., McGraw-Hill, New York, 1953.
20. Epstein, B., *Partial Differential Equations*, McGraw-Hill, New York, 1962.
21. Friedlander, F. G. *Sound Pulses*, Cambridge University Press, London, 1958.
22. Friedman, B., *Principles and Techniques of Applied Mathematics*, Wiley, New York, 1956.
23. Gantmacher, F. R., *The Theory of Matrices*, 2 vols. Chelsea, New York, 1959.
24. Garabedian, P. R., *Partial Differential Equations*, Wiley, New York, 1964.
25. Gelfand, I. M., and G. E. Shilov, *Generalized Functions and Operations*, vol. 1, Moscow, 1959 (tr. Academic, New York, 1964).
26. Goldstein, H., *Classical Mechanics*, Addison-Wesley, Reading, Mass., 1950.
27. Guillemin, E. A., *Theory of Linear Physical Systems*, Wiley, New York, 1963.
28. Hadamard, J., *Lectures on Cauchy's Problem*, Yale University Press, New Haven, 1923 (reprinted Dover, New York, 1952).
29. Heaviside, O., *Electromagnetic Theory*, Dover, New York, 1950.

30. Helwig, G., *Partial Differential Equations*, Blaisdell, New York, 1964.
31. Hobson, E. W., *Spherical and Ellipsoidal Harmonics*, Cambridge University Press, London, 1931.
32. Hoffman, K., and R. Kunze, *Linear Algebra*, Prentice-Hall, Englewood Cliffs, N.J., 1961.
33. Howarth, L., *Modern Developments in Fluid Dynamics*, Vol. 1—*High Speed Flow*, Oxford University Press, London, 1938.
34. Jahnke, E., and F. Emde, *Tables of Functions with Formulae and Curves*, Dover, New York, 1945.
35. Jeffreys, H., and B. Jeffreys, *Methods of Mathematical Physics*, 3rd ed., Cambridge University Press, London, 1956.
36. Karman, T. von, and M. A. Biot, *Mathematical Methods in Engineering*, McGraw-Hill, New York, 1940.
37. Kellogg, O. D., *Foundations of Potential Theory*, Springer, Berlin, 1929 (reprinted Dover, New York, 1953).
38. Lamb, H., *Hydrodynamics*, 6th ed., Cambridge University Press, London, 1932 (reprinted Dover, New York, 1944).
39. Lanczos, C., *Linear Differential Operators*, van Nostrand, London, 1961.
40. Love, A. E. H., *The Mathematical Theory of Elasticity*, 4th ed., Cambridge University Press, London, 1927
41. Lovitt, W. V., *Linear Integral Equations* McGraw-Hill, New York, 1924 (reprinted Dover, New York, 1950).
42. MacMillan, W. D., *The Theory of the Potential*, Dover, New York, 1958.
43. MacRobert, T. M. *Spherical Harmonics*, Methuen, London, 1927.
44. Milne, W. E., *Numerical Solution of Differential Equations*, Wiley, New York, 1953.
45. Morse, P. M., and H. Feshbach, *Methods of Theoretical Physics*, McGraw-Hill, New York, 1953.
46. Nering, E. D., *Linear Algebra and Matrix Theory*, Wiley, New York, 1963.
47. Petrowsky, I. G., *Lectures on Partial Differential Equations*, Interscience, New York, 1954.
48. Lord Rayleigh, *The Theory of Sound*, 2nd ed., Macmillan, London, 1896, 2 vols. (reprinted Dover, New York, 1944).
49. Richtmyer, R. D., *Difference Methods for Initial Value Problems*, Interscience, New York, 1958.
50. Sagan, H., *Boundary and Eigenvalue Problems in Mathematical Physics*, Wiley, New York, 1961.
51. Schwartz, L., *Théorie des distributions*, 2 vols., Hermann et Cie., Paris, 1950, 1951.
52. Schiff, L. I., *Quantum Mechanics*, McGraw-Hill, New York, 1949.
53. Sneddon, I. N., *Fourier Transforms*, McGraw-Hill, New York, 1949.
54. Sommerfeld, A. J. W., *Partial Differential Equations in Physics*, Academic, New York, 1949.
55. Stratton, J. A., *Electromagnetic Theory*, McGraw-Hill, New York, 1941.
56. Synge, J. L., and A. Schild, *Tensor Calculus*, University of Toronto Press, Toronto, 1952.
57. Temple, G., *Fluid Dynamics*, Oxford University Press, London, 1958.
58. Tolstov, G. P., *Fourier Series*, Prentice-Hall, Englewood Cliffs, N.J., 1962.
59. Tranter, C. J., *Integral Transforms in Mathematical Physics*, Wiley, New York, 1951.
60. Tychonov, A. N., and A. A. Samarskii, *Equations of Mathematical Physics*, Macmillan, New York, 1963.
61. Watson, G. N., *Bessel Functions*, Macmillan, 2nd ed., New York, 1944.
62. Webster, A. G., *Partial Differential Equations of Mathematical Physics*, Boston, 1927 (reprinted Dover, New York, 1955).
63. Weinberger, H. F., *A First Course in Partial Differential Equations*, Blaisdell, New York, 1965.
64. Whittaker, E. T., and G. N. Watson, *A Course of Modern Analysis*, 4th ed., Cambridge University Press, London, 1927.

Index

Action, 26, 181
Acoustic wave, 212
Addition theorem, 325, 356, 358
Additive, 10
Adiabatic, 211
Adjoint, 14
  self-adjoint, 14
Affine, 50
Algebraic equations, 3, 283
Analytic, 52, 134
Annulus, 163
Antianalytic, 52
Asymptotic, 243, 317
Asymptotic series, 317
Average value, 280

Basis, 4, 268
  orthogonal, 6
  orthonormal, 6
Basis vector, 4
Bending moment, 188
Bessel equation, 300
Bessel function, 299 ff.
  imaginary type, 80, 328
  modified, 327
Bicharacteristic, 398
Biharmonic function, 205
Bilinear series, 266, 267, 289, 296, 379
Boundary value problem, 136

Capacitance, 34
Cartesian space, 2
Cauchy integral, 165
Cauchy-Riemann equations, 163, 318
Characteristic, 56, 373
Characteristic cone, 375

Characteristic equation, 15
Characteristic function, 84
Characteristic lines, 56
Characteristic plane, 373
Characteristic surface, 373
Characteristic value, 15
Circle eigenfunctions, 309
Circular harmonics, 141
Clamped end, 188
Closed, 293
Commutation relation, 223
Compact, 124
Complete, 246
Complex derivative, 161
Complex integral, 161
Complex variable, 159
Condensation, 212
Conductivity, 97, 216, 364
Cone, 375
Conjugate, Hermitian, 14
Conjugate function, 165
Conjugate matrix, 14
Conjugate observable, 223
Continuity, 68
  equation of, 208
Contour, 163
Convergence, 91
  of distributions, 292
  mean square, 245
Convolution, 75, 117, 125
Coordinates, generalized, 30
Correctly set, 100
Current vector, 215
Curvature, 187
  principal, 191
Cylinder, 310, 326
Cylindrical functions, 298

δ_{kl}, 7
d'Alembert formula, 55
d'Alembertian, 78, 388
Damped wave equation, 78, 327, 388
Damping, 34
Degree, 363
Delta, 7, 40, 68
Derivative, 69
Descent, 386
Diagonal matrix, 36
Diagonalize, 128
Difference equation, 64, 129
Differential equation, 19, 24, 50
Diffracted wave front, 406
Diffraction, 323, 357, 402
Dimension, 4
Dirac distribution, 40, 68, 249, 267, 380
Dirichlet, 40
Dirichlet integral, 230
Dirichlet principle, 138
Dirichlet problem, 135, 153, 174, 229, 243
Displacement, 214, 215
Dissipation function, 33
Distribution, 39
   Dirac distribution, 40, 68
Divergence, 195
Divisor of zero, 14
Domain of dependence, 57, 121, 291, 376, 380, 396
Duhamel's rule, 77, 80, 118

Eigenfunction, 47, 84, 306, 346
Eigenfunction expansion, 244, 268, 289
Eigenvalue, 15, 84, 232, 233, 286
Eigenvector, 15
Eikonal, 397
Elastic waves, 201, 392
Electric field, 214
Electrode, 364
Electrodynamics, 385
Electromotive force (emf), 34
Elliptic, 52, 407
Energy, 32, 209
Entropy, 210
Enumeration function, 242
Equilibrium, 103
Eulerian derivative, 209
Even, 119
Expected mean, 222
Exponential matrix, 35
Exterior expansion, 362
Exterior problem, 145

Finite difference equation, 173

Finite difference pattern, 64, 129, 173
Finite differences, 64, 129
Force, 26, 32
   generalized, 32
Form, 10
Fourier-Bessel series, 310
Fourier coefficient, 246
Fourier cosine transform, 110, 157
Fourier double integral, 110
Fourier double series, 185
Fourier expansion, 246, 280
Fourier integral, 109, 113, 154, 201, 295
Fourier law, 97
Fourier series, 87, 91, 294, 299, 411
   cosine series, 90
   half-range series, 90
   sine series, 90
Fourier sine transform, 111, 155, 413
Fourier transform of distribution, 124
Free end, 188
Friction, 33, 34
Function, generalized, 68
   symbolic, 68
Functional, 10, 67, 278
   linear functional, 10, 67
Future, 57, 376

Gamma function, 301
Gauss, 177
Gauss profile, 115
Gauss theorem, 196
Generalized function, 68
Generating function, 300, 361
Gradient, 195
Gram-Schmidt process, 8
Green's formula, 201, 381
Green's function, 44, 74, 198, 264 ff., 270 ff., 379 ff.
Green's matrix, 40, 395
Grid, 64, 129, 243

$H(t)$, 42
$H_s^{(1)}(z)$, $H_s^{(2)}(z)$, 321
Half-plane problem, 153
Half-space, 272
Hamilton's principle, 29, 181
Hankel function, 321
Hankel transform, 312, 333, 416
Harmonic, 83, 134, 143
Harmonic function, 134, 138
Harmonic oscillator, 32, 226
Heat equation, 288
Heat kernel, 114, 289
Heaviside function, 75, 116

Heisenberg, 223
Helmholtz equation, 229, 246, 274
Helmholtz theorem, 200
Hermitian, 14
Hermitian operator, 221
Hertz, 218, 219
Hilbert space, 2, 88
  complex, 12, 221
Hinged end, 188
Homogeneous, 10, 119, 134, 143
Hoop, 336
Huygens' premise, 385
Hyperbolic, 51, 407
Hyperplane, 11

$I_s(z)$, 328
Idempotence, 9
Image, 119, 272
Incident wave front, 406
Inductance, 34
Induction, 214
Inner product, 5
Integral equation, 284
Interior expansion, 362
Inverse operator, 268, 284
Inverse point, 273
Irrotational, 195
Isentropic, 211
Isotropic, 134

$J_s(z)$, 299
Jordan form, 37

$K_s(z)$, 328
Kelvin, 363
Kernel, 268
Kinetic energy, 183
Kronecker delta, 7

Lagrange, 28
Lagrange density, 183, 191
Lagrange equation, 182
Laplace difference equation, 173
Laplace equation, 133
Laplace integral, 343
Laplace operator, 134, 195
Laplace transform, 167, 415
  one-sided, 169, 415
Lattice, 64, 129, 243
Legendre equation, 344, 368
  associated equation, 344
Legendre formula, 305
Legendre functions, 341
  associated functions, 344

Legendre polynomials, 343, 348, 349
Line source, 386
Linear dependence, 3
Linear equation, 3, 104
Linear functional, 10, 68
Linear transformation, 12
Liouville, 22, 43
Lorentz, 385

Magnetic field, 215
Mass, 34
Matrix, 13
  fundamental, 23
  Hermitian, 14
  nilpotent, 38
  orthogonal, 15
  response, 39
  symmetric, 14
  transposed, 14
  unitary, 15, 127
Maximum-minimum principle, 239
Maximum principle, 97, 135, 173
  converse of, 145
Maxwell, 215
Maxwell equations, 214, 390
Mean square convergence, 245, 255
Mean value property, 144
Membrane, 180
Minimum problem, 231, 235
Mixed problem, 150
Mode, 86
  normal, 86, 213
Momentum, 26
Monotonic, 241

Neumann function, 277, 281
Neumann problem, 139, 156, 243, 272
Newton's law of cooling, 140
Newton's laws of motion, 26
Nilpotent, 38
Nodal line, 244
Nonintegral order, 367
Nonsingular, 14
Norm, 2, 12
Null vector, 377

Observable, 221
Odd, 118
Operator, 44
  differential, 84
  linear, 267
Orthogonal, 6, 234, 280
Orthogonal matrix, 17, 269, 287
Orthonormal, 6, 235, 266

Oscillator, 33
  harmonic, 32, 226

$P_n(x)$, $P_n^m(x)$, 341
Parabolic, 52, 95, 288
Parseval's theorem, 249
Partial differential equation, 49
Past, 376
Path, 161, 319
  steepest descent, 319
Picard, 19
Plane wave expansion, 360
Plate, 187
Poisson, 382
Poisson's formula, 144, 274
Polar form, 160
Pole, 280
Potential, 212
Potential energy, 182, 191
Power distribution, 72
Present, 376
Pressure, 207
Pressure waves, 204, 393
Principal axes, 18
Probability amplitude, 222
Product of a distribution and a function, 73
Product region, 255
Projection, 8, 280
  orthogonal, 8, 286

Quantum, 220

R^n, 2
Radiation condition, 275, 324
Radiation equation, 97
Range, 14
Rate of strain, 207
Ray, 397
Rayleigh quotient, 239, 247
Rayleigh Ritz method, 250
Rectangular harmonics, 148
Reflected wave, 406
Reflection, 59, 401
Region of influence, 57
Relaxation, 177
Representation, 265, 291, 381
Residue theorem, 164
Resistance, 34
Resolvent, 284
Response matrix, 39
Retarded potential, 385
Retrograde cone, 375
Riemann, 43, 92

Riemann mapping theorem, 271
Robin function, 278
Rod, 187
Rodrigues' formula, 344
Rotation, 145
Rotational waves, 205

Saddle point, 318
Scalar, 2
Schläfli integral, 300, 329
Schrödinger equation, 118, 225
Schwartz, 67, 80
Schwarz, 5, 224
Secondary source, 402
Sector eigenfunctions, 309
Separation constant, 82
Separation of variables, 81, 254
Series expansions, 260
Shadow zone, 406
Shear, 188
Shear waves, 205, 393
Similar, 17
Simple structure, 36
Single-valued, 143
Singular support, 71
Singularity, 160, 374
Solenoidal, 196
Solid harmonic, 363
Sommerfeld, 275
Source, 119, 265
Spacelike, 377
Spherical eigenfunctions, 346, 353
Spherical harmonics, 342, 349
  solid harmonic, 363
  surface harmonic, 342
Stable, 131
State, 131
  equation of state, 211
Stationary, 30, 103
Steady state, 122
Stokes' rule, 76
Stokes' theorem, 197, 383
Strain, 201
Stress, 202
String, 53
Strip, 151
Sturm-Liouville, 47, 261
Subharmonic, 135
Subspace, 8
Superposition principle, 21, 151
Support, 71, 74
Surface harmonics, 342, 349
Symbolic function, 68, 74, 293
Symmetric, 265

System, 19
    differential, 19, 106, 113

Telegraph equation, 78
Template, 64
Test function, 67, 128
Timelike, 377
Trace, 23
Transformation, 12
    inverse, 14
    linear, 12
Transient, 102
Transmission problem, 364
Transport equation, 399
Transpose, 14
Tube, 400

Uncertainty principle, 223
Unit, 126
Unitary, 15, 127
Unstable, 130

Vanish, 70
Variation, 29, 181
    first variation, 29, 181
Variational derivative, 181
Variational principle, 29, 182, 204
Variational property, 239

Vector, 1, 194
    basis vector, 4
Vector field, 1
Vector potential, 199
Vector space, 2
Velocity, 54, 204, 376
Vibration, 180
    lateral vibration, 187
Viscosity, kinematic, 209
Viscous, 207
Volterra type, 79
Voluminal waves, 204

Wave, 53, 81, 372
Wave equation, 54, 74, 372, 386
Wave front, 397
Wave function, 223
    standing wave, 81
    traveling wave, 54
Weak convergence, 293, 379
Weber formulas, 337
Wedge, 402
Weierstrass, 230
Wronskian, 46, 302, 367

$Y_s(z)$, 303

Zonal harmonics, 354
Zone, 354